# Spectroscopy, Dynamics and Molecular Theory of Carbon Plasmas and Vapors

## Advances in the Understanding of the Most Complex High-Temperature Elemental System

# Spectroscopy, Dynamics and Molecular Theory of Carbon Plasmas and Vapors

## Advances in the Understanding of the Most Complex High-Temperature Elemental System

*Foreword by Sir Harold Kroto (Nobel Laureate, Chemistry, 1996)*

editors

# László Nemes
Hungarian Academy of Sciences, Hungary

# Stephan Irle
Nagoya University, Japan

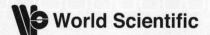

**World Scientific**

NEW JERSEY · LONDON · SINGAPORE · BEIJING · SHANGHAI · HONG KONG · TAIPEI · CHENNAI

*Published by*

World Scientific Publishing Co. Pte. Ltd.

5 Toh Tuck Link, Singapore 596224

*USA office:* 27 Warren Street, Suite 401-402, Hackensack, NJ 07601

*UK office:* 57 Shelton Street, Covent Garden, London WC2H 9HE

**British Library Cataloguing-in-Publication Data**
A catalogue record for this book is available from the British Library.

ISBN-13 978-981-283-764-6
ISBN-10 981-283-764-7

Printed in Singapore by B & Jo Enterprise Pte Ltd

# Foreword

The spheroidal pure carbon cage molecules, the Fullerenes, and their elongated cousins, the Carbon Nanotubes, have opened up totally new and unexpected fields of scientific investigation which promise revolutionary advances in electrical, magnetic and tensile strength behaviour. These advances have catalysed fundamental research in Nanoscience and its strategic twin Nanotechnology (N&N) and in the next few decades this research promises to lead to completely new and revolutionary technologies with socio-economic impact analogous to the advances of the last half of the 20th century which has seen phenomenal breakthroughs in advanced materials applications, lasers, computers and the Internet etc. From a fundamental point of view N&N is not really new at all but rather a new perspective on the way chemistry can impact other disciplines.

The fact that these breakthroughs are taking place at the beginning of the 21st Century indicates that we still have a lot to learn – particularly about carbon! Interestingly from the point of view of this compendium, is the fact that the major breakthroughs in N&N were stimulated by studies of the aggregation behaviour of small carbon species in the gas phase. There are many key areas of carbon behaviour that now need to be addressed and they are addressed in this volume. The articles promise to make a positive contribution to our understanding many problems and point the way for the next generation of researchers to successfully address many of these key issues: a) Studies of small carbon species from diatomic molecules to small carbon clusters in the gas phase are clearly vital if we are to understand carbon nanotube production processes. b) The only detailed way of studying such species is by spectroscopy and this allows the exploration of kinetic processes in plasmas and vapors. c) Modern lasers are ideal for spectroscopy and are also excellent for creating sample carbon vapors under controlled conditions. d) There is no doubt that we need an intrinsic understanding of the way that carbon clusters and subsequently forms condensed phase carbon nanostructures — in particular thin films. e) The fact that fullerenes are produced whenever soot is formed and furthermore is now being produced commercially by a combustion technique suggests that soot formation is still not that well understood. Indeed the fact that $C_{60}$ was not detected by any soot researchers even though it is a key constituent combustion produced carbon aggregates provides significant food for thought! f) It is of course always important that theory and experiment move forward in a synergistic and complementary way so electronic and vibrational properties probed with modern computer simulation methods promise deep insights into carbon dynamics.

The articles in this volume highlight most of the important fundamental research areas as well as the technological challenges that must be confronted and overcome if the paradigm-shifting new approaches that carbon N&N promises are to be realised. In particular the articles provide the information needed by students and specialists if they are to make progress in advancing our fundamental understanding and developing the applications of accurately nanostructured carbon materials.

Sir Harold Kroto FRS
The Florida State University
March 2011

# Preface

Nanotechnology and nanosciences are buzzwords of today. There are great expectations of brand new technologies and methods utilizing structures of size down to the molecular level, devices in a broad scale ranging from computers to medical devices and pharmaceuticals. The origin of such thoughts can probably be traced back to fiction novels written long ago, but perhaps it is not a mistake to date modern developments in the nanofields using the relatively recent discoveries of carbon nanostructures such as fullerenes, nanotubes, and graphenes. Nowadays there are new nanostructures built not from carbon but other chemical elements, such as silicon and gold to name just a few, and the domain of supramolecular organic chemistry is reaching far into the nanometer scale.

Nonetheless, nanostructures built purely and exclusively from carbon cover all possible dimensionalities, are the best studied nanscale materials around, and are among the most advanced candidates for application of nanotechnology in molecular electronics, sensors, and energy sciences. This book is a compendium of experimental and theoretical approaches for the study of spectroscopic and thermodynamic properties and the dynamics of carbon systems. The reader will soon discover that despite the chemical simplicity in terms of the atomic building blocks the formation as well as molecular and electronic structures of such nanomaterials are bewilderingly complex. We believe the present book will help the reader to appreciate the basic problems and will serve as thought provoker for research scientist and engineers in related fields.

This book is two-tiered, as it contains theoretical and experimental approaches. These are often intermingled as theoretical chapters contain references to experimental methods and the experimental chapters often contain theoretical considerations. The theoretical approaches covered in this book deal with the development and application of classical and quantum chemical potential energy functions governing structure formation and dynamics of carbon vapors. Fullerene and carbon nanotube formation, as well as carbon gasification reactions are the most prominent applications featured in this compendium. It is very important to gauge the validity of the potential energy surfaces; this can be achieved by comparison with high-level *ab initio* benchmark calculations, as well as by simulating the evolution of IR and Raman spectra for systems containing more and more carbon atoms.

There are various possibilities for the experimenter to study carbon vapors and carbon plasmas. The majority of such experiments use spectroscopic methods. Electronic spectroscopy is a relatively simple and convenient tool, it is non-invasive and provides a multitude of physical and chemical data. Different chapters overlap in their coverage of spectroscopic applications, with most of the overlap being found in emission spectroscopic studies. Plasmas are very complicated and a special position is enjoyed by laser induced plasmas both in materials science applications and in basic spectroscopic research. By no means, however, are laser-induced plasmas unique to carbon vapors, one special and very important example is astronomy. We trust the knowledge concentrated in this book may help astronomers in their spectroscopic investigations of various stellar and interstellar objects (such as carbon stars and interstellar matter).

We believe this book will serve not only a specialist reading but also a support text for education in physical sciences at the College and University level. It will prove to students of science that elementary processes in high temperature atomic media are very good examples of complicated mechanisms that need fundamental knowledge, much of which is still open to studies and need fresh thought and innovative experimental approaches. There is absoutely no doubt that carbon nanostructures will play an important role for commercial applications of nanotechnology in the near future. We are pleased to release this book for the international readership and would like to thank World Scientific Publishers for the wonderful opportunity in its production.

László Nemes                                                    Stephan Irle
Budapest, Hungary                                          Nagoya, Japan

# Contents

*Foreword*      v

*Preface*      vii

**Experimental**      **1**

Chapter 1      Spectroscopy of Carbon Nanotube Production      3
Processes
*B. A. Cruden*

Chapter 2      Spectroscopic Studies on Laser-Produced Carbon      55
Vapor
*K. Sasaki*

Chapter 3      Kinetic and Diagnostic Studies of Carbon      77
Containing Plasmas and Vapors Using Laser
Absorption Techniques
*J. Röpcke, A. Rousseau and P. B. Davies*

Chapter 4      Spectroscopy of Carbon Containing Diatomic      113
Molecules
*J. O. Hornkohl, L. Nemes and C. Parigger*

Chapter 5      Optical Emission Spectroscopy of $C_2$ and $C_3$      167
Molecules in Laser Ablation Carbon Plasma
*N. A. Savastenko and N. V. Tarasenko*

Chapter 6      Intra-Cavity Laser Spectroscopy of Carbon Clusters      199
*S. Raikov and L. Boufendi*

Chapter 7      Dynamics of Laser-Ablated Carbon Plasma for Thin      223
Film Deposition: Spectroscopic and Imaging
Approach
*R. K. Thareja and A. K. Sharma*

Chapter 8      Laser Spectroscopy of Transient Carbon Species in      255
               the Context of Soot Formation
               *V. Nevrlý, M. Střižík, P. Bitala and Z. Zelinger*

Chapter 9      Developing New Production and Observation             283
               Methods for Various Sized Carbon Nanomaterials
               from Clusters to Nanotubes
               *T. Sugai*

**Theoretical**                                                     **315**

Chapter 10     Potential Model for Molecular Dynamics of Carbon      317
               *A. M. Ito and H. Nakamura*

Chapter 11     Electronic and Molecular Structures of Small- and     343
               Medium-Sized Carbon Clusters
               *V. Parasuk*

Chapter 12     Vibrational Spectroscopy of Linear Carbon Chains      375
               *C.-P. Chou, W.-F. Li, H. A. Witek and*
               *M. Andrzejak*

Chapter 13     Dynamics Simulations of Fullerene and SWCNT           417
               Formation
               *S. Irle, G. Zheng, Z. Wang and K. Morokuma*

Chapter 14     Mechanisms of Carbon Gasification Reactions           445
               Using Electronic Structure Methods
               *J. F. Espinal, T. N. Truong and F. Mondragón*

*Index*                                                              503

# Experimental

## Chapter 1

# Spectroscopy of Carbon Nanotube Production Processes

Brett A. Cruden

*University Affiliated Research Center,*
*NASA Ames Research Center, Moffett Field, CA, U.S.A.*
*E-mail address: Brett.A.Cruden@nasa.gov*

Carbon Nanotubes (CNTs) have recently generated a great deal of interest in the scientific community due to their exceptional electrical, mechanical and thermal properties, which are creating applications in structural composites, sensors, microelectronics, etc. CNTs are generally produced by one of four primary techniques — arc discharge, laser ablation, chemical vapor deposition (CVD), or thermally driven gas phase processes. Of these processes, two (laser ablation and arc discharge) are inherently plasma-based approaches, CVD may be driven by glow discharge plasmas, and thermal approaches include flame (low density/temperature plasma) synthesis. This chapter will review efforts in spectroscopic studies of these different types of plasmas. Approaches employed include optical emission, absorption (both FTIR and broadband), laser scattering, laser induced fluorescence, incandescence and luminescence, and Raman spectroscopies. These approaches have been used to characterize properties ranging from atomic species densities to neutral temperatures to cluster formation. The studies have shed a great deal of light on the CNT production process, however much is still unknown about the specifics of the CNT growth process.

# 1. Introduction

In recent years, carbon nanotubes (CNTs) have been widely touted as a promising new material for structural, thermal and electrical applications.[1, 2] CNTs possess Young's modulus of greater than 1 TPa,[3–7] larger than that of diamond; and a strength-to-weight ratio that far exceeds that of steel and aluminum. This makes them of interest for structural applications and creating composite materials. The thermal conductivity of single-walled nanotubes (SWNTs) is also greater than that of diamond at 3000 W/m-K,[8–10] making them suitable for heat removal applications. Electrically, CNTs demonstrate high current carrying capacities, as high as $10^{10}$ A/cm$^2$.[11] Current transport in CNTs is known to be ballistic,[12, 13] i.e. conductance is limited only by quantum mechanics and is independent of length on micron scales. Thus, CNTs show a great potential in use as wiring, especially in small-scale (microelectronic) circuits. Further, depending on the nanotube's chirality, it can be semiconducting or metallic. While the metallic nanotubes are promising as wiring, the semiconducting nanotubes retain a high carrier mobility and thus show promise as next generation transistor materials.[14, 15] The combination of these properties make CNTs very promising as a new multifunctional material.

A variety of mechanisms exist for producing CNTs. In general, all methods involve a vaporized carbon source and a transition metal catalyst such as iron, cobalt, or nickel. The first method used for CNT production was an arc-discharge set-up where the high temperature produced in an electric arc converted the graphitic electrodes into carbon nanotubes.[16] Metals present in the arc system acted as catalysts for the conversion. Later experiments performed using a high powered laser on a metal/graphite catalyst proved to have similar effects.[17] More recent efforts have focused on producing carbon nanotubes using thermal processes in flowing chemical reactors, including the high pressure CO (HiPCO),[18] floating catalyst[19] and other hydrocarbon/metallorganic reactor processes.[20–24] At the same time, chemical vapor deposition (CVD) has emerged as a viable approach to preparing CNTs on substrates.[25] The CVD methods for carbon nanotubes may be divided into two general categories — thermal and plasma based CVD.[26]

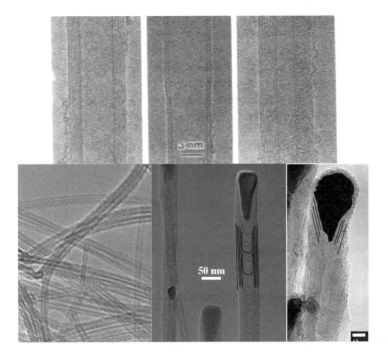

Fig. 1. TEM images showing different morphologies of carbonaceous nanostructures. (top) Various MWNTs. Reprinted by permission from Macmillan Publishers Ltd: Nature, Ref. [16], Copyright 1991. (bottom, from left to right) SWNTs from thermal CVD. Bamboo-style MWNTs. Stacked-cone type nanofibers. Lines are added to emphasize the fiber structure.

The differing morphologies of carbon nanotubes and the related nanofibers are demonstrated in Fig. 1. The first four images, showing nanotubes, depict sidewalls that are parallel to one another along the entire length of the tube. The primary difference in these images is the number of walls comprising each structure. The final two images show two different types of nanofibers. The bamboo-type morphology has several "compartments" on the inside of the nanotube, where the graphene sheets are bent inward and cross the inner radius of the tube. The stacked-cone nanofibers have graphene sheets that are at an angle to the tube axis, thus the fiber takes the form of several stacked cones. These types of structures, when grown in plasma CVD, are also characterized by having an elongated catalyst particle at the tip of the fiber. While these fibers possess vertical alignment, the difference in morphology from idealized carbon nanotubes may be expected to significantly degrade the ideal properties.

In the arc discharge technique, high temperatures, 4000°C or greater, are generated by an arc between two carbon electrodes, allowing carbon to be sublimated off the anode.[27] A layer of the carbon-based byproducts is deposited on the chamber walls, usually as a tangled "felt" of carbon nanotubes and contaminants. A typical set-up consists of two water-cooled carbon electrodes separated by a few millimeters in a He atmosphere. A voltage of 10-35 V is applied to achieve a current of 60-100 A. A stable plasma is maintained by moving the anode slowly as it is consumed. To favor single-walled nanotubes over multi-walled, a Ni/Y catalyst is incorporated into the anode. The yield and properties of the SWNT produced tend to vary with experimental set-up, but yields as high as 30-50% and production as high as 1.2 g/min has been reported.[28]

Laser vaporization techniques use a pulsed Nd:YAG or continuous $CO_2$ laser to vaporize a carbon target in a temperature controlled quartz tube furnace.[29] Soot or a filamentous material is created and condenses on a water cooled collector at the end of the tube. A typical set-up includes a graphite target inside a furnace at 1200°C with an inert gas flow at 500 Torr. In some cases, the target is double pulsed with the laser, where the second pulse is timed to break up the material ejected by the first pulse, reducing the amount of amorphous material in the soot and enabling more nanotubes to form. The pulses are approximately 40 ns apart, at an energy of 100-300 mJ/pulse, with a frequency of 10-30 Hz and a wavelength of 1064 or 532 nm. To favor single-walled nanotubes over multi-walled, catalyst metals, Co-Ni or Co-Pt, are incorporated into the target. Production rates of laser vaporization are higher than that of arc discharge methods, and can be as high as 1 g/day with yield in the range of 60-90%.[28] Rate improvement has been seen using a subpicosecond pulsed free electron laser focused onto a rotating target in a high temperature furnace. Rates as high as 45 g/hr are possible with this improved method.[28]

The thermal CVD approach has been employed to produce both single-walled and multi-walled carbon nanotubes under varying conditions. The plasma enhanced CVD (PECVD) methods have been demonstrated as producing single-walled nanotubes, multi-walled nanotubes and another class of nanostructured carbon characterized as "vertically aligned carbon nanofibers", "multi-walled nanofibers" or "bamboo"-like carbon nanotubes as discussed above. The majority of literature on plasma enhanced methods pertains to this final class of structures, though many authors do not take the care to distinguish these structures from carbon nanotubes. Various sources for CNT growth by PECVD have been explored, including direct current (dc) plasmas [30–35], microwave plasma[36–40] and rf/inductive plasmas.[41, 42] In some cases, a hot filament has been employed to improve heat input and aid dissociation.[30, 43] The chemistries for PECVD generally consist of some hydrocarbon precursor (methane, ethylene, acetylene) to provide growth species and some etchant to prevent amorphous carbon formation (hydrogen, ammonia). Some authors have claimed to perform low temperature growth using the PECVD approach, however generally have not properly accounted for heat transport issues between the plasma, the substrate and electrode equipment. The majority of works claiming nanotube production are in reality producing nanofibers, though a downstream plasma has been demonstrated to successfully produce SWNTs.[44]

Among the thousands of papers published on carbon nanotubes to date, a relatively small amount of papers actually concentrate on characterization of the CNT production environment. It is the intent of this chapter to review the work that has been done to date on spectroscopy of plasma systems that are used for CNT synthesis. This chapter is organized by the types of plasmas employed rather than by specific diagnostic technique. Plasma regimes are often broadly characterized by physicists by two characteristic parameters — electron density and temperature, as shown in Fig. 2. As discussed, there are three primary types of plasmas encountered in CNT synthesis: arc discharges, laser plumes, and glow discharges. These are outlined in Fig. 2. Additionally, there is a more limited amount of work on CNT synthesis in flames, another form of low energy plasma, which will be discussed briefly.

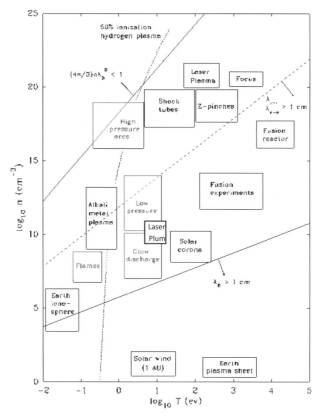

Fig. 2. Different plasma regimes. The types of plasmas used for CNT growth are highlighted. Modified from [45].

## 2. Arc Discharge

Arc discharges have been used extensively for the production of $C_{60}$ and other fullerenes. Adaptation of these approaches to nanotube synthesis was realized by introducing transition metal catalysts (e.g. Fe, Co, Ni) into the electrodes. Typical synthesis conditions use an inert (He or Ar) atmosphere at pressures from 0.1 to 100 kPa. An example experimental set up is given in Fig. 3. Graphitic electrodes with a gap of an order of a mm are subjected to voltages on the order of 20-50 V and carry currents of 50-100 A. While these types of discharges have been studied the longest for nanotube growth, there are only a handful of studies of the spectroscopic properties of these plasmas. Primarily, these studies involve analysis of optical emission of the $C_2$ Swan bands.[46–49] This includes general intensity measurements, temperature analysis and quantitative density measurement through $C_2$ self-absorption.[46, 47] In some cases, atomic lines were observed and characterized, including carbon, nickel and iron.[48, 50] These works are summarized in Table 1 below. This review does not include spectroscopic studies on similar arc plasmas that are used for fullerene production, of which there are several.[46]

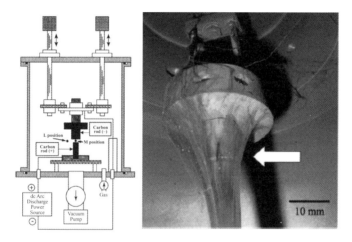

Fig. 3. Example experimental apparatus for arc discharge synthesis and image of collected CNTs. Reprinted from Diamond and Related Materials, Vol 14, Guo, et al., "Spectroscopic study during single-wall carbon nanotubes production by Ar, H-2, and H-2-Ar DC arc discharge", p. 887, Copyright 2005, with permission from Elsevier. (Ref. [50]).

Table 1. Summary of spectroscopic studies of arc discharges for CNT Production.

| Pressure | Current | Voltage | Ambient | Group | Ref |
|---|---|---|---|---|---|
| 17-88 kPa | 54-75 A | 21 | He, Kr/Xe/Ar-$H_2$ | Warsaw Univ | [46, 47, 51] |
| 13-39 kPa | 40-60 A | 24 | He | Osaka Prefecture Univ | [49, 52] |
| 3-13 kPa | 30-60 A | Not | $H_2$, $CH_4$, He, | Meijo Univ | [53] |
| 26 kPa | 30-70 A | Given | Ar, $H_2$, Ar-$H_2$ | | [50] |
| 60 kPa | 80 A | 32-45 | He | CEMES | [48] |

Byszweski, et al. were one of the first to report spectroscopy studies of a CNT arc plasma.[47] Their analysis utilized the $C_2$ Swan bands to ascertain both temperature and $C_2$ density. Density was extracted from the Swan bands by calculating their self-absorption. This is made possible by the close spacing of lines near the band head resulting in a much larger self-absorption than is observed at the band tail. Mathematically, this is given by:[54]

$$I_{J'J''} = \int \frac{\varepsilon_{J'J''}(v)L}{\kappa(v)L}\left(1-e^{-\kappa(v)L}\right)dv$$
$$\kappa(v) = \sum \kappa_{J'J''}(v)$$

(1)

where $I_{J'J''}$ is the intensity of the line originating from the transition between rotational states $J'$ and $J''$, $\varepsilon_{J'J''}(v)$ and $\kappa_{J'J''}(v)$ are the line emission and absorption intensities, and L is the length of the plasma column. The second of these equations demonstrates how the self-absorption would be increased significantly more at the band head where lines are overlapping. A Beer's Law type relationship is obtained considering that $\kappa(v)$ depends linearly on $C_2$ density, allowing extraction of density by considering the relative intensities of the band head to the band tail. These calculated $I_{J'J''}$ may be summed with a Boltzmann weighting and appropriately broadened line functions to produce a composite spectrum as shown in Fig. 4 for different assumed $C_2$

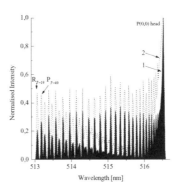

Fig. 4. Simulated spectra for the (0,0) Swan band transition with (1-solid) small and (2-dashed) large self-absorption. Line integrated $C_2$ densities are $10^{12}$ and $10^{16}$ cm$^{-2}$, respectively. Reprinted with permission from Ref. [54], copyright Institute of Physics.

densities. It is therefore possible to perform a fit to the band shape to determine both temperature and density. Alternatively, temperature may be extracted by taking a Boltzmann plot on the peaks that are removed from the band head and are not significantly altered in relative intensity by self-absorption.

Some of the results from this analysis are shown in Fig. 5 below.[47] The sample $C_2$ spectra of Fig. 5(a) shows both evidence of self-absorption ($C_2$ density extracted from the two spectra differ by about an order of magnitude) and an average temperature range of 4000-5500 K. While this temperature is the rotational temperature of the excited state of the $C_2$ molecule, it is believed to be representative of the actual neutral temperature of the arc on the basis of prior work comparing its temperature to that of CN and $N_2$.[55] The temperature trend versus position shows a peak temperature near the plasma center of about 5500 K that decreases monotonically away from the center. The temperature trend is slightly flatter at lower pressure, presumably related to a larger spatial extent of the plasma due to lower ambipolar diffusivity. The $C_2$ density trend is peaked near the center at around $2 \times 10^{16}$ cm$^{-3}$ and falls off with distance. As with temperature, this peak value extends over a larger distance at low pressure. An Abel transform of the data gives an estimated density profile versus radius, and again shows a more peaked profile at high pressure. With these densities and temperatures, the authors were able to simulate the composition of C species in the discharge. The calculation suggested that atomic C was the major constituent of the plasma, with density in excess of $10^{17}$ cm$^{-3}$. The simulation also found that below 4000 K, the C and $C_2$ species densities decreased in favor of a predominantly $C_3$ containing plasma, and further condensation of larger C species at even lower temperature. The authors concluded that the density and temperature range was the appropriate range for nucleation of carbon nanotubes and that this target temperature and density is what drives the pressure/current requirements for arc synthesis of CNTs. An additional finding by Lange and Huczko was that the catalyst did impact the $C_2$ formation, with more $C_2$ being produced with Co/Ni catalyst in comparison to Fe catalyst.[46] Neither the reason for this nor the significance to CNT production was discussed. A similar trend has been observed in laser ablation plasmas where Co increases the peak production of $C_2$ in the high temperature region, but reduces its concentration in the cooler zones.[56]

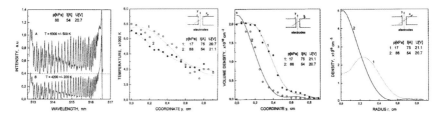

Fig. 5. (a) Sample C2 emission spectrum from arc discharge at two different heights. (b) Temperature fit from Swan bands as a function of position at two different pressure conditions. (c) Volume density at different pressure conditions and (d) volume density versus radial position as determined from Abel transform. Reprinted from J Phys Chem Solids, Vol 58, Byszewski, et al., "Fullerene and Nanotube Synthesis. Plasma Spectroscopy Studies", p. 1679, Copyright 1997, with permission from Elsevier. (Ref. [47])

Later, the same group also characterized emission characteristics in different rare gas mixtures atmospheres (i.e. Kr and Xe with $H_2$ instead of Ar-$H_2$) and found significant improvements in SWNT quality, with Kr being somewhat better than Xe.[51] Characterization of $C_2$ temperatures and densities found small differences between these three gas compositions at the electrode periphery where SWNTs are collected. Some modulation of temperature and $C_2$ density was observed in the center of the arc, though the trend reversed when the grain size of the electrode was changed. One condition that produced no SWNTs had substantially (almost 2x) increased $C_2$ concentration in the center. The authors used these observations to conclude that $C_2$ was not an important precursor to CNT growth and that the growth is driven by larger particulates removed from the electrode by the arc. They attributed the role of Kr and Xe as increasing residence time of the particulates.

Akita, et al., used an indirect method to extrapolate temperature from the $C_2$ Swan band intensities.[52] They calculated, based on Saha's equation, charge neutrality and an equation of state, what the composition (i.e. [C], [C$^+$] = $n_e$, [C$_2$], [C$_3$]) of a carbon containing plasma would be as a function of temperature. They then used this to estimate the relative emission intensity of $C_2$ as a function of temperature. Their calculation predicted $C_2$ intensity to pass through a maximum at 5500 K. Abel inversion of the spectroscopic signal intensity of $C_2$ revealed a profile that was peaked slightly off-center of the discharge. Assuming that the temperature increases monotonically toward the center would lead to the conclusion that the peak corresponds to a temperature of 5500 K and the temperature in the center is on the high end of the emission intensity-temperature curve. Using this assumption and the calculated curve as a calibration, they determined the temperature at the center of the arc to be as high as 7000 K. The model allowed them to also obtain density of the four species and temperature as a function of position, with atomic C being the predominant species within the inner 1.5-2.5 mm of radius in all cases. The temperature was reduced to below 5500 K by reducing pressure, though decreasing arc current or introducing active cooling to the electrode increased the peak (center) temperature to 6000 K. Both of these conditions gave improved yield in comparison to the other conditions. However, the seemingly counterintuitive result of increasing temperature under these conditions suggests that the model and methodology has oversimplified the physics of the arc plasma. Particularly, self-absorption was neglected and any diffusional effects are absent from the 0-d model. Note that a similar peak in spatial profile is observed in Fig. 5(d) but does not correlate with a high temperature as inferred through this approach. The ultimate conclusion of this work was that a plasma temperature of 6000 K, being composed predominantly of atomic C was preferred for CNT growth, consistent with the work of Byszewski, et al.[47]

In a later study by the same group, they introduced a flow of helium directly into the plasma to "cool" the gas.[49] It was found that an optimum helium flow existed, with the greatest production of CNTs occurring for flows on the order of 6-8 slm. After subtracting a background signal due to blackbody radiation from the cathode (estimating $T = 2000K$), they measured the relative intensities of the different vibrational modes of the Swan bands. The spectra were noisy and poorly resolved such that they did not attempt to extract an actual temperature but instead looked at the relative peak intensities corresponding to $\Delta v = 0, \pm 1$, and found that the ratio of peaks attributable to $[\Delta v = 0]/[\Delta v = -1]$ went through a maximum at the flow rate corresponding to optimum CNT production. They interpreted this as an indication that the temperature of the plasma was highest at this flow rate. It was surmised that intermediate rates of cooling by He flow caused a shrinkage of the plasma volume, increasing its core temperature, while high flow rates actually cooled the electrode temperature and reduced thermionic electron emission. However, the spectroscopic interpretation of this data is questionable as the emission spectrum does not resemble the typical Swan band spectra in its shape and could instead be consistent with other bands such as the high pressure $C_2$ bands or CN red band impurities. Even if the assignments were correct, this relative change in the band head intensities would not necessarily imply the temperature trend surmised unless sufficient analysis were performed to determine the actual vibrational upper states and the impact of self-absorption, if any.

In 1999, Zhao et al characterized optical emission spectra from arc discharges using $H_2$, $CH_4$ and He as carrier gases.[53] In all three gases, they characterized the 247.9 nm C atomic line as the strongest emitter, in addition to strong $C_2$ Swan bands, and weaker C+ lines (283.7, 657.8, 723.6 nm). In $H_2$ and $CH_4$, CH and atomic H was also observed. They observed a preference for fullerene formation over CNTs in the He discharge and thus surmised that H species were responsible for dissociating or otherwise discouraging the formation of fullerenes, leading to preferential CNT production. Later, Guo, et al., studied the variations in Ar, $H_2$ and Ar-$H_2$ mixture on the arc discharge.[50] Only the Ar-$H_2$ mixture produced high yield of SWNTs, and the corresponding optical emission spectra showed high intensities of both $C_2$ radical and Fe atoms. In $H_2$ alone the $C_2$ intensity was reduced and Fe was not observed, while in Ar, Fe but not $C_2$, was observed. It was interpreted that this gas composition therefore resulted in Fe and C densities that were appropriate for CNT growth. Other observations of note in this work included intense CH emission upon ignition that disappeared after a few seconds and an initially large Fe/$C_2$ ratio in the center of the discharge that decreased to a constant value within the first 30 s of operation. Away from the center of the discharge, these temporal trends were not observed.

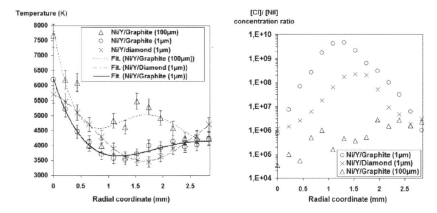

Fig. 6. Trace of temperature (as measured from $C_2$ Swan bands) and C/Ni concentration ratio (from 909.4 and 349.3 nm lines) versus radial position in an arc discharge with differing anode compositions. Reprinted from Carbon, Vol 45, Mansour, et al., "A significant improvement of both yield and purity during SWCNT synthesis via the electric arc process", p. 1651, Copyright 2007, with permission from Elsevier. (Ref. [48])

Mansour, et al., recently reported characterization of an arc discharge process involving heterogeneous anodes constructed of Ni/Y catalysts and graphite or diamond powders.[48] They found that the ratio of CNTs produced over amorphous carbon was greatly enhanced by small-grain (~1 μm) graphite or diamond powders over large-grain graphite. This trend was understood in part by examining optical emission characteristics of the plasma, including temperature fits of the $C_2$ Swan bands and looking at atomic C/Ni emission ratios. This data is reproduced in Fig. 6 above. The large grain graphitic electrode resulted in variable emission characteristics during the run, implying a lack of control over synthesis conditions due to anode inhomogeneities. This produced an irregular temperature profile along the radial direction. A highly peaked temperature in the center of the plasma (~8000 K) was measured relative to the other electrodes (~6000 K). This condition also gave much lower densities of atomic C, resulting in a low C/Ni concentration ratio. The small-grain graphitic electrode gave even larger atomic C production than the diamond electrode and displayed a correspondingly higher SWNT production. The authors credited the smaller atomic C concentration to producing higher plasma temperatures, which appears to be consistent with the radial profiles and anode composition trends. However, this interpretation is inconsistent with the models of Byszewski or Akita, which predicted greater C fraction at high temperatures.[47, 52] The authors attributed the overall trends to the thermal conductivity of the electrodes dissipating thermal energy from the plasma.

What these studies hold in common is that optical emission spectra are compared over a limited range of variations and the spectra that correlates to the best growth condition is presumed to represent an optimum condition for nanotube growth. The resulting characterization and conclusion is very specific for the particular set-up of and variations studied by the individual researcher. The conclusions drawn are neither necessary nor sufficient for understanding nanotube growth in a general sense. Identification of commonalities across these widely differing systems and conditions, however, may allow for some conclusions that approach this. One common conclusion is the necessity for formation of significant quantity of atomic carbon. Two independent studies estimated the required density of atomic carbon to be on the order of $10^{17}$ cm$^{-3}$ for SWNT formation.[47, 48] This was concluded in one case by analyzing intensity of the atomic C emission and the other by modeling the carbon chemistry using the measured $C_2$ temperature and density. Unfortunately, the two studies appear to disagree on the relationship of atomic carbon density and plasma temperature. There also appears to be a universally characterized temperature range of 4000-6000 K in the arc for CNT formation. Under these temperature and density conditions, the $C_2$ Swan bands are always observed. These conditions may be necessary, but not sufficient, as at least one case exists where a carbon density and temperature of this magnitude did not lead to CNT production in the presence of catalyst material.[51] The absence of atomic lines in some studies but not others is not well explained. Atomic carbon in particular should be expected in all spectra based on the above conclusions. It might be the case that authors not observing atomic C have simply not surveyed far enough into the UV or IR. Similar reasons might exist for the lack of observation of catalyst atoms in many studies.

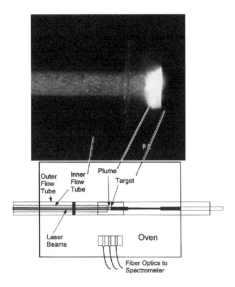

Fig. 7. A typical laser ablation set up. Reprinted from Chemical Physics Letters, Vol. 302, S. Arepalli and C.D. Scott, "Spectral measurements in production of single-wall carbon nanotubes by laser ablation", p. 139, Copyright 1999, with permission from Elsevier. (Ref. [57]).

## 3. Laser Plumes

The laser ablation technique uses a high energy laser to remove material from a graphite/Ni/Co target to induce nanotube formation. Two types of lasers have been employed, pulsed Nd:YAG or continuous $CO_2$ lasers. The ablation is usually performed inside an oven at 1200°C, though lower temperatures have also been found to produce CNTs. Typical energies are on the order of 150 mJ/pulse (Nd:YAG) or 1 kW ($CO_2$) and are focused onto an area a few mm in diameter. A typical set-up for the process is shown in Fig. 7. The laser ablation process was summarized in a recent review by Arepalli.[58] Processes examined spectroscopically are given in Table 2. Amongst the different techniques for CNT growth, laser ablation is by far the most extensively characterized approach spectroscopically. This may arise from a predisposition of laser users toward spectroscopy that is absent among the other synthesis communities. This characterization has also led to a greater understanding of the CNT production process in laser plumes as opposed to the other techniques. While emission spectroscopy is used to characterize laser plumes, there is a much wider range of other spectroscopic approaches being used. Many of these techniques are laser based, including laser induced fluorescene (LIF) and incandescence (LII), coherent anti-Stokes Raman scattering (CARS) and laser scattering/shadowgraphy. Absorption spectroscopy has also been applied. These results are summarized below.

Table 2. Summary of different spectroscopic characterizations of the laser ablation process.

| Laser | Duration/ Duty Cycle | Energy/ Power | Group | Diagnostics | Ref. |
|-------|---------------------|---------------|-------|-------------|------|
| 2 x Nd:YAG | 10 Hz (50 ns apart) | 300 mJ | NASA JSC | Emission LIF ($C_2$, Ni) | [57, 59, 60] |
| Nd:YAG | 8 ns | 140 mJ | Oak Ridge | Emission Absorption LIL ($Co,C_3$) Scattering | [29, 61, 62] |
| Nd:YAG | 3 ns | 1.7 $J/cm^2$ | NEC | Emission Scattering | [63] |
| $CO_2$ | Continuous | 1200 W | ONERA | CARS LII/LIF | [64] |
| $CO_2$ | Continuous | 130 $kW/cm^2$ | NEC | Emission Scattering | [65, 66] |

Arepalli and Scott first characterized the laser ablation process spectroscopically in 1999, with characteristic spectra as shown in Fig. 8.[57] The spectrum is dominated by the $C_2$ Swan bands and a continuum emission centered near 400 nm that was tentatively assigned to the $C_3$ $A^1\Pi_u$-$X^1\Sigma_u^+$ comet head system. There is also a continuum underlying the entire spectrum that slopes upward with increasing wavelength, characteristic of black-body radiation from particulates within the laser plume. These three contributors display separate temporal dependencies, with the $C_3$ peak increasing rapidly to its peak value at 100 ns, passing through a local minimum and maximum at 200 and 400 ns, then exponentially decaying. The $C_2$ spectra approaches its maximum value in less than 1 µs, then follows an exponential decay. The exponential decay was divided into three regions with differing decay constants over the first 6, 20 and 80 μs. The blackbody background rises in the first 1 μs, disappears and then reappears again after 30 μs.[60] This probably represents an initial formation of ablated particulate matter that is dissociated then recombines as the plume cools. The other trends suggest a rapid $C_3$ production whose intensity is initially modulated by heating and cooling of the plume within the first microsecond, while the $C_2$ production is slow enough that this modulation is not observed. The three recombination phases were not well understood but represent a combination of reaction and cooling. It is probable that the longer time recombination is that of $C_2$ with larger particulates with the two early recombinations involving recombination with atomic C, Co and small carbon clusters (i.e. self-recombination and $C_2$-$C_3$ recombinations).

Temperature measurements were attempted in this plume in several different fashions. The vibrational temperature of the $C_2$ Swan band was fit by integrating and ratioing the intensities of (1,0) and (0,0) bands. When a two pulse ablation scheme was used, the resultant temperature was found to decay from 3800 K to 3200 K over 25 μs. This final temperature agreed fairly well with the blackbody continuum emission (~2800 K) observed at long times.[60] With a single pulse ablation, the

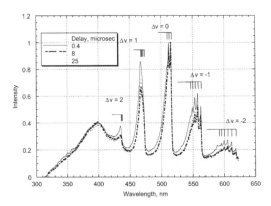

Fig. 8. Sample emission spectra from laser ablation plume. The spectra is normalized to the $C_2$ (0,0) peak at 516 nm. Reprinted from Chemical Physics Letters, Vol. 302, S. Arepalli and C.D. Scott, "Spectral measurements in production of single-wall carbon nanotubes by laser ablation", p. 139, Copyright 1999, with permission from Elsevier. (Ref. [57]).

peak temperature was near 2500 K. However, it was noted that this estimate did not well describe the entire rovibrational structure of the Swan bands, suggesting that the excited $C_2$ was not in thermal equilibrium. The authors attributed this to chemiluminescent C-C reaction or photodissociation of larger clusters. To check this, the authors performed LIF on $C_2$, exciting the (0,1) transition at 473 nm and collecting fluorescence at longer wavelengths.[60] Fitting the bands produced under LIF gave rotational temperatures of approximately 1500 K with a single pulse (the two pulse case could not be measured as the LIF was performed with the second laser). With the oven off, they measured a temperature of 300 K, indicating that this temperature reflects external heating and not thermal energy imparted by the laser. It is not clear from the data reported if this temperature discrepancy is due to a thermal non-equilibrium between rotational/vibrational modes or if it is due to a non-equilibrium between excited and ground states.

De Boer, et al., later followed up on these experiments by performing LIF on the Ni atom in the same reactor system.[59] The ablator target contained 1% Ni and 1% Co, which served as CNT growth catalysts. In these experiments, in addition to the two ablation lasers, a third laser was tuned into the 224.2-226.2 nm region to excite Ni atoms into the 4p levels. Since the process probed several low-lying electronic states, it was possible to estimate electronic temperatures by examining relative intensities for different excitations, however their estimate became insensitive to temperatures above ~1500 K. Nevertheless, the approach yielded a Ni electronic temperature between 1000-2000 K, similar to the $C_2$ rotational temperature. By probing at different locations and times, the authors were able to estimate a Ni atom velocity of 10 m/s, as compared to the $C_2$ velocity of around 50 m/s. The concentration of Ni in a particular location was observed to decay with a half life constant of ~370 μs, indicating little recombination. This observation was supported by an approximate calculation of Ni density based on fluorescence intensity, which resulted in a density of $10^{14}$ cm$^{-3}$. Considering the probe volume and the amount of Ni ablated from the target, this calculation suggested that the Ni was almost entirely in atomic form within the laser plume. These observations led to the conclusion that carbon cluster formation precedes formation of any catalyst nanoparticles.

Kokai, et al., also studied the laser ablation process with a pulsed Nd:YAG laser with emission and light scattering.[63] In their emission spectra they identified atomic lines of C (248 nm), Co (340.5 and 345.5 nm), Ni (341.5 and 346.2 nm), Ar (764 nm) and the $C_2$ Swan band (516.5 nm). These species comprise carbon clusters and catalyst atoms necessary for carbon nanotube formation in addition to electronically excited Ar carrier gas (present at 600 Torr). In contrast to the work of Arepalli, et al.,[57] no $C_3$ continuum was identified in spite of similar laser energy densities. The atomic lines showed significant broadening and wavelength shift at short times, with the 248 nm C line FWHM changing from 1.1 nm at 100 ns to 0.12 nm at 1 μs. This was attributed to Stark broadening and shift, though further quantification of the broadening was not attempted. A continuum emission attributed to free-free electron interactions (Bremsstrahlung) was also detected, peaking in the first 100 ns and vanishing within 400 ns. The atomic emission lines were all found to have two peaks at ~150 and 600 ns, then fully decayed at one to several μs. The first peak, having coincided with the Bremsstrahlung radiation, was attributed to electron impact excitation while the second peak was tentatively attributed to atom-atom excitation with backward moving atoms. The $C_2$ Swan bands, on the other hand, exhibited a comparatively long lifetime of up to 70 μs. This long emissive lifetime is suggestive of the chemiluminscent reactions reported by Arepalli,[57] but might also be due to ground state collisional excitation enabled over longer time scales by the relatively low excitation threshold of the Swan system. Spatially resolved images of these emission peaks showed separate regions corresponding roughly with the excitation energy of the emitting state. The peak emission occurred at increasing distances from the target as $C_2$ (2.4 eV) < Co (4.0 eV) < C (7.7 eV) < Ar (13.2 eV), suggesting higher energy atoms/electrons responsible for excitation moving further from the target in the same amount of time. The decay time for each of these species also progressed in the reverse of this order, indicative of the cooling of electrons, ions, metastables and/or high energy neutrals responsible for excitation. Laser scattering images showed the formation of particulate matter at later times and further distances than the atomic or $C_2$ emissions. The clustering of particulate matter in vortices with lifetimes

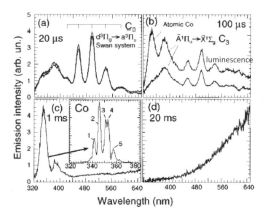

Fig. 9. (a-b) Emission and (b-d) Laser induced luminescence (LIL) from the laser plume at different times. The $C_2$ Swan band and $C_3$ emission are identified at short times. (<1 ms) At longer times, $C_3$, atomic Co and blackbody emission are identified via LIL. Reprinted with permission from Ref. [61], A. A. Puretzky, D. B. Geohegan, X. Fan and S. J. Pennycook, Appl. Phys. Lett. 76, 182 (2000). Copyright 2000, American Institute of Physics.

of around 1 s were observed, suggesting that CNT growth may be occurring within the vortices for up to 1s after ablation.

Puretzky, et al., performed spectroscopic characterization of similarly formed laser plumes using emission spectroscopy and laser induced luminescence as shown in Fig. 9.[29, 61] The emission spectroscopy showed similar characteristics to that of Arepalli, with significant $C_2$ Swan band contribution and a continuum near 380 nm attributable to $C_3$. Using a 308 nm XeCl laser, they were able to induce luminescence in the plume and identify the presence of atomic Co, $C_3$ and blackbody radiation from particulate matter. If any induced luminescence of $C_2$ was present, it was not easily distinguished from the emission signal. The baseline shift between the emission and luminescence spectra can be attributed to Rayleigh scattering and was used to track the formation of particulates and shape/motion of the laser plume. At short times (<100 μs), the luminescence signal was overwhelmed by emission, though at longer time emission was sufficiently weak that luminescence dominated the signal. By temporally monitoring the Co intensity, they were able to track the relative atomic Co density, finding it to peak at around 800 μs and decay over 2 ms. While the signal attributable to $C_3$

was not tracked directly, the relative intensity of the 380 nm to 345 nm peaks is decreasing over the first ms, suggesting a faster decay. The onset of blackbody luminescence begins around 1 ms, coinciding with the disappearance of the $C_3$ carbon clusters, suggesting that particulate matter is primarily carbonaceous in nature.

In later studies, Puretzky, et al., added optical absorption/scattering and pyrometry to more closely track the formation process.[62] For optical absorption, a pulsed Xe lamp was used to measure extinction spectra at different times. The absorption spectra showed an increasing absorbance at lower wavelength that is indicative of particulate scattering. The authors applied Mie theory and Rayleigh scattering/ absorbance to estimate particulate size from these spectra at different times and oven temperatures. Particulate diameters of 20 nm were found after 1 ms, with larger particles forming at lower temperature. Evidence of atomic Co and Ni absorption was also seen but not quantified. The

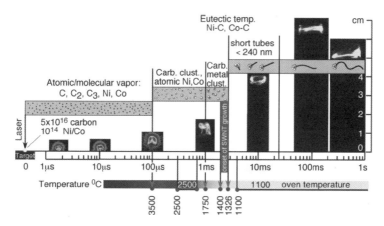

Fig. 10. Graphical summary of the CNT growth process in laser ablation as surmised by Puretzky, et al. on the basis of the body of spectroscopic studies. By tracking atomic, particulate densities and temperatures they were able to estimate the trends shown graphically here. The x-axis shows time on a logarithmic scale following the ablation event while the y-axis shows position within the furnace. Images shown are the scattering images of the plume. The different species present in the plume are depicted pictorially on the plot, while the temporal temperature trend is shown below. Reprinted from Applied Surface Science, Vol. 197, A.A Puretzky, et al., "Time-resolved diagnostics of single wall carbon nanotube synthesis by laser vaporization", p. 552, Copyright 2002, with permission from Elsevier. (Ref. [62])

pyrometry process simply collected emission spectra and fit the curve to blackbody radiation to extract temperatures. They were able to establish a temperature of near 3500°C at short times that decayed to the oven temperature of 1100°C within 4 ms. The authors noted that SWNT yield is greatest near the eutectic temperature of Ni/C and Co/C so surmised that nucleation begins near these temperatures. Fast extraction of sample matter from the oven at differing residence times was used to check the CNT growth rate and determine a general model for SWNT formation steps in the laser process as shown in Fig. 10 above.

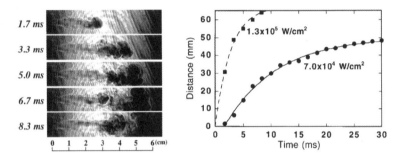

Fig. 11. (left) Shadowgraph of plume development (right) estimated plume propagation. With kind permission from Springer Science+Business Media: Applied Physics A, "Growth dynamics of carbon-metal particles and nanotubes synthesized by CO2 laser vaporization", Vol. 69, 1999, p. S232, Kokai, et al., Figs. 5&6. (Ref. [65])

Also in 1999, Kokai, et al., began reporting spectroscopic data on the laser ablation process using a continuous 1 kW $CO_2$ laser.[65, 66] Their spectroscopic data characterized a spectrum that was dominated by blackbody radiation at a temperature of around 3500 K during the ablation process. The blackbody radiation was attributable to particulate matter and was found to have lifetimes of up to 1 s when ablation was performed at an oven temperature of 1200°C, indicating that particulate matter remained suspended for a relatively long time. At short distances from the target, the $C_2$ Swan bands were also observed, but vanished approximately 4 cm from the target. This indicates that $C_2$ is formed and electronically excited for up to 3 ms after ejection from the surface but cools and condenses at longer time scales. The blackbody emission intensity was found to be greatest as the $C_2$ concentration began to decrease, indicating a combination of high temperature and particulate density coinciding with disappearance of $C_2$. Shadowgraphy of the plume was examined by imaging a HeNe laser across the system as shown in Fig. 11, which allowed visualization of particulate formation and estimation of plume propagation. The propagation was found to follow a classical drag-force model which predicts an exponentially decreasing velocity.

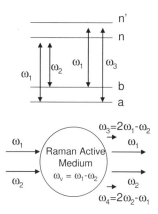

Fig. 12. Diagram of CARS process. (top) shows energy levels exciting from two lower energy vibrational states and the resultant CARS scattered frequency. (bottom) shows a schematic diagram of interacting beams with both Stokes and Anti-Stokes scattered frequencies. Adapted from Prog. Quantum Electron., Vol. 7, S.A.J Druet and J.P.E. Taran, "CARS Spectroscopy", p. 1-72, Copyright 1981, with permission from Elsevier. (Ref. [67]).

Dorval, et al., employed a host of diagnostics to help understand the continuous $CO_2$ laser ablation process.[64] Temperature of the system was measured using coherent anti-Stokes Raman scattering (CARS) on an added $H_2$ tracer gas. The CARS measurement method uses two pump lasers with a frequency difference equal to the vibrational (Raman) frequency of the probe molecule. The process is shown schematically in Fig. 12. The vibrational mode inelastically scatters the higher energy ($\omega_1$) pump beam and through coherent interaction with the pump and Stokes ($\omega_2$) beam, scattered light at two Raman shifted frequencies is produced. The frequency $\omega_3 = 2\omega_1-\omega_2$ is the anti-Stokes frequency, which is usually more intense than the Stokes frequency, $\omega_4 = 2\omega_2-\omega_1$. The corresponding energy level diagrams are shown in Fig. 12. Note that the pump laser energies don't necessarily correspond to an actual electronic transition of the molecule – Raman scattering always proceeds through excitation to a "virtual" energy level, though may be strengthened through a resonant interaction when a real energy level exists at or near that frequency. Temperature can be deduced by tuning the Stokes laser to excite different rotational lines, whose intensities can then be analyzed via Boltzmann plot to produce a ground state rotational temperature. In the typical set up, the pump laser is split into two beams that are focused into a small volume along with the Stokes laser and the CARS signal may be observed at an off-angle from the three beams. In Dorval's work, they used a frequency-doubled Nd:YAG (532 nm) and tunable dye laser (683 nm) as the pump and Stokes laser, respectively. Temperature obtained in this fashion was around 3200 K near the target, consistent with the target temperature measured by pyrometry. This temperature decayed with distance from the target, consistent with a finite element model of heat transport based on Navier-Stokes equations. A fast cooling region was identified in the first 3 mm, where the temperature dropped to 2100 K, with a slower cooling to the C-Ni and C-Co eutectic temperature of 1600 K at 7 mm. Beyond 7 mm, the temperature decayed slower still, approaching 1000 K as far as 20 mm out from the target. The authors postulated these regions could define a vapor-liquid-solid (VLS) mechanism as they correspond to gas, liquid and vapor phases of metal-carbon nanoparticles.

Dorval, et al. also used laser excitation to perform Laser Induced Incandescence (LII) to track soot volume fraction and LIF to track metal atom and $C_2$ density.[64] These techniques are similar to those employed by others.[59, 61–63] For incandescence, excitation was performed by a 532 nm doubled Nd:YAG laser transformed into a sheet across the plume and for fluorescence was used to pump a dye laser tuned around 248 nm. The incandescence data obtained was primarily qualitative and showed trends such as more confined and denser nanoparticle production in Ar as opposed to He ambient and an order of magnitude increase in density as target temperature increased from 3300 to 3500 K. The primary atomic emission lines for Ni and Co were observed at 338 and 350 nm, respectively, and were also used for fluorescence. Impurities were also observed in emission (Ca, 423 nm) and fluorescence (Fe). The intensity of Co lines excited at 248.46 and 248.62 nm was used to estimate electronic temperature of 2000 K at 4 mm from the target, which is similar to the temperature obtained by CARS. $C_2$ fluorescence was excited at 437.17 nm, corresponding to the (3,1) vibrational transition of the Swan bands, with fluorescence detected at 469 nm. These spatial profiles are collected in Fig. 13 along with other data obtained by this group. The Ni and Co signals decrease with increasing target temperature, while the $C_2$ signals increase. The spatial profiles of Ni and, to a lesser extent, Co are fairly flat over several mm, while the $C_2$ intensity is sharply peaked within 1-2 mm. These are consistent with other observations of a much faster quenching of $C_2$ species in comparison to that of the metal atoms. Dorval noted that optimum CNT production occurred for substrate temperatures near 3500 K where the $C_2$ concentration was higher and better overlapped spatially the maximum metal atom concentration.

Cau, et al. expanded on Dorval's work by performing a wider range of experimental conditions and comparing it with a detailed kinetic model of metal and carbon clustering.[56] They also performed LIF on $C_3$, exciting at 433 nm and observing fluorescence at 405 nm, corresponding to different rovibronic energies of the $A^1\Pi_u$-$X^1\Sigma_g^+$ comet band. They again observed a faster disappearance of Co relative to Ni and showed with the kinetic model that a faster clustering of Co could be expected assuming a higher dimer bond energy.[56] They also observed that the Ni clustering was enhanced by the presence of Co, though the reverse was not true. This was explained in the model by adjusting the He dilution rate. The reasons why this might be the case were not clear, though it was also noted that Co-C clustering may occur but was not included in the model. It might also indicate possible formation of Ni-Co alloy clusters, which also was not included in the model. They found that the presence of Co resulted in a fast decay of $C_2$ density, but did not alter the $C_3$ profile significantly. This points to possible Co-$C_2$ clustering processes. The short lifetime of $C_3$ suggests its role is primarily in soot particle formation.

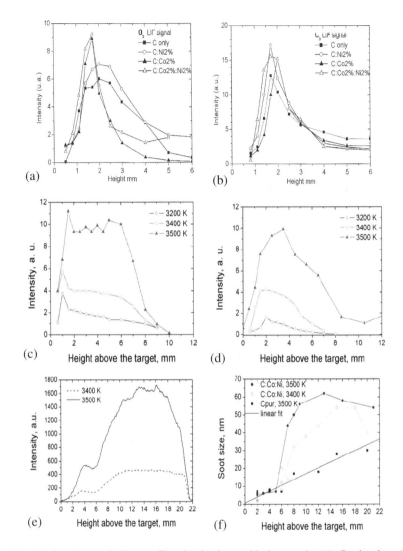

Fig. 13. Species evolution profiles in the laser ablation work. (a) $C_2$ density, (b) $C_3$ density. Reprinted with permission from Ref. [56], Cau, et al., "Spatial evolutions of Co and Ni atoms during single-walled carbon nanotubes formation: Measurements and modeling," J. Nanosci Nanotech., Vol. 6, p. 1298. Copyright 2006, American Scientific. (c) Ni atom density, (d) Co atom density, (e) particulate density and (f) particulate diameter. Reprinted with permission from Ref. [68], Cau, et al., "Synthesis of single-walled carbon nanotubes with the laser vaporization method: Ex situ and in situ measurements", Phys Status Solidi B, Vol. 243, p. 3063, Copyright 2006, Wiley-VCH Verlag GmbH & Co. KGaA.

Cau, et al. also quantified the LII data and LIF data at extended ranges beyond the end of the target.[68] The results, shown in Fig. 13, confirm that particle formation first coincides with the disappearance of $C_2$ species and results in the formation of particles approximately 6 nm in diameter. Up to 6 mm away from the target, the density and diameter of particles remains fairly constant, as do the concentration of metal atoms. As the Ni atom concentration begins to decrease, the density and diameter of particles increase significantly, up to 4 times as much in volume fraction and up to approximately 60 nm in diameter. The point of this transition is close to the point where the gas has cooled to the eutectic temperature.

The conclusion of the laser ablation studies is best told by Fig. 10, while Fig. 13 demonstrates many of these measurements directly in one particular system. What is notable about all these studies, as opposed to other types of CNT growth studies, is that the spectroscopic data and conclusions of the different studies are more or less in agreement with each other. Several different studies have monitored the formation of low molecular weight carbon species, $C_2$ and $C_3$, and find that they are quenched quickly relative to metal atoms such as Ni and Co. The formation of particulate matter has been tracked by scattering and is seen to coincide temporally and spatially with the disappearance of $C_2$ species. Disappearance of metal atoms occurs at later times but has been correlated with a further increase in particle density and mean diameter. Temperature measurement data suggest a temperature on the order of 3000-4000 K initially, which then cools down toward the oven temperature. Nucleation of mixed metal carbon nanoparticles occurs roughly at the eutectic temperature for metal-carbon mixtures in the range of 1400-1700 K. This suggests possible mechanisms related to nucleation of nanodomains within the clusters that spawn nanotube growth. It is notable that the particles observed here are all significantly larger than the CNTs themselves, in contrast to numerous studies in other systems suggesting a much closer correlation of nanoparticle and CNT diameter.[69]

## 4. Glow Discharge

The plasma regime commonly applied for materials processing, such as etching and deposition, is classified as "glow discharge" and pertains to conditions of relatively low electron temperature (a few eV) and density ($10^{8-12}$ cm$^{-3}$). Being the common regime for materials processing and particularly PECVD processes, this type of plasma has been widely adapted to carbon nanotube growth. The types of plasmas used for this type of processing are various, being excited by RF (both capacitively and inductively), microwave or DC power sources, but are all within the similar plasma regime. There are a few review papers which have summarized the work done in PECVD of carbon nanotubes and nanofibers.[26, 70] One of these reviews intended to engage the community of low temperature plasma processing specialists to apply the variety of diagnostic techniques used for studying microelectronics plasmas, however sophisticated spectroscopic techniques have yet to be employed on these systems.

In this section, all these types of glow discharge plasmas are grouped and discussed together, often without differentiating results based on the type of glow discharge employed. However, there are some important differences which do affect spectroscopic data and may affect growth mechanisms. In dc plasmas, the majority of energy in the plasma is lost near the cathode, where ions and electrons are accelerated across potentials of several hundred volts. Reaction chemistries may occur in the negative glow region where the accelerated electrons may create dissociation, ionization and excite species into emitting states. Studies of optical emission in dc plasmas are therefore primarily analyzing this region with high energy electrons at low densities and low degrees of precursor dissociation. The relative amount of energy lost in these collisions is low relative to microwave and ICP discharges. In the ICP and microwave discharges, the alternating current fields applied move electrons back and forth in the discharge, increasing their energy and inducing more chemistry-driving collisions. The ions, being of higher inertia, are not in general able to respond to the quickly varying ac fields and are moved only by static dc bias fields set up to maintain quasineutrality in the plasma. This acceleration occurs primarily at the "sheaths" (i.e. plasma/electrode boundaries.) The sheath potentials are small relative to the dc plasma, and are usually larger in ICPs than in Microwave plasmas, particularly when some capacitive component is present in the ICP.

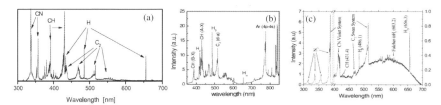

Fig. 14. Sample Optical Emission spectra from different glow discharges. (a) DC $C_2H_2/NH_3$ plasma. With kind permission from Springer Science+Business Media: Applied Physics A, "Dc plasma-enhanced chemical vapour deposition growth of carbon nanotubes and nanofibres: in situ spectroscopy and plasma current dependence", Vol. 88, 2007, p. 261, Jonsson, et al., (Ref. [71]), (b) Inductively coupled $CH_4/H_2/Ar$ plasma. Reprinted with permission from Ref. [72], I. B. Denysenko, et al., J. Appl. Phys. 95, 2713 (2004). Copyright 2004, American Institute of Physics. (c) Microwave $CH_4/N_2$ plasma. Reprinted from Carbon, Vol 41, Wang, et al., "Optical emission spectroscopy study of the influence of nitrogen on carbon nanotube growth", p. 1827, Copyright 2003, with permission from Elsevier. (Ref. [73])

A summary of spectroscopic studies published in the literature is given below in Table 3. The table is not meant to be exhaustive but includes many of the studies done in this area. Not included in this list are studies of these plasmas by Mass Spectrometry, of which there are several.[72, 74–76] The studies span three major types of plasma excitation methods (ICP, DC and Microwave) and to a smaller extent, RF or microwave excited atmospheric pressure glow discharges. Primarily, either methane or acetylene is the carbon precursor of choice, with a few variants. Most studies have employed either hydrogen or ammonia as a reducing gas to control amorphous carbon buildup, though a few authors have not used any sort of reducing gas. Argon is often also added to the system, usually to help stabilize the glow discharge and aid in igniting the plasma. By far, the dominant technique used to characterize the gas is optical emission spectroscopy (OES). Some authors have used OES to extract or estimate electron or neutral temperatures, though most have used it as a qualitative indicator of plasma chemistry. There are only two instances of absorptive spectroscopic approaches in the UV and IR. These results will be summarized below.

Table 3. Summary of spectroscopic studies of different glow discharges for CNT and CNF synthesis

| Plasma Type | Carbon Source | Reducer/ Diluent | Group | Spectroscopy |
|---|---|---|---|---|
| ICP | $C_2H_2$ | $H_2$ | Oak Ridge National Lab[42] | OES |
| ICP | $CH_4$ | $H_2$/Ar | Nanyang Tech Univ[72] | OES |
| ICP | $C_2H_2$ | $H_2$/Ar | Tsing Hua Univ[77] | OES |
| ICP | $CH_4$ | $H_2$ | NASA Ames[78] | OES |
|  |  |  |  | UV Absorption |
| ICP | $CH_4$ | $H_2$/$O_2$ | Stanford Univ[79] | OES |
| ICP | $CH_4$ | $H_2$/Ar | Univ of Michigan[80] | OES |
| DC | $C_2H_2$ $C_3H_4$ CO | $NH_3$ | Sungkyunkwan Univ[81] | OES |
| DC | $CH_4$ | $H_2$ | Kyoto Inst Tech[35] | OES |
| DC | $C_2H_2$ | $NH_3$ or $H_2$ | Goteberg[71] | OES |
| DC | $C_2H_2$ $CH_4$ $C_2H_4$ | $NH_3$ | Cambridge[82] | OES |
| DC | $C_2H_2$ | $NH_3$ | Kyung Hee Univ.[83] | OES |
| Pulsed DC | $C_2H_2$ | $NH_3$ | Univ. Barcelona[84] | OES |
| μWave | $CH_4$ | $H_2$/$NH_3$ | KAIST[85] | OES |
| μWave | $CH_4$ | $N_2$ | CAS[73] | OES |
| μWave | $C_2H_2$ | $NH_3$ | IIT Delhi[86] | OES |
| μWave | $CH_4$ | Ar | Heriot-Watt[87] | OES |
| μWave | $CH_4$ | $CO_2$ | Chiao-Tung Univ[88] | OES |
| μWave | $CH_4$ | $H_2$ | Purdue Univ[89] | OES |
| Atmospheric | $C_2H_2$ | He | UT Dallas[90] | FTIR |
| Atmospheric | $CH_4$ | Ar/$H_2$ | Masaryk Univ.[91] | OES |

Optical emission spectroscopy (OES) is the most frequently employed analysis approach for glow discharges. Typical optical emission spectroscopy traces are shown in Fig. 14. The exact species observed depend on gas composition and condition. The atomic hydrogen Balmer series is almost always observed ($\alpha$, $\beta$ and sometimes $\gamma$ peaks) as a decomposition product of $NH_3$, $H_2$ or hydrocarbon precursors. The CH radical is commonly observed near 431 nm though in many cases this emission is weak. Chemistries utilizing ammonia display NH radical emission at 335 nm as a direct dissociation product and also usually significant signal from the CN violet system (388 nm) as a recombination product of the hydrocarbon/ammonia precursors. In DC plasmas where CN was positively identified, peaks around 356-358 nm have been assigned to both $N_2$[82] and $N_2^+$[81, 83, 84], however these assignments are likely incorrect as this region is also assignable to the CN violet system and other bands from the $N_2/N_2^+$ systems were absent.[92] The presence of the $N_2$ 2nd positive band from ammonia containing discharges in microwave plasmas is more definitive.[86] This is explained by the greater degree of ionization in microwave discharges as compared to dc discharges inducing greater dissociation of precursors and subsequent formation of $N_2$ via recombination. The Swan band of the $C_2$ dimer is also observed in many, but not all, plasma systems. In plasmas employing $H_2$, a multitude of peaks attributable to various $H_2$ systems may be observed.

One of the first studies to utilize OES in CNT plasmas looked at the spatial variation in the atomic hydrogen line intensities to identify the location of the plasma sheath, but did not attempt to ascertain any of the plasma chemistry from the spectra.[35] One of the first studies to employ OES for understanding plasma chemistry tracked the strength of the H-$\alpha$ Balmer line on an RF inductively coupled plasma of methane and hydrogen. The authors observed that its intensity increased with power and decreased with argon dilution, which correlated with a transition between CNT and CNF growth.[41] It was therefore proposed on the basis of the mechanisms of Nolan et al.[93] that atomic hydrogen played a role in this transition by satisfying the edge chemistry of the carbon nanofiber structure. While the authors later qualified this conclusion with the recognition that other quantities in the system are changing concurrently, especially substrate temperature,[78] this mechanism has been supported by other authors.[70, 71]

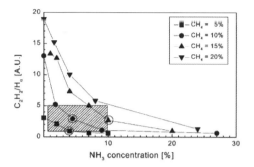

Fig. 15. Plot depicting metric of $C_2H_2$ intensity in mass spec ratioed against H $\alpha$ line intensity. CNT growth was only observed in the region marked with a shaded rectangle. Circled points displayed higher density growth. Reprinted from Diamond and Related Materials, Vol 11, Woo, et al., "In situ diagnosis of chemical species for the growth of carbon nanotubes in microwave plasma-enhanced chemical vapor deposition", p. 59, Copyright 2002, with permission from Elsevier. (Ref. [85])

When $NH_3$ was substituted for $H_2$, Woo, et al., observed CN and H$\alpha$ emission to increase in a similar fashion.[85] They compared the emission spectra to mass spectrometry data and found the stable byproduct HCN to follow the same trend. Varying $CH_4$ content, on the other hand, resulted in little change in H intensity, indicating that the H atom is primarily derived from $NH_3$ and $H_2$ dissociation. They also compared H$\alpha$ emission to $C_2H_2$ density measured in the mass spectrometer and used this ratio as a metric for CNT growth conditions. They found that a low ratio of $C_2H_2$/H$\alpha$ was required to obtain growth, as shown in Fig. 15. However, when the $NH_3$ concentration became sufficiently large, CNT growth was not observed at all and this metric did not apply. Starting from a gas mixture of $C_2H_2$/$NH_3$, Srivastava, et al., observed that substituting $H_2$ for $NH_3$ reduced and eliminated CN emission while only slightly reducing the strength of the H$\alpha$ line.[86] Substituting $N_2$ for $NH_3$, on the other hand, increased CN and eliminated H$\alpha$. This is consistent in showing that the H atom is derived most readily from $NH_3$, less so from $H_2$, and insignificantly from hydrocarbons. In Srivastava's work, the highest growth rates were obtained with mixtures of $H_2$/$NH_3$ or $N_2$/$NH_3$, where the H atom emission was reduced somewhat and the CN intensity was approximately the same. However, both of these conditions resulted in reduced G/D ratio in Raman spectroscopy, and the degree of cone structure in the CNF was reduced as the precursors tended toward less N content. The results indicate the H atom may act as an etchant while the N atom is more important in creating the nanofiber cone structure. This growth regime differed from that identified by Woo, perhaps because of differences in hydrocarbon precursor (acetylene vs. methane). In mixtures of $CH_4$ with pure $N_2$, the CN and $C_2$ intensities were found to peak at low $CH_4$ concentrations (~5%).[73] A substantial decrease in overall emission intensity with increasing $CH_4$ concentration suggests that there is a more efficient ionization of $N_2$ and N relative to $CH_4$ derived species.

In mixtures with only $NH_3$, the cyanogen radical intensity generally displays the same trend as H atom emission. Lee, et al., observed both to increase with increasing $NH_3$ fraction in $C_2H_2/NH_3$ discharge.[81] The length of CNFs produced decreased with increasing $NH_3$ fraction and increased Hα emission, again pointing to the possible effect of H atom as an etchant. The trend with $C_3H_4$ in place of $C_2H_2$ was opposite in regards to effect of $NH_3$ addition, though the relationship between CNF growth and Hα/CN intensity was consistent. Lim, et al. observed identical trends with $NH_3$ percentage, emission, and CNT growth quality.[83] Using a triode just above the cathode, they observed increasing intensity of Hα and CN emission with increasing bias but did not see a reduced growth quality. This omission trend originates from the energetics of the electrons being accelerated in the triode and does not indicate a change in plasma chemistry. An increase in Hα and CN intensities relative to other species was also observed with increasing plasma current by Jonnson, et al.,[71] though again the growth quality did not correlate because the plasma power, and thus surface heating, was increasing concurrently. Thus, one must be careful in interpreting emission trends without deconvoluting the impact of the electron energetics on the emission, and considering the effect of other important, and correlated, properties in the plasma. This will be discussed in more detail below.

Several authors have observed a weak dependence on conditions for both $C_2$ and CH emission.[81, 83, 86] Jonsson, et al., observed that CH emission intensity was always proportional to $C_2$ and thus concluded that CH was a byproduct of $C_2H_2$ dissociation.[71] On the other hand, Caughmann, et al. and Lin, et al. both observed significant CH emission when only $H_2$ was present, indicating CH is formed by hydrogen reaction with carbon deposits on the wall.[42, 77] Lin et al. also observed much stronger dependence of $C_2$ intensity on operating parameters than for CH, whose intensity generally scaled with Ar emission intensity. In a microwave powered atmospheric torch, the $C_2$ intensity and, to a lesser extent, CH, was found to increase when a sample was placed into the torch stream.[91] This indicates that both of these species are not strongly quenched by surface reactions, but may in fact be surface reaction products under certain conditions. The differences in these studies may originate from the different forms of plasma ignition, i.e. dc versus inductively and microwave coupled discharges. The latter cases should be more efficient at dissociating $H_2$. Less gas phase recombination of atomic H in an ICP would be expected at the lower pressure conditions as well, making surface reactions more important.

A few works have looked at emission for O-containing species. In $CH_4/CO_2$ mixtures Chen, et al., observed two different regimes depending on gas ratio, where the emission spectra was either dominated by $CO/C_2$ or OH/CH, with the former case providing better conditions for CNT growth and the latter for diamond or amorphous carbon.[88] However, the emission line assignments in the work are questionable, and the appropriate conclusion to be drawn here is unclear. Zhang, et al., substituted $O_2$ for $H_2$ in an upstream ICP discharge and observed significant enhancements in growth rate and CNT quality.[79] Though they incorrectly assigned one of the CO $3^{rd}$ Positive bands to atomic oxygen, the presence of the OH radical at 306.4 nm was identified with little ambiguity. It is of interest to note that significant advances in thermally driven nanotube growth have evolved surrounding chemistries that might involve the OH radical, such as alcohols[94, 95] and trace amounts of water.[96] This might suggest that the OH radical itself plays a role in CNT growth chemistry, and it is speculated that the OH removes amorphous carbon that would otherwise deactivate the catalyst. Zhang, et al., however, dismissed this popular conclusion and instead surmised that the oxygen scavenges atomic hydrogen which is otherwise deleterious to CNT growth. Which mechanism is correct is still unknown and further study would be required to approach a definitive answer.

Plasma analysis via emission alone is potentially quite subjective and qualitative. This is because the intensity of emission from a plasma species, $I_A$, is described by a combination of the species density, $n_A$, and electron energy distribution function (EEDF), $f_e(E)$. If the major important processes in producing emission are electron-impact excitation from the ground state and radiative decay, the intensity is described as follows:

$$I_{i \to j} = S(\lambda_{ij}) q_{ij} n_A \int_{E_{th}}^{\infty} \sigma_A(E) f_e(E) \left( \frac{E}{2m_e} \right)^{\frac{1}{2}} dE \qquad (2)$$

where $\sigma_A(E)$ is the excitation cross-section, $E_{th}$ is the threshold energy for transition, $S(\lambda_{ij})$ is the spectrometer sensitivity at the emission wavelength, $\lambda_{ij}$, and $q_{ij}$ is the branching ratio for that transition. The species density therefore is only one factor determining emission intensity. One popular method to correct for the electron energy distribution function is to normalize the emission intensity by the emission from an inert (actinometer) gas, such as argon. In this case, Eq. (2) may be written as:

$$\frac{I_A}{I_{Ar}} = \frac{S(\lambda_A) q_A n_A \int_{E_{th,A}}^{\infty} \sigma_A(E) f_e(E) \left( \frac{E}{2m_e} \right)^{\frac{1}{2}} dE}{S(\lambda_{Ar}) q_{Ar} n_{Ar} \int_{E_{th,Ar}}^{\infty} \sigma_{Ar}(E) f_e(E) \left( \frac{E}{2m_e} \right)^{\frac{1}{2}} dE} = C(f_e) \frac{n_A}{n_{Ar}} \qquad (3)$$

This normalization thus relates density to intensity with a functional relationship to the electron energy distribution function. The approach may be used for quantitative comparison over a range of conditions if the functional $C(f_e)$ is a constant. This would be exactly true under two conditions: (i) if the actinometer and the species of interest have an identical energy dependence, i.e. $\sigma_{Ar}(E)/\sigma_A(E) = \text{constant}$, or (ii) if the shape, but not necessarily the magnitude, of the EEDF is invariant (e.g if the electron distribution is Maxwellian and the electron temperature is constant). These two cases are rarely realized, however it is usually found to be approximately true when the excitations have similar threshold energies (i.e. $E_{th,A} \approx E_{th,Ar}$).

In many of the studies discussed above, a crude form of actinometry has been applied by examining variations in the ratios of particular emitting lines instead of absolute intensities. Lin, et al., however, used actinometry to study the variation in H, $C_2$ and CH emission, while noting that the actinometry assumption was probably only valid for the Hγ line, whose excitation threshold is similar to that of the 750 nm Ar line.[77] In their studies, they found that H atom density decreased with increased $C_2H_2/H_2$ ratio and increased at higher power. The growth quality was generally best at the cases where H atom density was highest, though this is not necessarily due to the H atom. For instance, significant growth rate enhancement was observed for increased substrate bias, although the H atom density as measured by actinometry did not change. Collard, et al., performed spatially resolved Ar actinometry using the Hα line.[80] They noted the impact that $H_2$ molecular dissociation has on Hα emission intensity (see below) and showed that it represents a constant offset to the density under conditions of constant electron temperature, as they expected in their plasma system. The results showed a H atom density peaked in the center of the discharge, dropping to a minimum near the electrode edge, then increasing outside the plasma zone, especially at higher pressure. This increase outside the plasma zone is a result of the assumed constant electron temperature and not a physical increase in density.

More detailed analysis of emission spectra has been performed by Cruden and Meyyappan to extract information about degree of $H_2$ dissociation and electron temperature and density in the plasma.[78] Their study also checked some of the actinometry assumptions. The analysis considered several different collisional-radiative paths to generating H and Ar emission. This involved accounting for dissociative excitation (for $H_2$), direct excitation from the ground state, cascaded excitation through higher energy levels, radiation trapping (for H branching ratio and population of Ar via cascades), collisional quenching and radiative decay. Considering all of these processes result in three general equations to describe the emission intensities of various lines:

$$I_{n'\to n''} \sim \frac{S\left(\lambda_{n'n''}\right)}{\lambda_{n'n''}} A_{n'n''} n_{n'}$$

$$n_{H,n'} = \frac{k_{exc}^{n'}\left(T_e\right) n_e n_H + \sum_i k_{diss,i}^{n'}\left(T_e\right) n_e n_i}{\sum_{n''=2}^{n'-1} A_{n'n''} + A_{n'1}\Lambda_{n'}\left(n_H\right) + \sum_i k_q^{n',i}\left(T_g\right) n_i}$$

$$n_{Ar,n'} = \frac{k_{exc}^{Ar}\left(T_e\right) n_e n_{Ar} + \sum_{n'' \, cascade} \dfrac{br_{n'',n'}^{thick}}{1+\left(\dfrac{br_{n'',n'}^{thick}}{br_{n'',n'}^{thin}}-1\right)\Lambda_{n''}(n_{Ar})} k_{n''}^{Ar}\left(T_e\right)}{\sum_{n''} A_{n'n''} + \sum_i k_q^{n',i}\left(T_g\right) n_i} n_e n_{Ar}$$

(4)

The first equation describes a general relationship for emission intensity given the density of an excited state. The second and third equations are number balances for the density of the excited states for H and Ar, respectively. Cruden and Meyyappan compiled the various rate constants from a variety of sources, expressing them as a function of electron temperature, assuming a Maxwellian electron energy distribution. The reader is referred to their publication for further explanation.[78] The equations in (4) reduce to one equation for each individual atomic line. A three parameter fit ($n_H$, $n_e$ and $T_e$) was only able to adequately reproduce three of the lines across all conditions studied (Ar 763 nm, H$\alpha$ and H$\beta$). In some cases, the H$\gamma$ and 750 nm Ar line were also predicted well. The authors attributed the lack of fit of the other measured Ar lines to uncertainties in the quenching coefficients and a non-Maxwellian plasma temperature. Electron temperatures were predicted to be in the range of 2-4 eV and H atom mole fraction was around $10^{-3}$. These fits were compared against actinometry using both Ar and H$_2$ molecule as the actinometer gas. The results in Fig. 16 show that H atom actinometry may be qualitatively correct over certain conditions. It is also apparent that using the H$\gamma$ line rather than H$\alpha$ is slightly more accurate, though the low intensity of H$\gamma$ makes it difficult to observe reliably at low power. Comparison of the fit electron temperature versus electron temperature determined by ratioing and normalizing the H $\alpha$ and $\beta$ lines,

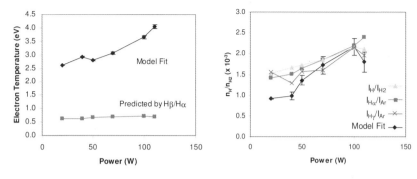

Fig. 16. Predicted electron temperature and ratio of H atom to $H_2$ molecule from a detailed collisional-radiative model. Also shown are comparisons to predictions via actinometry or analysis of Hβ to Hα ratio.

as done by Wilson, et al.,[87] underestimates the electron temperature and does not capture the entire trend.

A few authors have used the relative intensity of the hydrogen atomic Balmer series to estimate the electron, or electronic, temperature in the discharge. Lin, et al. used this ratio in a qualitative fashion and found it to be invariant with power, bias, and $C_2H_2$ flow rate in an inductively coupled discharge.[77] Wilson, et al. quantitated the α and β line ratios to determine an electronic temperature that was around 970°C in a microwave discharge.[87] This number is too low to be reasonable for a glow discharge, which typically displays a non-equilibrium of electronic temperature with the neutral temperature due to electron impact excitation. Thus, electronic temperatures should be closer to the electron temperature which is typically a few eV. Reasons for this underestimate have been explained by a few authors, albeit in different types of plasmas. Jonsson, et al. attempted a Boltzmann analysis of all three (α, β and γ lines) and found that they were not described by the Boltzmann relationship and therefore were not thermalized.[71] The detailed excitation model explained above helps understand this non-equilibrium, and also shows how the temperature is underestimated.[78] Under conditions of a capacitively coupled RF discharge, the dominant physics for H atom emission are dissociative excitation of $H_2$ and de-excitation by collisional quenching via neutral collision. Radiation trapping also

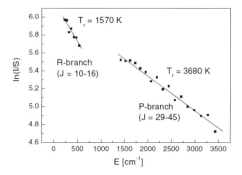

Fig. 17. Boltzmann plot of $C_2$, displaying a thermal non-equilibrium between low and high energy rotational states. Reprinted with permission from Ref. [91]. Copyright Institute of Physics.

significantly impacts the population of the $n = 2$ state responsible for the H-$\alpha$ line, but not so much for the higher states. Thus, the overall population is determined by a balance of kinetic processes and would not represent a thermal distribution.

Temperature measurement in glow discharges have also been attempted by analyses of the $C_2$ Swan rotational and vibrational bands in emission spectra. Jonsson, et al. used the ratio of the vibrational bands to extract a vibrational temperature on the order of 1-2 eV.[71] The large magnitude of this temperature indicates that the vibrational temperature is likely determined by electron impact energies and is nowhere near equilibration with the neutral temperatures. Zajickova, et al., used the $C_2$ Swan bands to estimate rotational temperature in a microwave plasma torch, finding temperatures in the range of 3200-3900 K.[91] They also observed the same $C_2$ self absorption effect reported for arc discharges,[47] and therefore realized different temperatures when analyzing spectra from lines near to and far away from the band head, as shown in Fig. 17. The magnitude of temperature measured is similar to that measured in OES of arc discharges,[47] laser plumes[57] and inductively coupled plasmas.[97] However, in the latter two cases, it has been shown that the excited state temperature of $C_2$ is not equilibrated with the ground state and appears to be "hotter" due to chemiluminescent reactions. In the case of the microwave torch, it is notable that analysis of

OH spectra from a pure Ar discharge with the same energy density gave temperatures of only 770 K. While the $CH_4/H_2$ containing discharge may expect a larger heating effect due to Frank-Condon heating,[98] this large disparity suggests the $C_2$ temperature may be overestimated, and the two different temperatures for different J ranges may be due to a non-thermal excitation process rather than self-absorption.

Analysis of $H_2$ spectra has also been attempted to quantify the neutral rotational spectra. Collard, et al., used a Boltzmann analysis of the $G^1\Sigma_g^+ \rightarrow B^1\Sigma_u^+$ (0,0) transition to measure rotational temperature in an inductively coupled GEC Reference Cell.[80] This system was chosen on the basis of several prior studies successfully estimating temperature from this band,[99–101] although several other studies have suggested this band to be perturbed and unreliable for temperature measurement.[102–104] The temperatures obtained were in the range of 400-520 K, and display monotonically increasing dependence on pressure and power. This temperature measurement appears low in comparison to measurements on inductively coupled plasmas of similar power density but differing composition (e.g. $CF_4$, $O_2$, $Cl_2$, Ar).[97, 98, 105–108] This is probably because they used the upper state rotational constants in analyzing the Boltzmann distribution rather than the lower state constants. At the pressures studied (<100 mtorr), the upper state will radiatively decay before it can thermalize, and thus the rotational distribution will reflect that of the ground state. The ground state rotational constants differ by about a factor of 2 from that of the upper state, so the correct temperature is more likely in the range of 800-1100 K. The dependence on pressure and power is as expected based on global heat balance considerations.

Cruden and Meyyappan used selected lines from the $H_2$ spectra with well-quantified line strengths to perform a temperature fit.[78] This included 25 different vibrational bands originating from 6 different electronic transitions. Some of these lines had to be discarded because they were not resolved from other known $H_2$ lines. An additional 1175 individual lines from $H_2$ were identified and included in the fit, though they did not yield any temperature information. A total of 21 different originating states were analyzed in this way, and significant scatter in the data was obtained. Eight of these states had previously been identified as unreliable for temperature measurement, and one state was clearly non-physical and thus discarded. Of the remaining 12 states, only 4 yielded consistent temperature information over the entire range of experimental conditions studied. The bands originated from the $d^3\Pi$ $(v=0)$, $J^1\Delta$ $(v=2)$, and $j^3\Delta$ $(v=1,3)$ states of $H_2$. They estimated a rotational temperature in the range of 800-1200 K, which was not strongly dependent upon experimental conditions, and mostly contained within rather large error bars of the fitting method. This temperature was somewhat larger than the estimated mean gas temperature based on the temperatures of all the surrounding surfaces. A global heat balance was performed, introducing a heating term for energy transfer via electron impact processes in the plasma and agreed somewhat with the predicted temperature trends.

Garg, et al. used a Boltzmann analysis of the Q-branch of the (0,0) Fulcher $\alpha$ ($d^3\Pi_u$-$a^3\Sigma_g^+$) transition of $H_2$.[89] Five of the first ten lines in this series were found to give a good Boltzman dependence while the others were overlapped and could not be independently measured. The method of Cruden and Meyyappan was also applied against some of this data, and yielded consistent results, even though it did not use the same line series.[109] The authors also introduced trace amounts of $N_2$ to the discharge and fit the rovibrational envelope of $N_2$ $C^3\Pi_u$-$B^3\Pi_g$ bands to check the measurement method. It is generally accepted that $N_2$ is a good thermometric species and should represent the true neutral temperature because of the absence of non-thermal means of populating the emitting state. Their results, shown in Fig. 18, indicate that the temperature measured by $N_2$ lies between the temperatures derived from $H_2$ assuming either the upper or ground (not lower) state rotational constants. In a low pressure plasma, emission occurs before the upper state undergoes any thermalizing collisions, and therefore its rotational population will reflect that of the ground state from which it is excited. At higher pressures, however, the upper state may equilibrate and display a thermal rotational population. At Garg's condition of 10 Torr, apparently the equilibrium lies somewhere in the middle of these two cases. The vibrational temperature of $N_2$ is always larger than the rotational temperature and shows an inverse correlation to the true neutral temperature.

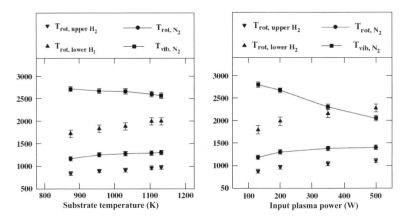

Fig. 18. Fit temperature trends in a microwave $CH_4/H_2$ discharge. The temperature of the thermometric species $N_2$ lies between that estimated from $H_2$ assuming rotational population determined by excited or ground state. Reprinted with permission from Ref. [89]. Copyright Institute of Physics.

Two types of absorption spectroscopy, ultraviolet[78] and infrared (FTIR),[90] have been employed on CNT producing plasmas. The former technique is useful for monitoring electronic transitions of specific species, such as $CH_3$, while the former looks at rovibrational transitions which may be assigned to specific molecules, such as $C_2H_2$. For methane plasmas, the methyl radical is observed with a well-characterized absorbance near 218 nm, as shown in Fig. 19. In this case, a density of approximately $10^{13}$ $cm^{-3}$ yielded an absorbance of 0.3%. This density was tracked versus pressure and plasma power, and also measured in the presence and absence of substrate heating. The $CH_3$ density increased nearly linearly with plasma power and decreased monotonically with increasing pressure (corresponding to an asymptotic increase in mole fraction up to 0.001 at 20 torr, where CNT growth was not observed). An interesting observation of the work was that the $CH_3$ density increased by approximately 50% when the substrate was heated to 900 C, in spite of any expansion of the gas expected due to the ideal gas law. This was attributed to the thermally driven hydrogen abstraction from methane, $CH_4 + H \rightarrow CH_3 + H_2$. This reaction was predicted to be the dominant method for methyl radical formation and methane decomposition in the plasma kinetics model of Delzeit, et al.[41]

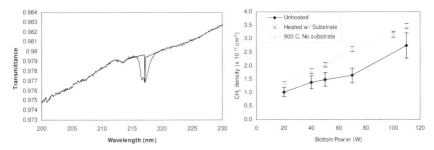

Fig. 19. UV Absorption spectra of $CH_3$ in a capacitively coupled $CH_4/H_2$ plasma. (left) sample spectrum, (right) density versus power with and without substrate heating. Reprinted with permission from Ref. [78], B. Cruden and M. Meyyappan, J. Appl. Phys. 97, 084311 (2005). Copyright 2005, American Institute of Physics.

Chandrashekar, et al. used FTIR absorption spectroscopy to study the chemistry of an atmospheric pressure plasma jet using acetylene and helium.[90] In this work, they examined the absorption attributable to the acetylene molecule and fit the rotational band envelope to determine ground state rotational temperature, and used Beer's Law to extract the density of $C_2H_2$ molecules. They found gas temperatures of approximately 350-400 K, which increased with increasing substrate temperature and plasma power, although this variation (~50 K) was within the measurement accuracy. The density of the acetylene was on the order of 1.2 x $10^{17}$ $cm^{-3}$, and decreased with both substrate temperature and plasma power, though in both cases this decrease was consistent with or less than the predicted density change based on the ideal gas law. While the dependence of these parameters is consistent with the range of substrate temperatures explored, the lack of strong dependence on plasma power is not well explained. The temperature rise in the gas would consume only $1/15^{th}$ of the total power. The authors proposed a somewhat larger fraction lost in decomposition, though the density change observed could be attributed to gas expansion rather than decomposition, with the remainder lost through some form of thermal parasitics. It is possible that the actual gas temperature and degree of decomposition is underestimated due to absorption by gas outside of the plasma zone,[110, 111] despite efforts of the minimize this effect. They found that the CNT growth rate and quality, as measured by Raman, both increased monotonically with these two parameters. This might suggest that a slight elevation in acetylene temperature increases its surface reactivity, though it seems more likely that some unmeasured carbon radical species play an important role in enhancing growth rate and quality.

The characterization of PECVD and other glow discharge approaches for carbon nanotube and nanofiber synthesis is complicated relative to the arc discharge and laser plume characterization as there are generally considerably more species present and the systems are generally further from equilibrium. The arc discharges and laser plumes are composed primarily of C and one or two metal atoms plus clusters of these, while the PECVD processes are typically based on hydrocarbons and contain etchants such as $H_2$ and $NH_3$. It is not generally possible to assume local thermal equilibrium (LTE) in the glow discharge and often not even possible to assume Maxwellian temperature distributions. The literature is therefore full of very qualitative observations about the state of the plasma. The few attempts made to quantitate the plasma parameters are riddled with ambiguity. Therefore there is little definitive knowledge gained from these studies and instead generalities, such as the role of H as an etchant are instead surmised. Much more detailed and focused study would be required to obtain a thorough understanding of these types of plasmas.

## 5. Flames

Flames, a low energy form of plasma, have recently seen increased interest in use for CNT synthesis. However, we are only aware of one study that employed spectroscopic analysis of the gas phase during synthesis.[112] Xu, et al., applied spontaneous Raman spectroscopy (SRS) to extract gas temperature and densities of $H_2$, CO, and $C_2H_2$. The approach directed a 532 nm pulsed Nd:YAG laser into the flame, and collected the scattered light into a gated ICCD detector. The $N_2$ anti-Stokes spectrum was fit against reference spectra to determine temperature while the other species densities were determined from the signal strengths at their expected locations (4160, 2145 and 1980 $cm^{-1}$, respectively). Other species, such as $CH_4$ (the carbon precursor), and

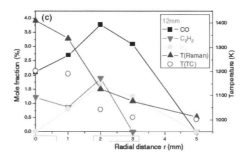

Fig. 20. Data collected from spontaneous Raman spectroscopy of a flame synthesis process. The blue box on the x-axis represents the approximate region where CNTs are produced for Fe and Ni based catalysts. Reprinted from Carbon, Vol 44, F.S. Xu, X.F. Liu and S.D. Tse., "Synthesis of carbon nanotubes on metal alloy substrates with voltage bias in methane inverse diffusion flames", p. 570, Copyright 2006, with permission from Elsevier. (Ref. [112])

$H_2O$ and $CO_2$, were monitored but no data was given. Figure 20 shows the spectral data collected at the optimum sampling height (12 mm). No CNTs were observed when sampled at 6 mm where the temperature is below 1100 K and no $H_2$ is observed, though CNTs were observed on Ni based probes between 9-15 mm where the temperature was generally in excess of this and $H_2$ was present. For Fe catalyst probes, higher temperatures (~1400 K or larger) were required and thus CNTs grew only in the radial center of the flame. The highest density of growth on Ni was obtained at $r = 3.25$ mm, where $C_2H_2$ density is below the detection limit, but CO density remains high. While the data is not conclusive regarding mechanisms, it suggests a minimum temperature required for producing CNTs, possibly related to the melting temperature of the catalysts, and a possible role of $H_2$ in reducing any oxide on the catalyst.

## 6. Conclusions

This chapter has presented a review of spectroscopic approaches to study different kinds of plasmas used for producing carbon nanotubes. Four general types of plasmas have been discussed, including arc discharge, laser ablation plumes, glow discharges and flames. The range of spectroscopic approaches employed varies from optical emission to laser induced incandescence, fluorescence and scattering to Raman spectroscopies (both coherent anti-stokes and spontaneous Raman spectroscopy) to absorption spectroscopies. Optical emission data has in some cases used very rudimentary interpretations (including erroneous peak assignments) and in other been subject to very detailed examination of processes including self-absorption, rotational/vibration/electronic temperature, and collisional-radiative modeling. The laser methods obtain spatially and temporally resolved data about various species concentrations and temperatures, while scattering approaches have been used to study particulate and cluster formation and sizes. Absorption spectroscopy in both the visible and the IR provide estimated species densities based on line-integrated data.

The arc discharge plasma has been studied primarily by observing Swan band spectra to determine $C_2$ concentration and density. In some cases, atomic emission has also been observed and quantified. Local thermal equilibrium is often obtained in arc discharges and allows application of Saha's equation to extract information about carbon atom and cluster densities with a limited amount of characterization data. These studies all arrived upon similar conclusions regarding the optimum carbon concentration and temperature for arc synthesis of CNTs, being around 4000-6000 K and a carbon atom concentration of around $10^{17}$ $cm^{-3}$. Optimum catalyst conditions for CNT growth, however, have not been ascertained via spectroscopy, nor have any detailed studies looked into particulate formation and what role it might play in the CNT growth. One study in particular has surmised, on the lack of strong correlation between spectroscopic data and CNT production rates, that direct removal of particulate matter from the cathodes may be important to growth.[51]

The laser plumes are the most thoroughly studied system for CNT growth and consequently the most well understood. A host of diagnostics used on laser ablation synthesis ovens has lead to a thorough understanding of different stages of growth. In particular, the time profile of catalyst atoms, low molecular weight carbon species and larger particulates have been monitored via laser induced fluorescence and scattering techniques. It is always found that the metal atom density declines following that of the $C_2$ and $C_3$ species and the disappearance of these carbon species coincide with the onset of nanoparticle formation. Larger nanoparticles begin to form as the metal species are lost, which presumably go on to nucleate nanotube growth once the plume has cooled to the eutectic point of the C-metal mixture. While this leads to a consistent picture elegantly summarized in Fig. 10, and kinetic models have shown good agreement with spectroscopic data,[56] there are still several unknowns. For instance, no study has yet tracked the evolution of C atoms, which may lead to different pathways than $C_2$ or $C_3$, and the actual coalescence process of metal atoms with the carbon clusters is still unknown.

The glow discharge and flame synthesis methods of CNT production are the most poorly understood of the CNT synthesis processes and also the least studied spectroscopically. In one sense, this lack of understanding may not be fully unraveled via spectroscopy because of the important role that the surface and substrate interactions play in carbon nanotube formation. The study is further complicated by the presence of a multitude of species including carbon, ammonia, cyanogen, hydrogen, etc. However, a great deal of useful spectroscopic analyses have been performed in plasmas used for PECVD and etching processes in the semiconductor industry. Coupled with surface science studies, a great deal is now known about how these processes proceed. Similar understanding is still not available in the PECVD of carbon nanotubes. Many of the studies have been far too qualitative to yield useful information. The results often appear to be conflicting as well, due to substantial differences between the different types of plasmas employed. A few quantitative studies have been performed, however there is a great deal more that could still be done.

## References

1. M. Meyyappan, Carbon Nanotubes Science and Applications. Boca Raton, FL: CRC Press, 2004.
2. J. Li, A. M. Cassell and B. A. Cruden, Vertically Aligned Carbon Nanostructures in Encyclopedia of Nanoscience and Nanotechnology (2008).
3. M. M. J. Treacy, T. W. Ebbesen and J. M. Gibson, Nature 381, 678 (1996).

4. A. Krishnan, E. Dujardin, T. W. Ebbesen, P. N. Yianilos and M. M. J. Treacy, Phys. Rev. B 58, 14013 (1998).

5. E. W. Wong, P. E. Sheehan and C. M. Lieber, Science 277, 1971 (1997).

6. P. Poncharal, Z. L. Wang, D. Ugarte and W. A. de Heer, Science 283, 1513 (1999).

7. M. Nakajima, F. Arai and T. Fukuda, IEEE Trans. Nanotech. 5, 243 (2006).

8. C. H. Yu, L. Shi, Z. Yao, D. Y. Li and A. Majumdar, Nano Lett. 5, 1842 (2005).

9. E. Pop, D. Mann, Q. Wang, K. Goodson and H. J. Dai, Nano Lett. 6, 96 (2006).

10. P. Kim, L. Shi, A. Majumdar and P. L. McEuen, Phys Rev Lett 87, 215502 (2001).

11. B. Q. Wei, R. Vajtai and P. M. Ajayan, Appl. Phys. Lett. 79, 1172 (2001).

12. S. J. Tans, M. H. Devoret, H. J. Dai, A. Thess, R. E. Smalley, L. J. Geerligs and C. Dekker, Nature 386, 474 (1997).

13. A. Javey, J. Guo, Q. Wang, M. Lundstrom and H. J. Dai, Nature 424, 654 (2003).

14. J. Appenzeller, J. Knoch, R. Martel, V. Derycke, S. J. Wind and P. Avouris, IEEE Trans. Nanotech. 1, 184 (2002).

15. A. P. Graham, G. S. Duesberg, R. V. Seidel, M. Liebau, E. Unger, W. Pamler, F. Kreupl and W. Hoenlein, Small 1, 382 (2005).

16. S. Iijima, Nature 354, 56 (1991).

17. A. Thess, R. Lee, P. Nikolaev, H. Dai, P. Petit, J. Robert, C. Xu, Y. H. Lee, S. G. Kim, A. G. Rinzler, D. T. Colbert, G. E. Scuseria, D. Tomanek, J. E. Fischer and R. E. Smalley, Science 273, 483 (1996).

18. P. Nikolaev, M. J. Bronikowski, R. K. Bradley, F. Rohmund, D. T. Colbert, K. A. Smith and R. E. Smalley, Chem Phys Lett 313, 91 (1999).

19. L. Ci, Y. Li, B. Wei, J. Liang, C. Xu and D. Wu, Carbon 38, 1933 (2000).

20. R. L. Vander Wal, Combust. Flame 130, 37 (2002).

21. B. C. Satishkumar, A. Govindaraj, R. Sen and C. N. R. Rao, Chem Phys Lett 293, 47 (1998).

22. H. M. Cheng, F. Li, G. Su, H. Y. Pan, L. L. He, X. Sun and M. S. Dresselhaus, Appl. Phys. Lett. 72, 3282 (1998).

23. R. Kamalakaran, M. Terrones, T. Seeger, P. Kohler-Redlich, M. Ruhle, Y. A. Kim, T. Hayashi and M. Endo, Appl. Phys. Lett. 77, 3385 (2000).

24. H. Neumayer and R. Haubner, Diam. Relat. Mat. 13, 1191 (2004).

25. H. Dai, J. Kong, C. Zhou, N. Franklin, T. Tombler, A. Cassell, S. Fan and M. Chapline, J Phys Chem B 103, 11246 (1999).

26. M. Meyyappan, L. Delzeit, A. Cassell and D. Hash, Plasma Sources Sci. Technol. 5, 205 (2003).

27. Y. Ando and S. Iijima, Jap. J. Appl. Phys. Pt. 2 32, L107 (1993).

28. C. T. Kingston and B. Simard, Anal. Lett. 36, 3119 (2003).

29. A. A. Puretzky, D. B. Geohegan, X. Fan and S. J. Pennycook, Applied Physics A-Materials Science & Processing 70, 153 (2000).

30. C. Taschner, F. Pacal, A. Leonhardt, P. Spatenka, K. Bartsch, A. Graff and R. Kaltofen, Surface & Coatings Technology 174-175, 81 (2003).

31. M. Chhowalla, K. B. K. Teo, C. Ducati, N. L. Rupesinghe, G. A. J. Amaratunga, A. C. Ferrari, D. Roy, J. Robertson and W. I. Milne, J Appl Phys 90, 5308 (2001).

32. V. I. Merkulov, D. H. Lowndes, Y. Y. Wei, G. Eres and E. Voelkl, Appl. Phys. Lett. 76, 3555 (2000).

33. M. Tanemura, K. Iwata, K. Takahashi, Y. Fujimoto, F. Okuyama, H. Sugie and V. Filip, J Appl Phys 90, 1529 (2001).

34. J.-h. Han, C. H. Lee, D.-Y. Jung, C.-W. Yang, J.-B. Yoo, C.-Y. Park, H. J. Kim, S. Ye, W. Yi, G. S. Park, I. T. Han, N. S. Lee and J. M. Kim, Thin Solid Films 409, 120 (2002).

35. Y. Hayashi, T. Negishi and S. Nishino, J Vac Sci Technol A 19, 1796 (2001).

36. L. C. Qin, D. Zhou, A. R. Krauss and D. M. Gruen, Appl. Phys. Lett. 72, 3437 (1998).

37. Q. Zhang, S. F. Yoon, J. Ahn, B. Gan and M.-B. Yu, J Phys Chem Solids 61, 1179 (2000).

38. S. H. Tsai, C. W. Chao, C. L. Lee and H. C. Shih, Appl. Phys. Lett. 74, 3462 (1999).

39. H. Hayashi, M. Koga, J. Kashirajima, K. Takahashi, Y. Hayashi and S. Nishino, Jpn J Appl Phys, Part 2 41, L1488 (2002).

40. C. Bower, O. Zhou, W. Zhu, D. J. Werder and S. Jin, Appl. Phys. Lett. 77, 2767 (2000).

41. L. Delzeit, I. McAninch, B. A. Cruden, D. Hash, B. Chen, J. Han and M. Meyyappan, J Appl Phys 91, 6027 (2002).

42. J. B. O. Caughman, L. R. Baylor, M. A. Guillorn, V. I. Merkulov and D. H. Lowndes, Appl. Phys. Lett. 83, 1207 (2003).

43. Z. F. Ren, Z. P. Huang, D. Z. Wang, J. G. Wen, J. W. Xu, J. H. Wang, L. E. Calvet, J. Chen, J. F. Klemic and M. A. Reed, Appl. Phys. Lett. 75, 1086 (1999).

44. Y. Li, D. Mann, M. Rolandi, W. Kim, A. Ural, S. Hung, A. Javey, J. Cao, D. Wang, E. Yenilmez, Q. Wang, J. F. Gibbons, Y. Nishi and H. Dai, Nano Lett. (2004).

45. J. D. Huba, "NRL Plasma Formulary," Naval Research Laboratory NRL/PU/6790-94-265 1994.

46. H. Lange and A. Huczko, New Diamond and Frontier Carbon Technology 11, 399 (2001).

47. P. Byszewski, H. Lange, A. Huczko and J. F. Behnke, J Phys Chem Solids 58, 1679 (1997).

48. A. Mansour, M. Razafinimanana, M. Monthioux, M. Pacheco and A. Gleizes, Carbon 45, 1651 (2007).

49. M. Nishio, S. Akita and Y. Nakayama, Thin Solid Films 464-65, 304 (2004).

50. Y. Guo, T. Okazaki, T. Kadoya, T. Suzuki and Y. Ando, Diam. Rel. Mat. 14, 887 (2005).

51. A. Huczko, H. Lange, M. Bystrzejewski, P. Baranowski, Y. Ando, X. Zhao and S. Inoue, J. Nanosci. Nanotech. 6, 1319 (2006).

52. S. Akita, H. Ashihara and Y. Nakayama, Jap. J. Appl. Phys., Pt. 1 39, 4939 (2000).

53. X. L. Zhao, T. Okazaki, A. Kasuya, H. Shimoyama and Y. Ando, Jap. J. Appl. Phys., Pt. 1 38, 6014 (1999).
54. H. Lange, K. Saidane, M. Razafinimanana and A. Gleizes, Journal Of Physics D-Applied Physics 32, 1024 (1999).
55. A. Huczko, H. Lange, A. Resztak and P. Byszewski, High. Temp. Chem. Process. 4, 125 (1995).
56. M. Cau, N. Dorval, B. Cao, B. Attal-Tretout, J. L. Cochon, A. Loiseau, S. Farhat and C. D. Scott, J. Nanosci. Nanotech. 6, 1298 (2006).
57. S. Arepalli and C. D. Scott, Chem. Phys. Lett. 302, 139 (1999).
58. S. Arepalli, J. Nanosci. Nanotech. 4, 317 (2004).
59. G. De Boer, S. Arepalli, W. Holmes, P. Nikolaev, C. Range and C. Scott, J. Appl. Phys. 89, 5760 (2001).
60. S. Arepalli, P. Nikolaev, W. Holmes and C. D. Scott, Applied Physics A-Materials Science & Processing 70, 125 (2000).
61. A. A. Puretzky, D. B. Geohegan, X. Fan and S. J. Pennycook, Appl. Phys. Lett. 76, 182 (2000).
62. A. A. Puretzky, D. B. Geohegan, H. Schittenhelm, X. D. Fan and M. A. Guillorn, Appl. Surf. Sci. 197, 552 (2002).
63. F. Kokai, K. Takahashi, M. Yudasaka and S. Iijima, J. Phys. Chem. B 104, 6777 (2000).
64. N. Dorval, A. Foutel-Richard, M. Cau, A. Loiseau, B. Attal-Tretout, J. L. Cochon, D. Pigache, P. Bouchardy, V. Kruger and K. R. Geigle, J. Nanosci. Nanotech. 4, 450 (2004).
65. F. Kokai, K. Takahashi, M. Yudasaka and S. Iijima, Applied Physics A-Materials Science & Processing 69, S229 (1999).
66. F. Kokai, K. Takahashi, M. Yudasaka, R. Yamada, T. Ichihashi and S. Iijima, J. Phys. Chem. B 103, 4346 (1999).
67. S. A. J. Druet and J. P. E. Taran, Prog. Quantum Electron. 7, 1 (1981).
68. M. Cau, N. Dorval, B. Attal-Tretout, J. L. Cochon, D. Pigache, A. Loiseau, N. R. Arutyunyan, E. D. Obraztsova, V. Kruger and M. Tsurikov, Phys. Status Solidi B-Basic Solid State Phys. 243, 3063 (2006).
69. A. G. Nasibulin, P. V. Pikhitsa, H. Jiang and E. I. Kauppinen, Carbon 43, 2251 (2005).
70. A. V. Melechko, V. I. Merkulov, T. E. McKnight, M. A. Guillorn, K. L. Klein, D. H. Lowndes and M. L. Simpson, J. Appl. Phys. 97, 41301 (2005).
71. M. Jonsson, O. A. Nerushev and E. E. B. Campbell, Applied Physics A-Materials Science & Processing 88, 261 (2007).
72. I. B. Denysenko, S. Xu, J. D. Long, P. P. Rutkevych, N. A. Azarenkov and K. Ostrikov, J Appl Phys 95, 2713 (2004).
73. E. G. Wang, Z. G. Guo, J. Ma, M. M. Zhou, Y. K. Pu, S. Liu, G. Y. Zhang and D. Y. Zhong, Carbon 41, 1827 (2003).

74. B. A. Cruden, A. M. Cassell, Q. Ye and M. Meyyappan, J Appl Phys 94, 4070 (2003).
75. B. A. Cruden, D. B. Hash, A. M. Cassell and M. Meyyappan, J. Appl. Phys. 96, 5284 (2004).
76. M. S. Bell, R. G. Lacerda, K. B. K. Teo, N. L. Rupesinghe, G. A. J. Amaratunga, W. I. Milne and M. Chhowalla, Appl. Phys. Lett. 85, 1137 (2004).
77. Y. Y. Lin, H. W. Wei, K. C. Leou, H. Lin, C. H. Tung, M. T. Wei, C. Lin and C. H. Tsai, J. Vac. Sci. Tech. B 24, 97 (2006).
78. B. Cruden and M. Meyyappan, J Appl Phys 97, 084311 (2005).
79. G. Y. Zhang, D. Mann, L. Zhang, A. Javey, Y. M. Li, E. Yenilmez, Q. Wang, J. P. McVittie, Y. Nishi, J. Gibbons and H. J. Dai, Proc. Natl. Acad. Sci. U. S. A. 102, 16141 (2005).
80. C. Collard, J. P. Holloway and M. L. Brake, Ieee Transactions On Plasma Science 33, 170 (2005).
81. T. Y. Lee, J.-h. Han, S. H. Choi, J.-B. Yoo, C.-Y. Park, T. Jung, S. Yu, W. K. Yi, I. T. Han and J. M. Kim, Diam. Rel. Mat. 12, 851 (2003).
82. S. Hofmann, B. Kleinsorge, C. Ducati, A. C. Ferrari and J. Robertson, Diam. Rel. Mat. 13, 1171 (2004).
83. S. H. Lim, H. S. Yoon, J. H. Moon, K. C. Park and J. Jang, Appl. Phys. Lett. 88 (2006).
84. J. Garcia-Cespedes, M. Rubio-Roy, M. C. Polo, E. Pascual, U. Andujar and E. Bertran, Diam. Rel. Mat. 16, 1131 (2007).
85. Y. S. Woo, D. S. Jeon, I. T. Han, N. S. Lee, J. E. Jung and J. M. Kim, Diam. Rel. Mat. 11, 59 (2002).
86. S. K. Srivastava, V. D. Vankar and V. Kumar, Thin Solid Films 515, 1552 (2006).
87. J. I. B. Wilson, N. Scheerbaum, S. Karim, N. Polwart, P. John, Y. Fan and A. G. Fitzgerald, Diam. Rel. Mat. 11, 918 (2002).
88. M. Chen, C.-M. Chen and C.-F. Chen, Thin Solid Films 420-421, 230 (2002).
89. R. K. Garg, T. N. Anderson, R. P. Lucht, T. S. Fisher and J. P. Gore, Journal of Physics D: Applied Physics 41, 095206 (2008).
90. A. Chandrashekar, J. S. Lee, G. S. Lee, M. J. Goeckner and L. J. Overzet, J. Vac. Sci. Technol. A 24, 1812 (2006).
91. L. Zajickova, M. Elias, O. Jasek, V. Kudrle, Z. Frgala, J. Matejkova, J. Bursik and M. Kadlecikova, Plasma Physics And Controlled Fusion 47, B655 (2005).
92. R. W. B. Pearse and A. G. Gaydon, The Identification of Molecular Spectra, Chapman and Hall, New York, New York (1976).
93. P. E. Nolan, D. C. Lynch and A. H. Cutler, J. Phys. Chem. B 102, 4165 (1998).
94. Y. L. Li, I. A. Kinloch and A. H. Windle, Science 304, 276 (2004).
95. Y. Murakami, S. Chiashi, Y. Miyauchi, M. H. Hu, M. Ogura, T. Okubo and S. Maruyama, Chem. Phys. Lett. 385, 298 (2004).
96. K. Hata, D. N. Futaba, K. Mizuno, T. Namai, M. Yumura and S. Iijima, Science 306, 1362 (2004).

97. B. A. Cruden, M. V. V. S. Rao, S. P. Sharma and M. Meyyappan, J Appl Phys 91, 8955 (2002).

98. M. W. Kiehlbauch and D. B. Graves, J Appl Phys 89, 2047 (2001).

99. H. N. Chu, E. A. Den Hartog, A. R. Lefkow, J. Jacobs, L. W. Anderson, M. G. Lagally and J. E. Lawler, Phys Rev A 44, 3796 (1991).

100. A. N. Goyette, W. B. Jameson, L. W. Anderson and J. E. Lawler, J Phys D: Appl Phys 29, 1197 (1996).

101. J. Laimer, F. Huber, G. Misslinger and H. Stori, Vacuum 47, 183 (1996).

102. A. Gicquel, K. Hassouni, Y. Breton, M. Chenevier and J. C. Cubertafon, Diam. Rel. Mat. 5, 366 (1996).

103. S. A. Astashkevich, M. Kaning, E. Kaning, N. V. Kokina, B. P. Lavrov, A. Ohl and J. Ropcke, J Quant Spectrosc Radiat Transfer 56, 725 (1996).

104. J. Ropcke, M. Kaning and B. P. Lavrov, J. Phys. IV 8, 207 (1998).

105. B. A. Cruden, M. V. V. S. Rao, S. P. Sharma and M. Meyyappan, Appl. Phys. Lett. 81, 990 (2002).

106. V. M. Donnelly and M. V. Malyshev, Appl. Phys. Lett. 77, 2467 (2000).

107. G. A. Hebner, J Appl Phys 80, 2624 (1996).

108. S. P. Sharma, B. A. Cruden, M. Rao and A. A. Bolshakov, J. Appl. Phys. 95, 3324 (2004).

109. B. A. Cruden, Unpublished.

110. B. A. Cruden, M. V. V. S. Rao, S. P. Sharma and M. Meyyappan, Rev. Sci. Instr. 73, 2578 (2002).

111. B. A. Cruden, M. V. V. S. Rao, S. P. Sharma and M. Meyyappan, Plasma Sources Sci. Technol. 11, 77 (2002).

112. F. S. Xu, X. F. Liu and S. D. Tse, Carbon 44, 570 (2006).

Chapter 2

# Spectroscopic Studies on Laser-Produced Carbon Vapor

Koichi Sasaki

*Department of Quantum Science and Engineering,*
*Faculty of Engineering, Hokkaido University, Kita 13, Nishi 8,*
*Kita-ku, Sapporo, Hokkaido 060-8628, Japan*
*E-mail address: sasaki@qe.eng.hokudai.ac.jp*

In this chapter, the growth process of carbon clusters in a plume produced by laser ablation of a graphite target in ambient He gas is investigated using spectroscopic diagnostic methods. The temporal variations of the spatial distributions of $C_2$ and $C_3$ densities are measured by laser-induced fluorescence imaging spectroscopy. The same method is also applied for evaluating the two-dimensional distribution of plume temperature. In addition, continuum optical emission is observed by optical emission spectroscopy. A scenario for the temporal evolution of carbon clusters in laser-produced carbon vapor is proposed based on the experimental observations.

## 1. Introduction

After the discoveries of fullerenes and nanotubes by Kroto *et al.* [Kroto *et al.* (1985)] and Iijima [Iijima (1991)], respectively, unique properties and potential applications of carbon clusters have led researchers to investigate their syntheses. Nevertheless, fundamental understanding of growth processes of carbon clusters is still insufficient. In order to develop an optimized synthesis method of clusters, we should have better understanding of the formation mechanisms. Although several models [Kroto and McKay (1988); Ozawa *et al.* (2002); Smalley (1992); Wakabayashi and Achiba (1992); von Helden *et al.* (1993); Hunter *et al.* (1993)] and review articles [Goroff (1996);

Irle *et al.* (2006)] on the gas-phase production processes of fullerenes have already been published, knowledge about the atomic-level formation mechanism of carbon clusters is still quite inadequate. A reason for the difficulty in obtaining deep understanding on the formation mechanism is the fact that experimental investigation on the fundamental aspect of the cluster formation is not an easy task.

Laser ablation of a graphite target in rare gas atmosphere is a synthesis method of carbon clusters [Guo *et al.* (1995); Yudasaka *et al.* (1998); Kataura *et al.* (1998); Kokai *et al.* (1999); Geohegan *et al.* (2001)]. Although the laser ablation technique is not suitable to the mass production, it is useful for investigating the growth processes of clusters [Kaizu *et al.* (1997); Shibagaki *et al.* (2000, 2001)]. The purpose of this chapter is to show an experimental investigation on initial growth processes of carbon clusters in carbon plumes produced by laser ablation of a graphite target.

Diagnostic techniques play essential roles in the fundamental investigation of growth processes. In experimental studies of laser ablation, the diagnostic technique should have high temporal and spatial resolutions since kinetics of laser ablation plumes include dynamic phenomena with transient properties. In this work, we employed spectroscopic diagnostics for investigating growth processes of carbon clusters. In particular, we used laser-induced fluorescence (LIF) imaging spectroscopy [Muramoto *et al.* (1997); Puretzki *et al.* (2000); Ikegami *et al.* (2001)] for visualizing the distributions of $C_2$ and $C_3$ radical densities in laser-ablation carbon plumes. The visualized density distributions were obtained as a function of time after the irradiation of the laser pulse for ablation. In addition, LIF imaging spectroscopy was also applied to the measurement of two-dimensional plume temperature. Based on the experimental results of the radical densities and the plume temperature, we obtained a scenario for the temporal evolution of carbon clusters in laser-produced carbon vapor.

## 2. Experimental Apparatus

### 2.1. *Laser ablation system*

Laser ablation of a graphite target was carried out in a vacuum chamber shown in Fig. 1. A graphite target was installed on a rotating holder. The vacuum chamber was evacuated below $5 \times 10^{-7}$ Torr using a turbo molecular pump. After the evacuation, He was injected into the chamber as ambient gas. Nd:YAG lasers at wavelengths of 1.06 $\mu$m and 266 nm were used for

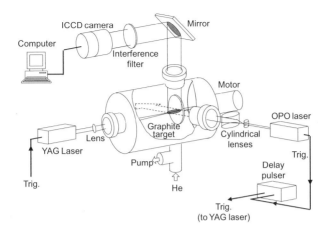

Fig. 1. Laser ablation system with laser-induced fluorescence imaging spectroscopy.

ablation. The YAG laser beam was focused onto the target surface using a lens, and the laser fluence on the target surface was estimated to be 2–3 J/cm$^2$. The duration of the YAG laser pulse was 10 ns. The angle between the YAG laser beam and the target surface was 90°.

## 2.2. *Optical emission spectroscopy*

A monochromator with a focal length of 25 cm and a photomultiplier tube was used for optical emission spectroscopy. An optical fiber was used to guide the optical emission to the monochromator. The entrance aperture (5 mm × 1 mm) of the optical fiber was imaged in front of the target using a lens with a focal length of 15 cm. The location of the entrance aperture was on the optical axis of the YAG laser beam at a distance of 5–10 mm from the target surface. In the time-domain measurement, the signal from the photomultiplier tube was recorded using a digital oscilloscope. In the measurement of the optical emission spectrum, the signal from the photomultiplier tube was gated and averaged using a boxcar integrator. For imaging spectroscopy, a CCD camera with a gated image intensifier was used for recording the image of the optical emission. The image intensifier was triggered at a delay time $t_D$ after the irradiation of the YAG laser pulse. Interference filters with various transmission wavelengths were placed in front of the CCD camera. The images taken by the CCD camera were averaged to eliminate the ambiguity due to the poor shot-to-shot reproducibility.

## 2.3. Laser-induced fluorescence imaging spectroscopy

To carry out the LIF measurement, as shown in Fig. 1, tunable laser pulses yielded from an optical parametric oscillator (OPO) were launched into the plume in front of the target. The tunable laser beam was arranged to have a planar shape using two cylindrical lenses. The width and the thickness of the planar tunable laser beam were approximately 27 and 1 mm, respectively. $C_2$ and $C_3$ radicals in the plume were excited by the tunable laser pulse. Fluorescence emissions yielded from excited $C_2$ and $C_3$ formed images on the planar laser beam. The images of the fluorescence were taken by a CCD camera with a gated image intensifier. Interference filters with high transmissions at the fluorescence wavelengths were installed in front of the camera to separate LIFs from stray lights and self-emissions of the plume. Since the LIF wavelengths were far from the excitation wavelengths, no stray light was mixed in the LIF images. Self-emissions of the plume at the same wavelengths as LIFs were observed. The self-emissions were intense just after the irradiation of the YAG laser pulse. In this case, we subtracted the images of the self-emissions from the LIF images to evaluate the density distributions of $C_2$ and $C_3$ at the ground states. In this way, we obtained the two-dimensional images of the $C_2$ and $C_3$ radical densities in the plume. The temporal variations of the density distributions were obtained by changing the delay time $t_D$ between the oscillations of the YAG and OPO lasers. The energy levels and the wavelengths used in the LIF imaging spectroscopy were summarized in Table 1 [Pearse and Gaydon (1976); Gausset et al. (1965); Takizawa et al. (2000)]. It is noted that $C_2$ radicals detected by the LIF scheme are at a metastable ($a^3\Pi_u$) state. However, since the energy separation between the $a^3\Pi_u$ state and the ground ($X^1\Sigma_g^+$) state is only 0.076 eV, the $a^3\Pi_u$ state is expected to have a large population and to have similar characteristics to the ground state [Suzuki et al. (1999)].

Table 1.  Energy levels and wavelengths used in the LIF imaging spectroscopy.

| Radical | Initial state | Excitation wavelength | Excited state | Fluorescence wavelength | Final state |
|---------|---------------|-----------------------|---------------|-------------------------|-------------|
| $C_2$ | $a^3\Pi_u(v''=0)$ | 516.52 nm | $d^3\Pi_g(v'=0)$ | 563.6 nm | $a^3\Pi_u(v''=1)$ |
| $C_3$ | $\tilde{X}^1\Sigma_g^+(000)$ | 405.13 nm | $\tilde{A}^1\Pi_u(000)$ | 426.4 nm | $\tilde{X}^1\Sigma_g^+(100)$ |

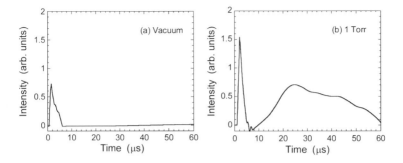

Fig. 2. Waveform of optical emission intensity at a wavelength of 650 nm.

## 3. Optical Emission from Laser-Produced Carbon Vapor [Sasaki *et al.* (2002)]

### 3.1. *Temporal variation of optical emission intensity*

Figure 2 shows the temporal variations of optical emission intensities from plumes produced in vacuum and in ambient He gas at 1 Torr. The wavelength and the fluence of the YAG laser pulse were 1.06 $\mu$m and approximately 2.2 J/cm$^2$, respectively. The measurement wavelength of the optical emission intensity was 650 nm. As shown in the figure, the optical emission intensity observed in vacuum had a spiky temporal variation just after the irradiation of the YAG laser pulse. On the other hand, the optical emission intensity observed in ambient He gas was composed of two components. One was a spiky emission which was similar to that observed in vacuum. The other was a delayed component which appeared $\sim$ 10 $\mu$s after the irradiation of the YAG laser pulse.

### 3.2. *Optical emission spectrum*

The spectral distribution of the spiky emission intensity observed in vacuum is shown in Fig. 3. In this measurement, the whole of the spiky emission just after the irradiation of the YAG laser pulse was gated using the boxcar integrator. The transmission wavelength of the monochromator was scanned slowly to obtain the spectrum. The wavelength resolution was approximately 1.5 nm. The wavelength dependence of the detection sensitivity was calibrated using a tungsten standard lamp. The Swan band emissions from C$_2$ [Pearse and Gaydon (1976)] were identified in the spectrum. In addition to the emissions from C$_2$, the spectrum had a continuum

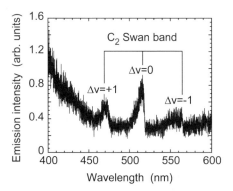

Fig. 3. Optical emission spectrum observed in vacuum.

component. The spectrum of the spiky emission was dependent on the laser fluence. At an intense laser fluence such as 3 $J/cm^2$, a lot of line emissions from C and $C^+$ were observed in the spectrum. On the other hand, the spectrum of the delayed ($> 10$ $\mu$s) optical emission observed in ambient He gas was similar to Fig. 2, and was roughly independent of the laser fluence.

A general explanation for the continuum emission from a laser ablation plume is blackbody radiation from clusters and particulates heated in the plume [Rohlfing (1988); Kokai et al. (1999)]. However, if we fit the spectrum shown in Fig. 3 with the Planck's radiation distribution function, the temperature of emitting particles is estimated to be higher than $3 \times 10^4$ K. Therefore, the spectrum shown in Fig. 3 cannot be explained by blackbody radiation. Another explanation for the continuum emission is the superposition of blackbody radiation and molecular band spectra from carbon clusters [Monchicourt (1991)]. Recently, it has been shown that carbon clusters $C_n$ with $n \leq 10$ have optical transition bands in a wide wavelength range [Jochnowitz and Maier (2008)]. Hence there is a possibility that the spectrum below 500 nm is reinforced with the superposition of a lot of band spectra of $C_n$.

### 3.3. Spatial distribution of delayed continuum emission

If the continuum emission is owing to the superposition of blackbody radiation and molecular band spectra of $C_n$, the delayed continuum emission observed at $t_D > 10$ $\mu$s may contain useful information on clustering reactions in the plume. Figures 4 shows spatial distribution of the optical

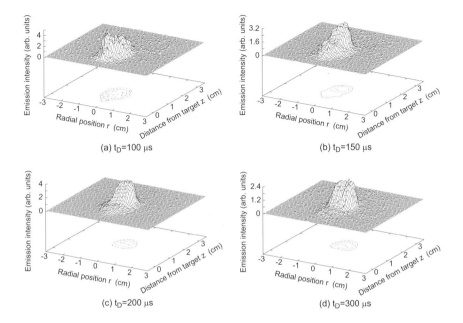

(a) $t_D$=100 μs　　　　(b) $t_D$=150 μs

(c) $t_D$=200 μs　　　　(d) $t_D$=300 μs

Fig. 4. Spatial distribution of optical emission intensity at a wavelength of 650 nm.

emission intensity observed at $t_D$ = 100, 150, 200, and 300 μs after the irradiation of the YAG laser pulse. The pressure of ambient He gas and the fluence of the YAG laser pulse were 5 Torr and 3 J/cm², respectively. The YAG laser pulse was irradiated at $r = z = 0$ cm in the figure ($r$ and $z$ stand for the radial position and the distance from the target, respectively). The images shown in Fig. 4 were taken using an interference filter with a transmission wavelength of 656.2 nm and a bandwidth of 10 nm. The gate width was 10 μs. Since there were no line emissions at $\lambda = 656.2 \pm 5$ nm, the images shown in Fig. 4 represent the spatial distributions of the continuum optical emission.

It is seen from Fig. 4 that the spatial distribution of the continuum optical emission was composed of two parts. One was the emission shown in Fig. 4(a), which had a peak at $z \simeq 1$ cm. This peak decreased monotonically with time, and disappeared at $t_D$ = 200 μs as shown in Fig. 4(c). The other was the emission shown in Figs. 4(c) and 4(d), which had a peak at $z \simeq 1.8$ cm. It is noted that the peak at $z \simeq 1.8$ cm increased at $t_D \geq 100$ μs. The increase in optical emission intensity in such a late phase of laser ablation is an interesting phenomenon. In addition, as will be descried later, it is known that the growth area of the continuum optical emission at $z \simeq 1.8$ cm corresponds to the decreasing area of the $C_2$ radical density. The spatial and temporal synchronizations between the decrease in the $C_2$ density and the growth of the continuum optical emission strongly suggest that the continuum emission is originated from the formation of clusters in the plume. Association reactions among C, $C_2$, and other species produce excited states of light clusters, and the superposition of molecular band spectra from the light clusters may contribute to the continuum emission. In addition, since heavy clusters and particulates produced by condensation reactions are probably at high temperatures, they may emit blackbody radiation.

## 4. Spatiotemporal Variations of $C_2$ and $C_3$ Radical Densities [Sasaki *et al.* (2002)]

### 4.1. $C_2$ and $C_3$ radical densities in vacuum

Figure 5 shows the distributions of $C_2$ radical density observed at $t_D = 1.5$ and 4 $\mu$s after the irradiation of the YAG laser pulse at 1.06 $\mu$m. Although the absolute $C_2$ density is unknown, the relative change in the $C_2$ density can be seen from the magnitudes of the vertical axes of the figures. The magnitudes of the vertical axes of all the figures showing the distribution of $C_2$ density are normalized by the maximum $C_2$ density observed in vacuum (The maximum $C_2$ density in vacuum was observed at $t_D = 0.6$ $\mu$s). It is seen from Fig. 5 that the density distribution of $C_2$ spreads rapidly after the irradiation of the YAG laser pulse. The flight speed of the peak position of the density distribution was approximately $1.5 \times 10^5$ cm/s. Because of the rapid expansion of the plume, the $C_2$ density at $t_D = 4$ $\mu$s was one-order of magnitude lower than that at $t_D = 1.5$ $\mu$s.

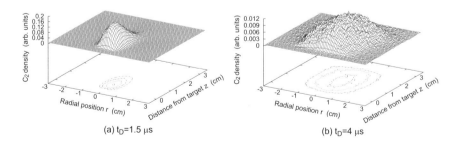

(a) $t_D=1.5$ $\mu$s                    (b) $t_D=4$ $\mu$s

Fig. 5.   Temporal variation of the density distribution of $C_2$ observed in vacuum.

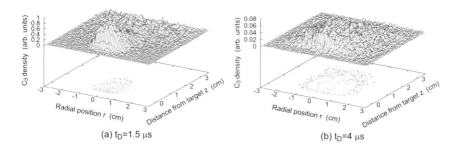

Fig. 6. Temporal variation of the density distribution of $C_3$ observed in vacuum.

The distributions of $C_3$ radical density at $t_D = 1.5$ and 4 $\mu$s are shown in Fig. 6. The maximum $C_3$ density in vacuum was observed at $t_D = 1.5$ $\mu$s. The magnitudes of the vertical axes of all the figures showing the distribution of $C_3$ density are normalized by the maximum $C_3$ density in vacuum at $t_D = 1.5$ $\mu$s. The vertical axes of Figs. 5 and 6 cannot be compared. The ratio of $C_2$ to $C_3$ densities have not been determined. Since the $C_3$ density observed in vacuum is low, Fig. 6 is poor in the signal-to-noise ratio. The rapid expansion of the density distribution was also observed in $C_3$. At $t_D = 4$ $\mu$s, the $C_3$ radical density was close to the noise level.

## 4.2. $C_2$ and $C_3$ radical densities in ambient He gas at 1 Torr

The distributions of $C_2$ radical density at $t_D = 4$, 20, and 100 $\mu$s observed in ambient He gas at 1 Torr are shown in Fig. 7. In ambient He gas at a pressure higher than 0.5 Torr, the expansion and the movement of the plume were restricted significantly, and the entire volume of the plume existed inside of the observation area for a long time. Comparing Fig. 7(a) with Fig. 5(b), it is known that the volume of the plume in ambient He gas was much smaller than that in vacuum at the same delay time. Because of the restricted expansion of the plume, the $C_2$ radical density observed in ambient He gas was much higher than that observed in vacuum. The density distribution of $C_2$ shown in Fig. 7(a) has a crescent shape with a sheer front and a gradually-decreasing tail. The crescent density distribution disappeared after $t_D \simeq 10$ $\mu$s. The movement of the plume was very slow after the disappearance of the crescent density distribution, and the density distribution of $C_2$ approached isotropic one as shown in Figs. 7(b) and 7(c). The slow expansion of the density distribution after $t_D = 10$ $\mu$s may be mainly due to diffusion.

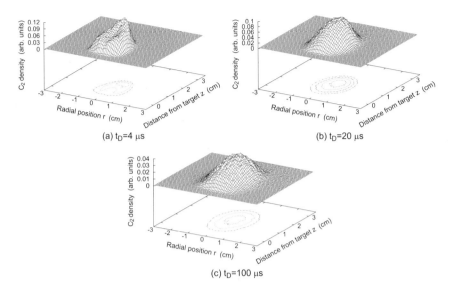

(a) $t_D$=4 μs

(b) $t_D$=20 μs

(c) $t_D$=100 μs

Fig. 7. Temporal variation of the density distribution of $C_2$ observed in ambient He gas at 1 Torr.

Figure 8 shows the distributions of $C_3$ radical density observed in ambient He gas at 1 Torr. The magnitudes of the vertical axes of Fig. 8 indicates that the $C_3$ density in ambient He gas was higher than the maximum $C_3$ density in vacuum. As shown in Fig. 8(a), the distribution of the $C_3$ density at $t_D = 20$ μs had two peaks. A peak was adjacent to the target surface, and the other peak was located near the leading edge of the plume. Comparing Fig. 8(a) with Fig. 7(b), it is seen that the peak near the leading edge roughly corresponds to the peak position of the $C_2$ density at the same delay time of $t_D = 20$ μs. At $t_D = 100$ μs, we observed the growth of the peak near the leading edge. On the other hand, the peak adjacent to the target decreased with time as shown in Figs. 8(b) and 8(c).

### 4.3.　$C_2$ and $C_3$ radical densities in ambient He gas at 5 Torr

In ambient He gas at 5 Torr, the movement and the expansion of the plume were restricted more significantly. In comparison with Fig. 7, the size of the density distribution of $C_2$ radical in 5 Torr was smaller than that in 1 Torr at the same observation time. In addition, the $C_2$ density in 5 Torr was higher than that in 1 Torr. The crescent density distribution due to the excitation of a shock wave was also observed in ambient He gas at 5 Torr

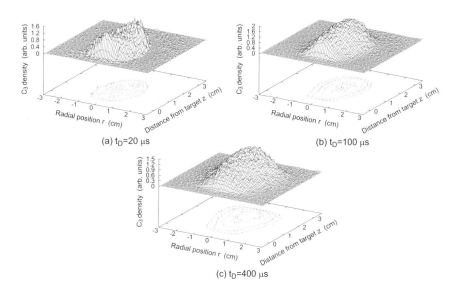

Fig. 8. Temporal variation of the density distribution of $C_3$ observed in ambient He gas at 1 Torr.

at $t_D \leq 8$ $\mu$s. At $t_D \leq 20$ $\mu$s, the peak in the density distribution of $C_2$ was positioned at the front area of the plume as shown in Figs. 9(a) and 9(b). On the other hand, at $t_D = 100$ $\mu$s, the peak in the density distribution was neighboring to the target. The $C_2$ density decreased at $t_D \geq 20$ $\mu$s. The decrease in the $C_2$ density was significant in the front area of the plume, resulting in the particular density distribution shown in Fig. 9(d).

The size of the density distribution of $C_3$ in 5 Torr was also smaller than that in 1 Torr as shown in Fig. 10. The peak $C_3$ density in 5 Torr was higher than that in 1 Torr at $t_D \leq 200$ $\mu$s. The double peak structure of $C_3$ radical density was also observed in He gas at 5 Torr as shown in Fig. 10(a). The growth of the peak in the front area of the plume was remarkable, while the peak adjacent to the target decreased monotonically. These different behaviors of the two peaks resulted in the density distribution of $C_3$ having the peak in the front area of the plume as shown in Figs. 10(c) and 10(d). It is noted that the growth area of the $C_3$ density corresponds to the decreasing area of the $C_2$ density. Comparing Fig. 10(d) with Fig. 9(d), it is known that the $C_2$ and $C_3$ radicals occupy different areas in the plume at $t_D = 400$ $\mu$s.

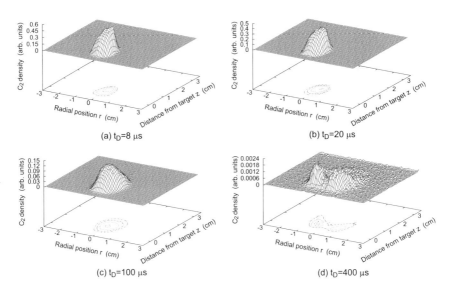

Fig. 9. Temporal variation of the density distribution of $C_2$ observed in ambient He gas at 5 Torr.

## 5. Temporal Change in the Total Numbers of $C_2$ and $C_3$

In order to examine the effect of reactions in the plume, we spatially integrated the density distributions shown in Figs. 5–10 to evaluate total numbers of $C_2$ and $C_3$ contained in the plume. The integration was carried out only when the entire volume of the density distribution was located in the observation area. The density distributions measured experimentally in both sides of $r \geq 0$ cm and $r \leq 0$ cm were averaged. The distribution thus obtained in $(r, z)$ plane was integrated under the assumption of the cylindrical symmetry. The temporal variations of the total numbers of $C_2$ and $C_3$ radicals are shown in Fig. 11. The magnitudes of the vertical axes of Figs. 11(a) and 11(b) are normalized by the maximum numbers of $C_2$ and $C_3$, respectively, observed in vacuum. In other words, the vertical axes show the degree of the enhancement in the numbers of $C_2$ and $C_3$ in gas phase, provided that the numbers of $C_2$ and $C_3$ ejected from the target directly are independent of the pressure of ambient He gas. It is noted that the quenching of the LIF emissions due to collision with particles ejected from the target is responsible for the small total numbers of $C_2$ and $C_3$ observed at $t_D \leq 5$ μs [Wakasaki et al. (2002)].

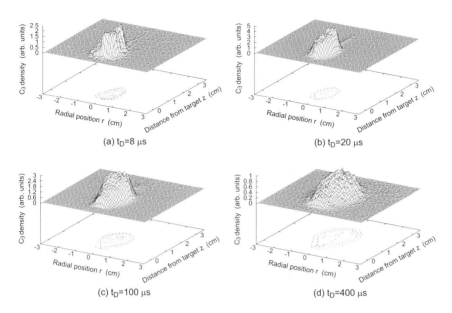

Fig. 10.  Temporal variation of the density distribution of $C_3$ observed in ambient He gas at 5 Torr.

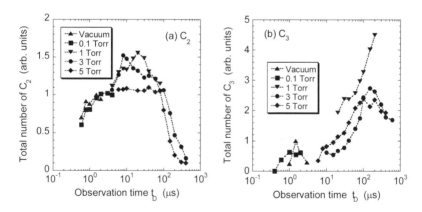

Fig. 11.  Temporal variation of the total number of (a) $C_2$ and (b) $C_3$ contained in the plume.

As shown in Fig. 11(a), the numbers of $C_2$ observed in ambient He gas at 1 and 3 Torr were 1.6 times bigger than that in vacuum at the maximum. The increase in the number of $C_2$ is probably due to a gas-phase reaction,

$$C + C + M \rightarrow C_2 + M. \tag{1}$$

The ambient He gas enhances this three-body reaction by increasing the density of M in two ways; one is the increase in the density of He and the other is the increase in the local density of carbon species. The latter effect is obtained by the fact that the expansion of the plume is restricted in ambient He gas. The peak in the total number of $C_2$ appeared at $t_D \simeq 8$ $\mu$s in 3 Torr, which was earlier than the peak time of $t_D \simeq 20$ $\mu$s observed in 1 Torr. This result is reasonable since reaction (1) is more efficient in ambient gas at a higher pressure. On the other hand, the total number of $C_2$ observed in 5 Torr was smaller than those in 1 and 3 Torr. However, the smaller number of $C_2$ in 5 Torr may not indicate the less efficient production of $C_2$. The production of $C_2$ is probably efficient in 5 Torr, but the loss of $C_2$ may also be significant. The loss of $C_2$ means the production of heavier carbon species $C_n$ with $n \geq 3$. The constant number of $C_2$ observed at $t_D = 7 - 60$ $\mu$s in 5 Torr may be attributed to the balance between the production and the loss of $C_2$. In the decreasing period of the total number, $C_2$ is consumed by the production of heavier carbon species.

As shown in Fig. 11(b), the enhancement in the total number of $C_3$ in ambient He gas was more significant than that of $C_2$. The quenching of the LIF emission due to collision with He is corrected in Fig. 8(b) [Wakasaki *et al.* (2002)]. In ambient He gas at 1 Torr, the total number of $C_3$ was 4.5 times bigger than that in vacuum. The increasing period of the total number of $C_3$ corresponded to the decreasing period of the total number of $C_2$. These results suggest the gas-phase production of $C_3$ due to a reaction,

$$C_2 + C + M \rightarrow C_3 + M. \tag{2}$$

This three-body reaction becomes more efficient in ambient He gas at a higher pressure. The production of $C_3$ from $C_2$ is supported by the fact that the increasing area of $C_3$ corresponds to the decreasing area of $C_2$ as shown in Figs. 7–10. The total numbers of $C_3$ observed in 3 and 5 Torr were smaller than that in 1 Torr, which may be due to the significant loss of $C_3$ to produce heavier carbon species.

## 6. Spatiotemporal Variation of Plume Temperature [Sasaki and Aoki (2008)]

### 6.1. *Evaluation of plume temperature*

Figure 12(a) shows a typical LIF excitation spectrum, which was obtained by scanning the wavelength of the OPO laser in the R branch range of the $a^3\Pi_u(v'' = 0) - d^3\Pi_g(v' = 0)$ transition of $C_2$ (Swan band). The

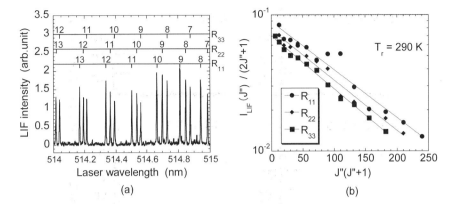

Fig. 12. (a) An example of the excitation spectrum of the R branch of the $a^3\Pi_u(v'' = 0) - d^3\Pi_g(v' = 0)$ transition of $C_2$. The rotational quantum numbers of the $a^3\Pi_u(v'' = 0)$ state are indicated in the figure. (b) The Boltzmann plot obtained by the spectrum shown in (a), indicating a rotational temperature of 290 K.

assignment of rotational quantum numbers of the $a^3\Pi_u(v'' = 0)$ state is indicated in the figure. The intensity of the OPO laser beam was strong enough for saturated excitation, and we recorded the LIF intensity after temporally integrating the pulsed fluorescence signal. Figure 12(b) shows the Boltzmann plot corresponding to Fig. 12(a), where $J''$ is the rotational quantum number of the $a^3\Pi_u(v'' = 0)$ state. The rotational temperature of $C_2$ was calculated to be 290 K from the average of the slopes of the fitted curves shown in Fig. 12(b). The rotational temperature is equal to the translational temperature of the plume in this experimental condition.

To obtain the 2D distribution of the plume temperature, we carried out the above measurement in the imaging mode with a planar laser beam. We took LIF images corresponding to the excitations from several rotational levels of the $a^3\Pi_u(v'' = 0)$ state and evaluated the rotational temperature of $C_2$ from the Boltzmann plot at each pixel of the ICCD camera. The temporal variation of the plume temperature was obtained by changing the delay time $t_D$ between the oscillations of the YAG and OPO lasers.

## 6.2. *Spatial distribution of plume temperature*

Figure 13 shows 2D distributions of the plume temperature observed at four delay times after the irradiation of the YAG laser pulse at 266 nm onto the graphite target installed in ambient He at 1 Torr. The plume temperature at $t_D = 10$ $\mu$s was $\sim 700$ K. Since the plume temperature at $t_D = 2$ $\mu$s

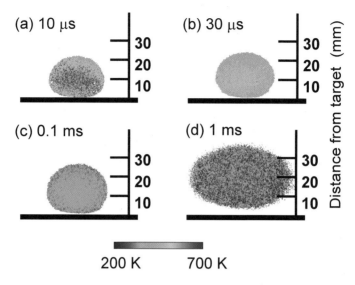

Fig. 13. Two dimensional distributions of the plume temperature observed in ambient He gas at 1 Torr.

evaluated by Saito *et al.* was ∼ 4000 K [Saito *et al.* (2003)], a rapid decrease in the plume temperature is expected in the initial stage of laser ablation. The plume temperature decreased at $t_D \leq 0.3$ ms, and reached ∼ 350 K at $t_D = 1$ ms. As the plume temperature decreased, the temperature at the center was slightly higher than that of the surrounding area.

We observed an interesting phenomenon when the pressure of ambient He was 5 Torr, at which the plume sizes became smaller (Fig. 14), compared with those at 1 Torr (Fig. 13), at the same delay times. At a gas pressure of 5 Torr, as shown in Fig. 14(a), we observed that the leading edge of the plume had a high temperature of 800 K at $t_D = 10$ μs. The thickness of the high-temperature region was as thin as 1 mm. The localized high temperature at the leading edge may be attributed to heating by the shock wave. At $t_D = 30$ μs, the main part of the plume heats up as shown in Fig. 14(b), which may be due to backward propagation of the shock wave. Following this, the plume temperature decreased with $t_D$ to ∼ 350 K, with the maximum spatial distribution at the center of the plume.

### 6.3. *Temporal variation of plume temperature*

Figure 15 shows the temporal decay of the averaged plume temperature after the irradiation of the YAG laser pulse. The averaged plume temperature

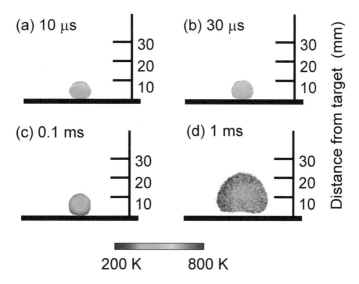

Fig. 14. Two dimensional distributions of the plume temperature observed in ambient He gas at 5 Torr.

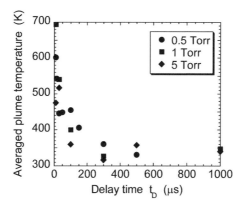

Fig. 15. Temporal decay of the averaged plume temperature.

was evaluated from the spatial integration of the 2D plume temperature. As shown in the figure, the decay curve of the plume temperature was roughly independent of the pressure of ambient He gas. The decay time constant in the initial phase ($t_D \leq 50$ $\mu$s) was approximately 0.1 ms. It became significantly gentler with $t_D$, and we observed a constant plume temperature of $\sim 350$ K at $t_D \geq 0.3$ ms.

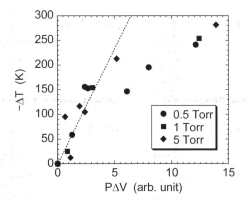

Fig. 16.  Relationship between the decrease in plume temperature ($\Delta T$) and the product of the ambient He pressure ($P$) and the increase in plume volume ($\Delta V$).

There are two candidates for the cooling mechanism of the plume. One is thermal conduction toward ambient He gas. The fact that plume temperature distribution has the maximum at the center of the plume suggests the contribution of thermal conduction [Figs. 13(b), 13(c), and 14(c)]. However, thermal conduction may not be the dominant cooling mechanism, since the decay curve of the plume temperature was independent of the pressure of ambient He gas, as shown in Fig. 15. The other mechanism for plume cooling is adiabatic expansion. In Fig. 16, we plotted the relationship between the decrease in plume temperature ($\Delta T$) and the product of the ambient He pressure ($P$) and increase in the plume volume ($\Delta V$). The plume volume was evaluated from the 2D distribution of the $C_2$ density measured by LIF imaging spectroscopy. According to Fig. 16, $\Delta T$ is proportional to $P\Delta V$ in the initial phase ($t_D \leq 50$ $\mu$s), which suggests that the decrease in plume temperature is determined by the work performed by the plume, indicating that the dominant cooling mechanism is adiabatic expansion. The thermal kinetics of the plume at $t_D \geq 50$ $\mu$s may not be adiabatic, resulting in the slowdown in $\Delta T$ as shown in Fig. 16.

## 7.  A Scenario for the Growth of Carbon Clusters

We discuss the growth processes of carbon clusters in a plume produced by laser ablation of a graphite target based on the temporal variations of the total numbers of $C_2$ and $C_3$ (Fig. 11) and the plume temperature (Fig. 15). At $t_D \leq 5$ $\mu$s, the plume moves from the irradiation point of the YAG laser pulse. The speed of the plume is decelerated by collision with ambient He gas, and the movement of the plume almost stops at $t_D \simeq$ 5 $\mu$s. The plume temperature in this period decreases very significantly from $\sim$ 4000 to $\sim$ 1000 K. After $t_D \simeq 5$ $\mu$s, the dynamics of the plume is governed by chemical reactions. At $5 \leq t_D \leq 10$ $\mu$s, $C_2$ radicals are produced by reaction (1), which is the first step of the cluster formation. The plume temperature in this period is less than 1000 K. The production of $C_3$ radicals by reaction (2) follows the production of $C_2$ at $10 \leq t_D \leq$ 200 $\mu$s. In this period, the total number of $C_2$ decreases. The plume temperature decreased from 700 to 400 K in this period. At $t_D \geq 200$ $\mu$s, the total numbers of $C_2$ and $C_3$ decrease, which may correspond to the formation of heavier clusters by reactions such as $C_2 + C_m + M \rightarrow C_{m+2} + M$, $C_3 + C_n + M \rightarrow C_{n+3} + M$, and $C_m + C_n + M \rightarrow C_{m+n} + M$. In this period, we observed continuum optical emission as shown in Fig. 4, which may be originated from the superposition of molecular band spectra of $C_n$ and high-temperature particulates produced by condensation reactions. In addition, as described in a previous paper [Kawashima *et al.* (1999)], we have carried out the detection of heavy carbon clusters in the plume using a laser photoionization technique. As a result, we have observed the increase in the cluster signal at $0.1 \leq t_D \leq 4$ ms. Therefore, the speculation from the present experimental results that heavy clusters are mainly formed at $t_D \geq 200$ $\mu$s is consistent with the previous photoionization diagnostics of the plume. Comparing the above observations with the plume temperature shown in Fig. 15, it is known that heavy carbon clusters grow at a low plume temperature of $\sim$ 350 K.

It is noted that the time scale described here is probably dependent on experimental conditions. For example, if the pressure of ambient gas is higher, the size of the plume may be smaller due to the confinement effect of the ambient gas, which may result in the faster growth of clusters. An important point shown by the present work is the sequential growth of clusters under the decreasing plume temperature.

Finally, we briefly discuss the role of ambient gas for the cluster formation. It is said generally that the role of ambient gas is plume cooling, which enhances the cluster formation by providing a low-temperature reaction environment. However, according to the experimental result shown in Fig. 15, plume cooling is not enhanced by a high ambient gas pressure. This is because plume cooling is governed by the work performed by the plume, and the expansion volume of the plume is related to pressure as $\Delta V \propto P^{-1}$. Therefore, the role of a high ambient gas pressure in cluster formation is not plume cooling but tight confinement of the plume. The plume's small volume at a high ambient gas pressure enhances the collision frequency, resulting in efficient cluster formation.

## 8. Conclusions

In this chapter, we investigated the growth processes of carbon clusters in a plume produced by laser ablation of a graphite target in ambient He gas. We presented a scenario for the growth of carbon clusters based on the temporal variations of the $C_2$ and $C_3$ radical densities in the plume, the plume temperature, and the continuum optical emission intensity. The growth efficiency of carbon clusters are governed by confinement and cooling of the plume. The plume confinement is controlled by the species and the pressure of ambient gas. The plume cooling is determined by the work performed by the plume. We expect that the fundamental understanding of the plume dynamics shown in this chapter is utilized for developing an optimized synthesis method of carbon clusters.

## References

Kroto, H. W., Heath, J. R., O'Brien, S. C., Curl, R. F., and Smalley, R. E. (1985). $C_{60}$: Buckminsterfullerene, *Nature* **318**, pp.162-163.

Iijima, S. (1991). Helical microtubules of graphitic carbon, *Nature* **354** (1991) pp.56-58.

Kroto, H. W. and McKay, K. (1988). The formation of quasi-icosahedral spiral shell carbon particles, *Nature* **331** (1988) pp.328-331.

Ozawa, M., Goto, H., Kusunoki, M., and Osawa. E. (2002). Continuously growing spiral carbon nanoparticles as the intermediates in the formation of fullerenes and nanoonions, *J. Phys. Chem. B* **106** 7135-7138.

Smalley, R. E. (1992). Self-assembly of the fullerenes, *Acc. Chem. Res.* **25** 98-105.

Wakabayashi, T., and Achiba, Y. (1992). A model for the $C_{60}$ and $C_{70}$ growth mechanism, *Chem. Phys. Lett.* **190** 465-468.

von Helden, G., Gotts, N. G., and Bowers, M. T. (1993). Experimental evidence for the formation of fullerenes by collisional heating of carbon rings in the gas phase, *Nature* **363** (1993) 60-63.

Hunter, J., Fye, J., and Jarrold, M. F. (1993). Annealing $C_{60}^{+}$: Synthesis of Fullerenes and Large Carbon Rings, *Science* **260** 784-786.

Goroff, N. S. (1996). Mechanism of fullerene formation, *Acc. Chem. Res.* **29** 77-83.

Irle, S., Zheng, G., Wang, Z., and Morokuma, K. (2006). The $C_{60}$ formation puzzle "solved": QM/MD simulations reveal the shrinking hot giant road of the dynamic fullerene self-assembly mechanism, *J. Phys. Chem. B* **110** 14531-14545.

Guo, T., Nikolaev, P., Thess A., Colbert, D. T., and Smalley, R. E. (1995). Catalytic growth of single-walled nanotubes by laser vaporization, *Chem. Phys. Lett.* **236** 49-54.

Yudasaka, M., Komatsu, T., Ichihashi, T., Achiba, Y., and Iijima, S. (1998). Pressure dependence of the structures of carbonaceous deposits formed by laser ablation on targets composed of carbon, nickel, and cobalt, *J. Phys. Chem. B* **102** 4892-4896.

Kataura, H., Kimura, A., Ohtsuka, Y., Suzuki, S., Maniwa, Y., Hanyu, T., and Achiba, Y. (1998). Formation of thin single-wall carbon nanotubes by laser vaporization of Rh/Pd-graphite composite rod, *Jpn. J. Appl. Phys.* **37** L616-L618.

Kokai, F., Takahashi, K., Yudasaka, M., Yamada, R., Ichihashi, T., and Iijima, S. (1999). Growth dynamics of single-wall carbon nanotubes synthesized by $CO_2$ laser vaporization, *J. Phys. Chem.* **103** 4346-4351.

Geohegan, D.B., Schittenhelm, H., Fan, X., Pennycook, S.J., Puretzky, A.A., Guillorn, M.A., Blom, D.A., and Joy, D.C. (2001). Condensed phase growth of single-wall carbon nanotubes from laser annealed nanoparticulates, *Appl. Phys. Lett.* **78** 3307-3309.

Kaizu, K., Kohno, M., Suzuki, S., Shiromaru, H., Moriwaki, T., and Achiba, Y. (1997). Neutral carbon cluster distribution upon laser vaporization, *J. Chem. Phys.* **106** 9954-9956.

Shibagaki, K., Kawashima, T., Sasaki, K., and Kadota, K. (2000). Formation of positive and negative carbon cluster ions in the initial phase of laser ablation in vacuum, *Jpn. J. Appl. Phys.* **39** 4959-4963.

Shibagaki, K., Sasaki, K., Takada, N., and Kadota, K. (2001). Synthesis of heavy carbon clusters by laser ablation in vacuum, *Jpn. J. Appl. Phys.* **40** L851-L853.

Muramoto, J., Nakata, Y., Okada, T., and Maeda, M. (1997). Observation of nano-particle formation process in a laser-ablated plume using imaging spectroscopy *Jpn. J. Appl. Phys.* **36** L563-L565.

Puretzky, A.A., Geohegan, D.B., Fan, X., Pennycook, S.J. (2000). In situ imaging and spectroscopy of single-wall carbon nanotube synthesis by laser vaporization *Appl. Phys. Lett.* **76** 182-184.

Ikegami, T., Ishibashi, S., Yamagata, Y., Ebihara, K., Thareja, R.K., and Narayan, J. (2001). Spatial distribution of carbon species in laser ablation of graphite target, *J. Vac. Sci. Technol. A* **19** 1304-1307.

Pearse, R. W. B. and Gaydon, A. G. (1976). *The identification of molecular spectra* (John Wiley & Sons, New York).

Gausset, L., Herzberg, G., Lagerqvist, A., and Rosen, B. (1965). Analysis of 4050-A Group of $C_3$ Molecule, *Astrophys. J.* **142** 45.

Takizawa, K., Sasaki, K., and Kadota, K. (2000). Characteristics of $C_3$ radicals in high-density $C_4F_8$ plasmas studied by laser-induced fluorescence spectroscopy, *J. Appl. Phys.* **88** 6201-6206.

Suzuki, C., Sasaki, K., and Kadota, K. (1999). Formation of $C_2$ radicals in high-density $C_4F_8$ plasmas studied by laser-induced fluorescence, *Jpn. J. Appl. Phys.* **38** 6896-6901.

Sasaki, K., Wakasaki, T., and Kadota, K. (2002). Observation of continuum optical emission from laser-ablation carbon plumes, *Appl. Surf. Sci.* **197-198** 197-201.

Rohlfing, E. A. (1988). Optical-emission studies of atomic, molecular, and particulate carbon produced from a laser vaporization cluster source, *J. Chem. Phys.* **89** 6103-6112.

Monchicourt, P. (1991). Onset of carbon cluster formation inferred from light-emission in a laser-induced expansion, *Phys. Rev. Lett.* **66** 1430-1433.

Jochnowitz, E. B. and Maier, J. P. (2008). Electronic Spectroscopy of carbon chains, *Annu. Rev. Phys. Chem.* **59** 519-544.

Sasaki, K., Wakasaki, T., Matsui S., and Kadota., K. (2002). Distributions of $C_2$ and $C_3$ radical densities in laser-ablation carbon plumes measured by laser-induced fluorescence imaging spectroscopy *J. Appl. Phys.* **91** 4033-4039.

Wakasaki, T., Sasaki, K., and Kadota, K. (2002). Collisional quenching of $C_2(d^3\Pi_g)$ and $C_3(\tilde{A}^1\Pi_u)$ and its application to the estimation of absolute particle density in laser-ablation carbon plumes, *Jpn. J. Appl. Phys.* **41** 5792-5796.

Sasaki., K. and Aoki, S. (2008). Temporal variation of two-dimensional temperature in a laser-ablation plume produced from a graphite target, *Appl. Phys. Express* **1** 086001.

Saito, K., Sakka, T., and Ogata, Y. H. (2003). Rotational spectra and temperature evaluation of $C_2$ molecules produced by pulsed laser irradiation to a graphite-water interface, *J. Appl. Phys.* **94** 5530-5536.

Kawashima, T., Sasaki, K., and Kadota, K. (1999). Diagnostics of laser-ablated carbon plumes by photoionization using a tunable laser, *Appl. Phys. A* **69**[Suppl.] S767-S770.

Chapter 3

# Kinetic and Diagnostic Studies of Carbon Containing Plasmas and Vapors Using Laser Absorption Techniques

J. Röpcke,[1,*] A. Rousseau[2] and P. B. Davies[3]

[1]*INP-Greifswald, 17489 Greifswald, Felix-Hausdorff-Str. 2, Germany*
[2]*LPP, Ecole Polytechnique, UPMC, Université Paris-Sud 11, CNRS, Palaiseau, France*
[3]*University of Cambridge, Cambridge CB2 1EW, U.K.*
[*]*E-mail address: roepcke@inp-greifswald.de*

Within the last decade mid infrared absorption spectroscopy between 3 and 20 μm, known as Infrared Laser Absorption Spectroscopy (IRLAS) and based on tunable semiconductor lasers, namely lead salt diode lasers, often called tunable diode lasers (TDL), and quantum cascade lasers (QCL) has progressed considerably as a powerful diagnostic technique for *in situ* studies of the fundamental physics and chemistry of molecular plasmas and vapors containing carbon. The increasing interest in processing plasmas and vapors containing hydrocarbons, fluorocarbons and organo-silicon compounds has lead to further applications of IRLAS because most of these compounds and their decomposition products are infrared active. IRLAS provides a means of determining the absolute concentrations of the ground states of stable and transient molecular species, which is of particular importance for the investigation of reaction kinetics. Information about gas temperature and population densities can also be derived from IRLAS measurements. A variety of free radicals and molecular ions have been detected, especially using TDLs. Since plasmas and vapors with molecular feed gases are used in many applications such as thin film deposition, semiconductor processing, surface activation and cleaning, and materials and waste treatment, this has stimulated the adaptation of infrared spectroscopic techniques to industrial requirements. The recent development of QCLs offers an attractive new option for the monitoring and control of industrial plasma and vapor processes as well as for highly time-resolved studies on the kinetics of plasma processes.

The aim of the present contribution is threefold: (i) to review recent achievements in our understanding of molecular phenomena in plasmas and vapors containing carbon, (ii) to report on selected studies of the spectroscopic properties and kinetic behavior of radicals, and (iii) to describe the current status of advanced instrumentation for QCLAS in the mid infrared.

## 1. Introduction

Low-pressure, non-equilibrium molecular plasmas and vapors are of increasing interest not only in fundamental research but also in plasma and vapor processing and technology. Molecular plasmas and vapors are used in a variety of applications such as thin film deposition, semiconductor processing, surface activation and cleaning, and in materials and waste treatment. The investigation of plasma physics and chemistry *in situ* requires detailed knowledge of plasma parameters, which can be obtained by appropriate diagnostic techniques. The need for a better scientific understanding of plasma physics and chemistry has stimulated the improvement of established diagnostic techniques and the introduction of new ones. Methods based on traditional spectroscopy have become amongst the most important because they provide a means of determining the population densities of species in both ground and excited states. The spectral line positions provide species identification while line profiles are often connected with gas temperature while relative intensities provide information about population densities. An important advantage of AS over Optical Emission Spectroscopy (OES) methods is that only relative intensities need to be measured to determine absolute concentrations, avoiding the problems of complete instrument calibration inherent in the OES methods. Absorption spectroscopy has been applied right across the spectrum from the Vacuum Ultra Violet (VUV) to the Far Infra-Red (FIR). Continuously emitting lamps (e.g. the Xe-lamp for the VIS and NIR, and the $D_2$-lamp for the UV) and tunable narrow-band light sources (e.g. tunable dye lasers, diode lasers) can be used as external light sources.

In the case where an external light source has much higher intensity than that of the plasma itself, the absorption of radiation can be described by the Beer-Lambert law which is,

$$I_v(l) = I_v(0) \exp(-\kappa(v)l).$$

$I_v$ (0) and $I_v$ (l) are the fluxes of the radiation entering and leaving the plasma, $l$ is the length of the absorbing (homogeneous) plasma column and $\kappa(v)$ is the absorption coefficient. Figure 1 illustrates this situation [1].

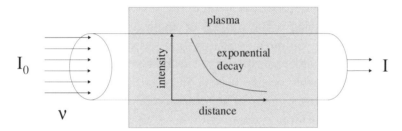

Fig. 1. Absorption of external radiation in a plasma according to the Beer-Lambert Law [1].

With the development of tunable, narrow band light sources such as tunable dye lasers and infrared diode lasers, these have been substituted for continuous light sources in AS experiments. These narrow band laser sources have the advantage of high spectral intensity, narrow bandwidth, and continuous tunability over the absorption profile.

The increasing interest in processing plasmas and vapors containing hydrocarbons, fluorocarbons or organo-silicon compounds has led to further applications of infrared AS techniques because most of these compounds and their decomposition products are infrared active. Fourier Transform Infrared (FTIR) spectroscopy has been used for *in situ* studies of methane plasmas for a number of years, but it is generally insufficiently sensitive for detecting free radicals or ions in processing plasmas. TDLAS is increasingly being used in the spectral region between 3 and 20 μm for measuring the concentrations of free radicals, transient molecules and stable products in their electronic ground states. TDLAS can also be used to measure neutral gas temperatures [2] and to investigate dissociation processes of molecular low temperature plasmas [3–6]. The main applications of TDLAS until now have been for investigating molecules and radicals in fluorocarbon etching plasmas [2,5,7], in plasmas containing hydrocarbons [6,8–14,19,22] and nitrogen, hydrogen and oxygen [17,18,20,21]. A wide variety of low molecular weight free radicals and molecular ions has been detected by TDLAS in purely spectroscopic studies e.g. $Si_2^-$ [15] and $SiH_3^+$ [16] in silane plasmas. Most of these spectroscopic results have yet to be applied in plasma diagnostic studies.

Molecular plasmas and vapors are increasingly being used not only for basic research but also, due to their favorable properties, for materials processing technology. These fields of application have stimulated the development of infrared spectroscopic techniques for industrial requirements. In order to exploit the capabilities of infrared TDLAS for effective and reliable on-line plasma diagnostics and process control in research and industry, compact and transportable tunable infrared multi-component acquisition systems (IRMA, TOBI) have been developed [17,18]). These systems are mainly focused on (i) high speed detection of stable and transient molecular species in plasmas under non-stationary excitation conditions and (ii) on sensitive (sub-ppb) trace gas detection with the aid of multi-pass absorption cells.

The main disadvantage of TDLAS systems, based upon lead salt diode lasers, is the necessary cryogenic cooling of the lasers (and also of the detectors), because they operate at temperatures below 100 K. Systems based upon lead salt diode lasers are typically large in size and require closed cycle refrigerators and/or cryogens like liquid nitrogen. The recent development and commercial availability of quantum cascade lasers (QCL) offers an attractive new option for infrared absorption spectroscopy.

The present chapter is intended to give an overview of recent achievements which have led to an improved understanding of phenomena in non-equilibrium molecular plasmas and vapors based on the application of IRLAS techniques. The chapter is divided in four main sub-chapters: In sub-chapter 2 special attention is devoted to recent studies of plasma chemistry and reaction kinetics in gas discharges containing hydrocarbons, nitrogen, oxygen and hydrogen. Sub-chapter 3 describes the gas-phase characterization in diamond hot-filament CVD. Sub-chapter 4 concerns recent results of spectroscopic properties and kinetic behavior of selected radicals, which are of special importance for reaction kinetics and chemistry in molecular processing plasmas. The current status of advanced spectroscopic instrumentation is described in sub-chapter 5.

## 2. Plasma Chemistry and Reaction Kinetics

### 2.1. *General considerations*

Low temperature plasmas, in particular microwave and Radio Frequency (RF) plasmas, have high potential for applications in plasma technology. In molecular low temperature plasmas, the species and surface conversion is frequently governed by high degrees of dissociation of the precursor molecules and high amounts of chemically active transient and stable molecules present. For further insight into plasma chemistry and kinetics a challenging subject is to study the mainly electron induced plasma reactions leading to entire series of different chemical secondary reactions involving the whole group of substances making up the source gas molecules. Hydrocarbon precursors are of special importance, since they are used in a variety of Plasma Enhanced Chemical Vapour Deposition (PECVD) processes to deposit thin carbon films. In all cases, the monitoring of transient or stable plasma reaction products, in particular the measurement of their ground state concentrations, is the key to improved understanding of fundamental phenomena in molecular non-equilibrium plasmas which can in turn be applied to many other aspects of plasma processing.

Transient molecular species, in particular radicals, influence the properties of nearly all molecular plasmas, both in the laboratory and in nature. They are of special importance in several areas of reaction kinetics and chemistry. The study of the behaviour of radicals together with their associated stable products provides a very effective approach to understand phenomena in molecular plasmas. Radicals containing carbon and oxygen are of special interest for fundamental studies and for applications in plasma technology.

Although the methyl radical ($CH_3$) is acknowledged to be one of the most essential intermediates in hydrocarbon plasma chemistry, only a few methods are available for its detection *in situ*. Sugai *et al.* and Zarrabian *et al.* employed the technique of threshold ionisation mass spectrometry to detect the methyl radical in electron cyclotron resonance plasmas containing methane [23–25]. Based on cavity ring-down spectroscopy with ultraviolet radiation $CH_3$ concentration measurements have been performed in a hot-filament reactor [26–28]. Most of the measurement techniques for detecting the methyl radical are based on absorption spectroscopy either with 216 nm ultraviolet radiation or in the infrared near 606 cm$^{-1}$. For example, the ultraviolet absorption of $CH_3$ at 216 nm was used for number density measurements by Child *et al.* and Menningen *et al.* and in different CVD diamond growth environments, hot-filament, dc and microwave plasmas [29–31]. In 2003 Lombardi *et al.* performed a comparative study to detect methyl radicals using both broadband ultraviolet absorption and TDLAS [32].

The infrared TDLAS technique has proven to be highly useful because it can also be used to measure the concentrations of related species provided they are IR active. Already in 1990 Wormhoudt demonstrated this flexibility by measuring $CH_3$ and $C_2H_2$ in a $CH_4$-$H_2$ RF plasma using a long path plasma absorption cell [33]. Actually, TDLAS is probably the best method for detecting the methyl radical for several reasons. The $v_2$ out-of-plane bending mode is not only intense but has many lines between 600 and 650 cm$^{-1}$. It is then possible to derive rotational and vibrational temperatures from their relative intensities. The ($J = K$) Q-branch lines of the $v_2$ fundamental band near 606 cm$^{-1}$ are particularly useful because several of them lie within 0.5 cm$^{-1}$ of each other i.e. within a laser spectral mode. Rotational temperatures in the plasma are therefore easily measured from them. For more than a decade the study for quantifying the concentrations of methyl radicals via the determination of the line strength of the

Q (8,8) line of methyl at 608.3 cm$^{-1}$ by Wormhoudt and McCurdy has been highly important [34]. In 2005 using the decay of the methyl radical in the off-phase of a pulsed plasmas new, precise measurements of the transition dipole moment of the $\upsilon_2$ fundamental band have been performed [35] (see sub-chapter 4.1.1).

One of the most successful applications of TDLAS is for studying the decomposition of hydrocarbons in a variety of PECVD processes. Systematic TDLAS measurements of several different hydrocarbons, including methyl, in a 20 kHz methane plasma in a parallel plate reactor were reported by Davies and Martineau [3,36]. Goto and co-workers have published numerous studies of methyl and methanol concentrations in RF and Electron Cyclotron Resonance (ECR) plasmas under different conditions e.g. investigating the influence of rare gases on the plasma. They have also combined IR absorption with emission spectroscopy, and investigated the effect of water vapor on the methyl radical concentration in argon/methane and argon/methanol RF plasmas using TDLAS [4,10,11,37]. Kim *et al.* measured $CH_3$, $C_2H_2$ and $CH_3OH$ concentrations in methanol/water RF plasmas by TDLAS and found that methanol was almost completely dissociated even at medium applied power levels [38]. In 1999 a group of eleven species, $CH_4$, $C_2H_2$, $C_2H_4$, $C_2H_6$, CO, $CO_2$, $CH_3$, $H_2O$, $CH_2O$, $CH_3OH$, HCOOH, were detected in $O_2$-$H_2$-Ar microwave plasmas with small admixtures of methane or methanol by TDLAS [6]. Busch *et al.* monitored the densities and temperatures of $CH_4$, $O_2$, $CH_3$, CO and $CO_2$ and studied aspects of the chemistry in a capacitively coupled RF discharge in 2001 [13].

## 2.2. *Molecular microwave plasmas containing hydrocarbons*

### 2.2.1. *$H_2/N_2/Ar$ plasmas with admixtures of $CH_4$ or $CH_3OH$*

In recent years several types of microwave discharge containing hydrocarbons as precursor gases have been at the centre of interest. The most recent applications of TDLAS for plasma diagnostic purposes include studies in which many different species have been monitored under identical plasma conditions [32,39]. This experimental data has frequently been used to model plasma chemical phenomena.

In 2003 Hempel *et al.* studied hydrocarbon plasmas with admixtures of nitrogen in a planar microwave reactor [40] using a tuneable diode laser (TDL) spectrometer [39]. The interest in such plasmas is based on various applications including deposition of diamond layers [41–43] and of hydrogenated carbon nitride films [44–47], detoxification of combustion gases [48], conversion to higher hydrocarbons [49], studies of astronomical objects such as interstellar clouds and stellar atmospheres [50,51]. Such types of plasma are also gaining importance in fusion physics, since they are representative of the edge discharges observed in the proximity of the carbon surfaces of the tokamak divertors [52].

Figure 2 shows the experimental arrangement. Hempel and co-workers used TDLAS to detect the methyl radical and nine stable molecules, $CH_4$, $CH_3OH$, $C_2H_2$, $C_2H_4$, $C_2H_6$, $NH_3$, $HCN$, $CH_2O$ and $C_2N_2$, in $H_2$-Ar-$N_2$ microwave plasmas containing up to 7 % of methane or methanol, under both flowing and static conditions. The degree of dissociation of the hydrocarbon precursor molecules varied between 20 and 97 %. The methyl radical concentration was found to be in the range $10^{12}$ to $10^{13}$ molecules cm$^{-3}$. It was established by analyzing the temporal development of the molecular concentrations under static conditions that $HCN$ and $NH_3$ are the final products of plasma chemical conversion. The fragmentation rates of methane and methanol and the respective conversion rates to methane, hydrogen cyanide and ammonia were determined for different relative proportions of hydrogen to nitrogen.

The novel experimental aspect introduced by Hempel *et al.* was the installation of multiple pass optics directly within the plasmas reactor to achieve higher sensitivity. Twenty four passes were realised with the White cell arrangement, leading to an optical length inside the reactor of about 36 m [53]. Figure 2b shows a ray diagram of the alignment within the White cell [54].

In fact, it is sometimes possible to detect the IR spectra of more than one species in a single laser mode using TDLAS, as shown in Fig. 4. As an example of the experimental results Fig. 3 gives an overview of the mass balance and degree of dissociation, as well as the product concentrations which range over five orders of magnitude in a methanol containing discharge under flowing conditions as a function of the nitrogen flow rate. A key objective of this type of study is to be able to model the chemistry of the plasma, for which it is necessary to monitor as many plasma species as possible.

In an earlier paper chemical modelling was successfully used to predict the concentrations of molecular species in methane plasmas in the absence of oxygen and the concentration trends of the major chemical products as oxygen were added [13,20,21].

In the work of Hempel *et al.* chemical modelling of the methane plasma with admixtures of nitrogen under static conditions was performed to predict the concentrations of those gaseous species which had been detected so far. A total of 145 reactions for 22 gaseous species were included in the model leading to relatively close agreement of experimental and calculated concentrations, and to improved knowledge of the main chemical reaction pathways and plasma chemical processes [39].

The rate coefficients for the electron collision processes have been determined by solving the time-dependent Boltzmann equation for given values of the reduced electric field, microwave frequency and mixture composition up to the establishment of the steady state. This electron kinetic equation has been solved by means of the multiterm method described by Loffhagen and Winkler [57]. Respective cross sections for

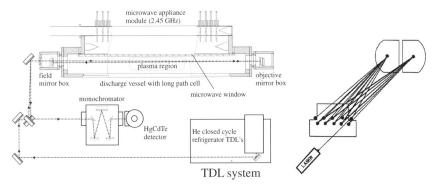

Fig. 2a. Experimental arrangement of the planar microwave plasma reactor (side view) with White cell multiple pass optical arrangement and TDL infrared source. The laser beam path is indicated by dotted lines [39]. 2b. White cell with field mirror and objective mirrors showing the laser beam path [39].

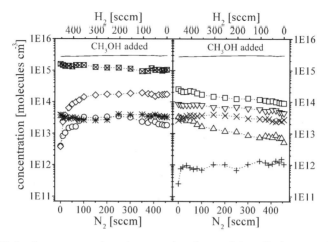

Fig. 3. Molecular concentrations in a methanol containing discharge under flowing conditions as a function of the nitrogen flow rate (⊠ - CH₃OH, ◇ - HCN, ○ - NH₃, * - CH₃, □ - CH₄, × - CH₂O, △ - C₂H₂, + - C₂H₄, ▽ - C₂H₆) [39].

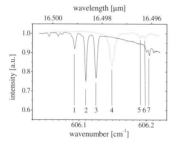

Fig. 4. TDL absorption spectra of some methyl and methanol lines in a H₂-Ar-N₂-CH₃OH microwave discharge (1,3,7 – CH₃; 2 – CH₃OH; 4,5,6 – N₂O). The dotted lines due to N₂O are from a reference gas cell placed in the beam path [39].

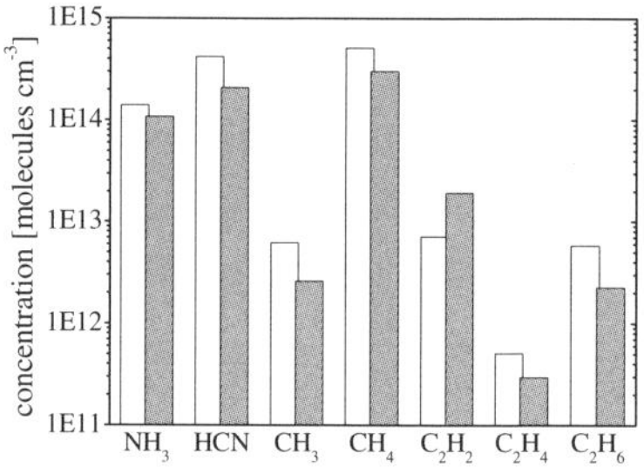

Fig. 5. Comparison of species concentrations in a representative H₂-N₂-Ar-CH₄-plasma (white – measured by TDLAS, grey – calculated) [39].

electron impact collisions for hydrogen, argon, oxygen and methane were taken from the established literature [55,58,59]. A comparison of modelled and experimental species concentrations in a representative H₂-N₂-Ar-CH₄-plasma is presented in Fig. 5 showing good agreement between them.

## 2.2.2. *N₂/O₂/Ar plasmas with admixtures of CH₄*

Although hydrogen and hydrocarbon containing plasmas with admixtures of oxygen and nitrogen have been extensively studied [1,39] there is still a lack of experimental data concerning the absolute densities of radicals in these discharges. The hydroxyl radical is known to be one of the main oxidising radicals. So far only a few studies have been reported on absolute OH concentrations in plasmas. Mostly, they were measured in the UV spectral region. In the present article quantitative measurements of OH in plasmas by means of TDLAS at 530 cm⁻¹ are reported.

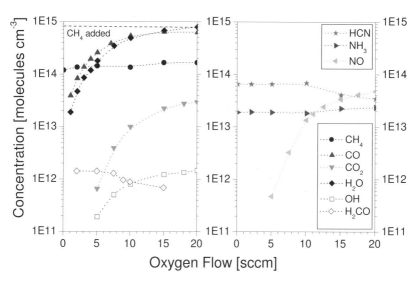

Fig. 6. Most abundant molecular species and their concentrations in an Ar/CH₄/N₂/O₂-plasmas (420 sccm Ar, 10 sccm CH₄, 10 sccm N₂, (0 ... 20) sccm O₂; 1.5 mbar) [62].

The measurements were performed in a planar microwave reactor in Ar/CH₄/N₂/O₂-discharges. The reactor was equipped with multi pass cell optics (White cell) in order to increase the absorption length (Fig. 2). A mirror spacing of 1.5 m and 40 m passes gave an effective absorption length of 60 m. Details on the experimental setup, data acquisition and data processing can be found elsewhere [6,40,60,61]. The pressure was kept constant at 1.5 mbar during the experiments. With a typical input power of 1.5 kW and 420 sccm Ar two different gas mixtures were used: (0 ... 20) sccm $H_2$, 20 sccm $N_2$ + $O_2$ and 10 sccm $CH_4$, 10 sccm $N_2$, (0 ... 20) sccm $O_2$. In both cases the oxygen content in the discharge was varied. Apart from the precursor molecule, the most abundant species in a methane containing plasma were determined to be $H_2O$ and CO (Fig. 6). When comparing the amount of $CH_4$ added to the discharge and the measured CO concentration values it turns out that methane is mainly converted into CO and only to a lesser extent into HCN and $CO_2$. However, the $CO_2$ selectivity, i.e. $[CO_2]$ / $([CO] + [CO]_2)$, is increasing with a higher amount of oxygen in the discharge.

In order to elucidate the underlying reactions one of the well known intermediate molecules for the conversion of $CH_4$, i.e. formaldehyde ($H_2CO$), and the concentration of OH were measured, too. For higher oxygen flows less $H_2CO$ can be found in the discharge whereas the concentration of the OH radical is increasing.

The behavior of the carbon containing molecules mentioned above can be understood qualitatively in terms of a model which was developed for a $H_2/CH_4/O_2$ plasma [6,55]. Although the bath gas was changed (Ar instead of $H_2$) the major reactions for the conversion of $CH_4$ should still be valid since mainly O or OH are involved. The prediction of the model calculation for the OH concentration is $5 \cdot 10^{11}$ cm$^{-3}$ which agrees well with the values obtained in this study.

$$CH_4 \xrightarrow{e,O} CH_3 \xrightarrow{O,OH} H_2CO \xrightarrow{OH} HCO \xrightarrow{O} CO \xrightarrow{O,OH} CO_2$$

After the dissociation of $CH_4$ into $CH_3$ the main conversion path is terminated at $H_2CO$ at low oxygen flows. With a higher oxygen content more radicals (O, OH) are available which first leads to a conversion into CO and finally into $CO_2$. Formaldehyde as the intermediate molecule is further converted leading to the observed maximum in $[H_2CO]$ at lower oxygen flows. The more $CO_2$ is produced with higher oxygen flows — as the final product of the $CH_4$ conversion path — the more the $CO_2$ selectivity increases. However all the measurements were performed in a discharge regime which is still oxygen poor [6]. As a result CO remains the most abundant molecule due to an incomplete conversion of $CH_4$ [62].

## 3. Gas-Phase Characterization in Diamond Hot-Filament CVD

In CVD diamond coating, hot-filament reactors are the most suitable ones when it comes to growing layers on complex geometries or up-scaled industrial coating processes with high numbers of substrates to be coated. Even though the precursor gases are often solely hydrogen and methane, the gas composition within the active coating zone becomes abundant of species. Therefore many works have been done for characterizing the gas-phase conditions and kinetics by numerous interdependent gas-phase reactions. Theoretical modeling, such as in [63–68] has been done to reveal gas-phase processes. Modeling results are dependent on input border conditions, being different for every reactor and filament–substrate arrangement.

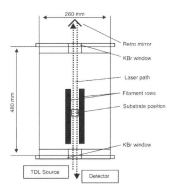

Fig. 7. Photo showing the IR-TDLAS-system IRMA consisting of optical bench with integrated He-cooling system (left) and data processing unit (right) positioned in front of the HF-CVD coating reactor (in the back). The arrows indicate the way of the laser beam [74].

Fig. 8. Top view of the CVD reactor chamber with filament substrate arrangement and position of the laser beam path striving directly above the substrates surface. The shown set up with two paths of the laser through the chamber results in an absorption length of 960 mm window to window [74].

For experimental gas-phase diagnostics, sumptuous optical techniques such as REMPI, CRDS, CARS, LIF have been applied [28,69–71]. Each of these techniques has its advantages and drawbacks, as e.g. elucidated in [72]. Mostly, though, only one single species of the gas-phase can be measured at a time. The improvement of process effectiveness, reliability and reproducibility in hot-filament reactors requires investigation of process parameter effects on as many gas species as possible. For this purpose a compact infrared tunable diode laser absorption spectroscopy (IRMA) system has been used, that is capable of measuring absolute molecular gas concentrations of up to 8 species simultaneously in real time (Figs. 7, 8).

Employing TDLAS the influence of adjustable process parameters on the most significant carbon containing species within our HFCVD coating unit was characterized. The coating conditions ranged in typical parameter windows applied for industrial diamond coating processes. The IRMA system [29] was applied for process monitoring by measuring up to 3 species concentrations ($CH_4$, $C_2H_2$, CO) simultaneously. $CH_4$ and $C_2H_2$ were found to be the most abundant carbon species, which is in good agreement with HF-CVD gas-phase characterization in literature. $CH_4$ was measured to be in the magnitude of $10^{14}$ molecules/cm$^3$, $C_2H_2$ concentrations ranged one order of magnitude lower. Measuring significant concentrations of CO leads to the conclusion of the presence of air leakage, which for coating units of this size or industrially employed larger ones is generally expected to some extent. It was possible to measure absolute concentration of methyl radicals, considered to be the major diamond growth species.

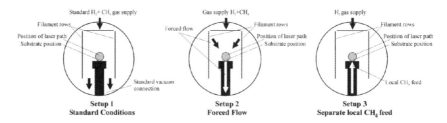

Fig. 9. Front view of the CVD reactor in the three different applied operating states. The conditions in setup 1 with a mixed $H_2/CH_4$ gas feed and a vacuum connection far from the substrate resembles a widespread setup of common HFCVD diamond coating units. The forced flow arrangement (setup 2) and the possibility for separate $CH_4$ feeding (setup 3) are employed to improve the local gas phase above the substrate [73].

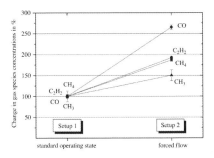
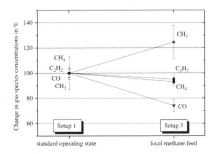

Fig. 10. Application of a forced flow towards the substrate causes a local increase in all measured carbon species at 25 mm distance from the filaments (substrate center position). The concentrations of $CH_4$, $C_2H_2$ and CO were measured simultaneously at a process pressure of 7 mbar, 1000 sccm gas flow and 1% methane content. $CH_3$ was measured separately at the same conditions but at 1.5% $CH_4$ [73].

Fig. 11. Feeding methane locally at substrate position separately from hydrogen leads to more favorable diamond growth conditions. The $CH_4$ is dissociated more effectively, leading to a higher $CH_3$ growth species concentration. The formation of longer chained molecules like $C_2H_2$ and carbon traps like CO is alleviated. The concentrations of $CH_4$, $C_2H_2$ and CO were measured simultaneously at a process pressure of 7 mbar, 1000 sccm gas flow and 1% methane content. $CH_3$ was measured separately at the same conditions but at 1.5% $CH_4$ [73].

The possibilities to increase the diamond growth rate in HF-CVD beyond the conventional process parameter window by changing gas phase conditions at the substrate were evaluated (Fig. 9). For that, new features in the coating unit of the reactor were installed (i) a forced gas flow towards substrate surface and (ii) a separate $CH_4$ feeding locally at the substrate. Application of a forced gas flow showed a remarkable increase in the diamond growth rate of a factor 46 compared to standard coating setups. By lowering the methane content in the forced flow, diamond quality factors 495% were achieved. To depict the deposition results, diamond layers after 70 h deposition time are shown in Fig. 12 for each operating state by using ideal deposition parameters, respectively. It is obvious that the new operating states cause a significantly increased diamond growth rate. TDLAS showed an increase of all measured carbon containing species $CH_4$, $C_2H_2$, $CH_3$ and CO when applying the forced flow (Fig. 10). Estimations showed that the mass transport dominated by diffusion in the standard setup shifts to a convective gas transport in the forced flow setup. The induced laminar flow causes a more effective growth species transport to the substrate and leads to higher growth rates.

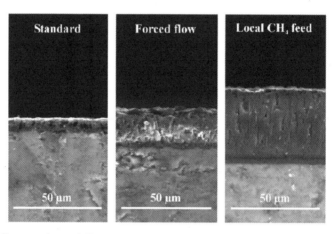

Fig. 12. Cross-sections of diamond layers deposited under setups 1–3 (left to right) after 70 h. The layers shown were grown by using ideal deposition parameters for the different setups, respectively. For standard conditions the methane content used was 0.8%, for forced flow (800 sccm) and local $CH_4$ feed conditions the methane content was 0.3%. The new operating states cause a significantly increased diamond growth rate [73].

The application of feeding methane locally at substrate position leads to exceptionally high growth rates (0.68 mm/h) at correspondingly high diamond quality (91%). For this, however, the methane content has to be lowered. This leads at the same time to a more homogenous deposition lateral on the surface. From the TDLAS gas phase measurements, a more effective precursor dissociation, a higher $CH_3$ density and a rise in the $CH_3/C_2H_2$ ratio above the substrate surface can be derived (Fig. 11). Having already proved beneficial effects on diamond infiltration (CVI) [5], the two applied concepts are considered as a promising HFCVD reactor extension. The results in the forced flow setup give hints for future reactor design concerning gas flow arrangements. The elevated growth rates when feeding $CH_4$ locally can lead to more economic batch production at even lower precursor consumption [73,74].

## 4. Kinetic Studies and Molecular Spectroscopy of Radicals

### 4.1. *Line strengths and transition dipole moment of CH$_3$*

#### 4.1.1. *The $\nu_2$ fundamental band*

This sub-chapter describes a new measurement of $\mu_2$ for the $\nu_2$ fundamental band of the methyl radical in order to resolve the differences between earlier experimentally measured values and between experiment and theory. The method used for determining the absolute methyl radical concentrations was the same as that used by Yamada and Hirota [8]. However, integrated intensities and many more methyl radical lines were used. Furthermore the kinetic conditions were more precisely specified and the temperature determined more exactly. The resulting value of $\mu_2$ is now in much better agreement with theory.

The methyl radical has no electric dipole allowed rotational transitions because of its $D_{3h}$ symmetry and so IR spectroscopy is one of the few suitable methods for its detection. The determination of methyl radical concentrations in terrestrial and astronomical sources using IR spectroscopy relies on the availability of accurate line strengths and transition dipole moments. The $\nu_2$ band of CH$_3$ is the strongest of its IR active fundamentals and particularly useful for quantitative measurements. The need for a more accurate and precise value of $\mu_2$ has been highlighted by the measurements of CH$_3$ in the atmospheres of Saturn [78], Neptune [79] and in the interstellar medium [80].

The experimental set-up of the planar microwave plasma reactor with the optical arrangements used for the methyl transition dipole moment study is comparable to that shown in Fig. 2. Details of the diode laser spectrometer, IRMA, and discharge absorption cell have been reported elsewhere [40,60]. The methyl radical was produced in mixtures of tertiary butyl peroxide ($[(CH_3)_3CO]_2$) and argon at total pressure of 1 mbar. Two kinds of experiments were performed: (a) time dependent measurement of the decay of the absorption coefficient when the discharge was turned off, to obtain absolute methyl concentrations and (b) measurements of the absorption coefficients of different rovibronic

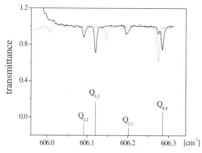

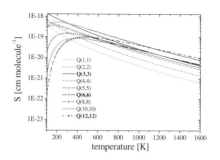

Fig. 13. Survey spectrum showing several ($J = K$) $Q$ branch lines of the $\nu_2$ fundamental of the $CH_3$ free radical around the band origin. The spectrum represented by the dashed line is a calibration spectrum from $N_2O$ and $CO_2$ [35].

Fig. 14. Line strengths, S, for different Q branch transitions of $CH_3$ as a function of temperature. Values calculated from the reference temperature values at 296K [35].

lines. In total ten lines were studied in the fundamental band, seven in the first hot band, and one from the second hot band. A survey spectrum of the $Q$-branch region of the $\nu_2$ fundamental band is shown in Fig. 13.

In order to derive accurate line strengths and the transition dipole moment, it is necessary to obtain the absolute concentration of the methyl radical and its temperature in the discharge. The decay method was the experimental approach for methyl radical concentration measurements. The plasma was switched on and off for periods of ten seconds and the decay of the methyl radical signal measured during the off period with ms time resolution [75,76]. The absolute concentration was obtained from the decay of the integrated absorption coefficient and the recombination rate constant. It is well known from numerous kinetic studies that the main loss channel under the conditions used here is self recombination via a three body reaction. Hence by measuring the integrated absorption coefficient as a function of time and knowing the value of the recombination rate constant the absolute concentration of the methyl radical can be obtained.

The rate constant $k_1$ for the self recombination reaction of methyl radicals has been extensively investigated in experimental and theoretical work [81–88]. The selected value for $k_1$ was based on the compilation of

Baulch *et al.* and was appropriate for the specific temperature and argon concentration [35,89]. The translational, rotational and vibrational temperatures of the methyl radical were measured. A near similarity of $T_{trans}$ and $T_{rot}$ was observed. Based on experimental results the vibrational temperature was found to be in equilibrium with the translational and rotational temperature within experimental uncertainties i.e. $T_{vib} = 600$ K. For details see [35,77].

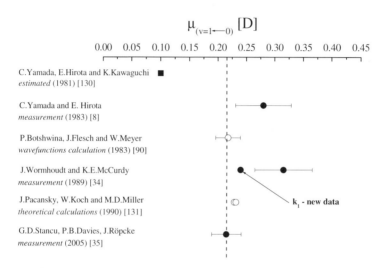

Fig. 15. Chronological summary of calculated and measured values of $\mu_2$ the transition dipole moment of the $\nu_2$ fundamental band of $CH_3$ [35].

Figure 14 shows as an example of the temperature dependence of the line strengths of several transitions from the lower energy levels. The gas temperature was found to be about 800 K near the deposition substrate, which was essentially the same as the temperature measured by a thermocouple probe. The line strengths of the nine $Q$ branch lines in the $\nu_2$ fundamental band of the methyl radical in its ground electronic state were used to derive a more accurate value of the transition dipole moment of this band: $\mu_2 = 0.215(25)$ Debye. Improved accuracy over earlier measurements of $\mu$ was obtained by integrating over the complete line profile instead of measuring the peak absorption and assuming a Doppler line width to deduce the concentration; and the derivation of more accurate temperatures by examining a large number of lines. In addition a more precise value for the rate constant for methyl radical recombination than available earlier was employed. The new value of $\mu_2$ is in very good agreement with high quality *ab initio* calculations. Figure 15 shows the chronological summary of calculated and measured values of $\mu_2$ the transition dipole moment of the $v_2$ fundamental band [35].

### 4.1.2. *The $v_2$ first hot band*

The infrared laser absorption spectra of the methyl radical recorded in electric discharges consist not only of fundamental band lines but also of weaker hot band lines suggesting the possibility of determining the transition dipole moments of the hot bands for comparison with theory. This sub-chapter describes the determination of the transition dipole moment of the first hot band of the $v_2$ band using the same experimental method as used for the fundamental band in 4.1.1.

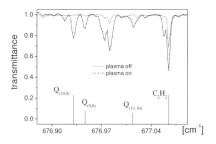
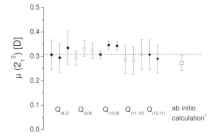

Fig. 16. Survey spectrum showing Q-branch lines from the first hot band. Several lines of stable species are observed both inside and outside the plasma [91].

Fig. 17. The transition dipole moment of the $v_2$ first hot band, measured values compared with the *ab intio* calculation. Multiple determinations based on five measured transitions are shown [91].

A survey spectrum of some of the Q branch lines of the first hot band is shown in Fig. 16. The required methyl recombination rate coefficient in the presence of Argon and at the experimental temperatures is available from the literature. The measured line strengths for the $v_2$ first hot band Q-branch transitions are given in [91]. The translational, rotational and vibrational temperatures are required to calculate the corresponding partition functions. Although the partition functions were deduced in the earlier study [35] on the fundamental band they were also evaluated here from the hot band transitions and using a more extended analysis.

The translational temperature was obtained by fitting the line profiles to a Gaussian function since the line width is determined by Doppler broadening at the pressures used here. The net broadening due to the intrinsic Doppler broadening of the line corresponded to $T = 630$ K. The rotational temperature was obtained using $Q(4,2)$, $Q(9,8)$, $Q(10,9)$, $Q(11,10)$, and $Q(12,11)$. Experimentally the vibrational temperature is obtained from measured integrated absorption coefficients of resolved vibration rotation transitions in the fundamental and the hot bands. The transition dipole moment of the first hot band was determined to be 0.31(6) D, Fig. 17. This value is in satisfactory agreement with the value of 0.27(3) D from a high precision *ab initio* calculation using the Self Consistent Electron Pairs (SCEP) method reported by Botschwina, Flesch and Meyer [90,91].

### 4.2. *Molecular spectroscopy of the CN radical*

The CN radical is of fundamental importance in laboratory spectroscopy and in astrophysics. Electronic emission spectra arising from the red $(A^2\Pi - X^2\Sigma^+)$ and violet $(B^2\Sigma^+ - X^2\Sigma^+)$ band systems excited in flames and discharges have been studied in the laboratory over decades while CN spectra have been detected in the atmospheres of stars and in the interstellar medium. Most recently the electronic band systems have been very extensively measured and analyzed in emission [92,93] using high resolution Fourier transform spectroscopy.

Rotationally resolved spectra of the fundamental band of the CN free radical in four isotopic forms have been measured using TDLAS [94]. The source of the radical was a microwave discharge in a mixture of isotopically selected methane and nitrogen diluted with argon. The lines were measured to an accuracy of $5 \times 10^{-4}$ cm$^{-1}$ and fitted to the formula for the vibration rotation spectrum of a diatomic molecule, including quartic distortion constants. The band origins of each of the isotopomers from the five parameter fits were found to be $^{12}C^{14}N$: 2042.42115(38) cm$^{-1}$, $^{13}C^{14}N$: 2000.08479(23) cm$^{-1}$, $^{12}C^{15}N$: 2011.25594(25) cm$^{-1}$, $^{13}C^{15}N$: 1968.22093(33) cm$^{-1}$ with one standard deviation from the fit given in parenthesis. Some of the lines showed a resolved splitting due to the spin rotation interaction. This was averaged for fitting purposes. The average equilibrium internuclear distance derived from the $v = 0$ and $v = 1$ rotational constants of the four isotopomers is 1.171800(6)Å which is in good agreement with the value determined from microwave spectroscopy. Figure 18 shows a stick diagram of all the lines measured in the four isotopic forms and their intensities calculated for a rotational temperature of 950 K [94].

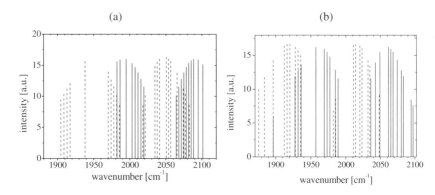

Fig. 18. Stick diagrams showing the detected absorption lines from the fundamental band of CN and their intensities at 950 K: (a) $^{12}C^{14}N$ and $^{13}C^{14}N$ (dotted lines) and (b) $^{12}C^{15}N$ and $^{13}C^{15}N$ (dotted lines) [94].

## 5.  Quantum Cascade Laser Absorption Spectroscopy for Plasmas Diagnostics and Control

### 5.1. General considerations

The recent development of pulsed QCLs and their commercial availability offer promising new possibilities for infrared absorption spectroscopy [95–97]. Pulsed QCLs are able to emit mid infrared radiation at near room temperature operation. Compared to lead salt laser systems, QCL systems are very compact mid infrared sources characterized by narrow line width combining single-frequency operation and considerably higher powers, i.e. tens of mW. The output power is sufficient to combine them with thermoelectrically cooled infrared detectors, which permits a decrease of the apparatus size and provides an opportunity to design compact cryogen-free mid infrared spectrometer systems. The positive features of quantum cascade laser absorption spectroscopy (QCLAS) opens up new fields of application in research and industry. Recently a compact quantum cascade laser measurement and control system (Q-MACS) has been developed (Figs. 19 and 20) for time-resolved plasma diagnostics, process control and trace gas monitoring which can be used as platforms for various applications of QCLAS [1,98].

Fig. 19. Q-MACS Basic with laser head, supply unit, connection cable and optical collimation module [103].

Fig. 20. New generation of compact QCLAS equipment with laser head in TO-8 configuration (left-hand side), reference gas cell (middle) and detector (right-hand side). The laser beam is guided by 2 Out of Axis Paraboloids [103].

Nowadays QCLAS has been used to detect atmospheric trace constituents or trace gases in exhaled breath. Furthermore it has already been successfully applied to the study of plasma processes, e.g. of microwave and RF discharges [99,100]. Of special interest for all hydrocarbon-based processes is the capacity to detect transient molecules like the $CH_3$ radical, as the supposed key growth species, by means of QCLAS [103].

The scan through an infrared spectrum is commonly achieved by two different methods. In the *inter* pulse mode a bias DC ramp is applied to a series of short laser pulses of a few ten nanoseconds [76,101]. Another option is the *intra* pulse mode, i.e. the scanning in single, longer pulses acquiring an entire spectrum [102]. Since this scan is performed in tens up to a few hundred nanoseconds a time resolution below 100 ns has become possible for quantitative in-situ measurements of molecular concentrations in plasmas for the first time. Therefore it fits very well to measurements of rapidly changing chemical processes.

### 5.2. *Trace gas measurements using optically resonant cavities*

The development of compact and robust optical sensors for molecular detection is of interest for an increasing number of applications, such as environmental monitoring and atmospheric chemistry [104–106], plasma diagnostics and industrial process control [1,100], combustion studies, explosive detection and medical diagnostics [107–111]. Absorption spectroscopy in the mid-infrared spectral region using lasers as radiation sources is an effective method for monitoring molecular species. In principle path lengths up to several kilometres can be achieved by using optical resonators for cavity ring-down spectroscopy (CRDS) [112], cavity enhanced absorption spectroscopy (CEAS) [113], or integrated cavity output spectroscopy (ICOS) [114,115]. The majority of cavity based methods have used sources of radiation in the ultraviolet and visible regions. For many years the infrared spectral range could not be employed either for CRDS or for the CEAS or ICOS techniques, because of the lack of suitable radiation sources with the required power and tunability, but this situation has now changed. Near infrared (NIR) applications have profited from developments in telecommunications where cheap and compact light sources became available [116,117] in the 1990s whereas similar lasers were not available in the MIR. Recent advances in semiconductor laser technology, in particular the advent of intersubband quantum cascade lasers (QCL) and interband cascade lasers (ICL) [95,97,118,119], provides new possibilities for highly sensitive and selective trace gas detection using MIR absorption spectroscopy in combination with multiple pass cells [90,120–126] and enabled sensitivities of $5 \times 10^{-10}$ cm$^{-1}$Hz$^{-1/2}$ to be accomplished but at the expense of large sample volumes [76]. The relatively high output power of the QCL permits the use of optical cavities with high finesse. With the help of such cavities the effective path length of the laser beam in the absorbing medium can essentially be increased to more than the 200 m limit usually available from conventional optical multi-pass cells [121,127] while keeping the sample and pumped volume small.

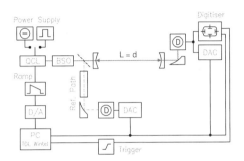

Fig. 21. Schematic diagram of the apparatus used for both CRDS and CEAS experiments. Beam shaping optics (BSO) provided efficient light transmission to the resonator. The detector (D) signals were recorded with data acquisition cards (DAC). The quantum cascade lasers (QCL) were driven either by a pulsed or a cw source and the tuning ramp for the laser was generated via the digital-to-analog-converter (D/A) [128].

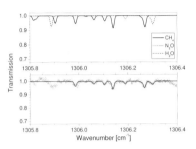

Fig. 22. Calculated spectrum (upper) for air containing $CH_4$, $N_2O$ and $H_2O$. Experimental conditions: $p = 2.2$ mbar, $L_{eff} = 1080$ m, $0.005$ cm$^{-1}$ instrumental broadening. The corresponding CEAS spectrum (lower) was observed with a TE cooled detector (open circles) and fitted (solid line) to determine the actual concentrations [128].

While pulsed QCLs working at room temperature have been commercially available for several years, room-temperature cw QCLs have only recently been introduced. Thermoelectrically cooled QCLs have been combined with high finesse optical resonators, (Fig. 21) in order to profit from their enlarged path lengths at reasonably small sample volumes in combination with the high absorption cross section in the infrared molecular fingerprint region. Two different approaches have been investigated. Firstly, pulsed QCLs at 7.42 μm and 8.35 μm were used to perform CRDS and ICOS experiments. The spectra measured at pressures between 100 mbar and 300 mbar were normally characterized by a broadening of the absorption lines and reduced absorption in comparison with theoretical expectations, which makes an absolute calibration necessary. The resulting decrease in sensitivity, i.e. to

~ $5 \times 10^{-7} \, \text{cm}^{-1}\text{Hz}^{-1/2}$, means that long path cell configurations of similar sample volumes have superior sensitivity to either CRDS or ICOS. It transpires that the frequency-down chirp inherent to pulsed QCLs sets the fundamental limit.

Due to the chirp the effective line width of the QCL is much broader than the narrow molecular absorption features and the rather fast chirp rate does not allow for an efficient build-up of the laser field in the cavity. CRDS using pulsed QCLs therefore has only a limited number of useful applications, e.g. for the determination of the reflectivity of the cavity mirrors in preliminary experiments or for the detection of complex and broad molecular absorptions at higher pressures. Secondly, a cw QCL at 7.66 μm has been combined with an unstabilized and unlocked cavity. With this straightforward arrangement, comprising only a TE cooled detector, an effective path length of 1080 m has been achieved. The main limit to sensitivity, of $2 \times 10^{-7} \, \text{cm}^{-1}\text{Hz}^{-1/2}$, was the cavity mode noise which may be reduced by off-axis alignment. Such mode noise is absent in conventional long path cell spectrometers. With a 20 s measurement interval detection limits for $N_2O$ and $CH_4$ of $6 \times 10^{8}$ molecules/cm$^3$ and $2 \times 10^{9}$ molecules/cm$^3$ respectively could be achieved at 2.2 mbar indicating that sub-ppb levels could easily be measured at higher pressures, see Fig. 22. Furthermore, the small sample volume of 0.3 l used here, and consequently the reduced pumping requirement, is an advantage over long path cells with much larger volumes than 0.5 l. The detection limits achieved here are relevant both for the detection of radicals in plasma chemistry and for trace gas measurements with field-deployable systems. Additionally, trace gas measurements may also be carried out without pre-concentration procedures. Radicals with small abundances in the gas phase might now be detectable via QCL based spectroscopy. The choice of an appropriate method depends on whether the important criterion is ultimate sensitivity or a more compact system. In the former case a multi pass cell spectrometer would be preferable because of its better signal to noise characteristics but this configuration would exclude the detection of processes on short time scales or certain types of *in situ* measurements. Sophisticated locked cavity schemes might overcome such restrictions at the expense of complex spectrometer geometries. To achieve a compact system a small volume cavity based spectrometer employing cw QCLs would be more appropriate. This configuration would also be of special interest for applications where the pressure cannot arbitrarily be chosen in order to adapt the absorption line width to the laser line width, e.g. in low pressure plasmas. Moreover, for these applications *in situ* measurements are essential because multi-pass cell sampling *ex situ* is not an option [128].

## 5.3. *In situ monitoring of plasma etch processes with a QCL arrangement in semiconductor industrial environment*

During the last forty years plasma etching has become a fundamental feature for processing integrated circuits. The optimization of the plasma chemistry in etch processes includes the identification of the mechanisms responsible for plasma induced surface reactions combined with the achievement of uniformity in the distribution of molecules and radicals for homogeneous wafer treatment.

In etch plasmas used for semiconductor processing concentrations of the precursor gas $NF_3$ and of the etch product $SiF_4$ were measured on-line and *in situ* using a new experimental arrangement, designated the Q-MACS Etch system, which is based on QCLAS. In addition, the etch rates of $SiO_2$ layers and of the silicon wafer were monitored including plasma etching endpoint detection. For this purpose the Q-MACS Etch system worked in an interferometic mode. The experiments were performed in an industrial dual frequency capacitively coupled magnetically enhanced reactive ion etcher (MERIE), which is a plasma reactor developed for dynamic random access memory (DRAM) technologies, Fig. 23. The quantum cascade laser system Q-MACS Etch consists of a pulsed infrared QCL source with the laser wavelength tunable in the range 1027–1032 $cm^{-1}$, optical components, detectors and data acquisition cards controlled by a PC, Fig. 24. On board the IR beam is split into two channels using two IR transparent ZnSe beam splitters. The main part of the beam is coupled into an IR fibre using an off axis parabolic mirror, and then collimated into the plasma reactor. In the second channel a reference spectrum of $C_2H_4$ is measured for line locking. Industrial requirements, such as (i) no open optical path and (ii) the availability of just one optical access port makes coupling the intrared beam into the reactor a challenging task. The solution realized in the Q-MACS Etch is based on the use of (a) mid infrared fibers and (b) internal reflections in the reactor chamber. The capability of Q-MACS Etch for $SiF_4$ concentration monitoring via reactor side access and for monitoring the etch rates via reactor top access is shown in Fig. 23.

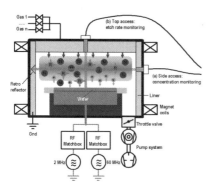 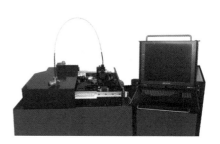

Fig. 23. Experimental arrangement of the MERIE plasma etching reactor: a) side access and b) top access monitoring of the Q-MACS Etch system [100].

Fig. 24. Picture of the Q-MACS Etch with the optical module and the separate control unit in a rack [129].

This first application of a quantum cascade laser arrangement for monitoring of industrial etch processes has opened up a challenging new option for control of demanding semiconductor production applications. Focused on sensitive and fast concentration measurements of key molecular components, while ensuring compactness, robustness and long term stability, this new class of process control equipment has the potential to become implemented into other fields of technology [100,129], in particular also for applications of carbon containing plasmas and vapors.

## 6. Summary and Conclusions

During the past few years a variety of phenomena in carbon containing plasmas and vapors in which many short-lived and stable species are produced have been successfully studied based on diode laser absorption techniques in the mid infrared spectral range, with which the present chapter is concerned. It has been possible to determine absolute concentrations of ground states using spectroscopy thereby providing a link with chemical modelling of the plasma, the ultimate objective being better to understand the chemical and reaction kinetic processes occurring in the plasma. The other essential component needed to reach this objective is to determine physical parameters of the plasma, as for example, temperatures, degrees of dissociation and dynamics of reaction kinetic processes, and the present article discusses methods for achieving this. The need for a better scientific understanding of plasma physics and chemistry has stimulated the application of IRLAS, which has been proven to be one of the most versatile techniques for studying molecular plasmas. Based on the recent development of quantum cascade lasers the further spread of this method of high resolution mid infrared spectroscopy to industrial applications has become a reality.

## Acknowledgements

This work was partly supported by (a) the Deutsche Forschungsgemeinschaft, Sonderforschungsbereich Transregio 24, and Transferbereich 36, (b) the Bundesministerium für Bildung und Forschung, FKZ 13N7451/8 and (c) the Deutscher Akademischer Austauschdienst and EGIDE as part of the French-German PROCOPE Collaboration Program (Project 04607QB).

Furthermore it was supported within the scope of a technology development by the EFRE fund of the European Community and by funding of the State Saxony of the Federal Republic of Germany.

The authors give sincere thanks to all present and former members of the laboratories involved in Greifswald, Paris and Cambridge for permanent support and a stimulating scientific climate. In particular, the authors are indebted to all co-authors of former papers whose contributions made the present chapter possible. K.-D. Weltmann is gratefully acknowledged for his continuous encouragement and general support.

# References

[1] Röpcke J, Lombardi G, Rousseau A and Davies P B, *Plasma Sources Sci. Technol.* **2006**, *15*, S148.

[2] Haverlag M, Stoffels E, Stoffels W W, Kroesen G M W and de Hoog F J 1996 *J. Vac. Sci. Technol.* A **14** 380.

[3] Davies P B and Martineau P M 1992 *Adv Mater.* **4** 729.

[4] Naito S, Ito N, Hattori T and Goto T 1995 *Jpn. J. Appl. Phys.* **34** 302.

[5] Haverlag M, Stoffels E, Stoffels W W, Kroesen G M W and de Hoog F J 1994 *J. Vac. Sci. Technol.* A **12** 3102.

[6] Röpcke J, Mechold L, Käning M, Fan W Y and Davies P B 1999 *Plasma Chem. Plasma Process.* **19** 395.

[7] Haverlag M, Stoffels E, Stoffels W W, Kroesen G M W. and de Hoog F J 1996 *J. Vac. Sci. Technol.* A **14** 384.

[8] Yamada C and Hirota E 1983 *J. Chem. Phys.* **78** 669.

[9] Naito S, Ikeda M, Ito N, Hattori T and Goto T 1993 E *Jpn. J. Appl. Phys.* **32** 5721.

[10] Naito S, Ito N, Hattori T and Goto T 1994 *Jpn. J. Appl. Phys.* **33** 5967.

[11] Ikeda M, Ito N, Hiramatsu M, Hori M and Goto T 1997 *J. Appl. Phys.* **82** 4055.

[12] Kroesen G M W, den Boer J H W G, Boufendi L, Vivet F, Khouli K, Bouchoule A and de Hoog FV 1996 *J. Vac. Sci. Technol.* A **14** 546.

[13] Busch C, Möller I and Soltwisch H 2001 *Plasma Sources Sci. Technol.* **10** 250.

[14] Serdioutchenko A, Möller I and Soltwisch H 2004 *Spectrochim. Acta* **60A** 3311.

[15] Liu Z and Davies P B 1996 *J. Chem. Phys.* **105** 3443.

[16] Davies P B and Smith D M 1994 *J. Chem. Phys.* **100** 6166.

[17] Rousseau A, Dantier A, Gatilova L V, Ionikh Y, Röpcke J and Tolmachev Y A 2005 *Plasma Sources Sci. Technol.* **14** 70.

[18] Ionikh Y, Meshchanov A V, Röpcke J and Rousseau A 2006 *Chem. Phys.* **322** 411.

[19] Zijlmans R A B, Gabriel O, Welzel S, Hempel F, Röpcke J, Engeln R, Schram D C 2006 *Plasma Sources Sci. Technol.* **15** 564.

[20] Gatilova L V, Allegraud K, Guillon J, Ionikh Y Z, Cartry G, Röpcke J and Rousseau A 2007 *Plasma Sources Sci. Technol.* **16** S107.

[21] Van Helden, J H, Wagemans W, Yagci G, Zijlmans R A, Schram D C, Engeln R, Lombardi G, Stancu G D and Röpcke J 2007 *J. Appl. Phys.* **101** 043305.

[22] Rousseau A, Guaitella O, Gatilova L, Hannemann M. and Röpcke J 2007 *J. Phys. D.* **40** 2018.

[23] Sugai H, Kojima H, Ishida A and Toyoda H 1990 *Appl. Phys. Lett.* **56** 2616.

[24] Sugai H and Toyoda H 1992 *J. Vac. Sci Technol.* A **8** 1193.

[25] Zarrabian M, Leteinturier C and Turban G 1998 *Plasma Sources Sci. Technol.* **7** 607.

[26] Zalicki P, Ma Y, Zare R N, Wah, E H, Dadamio J R, Owano T G and Kruger C H 1995 *Chem. Phys. Lett.* **234** 269.

[27] Zalicki P, Ma Y, Zar, R N, Wahl E H, Dadamio J R, Owano T G and Kruger C H 1995 *Appl. Phys. Lett.* **67** 144.

[28] Wahl E H, Owano T G, Kruger C H, Zalicki P, Ma Y and Zare R N 1997 *Diam. Relat. Mater.* **6** 476.

[29] Childs M A, Menningen K L, Chevako P, Spellmeyer N W, Anderson L W and Lawler J E 1992 *Phys. Lett. A* **171** 87.

[30] Menningen K L, Childs M A, Chevako P, Toyoda H, Anderson L W and Lawler J 1993 *Chem. Phys. Lett.* **204** 573.

[31] Menningen K L, Childs M A, Toyoda H, Ueda Y, Anderson L W and Lawler J E 1994 *Diam. Relat. Mater.* **3** 422.

[32] Lombardi G, Stancu G D, Hempel F, Gicquel A and Röpcke J 2004 *Plasma Sources Sci. Technol.* **13** 27.

[33] Wormhoudt J 1990 *J. Vac. Sci. Technol. A* **8** 1722.

[34] Wormhoudt J and McCurdy K E 1989 *Chem. Phys. Lett.* **156** 47.

[35] Stancu G D, Röpcke J and Davies P B 2005 *J. Chem. Phys.* **122** 014306.

[36] Davies P B and Martineau P M 1990 *Appl. Phys. Lett.* **57** 237.

[37] Ikeda M, Hori M, Goto T, Inayoshi M, Yamada K, Hiramatsu M and Nawata M 1995 *Jpn. J. Appl. Phys.* **34** 2484.

[38] Kim S, Billesbach D P and Dillon R. 1997 *J. Vac. Sci. Technol. A* **15** 2247.

[39] Hempel F, Davies P B, Loffhagen D, Mechold L and Röpcke J 2003 *Plasma Sources Sciences Technol.* **12** S98.

[40] Ohl A 1998 *J. Phys. IV* **8** 83.

[41] Vandevelde T, Nesladek M, Quaeyhaegens C and Stals L 1997 *Thin Solid Films* **308-309** 154.

[42] Vandevelde T, Wu T D, Quaeyhaegens C, Vlekken J, D'Olieslaeger M and Stals L 1999 *Thin Solid Films* **340** 159.

[43] Mutsukura N 2001 *Plasma Chem. Plasma Process.* **21** 265.

[44] Bhattacharyya S, Granier A and Turban G 1999 *J. Appl. Phys.* **86** 4668.

[45] Zhang M, Nakayama Y, Miyazaki T and Kume M 1998 *J. Appl. Phys.* **85** 2904.

[46] Dinescu G, De Graaf A, Aldea E, and van de Sanden M C M 2001 *Plasma Sources Sci. Technol.* **10** 513.

[47] De Graaf A, Aldea E, Dinescu G and Van de Sanden M C M 2001 *Plasma Sources Sci. Technol.* **10** 524.

[48] Penetrante B M, Hsiao M C; Bardsley J N; Merritt B T, Vogtlin G E, Kuthi A, Burkhart C P and Bayless J R 1997 *Plasma Sources Sci. Technol.* **6** 251.

[49] Kareev M, Sablier M and Fujii T 2000 *J. Phys. Chem.* **104** 7218.

[50] Coll P, Coscia D, Gazeau M C, De Vanssay E, Guillemin J C and Raulin F 1995 *Adv. Space Res.* **16** 93.

[51] De Vanssay E, Gazeau M C, Guillemin J C and Raulin F 1995 *Planet. Space Sci.* **43** 25.

[52] Tabares F L, Tafalla D, Tanarro I, Herrero V J, Islyaikin A and Maffiotte C 2002 *Plasma Phys. Control. Fusion* **44** L37.

[53]  White J U 1942 *J. Optical Soc. America* **32** 285.

[54]  Gemini Scientific Instr. 2001 Multi-Pass Optics Alignment Instructions, Manual.

[55]  Fan W Y, Knewstubb P F, Käning M, Mechold L, Röpcke J and Davies P B 1999 *J. Phys. Chem.* A **103** 4118.

[56]  Mechold L, Röpcke J, Duten X and Rousseau A 2001 *Plasma Sources Sci. Technol.* **10** 52.

[57]  Loffhagen D and Winkler R 1996 *J. Phys. D: Appl. Phys.* **29** 618.

[58]  Buckman S J and Phelps A V 1985 *J. Chem. Phys.* **82** 4999.

[59]  Phelps A V and Pitchford L C 1985 *Phys. Rev.* **31** 2932.

[60]  Röpcke J, Mechold L, Käning M, Anders J, Wienhold F G, Nelson D, and Zahniser M 2000 *Rev. Sci. Instrum.* **71** 3706.

[61]  McManus J B, Nelson D, Zahniser M, Mechold L, Osiac M, Röpcke J and Rousseau A 2003 *Rev. Sci. Instr.* **74** 2709.

[62]  Welzel, S., Rousseau, A., Davies, P. B. and Röpcke, J., *J. Phys.: Conf. Series* **86** (2007) 012012.

[63]  Goodwin D G, Gavillet G G. J Appl Phys 1990;68(12):6393–400.

[64]  Mankelevich Y A, Rakhimov A T, Suetin N V. Diam Rel Mat 1998; 7:1133–7.

[65]  Mankelevich YA, Suetin NV, Ashfold MNR, Smith JA, Cameron E. Diam Rel Mat 2001;10:364–9.

[66]  Dandy DS, Coltrin ME. J Appl Phys 1994;76(5):3102–13.

[67]  Tsang RS, May PW, Cole J, Ashfold MNR. Diam Rel Mat 1999;8:1388–92.

[68]  Lang T. Reaktionskinetik von Kohlenwasserstoffen in Wasserstoffplasmen. Dissertation. Technische Universita¨ t Wien, Wien, 1993.

[69]  Celii FG, Butler JE. J Appl Phys 1992;71(6):2877–83.

[70]  Ashfold MNR, May PW, Petherbridge JR, Rosser KN, Smith JA, Mankelevich YA, et al. Phys Chem Chem Phys 2001;3: 3471–85.

[71]  Smith JA, Cameron E, Ashfold MNR, Mankelevich YA, Suetin NV. Diam Rel Mat 2001;10:358–63.

[72]  Wahl EH. Stanford University report No. TSD-136 2001.

[73]  Hirmke, J., Glaser, A., Hempel, F., Stancu, G.D., Röpcke, J., Rosiwal, S. and Singer, R.F., *Vacuum* **81** (2007) 619.

[74]  Hirmke, J., Hempel, F., Stancu, G.D., Röpcke, J., Rosiwal, S. and Singer, R.F., *Vacuum* (2006) **80** 967.

[75]  Zahniser M S, Nelson D D and Kolb C E 2002 in *Applied Combustion Diagnostics*, Kohse-Hoinghaus K and Jeffries J (eds), Tallor and Francis, New York, 648.

[76]  Nelson D D, Shorter J H, McManus J B and Zahniser M S 2002 *Appl. Phys. B* **75** 343.

[77]  Stancu G D, Ph. D. Thesis, 2004 University of Greifswald, Germany.

[78]  Bézard B, Feuchtgruber H, Moses J I and Encrenaz T 1998 *Astro. and Astrophys.* **334** L41.

[79] Bezard B, Romani P N, Feuchtgruber H and Encrenaz T 1999 *The Astrophysical Journal* **515** 868.

[80] Feuchtgruber H, Helmich F P, van Dishoeck E F and Wright C M 2000 *The Astrophysical Journal* **535**, L111.

[81] Laguna G A and Baughcum S L 1982 *Chem. Phys. Lett.* **88** 568.

[82] Cody R J, Payne Jr W A, Thorn Jr R P, Nesbitt F L, Iannone M A, Tardy D C and Stief L 2002 *J. Phys. Chem.* A **106** 6060.

[83] Walter D, Grotheer H H, Davies J W, Pilling M J and Wagner A F 1990 *23$^{-rd}$ Symposium on Combustion* 107.

[84] Macpherson M T, Pilling M J and Smith M J C 1983 *Chem. Phys. Lett.* **94** 430.

[85] Macpherson M T, Pilling M J and Smith M J C 1985 *J. Phys. Chem.* **89** 2268.

[86] Slagle I R, Gutman D, Davies J W and Pilling M J 1988 *J. Phys. Chem.* **92** 2455.

[87] Wagner A F and Wardlaw D M 1988 *J. Phys. Chem.* **92** 2462.

[88] Callear A B and Metcalfe M P 1976 *Chem. Phys.* **14** 275.

[89] Baulch D L, Cobos C J, Cox R A, Frank P, Hayman G, Just Th, Kerr J A, Murrells T, Pilling M J, Troe J, Walker R W and Warnatz J 1994 *J. Phys. Chem. Ref. Data* **23** 980.

[90] P. Botschwina, J. Flesch, W. Meyer, Chem. Phys. 1983, **74**, 321.

[91] Stancu, G.D., Röpcke, J., and Davies, P.B J. Phys. Chem. A, 2008, **112** (28), 6285–6288.

[92] Rehfuss B D, Suh M H, Miller T A and Bondybey V E 1992 *J. Mol. Spec.* **151** 437.

[93] Prasad C V V, Bernath P F, Frum C and Engleman R 1992 J. Mol. Spec. **151** 459.

[94] Hübner M, Castillo M, Davies P B and Röpcke J 2005 *Spectrochim. Acta A Mol. Spectrosc.* **61** 57.

[95] Faist J, Capasso F, Sivco D L, Sirtori C, Hutchinson A L and Cho A 1994 *Science* **264** 553.

[96] Gmachl C, Sivco D L, Colombelli R, Capasso F and Cho A Y 2002 *Nature* **415** 883.

[97] Beck M, Hofstetter, D., Aellen, T., Faist, J, Oesterle, U., Ilegems, M., Gini, E. and Melchior, H., 2002 *Science* **295** 301.

[98] Hempel F, Glitsch S, Röpcke J, Saß S and Zimmermann H 2005 in: *Plasma Polymers and Related Materials* (M. Mutlu ed.), Hacettepe University Press 142.

[99] Cheesman A et al., 2006 *J. Phys. Chem. A* **110** 2821.

[100] Stancu G D, Lang N, Röpcke J, Reinicke M, Steinbach A and Wege S, *Chem. Vap. Deposition* **13** (2007) 351-360.

[101] Namjou K et al., 1998 *Opt. Lett.* **23** 219.

[102] Beyer T et al., 2003 *J. Appl. Phys.* **93** 3158.

[103] Welzel, S., Rousseau, A., Davies, P. B. and Röpcke, J., *J. Phys.: Conf. Series* **86** (2007) 012012.

[104] Nelson, D.D., McManus, B., Urbanski, S., Herndon, S. and Zahniser, M.S., Spectroch. Acta A 60(14), 3325-3335 (2004).

[105] Steinfeld, J.H. and Pandis, S.N., "Atmospheric Chemistry and Physics: From Air Pollution to Climate Change", Wiley, New York [u.a.], (1998).

[106] Brown, S.S., Chem. Rev. 103(12), 5219-5238 (2003).

[107] Gupta, H. and Fan, L.S., Ind. Eng. Chem. Res. 42(12), 2536-2543 (2003).

[108] Bauer, C., Geiser, P., Burgmeier, J., Holl, G., and Schade, W., Appl. Phys. B 85(2-3), 251-256 (2006).

[109] Silva, M.L., Sonnenfroh, D.M., Rosen, D.I., Allen, M.G., and O'Keefe, A., Appl. Phys. B 81(5), 705-710 (2005).

[110] Todd, M.W., Provencal, R.A., Owano, T.G., Paldus, B.A., Kachanov, A., Vodopyanov, K.L., Hunter, M., Coy, S.L., Steinfeld, J.I. and Arnold, J.T., Appl. Phys. B 75(2-3), 367-376 (2002).

[111] McCurdy, M.R., Bakhirkin, Y., Wysocki, G., Lewicki, R. and Tittel, F.K., J. Breath Res. 1(1), 014001 (2007).

[112] O'Keefe, A. and Deacon, D.A.G., Rev. Sci. Instrum. 59(12), 2544-2551 (1988).

[113] Engeln, R., Berden, G., Peeters, R. and Meijer, G., Rev. Sci. Instrum. 69(11), 3763-3769 (1998).

[114] O'Keefe, A., Scherer, J.J. and Paul, J.B., Chem. Phys. Lett. 307(5-6), 343-349 (1999).

[115] O'Keefe, A., Chem. Phys. Lett. 293(5-6), 331-336 (1998).

[116] Romanini, D., Kachanov, A.A. and Stoeckel, F., Chem. Phys. Lett. 270(5-6), 538-545 (1997).

[117] Berden, G., Peeters, R. and Meijer, G., Chem. Phys. Lett. 307(3-4), 131-138 (1999).

[118] Yang, R.Q., Hill, C.J., Yang, B.H., Wong, C.M., Muller, R.E. and Echternach, P. M., Appl. Phys. Lett. 84(18), 3699-3701 (2004).

[119] Yang, R.Q., Hill, C.J. and Yang, B.H., Appl. Phys. Lett. 87(15), 151109 (2005).

[120] Roller, C., Kosterev, A.A., Tittel, F.K., Uehara, K., Gmachl, C. and Sivco, D.L., Opt. Lett. 28(21), 2052-2054 (2003).

[121] Nelson, D.D., McManus, J.B., Herndon, S.C., Shorter, J.H., Zahniser, M.S., Blaser, S., Hvozdara, L., Muller, A., Giovannini, M. and Faist, J., Opt. Lett. 31(13), 2012-2014 (2006).

[122] Lewicki, R., Wysocki, G., Kosterev, A.A. and Tittel, F.K., Opt. Express 15(12), 7357-7366 (2007).

[123] Grossel, A., Zeninari, V., Parvitte, B., Joly, L. and Courtois, D., Appl. Phys. B 88(3), 483-492 (2007).

[124] Kosterev, A.A., Bakhirkin, Y.A. and Tittel, F.K., Appl. Phys. B 80(1), 133-138 (2005).

[125] Kosterev, A.A., Curl, R.F., Tittel, F.K., Gmachl, C., Capasso, F., Sivco, D.L., Baillargeon, J.N., Hutchinson, A.L. and Cho, A.Y., Appl. Opt. 39(24), 4425-4430 (2000).

[126] Tuzson, B., Zeeman, M.J., Zahniser, M.S. and Emmenegger, L., Infrared Phys. Tech. 51(3), 198-206 (2008).

[127]  McManus, J.B., Kebabian, P.L. and Zahniser, M.S., Appl. Opt. 34(18), 3336-3348 (1995).

[128]  Welzel, S., Lombardi, G., Davies, P.B., Engeln, R., Schram, D.C., and Röpcke, J. J. *Appl. Phys.*, 2008, **104** 093115.

[129]  Lang, N., Röpcke, J., Steinbach, A., Wege, S., 6. OPTAM, Düsseldorf 2008, VDI-Berichte 2047, 11-21 (2008).

[130]  Yamada, C., Hirota E. and Kawaguchi, K. J. Chem. Phys. 1981 **75** 5256.

[131]  Pacansky, J., Koch, W. and Miller, M. D., J. Amer. Chem. Soc. 1991 **113** 317.

## Chapter 4

# Spectroscopy of Carbon Containing Diatomic Molecules

James O. Hornkohl[*]

*University of Tennessee Space Institute, U.S.A.*

László Nemes

*Laboratory for Laser Spectroscopy, Chemical Research Center, Hungarian Academy of Sciences, Hungary*

Christian Parigger

*Center for Laser Applications, University of Tennessee Space Institute, U.S.A.*

The topic of this chapter is the use of a recorded diatomic spectrum as a diagnostic tool. In this application of diatomic spectroscopy, one attempts to infer the physical environment from a spectrum produced by that environment. In current practice, temperature is normally the state variable inferred from a recorded spectrum. A concise review of applied diatomic quantum theory is given with emphasis on how a line strength table (a line list of a particular spectrum which includes the line strengths) is created from accurate measurements of line positions. An algorithm is given for finding the molecular parameters in the upper and lower Hamiltonians by fitting calculated term value differences to measured line positions. Once the molecular parameters have been found, one can compute the line positions and Hönl-London factors even for those lines that were not included in the fitting (*i.e.*, lines that are not experimentally known with high accuracy.) Additionally using the electronic transition moment and the Franck-Condon factors, one can compute the line strength table for

---

[*]retired.

the spectrum. Several examples are given showing what type of synthetic spectra can be computed from the line strength table.

# 1. Introduction

The spectrum of a particular band system of diatomic carbon, known as the Swan spectrum, is one of the earliest known molecular spectra. According to Johnson,[1] W. Swan first recorded the band spectrum named after him in 1856. Good quality experimental measurements of the Swan system were possible several decades before a theoretical description of diatomic spectra was provided by quantum mechanics. In 1927 Johnson concluded that this spectrum was produced by $C_2H_2$ molecules, not $C_2$, while noting that others suggested that the Swan bands were spectra of CO or CH. However, Johnson and Asundi[2] noted that the newly developed quantum theory had conclusively shown that the Swan bands are produced by diatomic carbon molecules. Although study of the Swan bands has lead to significant developments in molecular spectroscopy, the continuing strong interest in the Swan spectrum is now mostly related to its use as a diagnostic tool in combustion and materials research for temperature measurement and in astrophysics to measure the relative abundances of the various carbon isotopes in stars.

In significant measure the utility of the Swan bands is its ubiquity. If carbon is present in significant quantity at flame temperatures or higher, one is nearly guaranteed of emission from Swan bands. If nitrogen or hydrogen are also present spectra of CN and CH are also likely.

The topic of this chapter is applied diatomic spectroscopy, the inference of the physical environment of the diatomic molecules whose spectrum has been recorded using carbon containing diatomic molecules as the examples. Particular emphasis will be placed on a theoretical quantity called the line strength and its use in computing synthetic spectra.

## 1.1. *Differences between atomic and diatomic spectra*

The single nucleus of an atom can interact with only the atom's electrons. The two nuclei in a diatomic molecule not only interact with the electrons but also with each other. The nuclei rotate and vibrate about the molecular center of mass, and because nearly all of the molecule's mass resides in the nuclei, these motions are mostly with respect to nuclei's center of mass. Atomic spectra are controlled mostly by the orbital and spin angular momenta of the electrons while nuclear spin produces only very small hyperfine structure. In a molecule the orbital angular momentum of the nuclei contributes significantly to the total angular momentum and makes small contributions to the molecule's energy eigenvalues. Vibration of the nuclei makes larger contributions to the energy eigenvalues. To a useful degree of approximation, one can speak of independent electronic, vibrational, and rotational motions of the molecule.

The electronic, vibrational, and rotational states of a molecule each exchange energy with collision partners at different rates. Thus, for example, in an electric arc heated low density supersonic wind tunnel, the slow relaxation of vibrational states in comparison to relaxation of rotational states produces a non-equilibrium state in the low density gas downstream of the expansion nozzle. A measurement of the distribution of energy among vibrational states often yields a Maxwell-Boltzmann distribution but at a temperature more nearly the gas temperature in the arc heater while the temperature inferred from the distribution of rotational states is more nearly the free stream temperature. The rotation and vibration of nuclear masses in a molecule greatly complicate its internal motions and its interactions with other molecules. Conversely, one wishing to infer from a spectrum the physical environment that produced the spectrum finds that molecular spectra have more to offer than atomic spectra. The diatomic molecule is the first step away from atoms to molecules.

## 1.2. *The line strength*

In the preface to their book on atomic spectra Condon and Shortley[3] define a quantity "called the *strength* of a line, which we find to give a more convenient theoretical specification of the radiation intensity than either of the Einstein transition probabilities." We follow their suggestion even though the Einstein coefficients for absorption, spontaneous emission, and induced emission are probably more familiar than the quantity called line strength. The oscillator strength and cross section are two additional quantities likely more familiar than line strength, but the line strength is the theoretical quantity in terms of which the more familiar quantities are defined. Hilborn[4] gives a very useful discussion and a convenient table concerning the mathematical relationships between these various quantities. Thorne, *et al.*,[5] give a comparable discussion and table.

A spectral line is produced when a molecule changes state through an interaction with the electromagnetic field. Given that a quantum system is in initial state $|i\rangle$ at time $t_0$, the probability of finding the system in state $|f\rangle$ at time $t$ is expressed as the absolute square of matrix elements of the evolution operator $U(t, t_0)$,

$$p_{f \leftarrow i} = |\langle f|U(t, t_0)|i\rangle|^2. \tag{1}$$

The total Hamiltonian is the sum of molecular Hamiltonian, the radiation Hamiltonian, and the operators, denoted here by the symbol $V$, representing the interactions between the molecule and radiation field,

$$H = H_{\mathrm{mol}} + H_{\mathrm{rad}} + V, \tag{2}$$
$$= H_0 + V. \tag{3}$$

As shown in many quantum texts the matrix elements of the evolution operator are proportional to matrix elements of the interaction operator. An example is seen in Fermi's golden rule for probability per unit time for a transition,

$$\frac{dp}{dt} = \frac{2\pi}{\hbar} |\langle f|V|i\rangle|^2 \rho_f \tag{4}$$

where $\rho_f$ is the density of final states. The absolute square of the matrix elements of the interaction operator summed over all transitions leading to the same spectral line defines the line strength,

$$S_{fi} \equiv \sum_f \sum_i |\langle f|V|i\rangle|^2. \tag{5}$$

In the absence of externally supplied electromagnetic fields, the sums are carried over the magnetic quantum numbers $m_f$ and $m_i$ to account for the statistical weights $2j_f + 1$ and $2j_i + 1$ of the final state having total angular momentum $j_f$ and initial state having total angular momentum $j_i$. The notations $\langle f|$ and $|i\rangle$ are overly simplified for a real system. More realistic notations are $\langle n_f j_f m_f|$ and $|n_i j_i m_i\rangle$ where $j$ and $m$ refer to the total angular momentum while $n$ refers to all other required quantum numbers. In this notation the equation for the line strength reads

$$S_{n_f j_f m_f \leftrightarrow n_i j_i m_i} = \sum_{m_j = -j_f}^{j_f} \sum_{m_i = -j_i}^{j_i} |\langle n_f j_f m_f|V|n_i j_i m_i\rangle|^2. \tag{6}$$

## 2. Diatomic Quantum Theory

The primary topic of this chapter is how to use a recorded diatomic spectrum to infer the physical state of the environment that produced the spectrum. In this section we summarize the essential quantum theory of the diatomic molecule.

### 2.1. *Diatomic eigenfunctions*

The system in question is composed of a diatomic molecule and the radiation field. Using the fundamental symmetries (conservation laws) for energy, linear momentum, and angular momentum, one can write an exact equation for the internal diatomic eigenfunction (*i.e.*, the eigenfunction remaining after motion of the total center of mass has been removed),

$$\langle \mathbf{r}_1 \mathbf{r}_2 \cdots \mathbf{r}_N \mathbf{r} | nJM \rangle = \sum_{\Omega=-J}^{J} \langle \mathbf{r}'_1 \mathbf{r}'_2 \cdots \mathbf{r}'_{N-1} \, r \, \rho \, \zeta | n \rangle \, D_{M\Omega}^{J^*}(\phi \, \theta \, \chi). \quad (7)$$

This equation is written for a molecule having $N$ electrons with spatial coordinates $\mathbf{r}_i$ for $i = 1, N$ in the laboratory coordinate system and $\mathbf{r}'_i$ in a coordinate system attached to the rotating molecule. Coordinates $r$, $\theta$, and $\phi$ are the spherical polar coordinates of the internuclear vector $\mathbf{r}$ in the laboratory coordinate system. Coordinates $\rho$, $\chi$, and $\zeta$ are the cylindrical coordinates of the $N$th electron in the molecule-fixed coordinate system. The scaler $r$ is the distance between the two nuclei, $\rho$ is the distance of the $N$th electron from the internuclear axis , and $\zeta$ is the distance of that same electron from a plane perpendicular to the internuclear axis and passing through the center of mass of the nuclei. The Euler angles serve the dual purpose of describing coordinate rotations $\mathbf{r} \leftrightarrow \mathbf{r}'$ and physical rotation of the molecule. All information about the total angular momentum of the diatomic molecule is contained in $D_{M,\Omega}^{J^*}(\phi \, \theta \, \chi)$, the complex conjugate of the rotation matrix element or Wigner $D$-function. The two nuclei lie on the $z'$ axis. The magnetic quantum number $M$ is for the laboratory $z$ component of the total angular momentum, and $\Omega$ is for the $z'$ component of the total angular momentum. The symbol $n$ represents all other required quantum numbers.

In proper spectroscopic notation the above equation would be written in terms of the quantum numbers $F$, $M_F$, and $\Omega_F$ associated with the total angular momentum including nuclear spin,

$$\mathbf{F} = \mathbf{R} + \mathbf{L} + \mathbf{S} + \mathbf{T} \tag{8}$$

$$= \mathbf{J} + \mathbf{T}, \tag{9}$$

but the total nuclear spin $\mathbf{T}$ responsible for hyperfine structure will be ignored here. In the above, $\mathbf{R}$ is the total orbital angular momentum of the two nuclei, $\mathbf{L}$ is the total electronic orbital angular momentum, and $\mathbf{S}$ is the total electronic spin. The total orbital angular momentum of the molecule is given by

$$\mathbf{N} = \mathbf{R} + \mathbf{L}. \tag{10}$$

Because Eq. (7) depends only upon fundamental symmetries, it is exact. This generality exists because the equation says nothing more than what happens to a quantum system when the direction of the $z$ axis is changed. The equation is of special significance to the diatomic molecule because the Wigner $D$-functions are perfect basis functions for diatomic states. Equation (7) gives exact diatomic states as a finite sum of these basis states. We will see below that the summation of basis functions is often much shorter than the summation of $2J + 1$ basis functions shown in Eq. (7).

### 2.2. *Diatomic parity*

The influence of the weak force on diatomic spectra is yet to be observed. From a practical standpoint only electromagnetic forces need be considered, and we may assume that diatomic parity is rigorously conserved.

The parity operator $P$ changes the sign of all signed spatial coordinates,

$$\langle \mathbf{r}_1 \mathbf{r}_2 \cdots \mathbf{r}_N \mathbf{r} | P | nJM \rangle = \langle -\mathbf{r}_1, -\mathbf{r}_2 \cdots -\mathbf{r}_N, -\mathbf{r} | nJM \rangle. \tag{11}$$

Scaler distances such as $r$ and $\rho$ are immune to the parity operator as are all spins. The Euler rotations are proper but the parity operation is an improper rotation. Transformations of the Euler angles can be chosen to invert the signs of two of the rotated cartesian coordinates but not all three. Thus the diatomic parity operator is the product of two parity operators, one a transformation of the Euler angles chosen to invert the signs of two molecule-fixed coordinates and the other which inverts the sign of the remaining molecule-fixed coordinate.

Table 1 shows the three transformations of Euler angles that invert the signs of two rotated coordinates. The first entry $\sigma_v(yz)$ is chosen because its sign factor is independent of $\Omega$ and can therefore be removed from the

Table 1. The three ways of using the Euler angles to change the signs of two of the three components of a rotated vector. This table shows the sign changes on the components $x'$, $y'$ and $z'$ of a coordinate vector $\mathbf{r}'(x'y'z')$ produced by three different Euler angle transformations, and the effect of the Euler angle transformations on $D^{J^*}_{M\Omega}(\alpha\beta\gamma)$.

| Group notation | Alternate notation | Euler angle transformation | Coordinate transformation | Effect on $D^{J^*}_{M\Omega}(\alpha\beta\gamma)$ |
|---|---|---|---|---|
| $C_2(x)$ | $\sigma_{\mathbf{v}}(yz)$ | $\alpha \to \pi + \alpha$ <br> $\beta \to \pi - \beta$ <br> $\gamma \to -\gamma$ | $x' \to x'$ <br> $y' \to -y'$ <br> $z' \to -z'$ | $(-)^{J+2M} D^{J^*}_{M,-\Omega}(\alpha\beta\gamma)$ |
| $C_2(y)$ | $\sigma_{\mathbf{v}}(xz)$ | $\alpha \to \pi + \alpha$ <br> $\beta \to \pi - \beta$ <br> $\gamma \to \pi - \gamma$ | $x' \to -x'$ <br> $y' \to y'$ <br> $z' \to -z'$ | $(-)^{J-\Omega} D^{J^*}_{M,-\Omega}(\alpha\beta\gamma)$ |
| $C_2(z)$ | $\sigma_{\mathbf{v}}(xy)$ | $\alpha \to \alpha$ <br> $\beta \to \beta$ <br> $\gamma \to \pi + \gamma$ | $x' \to -x'$ <br> $y' \to -y'$ <br> $z' \to z'$ | $(-)^{-\Omega} D^{J^*}_{M,-\Omega}(\alpha\beta\gamma)$ |

summation over $\Omega$ in Eq. (7). The operator $P_\Sigma$ is defined to be the operator that inverts the sign of $x'$. Application of the complete parity operator can then be written as

$$\langle \mathbf{r}_1\mathbf{r}_2 \cdots \mathbf{r}_N\mathbf{r}|P|nJM\rangle = P_\Sigma P_{\phi\theta\chi} \sum_{\Omega=-J}^{J} \langle \mathbf{r}'_1\mathbf{r}'_2 \cdots \mathbf{r}'_{N-1} \, r \, \rho\, \zeta|n\rangle \, D^{J^*}_{M\Omega}(\phi\,\theta\,\chi)$$

$$= p_\Sigma \, (-)^{J+2M} \sum_{\Omega=-J}^{J} \langle -\mathbf{r}'_1, -\mathbf{r}'_2 \cdots -\mathbf{r}'_{N-1},\, r,\, \rho,\, -\zeta|n\rangle \, D^{J^*}_{M,-\Omega}(\phi\,\theta\,\chi)$$

$$= p \sum_{\Omega=-J}^{J} \langle \mathbf{r}'_1, \mathbf{r}'_2 \cdots \mathbf{r}'_{N-1},\, r,\, \rho,\, \zeta|n\rangle \, D^{J^*}_{M\Omega}(\phi\,\theta\,\chi) \tag{12}$$

in which $p = \pm 1$ is the eigenvalue of the parity operator $P$. The minus sign on $-\Omega$ was dropped above because Eq. (7) is unchanged if the sign of $\Omega$ is changed. The parity eigenvalue is the product to two eigenvalues,

$$p = p_\Sigma \, p_{\phi\theta\chi}. \tag{13}$$

The parity operator $P$ has the real eigenvalue $p = \pm 1$. However, the eigenvalues of $P_{\phi\theta\chi}$,

$$p_{\phi\theta\chi} = (-)^{J+2M}$$
$$= \pm 1 \qquad J \text{ integer}$$
$$= \pm i \qquad J \text{ half-integer}, \tag{14}$$

are imaginary when $J$ is half-integer, and hence, $p_\Sigma$ must also be imaginary when $J$ is half-integer. If unity in the form of $i/i$ is inserted in Eq. (13), then for half-integer $J$

$$p = \frac{i}{i} p_\Sigma p_{\phi\theta\chi} \tag{15}$$

$$= i p_\Sigma \frac{(-)^{J+2M}}{i} = -i p_\Sigma (-)^{J-1/2} \qquad \text{or} \tag{16}$$

$$= \frac{p_\Sigma}{i} i (-)^{J+2M} = i p_\Sigma (-)^{J+1/2}. \tag{17}$$

Note from Eq. (16) that because $p$ is real and $(-)^{J\pm1/2}$ is real for half-integer $J$, the product $i p_\Sigma$ must be real for half-integer $J$. If one sets the convention to always use Eq. (16) instead of (17), then the $-i$ can be dropped from (16) and $p_\Sigma$ can be treated as if it were real. Thus, the diatomic parity eigenvalues can be written as

$$p = p_\Sigma (-)^J \qquad J \text{ integer,} \tag{18}$$

$$p = p_\Sigma (-)^{J-1/2} \qquad J \text{ half-integer.} \tag{19}$$

where $p_\Sigma = \pm1$ for both whole and half-integer $J$. The convention to always use $J - 1/2$ (and never $J + 1/2$) follows that for the $e/f$ parity designation. Brown, et al.,[6] recommend that for integer $J$

> *levels with parity $+(-1)^J$ be called e levels, and*
> *with parity $-(-1)^J$ be called f levels*

while for half-integer $J$ they recommend that

> *levels with parity $+(-1)^{J-1/2}$ are e levels, and*
> *levels with parity $-(-1)^{J-1/2}$ are f levels.*

It will be shown below that the parity eigenvalue $p$ is easily computed when one is using the Hund's case (a) or (b) basis. Parity eigenvalues are easy to compute because parity is a member of the complete set of commuting observables, and the orthogonal matrix that diagonalizes the Hamiltonian matrix computed in a Hund basis will also diagonalize the parity matrix computed in that same basis. Once the parity of a level is known, computation the $e/f$ parity is simple as shown in the following FORTRAN routine.

```fortran
      subroutine findef (J_half_integer, J, p, ef)
c Find the e/f parity designation.
c     J_half_integer is true if J is half-integer.
c     J is the total angular momentum quantum number.
c     p is the parity.
c     The character variable is returned 'e' or 'f'.
      double precision J
      logical J_half_integer
      integer p
      character*1 ef
      if (J_half_integer)  then
         Jtst = nint(J - 0.5d0)
      else
         Jtst = nint(J)
      end if
      if (mod(Jtst,2).eq.0)  then
         iJ = 1
      else
         iJ = -1
      end if
      if (iJ.eq.p)   then
         ef = 'e'
      else
         ef = 'f'
      end if
      return
      end
```

The parity input in the above routine might be, for example, one of the parity eigenvalues shown in Table 6.

## 2.3. *Homonuclear diatomics*

The states of a homonuclear diatomic molecule must be symmetric or antisymmetric under exchange of the two identical nuclei. When the spin of a nucleus is half-integer (*i.e.*, when the identical nuclei are fermions) the states must be antisymmetric under exchange of the identical nuclei. When the spin of a nucleus is integer (*i.e.*, when the identical nuclei are bosons) the states must be symmetric under exchange of the identical nuclei.

The exchange of two identical particles refers to the exchange of all physical properties, not the mere exchange of coordinates. The full diatomic nuclear exchange operator $\mathcal{E}_{ab}$ is the product of a nuclear parity operator $P_{ab}$ which inverts the signs of the spatial coordinates of the two identical nuclei, and the nuclear spin exchange operator $\mathcal{I}_{ab}$ that swaps the two spins,

$$\mathcal{E}_{ab} = P_{ab}\,\mathcal{I}_{ab}. \tag{20}$$

The nuclear parity operator is written as the product of the total parity operator and a new parity $P_{gu}$ that inverts the signs of all electronic coordinates,

$$P_{ab} = P\,P_{gu}. \tag{21}$$

The operator that exchanges two identical nuclei is therefore the product of four operators

$$\mathcal{E}_{ab} = P_{\Sigma}\,P_{\phi\theta\chi}\,P_{gu}\,\mathcal{I}_{ab}. \tag{22}$$

Using previous results for the eigenvalues of $P_{\Sigma}$ and $P_{\alpha\beta\gamma}$, the eigenvalue of $\mathcal{E}_{ab}$ becomes

$$e_{ab} = p_{\Sigma}\,(-)^{J}\,p_{gu}\,p_{\mathcal{I}} \qquad J \text{ integer} \tag{23}$$

$$e_{ab} = p_{\Sigma}\,(-)^{J-1/2}\,p_{gu}\,p_{\mathcal{I}} \qquad J \text{ half-integer.} \tag{24}$$

Finding the eigenvalues of the nuclear spin exchange operator $\mathcal{I}_{ab}$ is the next goal. Once an equation for these eigenvalues is found, the number of symmetric and antisymmetric states can be counted, thereby establishing the nuclear spin statistics.

The total nuclear spin operator, $T$, is defined,

$$\mathbf{T} = \mathbf{I}_a + \mathbf{I}_b, \tag{25}$$

where $\mathbf{I}_a$ and $\mathbf{I}_b$ are the nuclear spin operators for the two nuclei. The spin states of the identical nuclei are labelled $|I_a M_a\rangle$ and $|I_b M_b\rangle$. The states of total nuclear spin $|T M_T\rangle$ are built from the Clebsch-Gordan coefficient $\langle I_a M_a I_b M_b | T M_T\rangle$,

$$|T M_T\rangle = \sum_{M_a=-I_a}^{I_a} \sum_{M_b=-I_b}^{I_b} |I_a M_a\rangle\,|I_b M_b\rangle\,\langle I_a M_a I_b M_b | T M_T\rangle. \tag{26}$$

Taking into account a symmetry of the Clebsch-Gordan coefficient,

$$\langle j_2 m_2 j_1 m_1 | j m\rangle = (-)^{j_1+j_2-j}\langle j_1 m_1 j_2 m_2 | j m\rangle, \tag{27}$$

the eigenvalues of the nuclear spin exchange operator,

$$\mathcal{I}_{ab}|TM_T\rangle = (-)^{2I-T}|TM_T\rangle, \tag{28}$$

where $I = I_a = I_b$, are found to be

$$p_{\mathcal{I}} = (-)^{2I-T}. \tag{29}$$

For a specified value of $T$ the number of $|TM_T\rangle$ states is $2T + 1$. For $I_a = I_b = I$, the possible values of $T$ are $= 0, 1, \cdots, 2I$. Also, for integer $I$ the sign factor $(-)^{2I}$ in Eqs. (28) and (29) is positive while for half-integer $I$ the sign factor is negative. Therefore, for integer $I$, $p_{\mathcal{I}}$ is positive for even values of $T$ and negative for odd values of $T$, but this symmetry with respect to $T$ is reversed for half-integer $I$. Let $g_+$ be the number of states for which the eigenvalues $p_{\mathcal{I}}$, Eq. (29), are positive, and let $g_-$ be the number of states for which $p_{\mathcal{I}}$ is negative. Table 2 gives the values of $g_+$ and $g_-$ versus $I$. The ratio column gives the larger of $g_+/g_-$ or $g_-/g_+$. By deduction one finds

$$\left. \begin{array}{l} g_+ = (2I+1)(I+1) \\ g_- = (2I+1)I \end{array} \right\} I \text{ integer}, \tag{30}$$

$$\left. \begin{array}{l} g_+ = (2I+1)I \\ g_- = (2I+1)(I+1) \end{array} \right\} I \text{ half-integer}. \tag{31}$$

Thus, the ratio of the larger number of states to the smaller number (*i.e.*, the ratio of the nuclear spin statistical weights) is given by

$$R = \frac{I+1}{I}. \tag{32}$$

This equation fails when $I = 0$ because then states of only one nuclear spin symmetry exist.

Table 2.  The number $g_+$ of nuclear spin states having positive spin exchange symmetry and the number $g_-$ of nuclear spin states having negative exchange symmetry for various values of nuclear spin $I$.

| $I$ | $T$ | $2I - T$ | $g_+$ | $g_-$ | ratio |
|---|---|---|---|---|---|
| 0 | 0 | 0 | 1 | 0 | $\infty$ |
| 1/2 | 0,1 | 1,0 | 1 | 3 | 3 |
| 1 | 0,1,2 | 2,1,0 | 5+1=6 | 3 | 2 |
| 3/2 | 0,1,2,3 | 3,2,1,0 | 5+1=6 | 7+3=10 | 5/3 |
| 2 | 0,1,2,3,4 | 4,3,2,1,0 | 9+5+1=15 | 7+3=10 | 3/2 |
| 5/2 | 0,1,2,3,4,5 | 5,4,3,2,1,0 | 9+5+1=15 | 11+7+3=21 | 7/5 |
| 3 | 0,1,2,3,4,5,6 | 6,5,4,3,2,1,0 | 13+9+5+1=28 | 11+7+3=21 | 4/3 |

## 2.4. *Born-Oppenheimer approximation*

In the Born-Oppenheimer approximation the electronic-vibrational eigenfunction is approximated as the product of an electronic eigenfunction $\langle \mathbf{r}_1' \mathbf{r}_2' \cdots \mathbf{r}_{N-1}' \rho \zeta; r|n \rangle$ and vibrational eigenfunction $\langle r|v \rangle$,

$$\langle \mathbf{r}_1 \mathbf{r}_2 \cdots \mathbf{r}_{N-1}\, r\, \rho\, \zeta\, \phi\, \theta\, \chi | nJMv \rangle$$

$$\approx \sum_{\Omega=-J}^{J} \langle \mathbf{r}_1' \mathbf{r}_2' \cdots \mathbf{r}_{N-1}' \rho \zeta; r|n \rangle\, \langle r|v \rangle\, D_{M\Omega}^{J^*}(\phi\, \theta\, \chi). \tag{33}$$

The $r$ preceded by a semicolon in the electronic eigenfunction denotes that the electronic eigenfunction is a parametric function of the internuclear distance $r$. That is, the electronic Schrodinger equation must be repeatedly solved at fixed $r$ over a range of $r$ values. This makes the electronic eigenvalues functions of $r$.

The vibrational eigenfunction $\langle r|v \rangle$ should be more specifically designated $\langle r|v_n \rangle$ because the vibrational basis is different for each electronic electronic eigenfunction $\langle \mathbf{r}_1' \mathbf{r}_2' \cdots \mathbf{r}_{N-1}' \rho \zeta; r|n \rangle$. Traditionally, this dependence is not displayed.

The angular coordinates $\theta$ and $\phi$ are nuclear coordinates. However, their separation shown in Eq. (7) is exact, not part of the Born-Oppenheimer approximation.

## 2.5. *Hund's angular momentum coupling cases*

The total angular momentum (ignoring nuclear spin),

$$\mathbf{J} = \mathbf{R} + \mathbf{L} + \mathbf{S} \tag{34}$$

$$= \mathbf{N} + \mathbf{S}, \tag{35}$$

can be built from its components in various ways. Given the states $|NM_n\rangle$ and $|SM_S\rangle$, one can build the $|JM\rangle$ states using Clebsch-Gordan coefficients in the laboratory coordinate system,

$$|JM\rangle = \sum_{M_N=-N}^{N} \sum_{M_S=-S}^{S} |NM_N\rangle\, |SM_S\rangle\, \langle NM_N SM_S|JM\rangle \tag{36}$$

or in the rotating coordinate system,

$$|J\Omega\rangle = \sum_{\Lambda=-N}^{N} \sum_{\Sigma=-S}^{S} |N\Lambda\rangle\, |S\Sigma\rangle\, \langle N\Lambda S\Sigma|J\Omega\rangle, \tag{37}$$

where a Roman magnetic quantum number is referenced to laboratory co-ordinates and a Greek magnetic quantum number to rotating coordinates. Using the above and the inverse Clebsch-Gordan series,

$$D_{M,\Omega}^{J}(\phi\,\theta\,\chi) = \sum_{M_N=-N}^{N} \sum_{M_S=-S}^{S} \sum_{\Lambda=-N}^{N} \sum_{\Sigma=-S}^{S}$$

$$\times \langle N M_N S M_S | J M \rangle \langle N \Lambda S \Sigma | J \Omega \rangle D_{M_N\Lambda}^{N}(\phi\,\theta\,\chi) D_{M_S\Sigma}^{S}(\phi\,\theta\,\chi),$$ (38)

one can build $|JM\rangle$ in various ways from its components. Approximate physical models, known as Hund's angular momentum coupling cases, have been developed to describe the various ways in which the total is composed from its parts. We will be concerned here with the mathematical description of two Hund's cases.

### 2.5.1. *Hund's case (a)*

The Hund's case (a) basis is obtained when all terms in the summation over $\Omega$ in Eq. (33) but one are dropped and the result is normalized,

$$|a\rangle = \langle \mathbf{r_1 r_2} \cdots \mathbf{r}_{N-1}\, r\, \rho\, \zeta\, \phi\, \theta\, \chi | n v J M \Omega \Lambda S \Sigma \rangle$$ (39)

$$= \sqrt{\frac{2J+1}{8\pi^2}} \langle \mathbf{r'_1 r'_2} \cdots \mathbf{r'}_{N-1}\, \rho\, \zeta; r | n \rangle \langle r | v \rangle D_{M\Omega}^{J\,*}(\phi\,\theta\,\chi).$$ (40)

If $D_{M\Omega}^{J\,*}(\phi\,\theta\,\chi)$ is expanded in accord with the inverse Clebsch-Gordan series (38) one sees that $|S\Sigma\rangle$ is locked to $|J\Omega\rangle$ because $D_{M_N\Lambda}^{N\,*}(\phi\,\theta\,\chi)$ and $D_{M_S\Sigma}^{S\,*}(\phi\,\theta\,\chi)$ share the same Euler angles. The physical model for case (a) is strong coupling of the electronic spin to the orbital angular momentum. The internuclear component of $\mathbf{J}$ is the sum of the internuclear components of $\mathbf{L}$ and $\mathbf{S}$,

$$J_{z'} = L_{z'} + S_{z'}.$$ (41)

Their respective quantum numbers $\Lambda$ and $\Sigma$ and their sum $\Omega$ are quantum numbers of the case (a) basis.

### 2.5.2. *Hund's case (b)*

In Hund's case (b) the electronic spin states are independent of the electronic orbital angular momentum states. The $|JM\rangle$ states cannot be built using the inverse Clebsch-Gordan series because the spin states do not rotate in sync with the orbital rotations. The case (b) eigenfunction can be built by coupling $|NM_N\rangle$ states with $|SM_S\rangle$ states with the Clebsch-Gordan coefficient $\langle NM_N SM_S|JM\rangle$. The result is

$$|b\rangle = \langle \mathbf{r}_1 \mathbf{r}_2 \ldots \mathbf{r}_{N-1}\, r\, \rho\, \zeta\, \phi\, \theta\, \chi|nJMN\Lambda S\rangle = \sqrt{\frac{2N+1}{8\pi^2}} \sum_{M_N=-N}^{N} \sum_{M_S=-S}^{S}$$

$$\times \langle \mathbf{r}_1' \mathbf{r}_2' \ldots \mathbf{r}_{N-1}'\, \rho\, \zeta\, r|n\rangle\, \langle r|v\rangle\, \langle NM_N SM_S|JM\rangle\, |SM_S\rangle\, D_{M_N\Lambda}^{N*}(\alpha\beta\gamma). \quad (42)$$

There are additional Hund's coupling models but cases (a) and (b) are the most widely used.

The exact equation (7) contains a sum over $2J+1$ values of $\Omega$. Even for modest values of $J$ the number of terms can be large. A remarkable property of a diatomic state is that often only very few of these terms are required to accurately describe the state even for large $J$. A diatomic eigenfunction is approximated as a sum of Hund's basis states, but the number of terms required for high accuracy is often much fewer than indicated by Eq. (7).

### 2.5.3. *Parity in Hund's cases (a) and (b)*

The case (a) eigenfunctions constitute a complete basis, but they are eigenfunctions lacking the mathematical properties of angular momentum eigenfunctions [see Eqs. (69–74) and (80–83).] Parity and angular momentum are in perfect accord for the electromagnetic force. Therefore, because case (a) eigenfunctions are not angular momentum eigenfunctions they are also not parity eigenfunctions. This becomes apparent when the case (a) parity matrix elements are calculated,

$$p_{ij}^{(a)} = p_\Sigma\, (-)^J\, \delta(J_iJ_j)\, \delta(\Omega_i,-\Omega_j)\, \delta(\Lambda_i\Lambda_j)\, \delta(n_in_j). \quad (43)$$

Similar comments apply to the case (b) parity matrix.

$$p_{ij}^{(b)} = p_\Sigma\, (-)^{N_i}\, \delta(N_iN_j)\, \delta(\Lambda_i,-\Lambda_j)\, \delta(n_in_j) \quad (44)$$

### 2.5.4. *Case (a) ↔ case (b) transformations*

Transformations between cases (a) and (b) are accomplished with

$$|a\rangle = \sum_b |b\rangle\, \langle b|a\rangle \quad (45)$$

$$= \sqrt{\frac{2J+1}{2N+1}} \sum_N \langle \mathbf{r}_1 \mathbf{r}_2 \ldots \mathbf{r}_{N-1}\, r\, \rho\, \zeta\, \theta\, \phi\, \chi|nJMN\Lambda S\rangle\, \langle N\Lambda S\Sigma|J\Omega\rangle \quad (46)$$

$$|b\rangle = \sum_a |a\rangle \langle a|b\rangle \tag{47}$$

$$= \sqrt{\frac{2N+1}{2J+1}} \sum_\Omega \langle \mathbf{r}_1 \mathbf{r}_2 \ldots \mathbf{r}_{N-1}\, r\, \rho\, \zeta\, \theta\, \phi\, \chi | nJM\Omega\Lambda S\Sigma\rangle \langle N\Lambda S\Sigma | J\Omega\rangle$$

$$= \sum_\Omega (-)^{S+\Sigma} \langle \mathbf{r}_1 \mathbf{r}_2 \ldots \mathbf{r}_{N-1}\, r\, \rho\, \zeta\, \theta\, \phi\, \chi | nJM\Omega\Lambda S\Sigma\rangle \langle J, -\Omega S\Sigma | N, -\Lambda\rangle. \tag{48}$$

Molecular term values are the eigenvalues of the Hamiltonian matrix. The wavenumber of a spectral line is the difference between two term values. The intensity of a spectral line is controlled by the matrix elements of the operator responsible for the transition. In practice, application of quantum theory to diatomic spectroscopy often consists of computing matrix elements. We turn now to approximate computation of the Hamiltonian matrix in the Hund's case (a) basis.

## 3. The Diatomic Hamiltonian

Despite its being the simplest molecule, the diatomic Hamiltonian is quite complicated. Brown and Carrington[7] write a Hamiltonian having 32 terms some of which are themselves complicated. A much simpler model will be used here, one which does not attempt an accurate physical description of the molecule but instead incorporates semi-empirical parameters. The kinetic energy of the electrons $K_e$ and nuclei $K_n$, their Coulomb potential energy $V$, and interactions between the magnetic moments of the electrons and nuclei contribute to the Hamiltonian,

$$H = K_e + K_n + V + H_{fs} + H_{hfs}, \tag{49}$$

where $H_{fs}$, the fine structure Hamiltonian, represents the magnetic moments produced by the orbital and spin angular momenta of the electrons, and $H_{hfs}$, the hyperfine structure Hamiltonian, represents the magnetic moments produced by the orbital and spin angular momenta of the nuclei. The hyperfine structure term is smaller than the fine structure term by roughly the ratio of the electronic to nuclear masses. Except for the simplest molecules, *ab initio* calculations cannot provide sufficiently accurate solutions for the eigenfunction $\langle \mathbf{r}'_1 \mathbf{r}'_2 \cdots \mathbf{r}'_{N-1}\, \rho\, \zeta; r | n \rangle$ of Eq. (33). Therefore, in our minds eye we calculate the matrix elements $\langle n|H|n'\rangle$ by integrating over the coordinates $\mathbf{r}'_1 \mathbf{r}'_2 \cdots \mathbf{r}'_{N-1} \rho \zeta$,

$$
\langle n|H|n' \rangle = \int \int \cdots \int \psi_n^*(\mathbf{r}'_1 \mathbf{r}'_2 \cdots \mathbf{r}'_{N-1} \rho \zeta; r)\, H(\mathbf{r}'_1 \mathbf{r}'_2 \cdots \mathbf{r}'_{N-1}\, r\, \rho\, \zeta)
$$
$$
\times\, \psi_{n'}(\mathbf{r}'_1 \mathbf{r}'_2 \cdots \mathbf{r}'_{N-1}\, \rho\, \zeta; r)\, d\mathbf{r}'_1\, d\mathbf{r}'_2 \cdots d\mathbf{r}'_{N-1}\, d\rho\, d\zeta
$$
$$
= \left\{ -\frac{\hbar^2}{2\mu r^2}\frac{\partial}{\partial r}\left(r^2 \frac{\partial}{\partial r}\right) + \frac{\hbar^2}{2\mu r^2}\mathbf{R}^2(\theta\,\phi) + T_n + V(r) + H_{\text{fs}}(r\,\phi\,\theta\,\chi) \right\}\delta nn'
$$

(50)

and obtain a radial Schrodinger equation to which many angular momentum terms have been added. The constant $T_n$, called the electronic term value, is the constant part of $\langle n|H|n\rangle$. $V(r)$ is a function of $r$ because the eigenfunction $\psi_{n'}(\mathbf{r}'_1 \mathbf{r}'_2 \cdots \mathbf{r}'_{N-1} \rho \zeta; r)$ is a parametric function of $r$, $\mathbf{R}(\phi\theta)$ is the orbital angular momentum of the two nuclei, and $H_{\text{fs}}(r\,\phi\,\theta\,\chi)$ is the fine structure term, a function $r$, $\phi$, $\theta$, and $\chi$ because integration over these variables was not performed. A similar term representing the hyperfine structure should also be present, but hyperfine terms are ignored in the following.

A proper eigenvalue of the Hamiltonian should be a number, the energy eigenvalue (term value.) By only partially evaluating the Hamiltonian matrix elements we obtain a result that is a function of the variables ignored in the evaluations. The integrations over $\mathbf{r}'_1 \mathbf{r}'_2 \cdots \mathbf{r}'_{N-1} \rho \zeta)$ are not actually performed, but the result of doing them in principle is a tractable radial Schrodinger equation.

### 3.1. The rotational Hamiltonian

Many of the most interesting aspects of diatomic spectra come from the terms $\mathbf{R}(\theta\phi)$ and $H_{\text{fs}}(r\,\phi\,\theta\,\chi)$. The rotational Hamiltonian is defined as

$$
H_{\text{rot}} = \frac{\hbar^2}{2\mu r^2}\mathbf{R}^2(\theta\,\phi).
$$

(51)

The orbital angular momentum of the nuclei $\mathbf{R}(\theta\phi)$ seems simple enough until one notices that it does not explicitly appear in either the Hund's case (a) or (b) basis. Expressing $\mathbf{R}(\theta\phi)$ in terms of $\mathbf{J}$ yields

$$\mathbf{R}^2 = (\mathbf{J} - \mathbf{L} - \mathbf{S})^2 \tag{52}$$

$$= \mathbf{J}^2 + \mathbf{L}^2 + \mathbf{S}^2 - 2\mathbf{J} \cdot \mathbf{L} - 2\mathbf{J} \cdot \mathbf{S} + 2\mathbf{L} \cdot \mathbf{S} \tag{53}$$

$$= \mathbf{J}'^2 + \mathbf{L}'^2 + \mathbf{S}'^2 - 2\mathbf{J}' \cdot \mathbf{L}' - 2\mathbf{J}' \cdot \mathbf{S}' + 2\mathbf{L}' \cdot \mathbf{S}' \tag{54}$$

$$= \mathbf{J}^2 + \mathbf{L}^2 + \mathbf{S}^2 - (J_+L_- + J_-L_+) - 2J_zL_z \tag{55}$$
$$- (J_+S_- + J_-S_+) - 2J_zS_z + (L_+S_- + L_-S_+) + 2L_zS_z$$

$$= \mathbf{J}'^2 + \mathbf{L}'^2 + \mathbf{S}'^2 - \left(J'_+L'_- + J'_-L'_+\right) - 2J_{z'}L_{z'} \tag{56}$$
$$- \left(J'_+S'_- + J'_-S'_+\right) - 2J_{z'}S_{z'} + \left(L'_+S'_- + L'_-S'_+\right) + 2L_{z'}S_{z'}.$$

The appearance of $\mathbf{L}^2$ above requires that we go back to the integrations in Eq. (50). The electronic orbital angular momentum,

$$\mathbf{L} = \mathbf{L}(\mathbf{r}'_1 \mathbf{r}'_2 \cdots \mathbf{r}'_{N-1} \, \rho \, \zeta \, \chi), \tag{57}$$

is a function of the variables of integration in Eq. (50) and the third Euler angle $\chi$. The coordinates $\rho$, $\chi$, and $\zeta$ are cylindrical coordinates of a single electron (the electron excluded from the collection $\mathbf{r}'_1, \mathbf{r}'_2 \cdots \mathbf{r}'_{N-1}$) in the molecule-fixed coordinate system. The integration in Eq. (50) of $\mathbf{L}^2(\mathbf{r}'_1 \mathbf{r}'_2 \cdots \mathbf{r}'_{N-1} \, \rho \, \zeta \, \chi)$ yields a constant which we add to the constant $T_n$, and the $\chi$-dependent term

$$\langle n | \frac{-\hbar^2}{2m_e\rho^2} | n \rangle \frac{\partial^2}{\partial \chi^2} = \frac{1}{I_e} J_{z'}^2 \tag{58}$$

where $m_e$ is the electronic mass and $I_e$ is the expectation value of the moment of inertia of the $N$th electron. Because $m_e$ is small in comparison to reduced nuclear mass $\mu$, this terms adds a large but constant term to the energy eigenvalue when $\Lambda^2 \neq 0$. This is also added to the constant $T_n$. Thus, the net effect the integral of $\mathbf{L}^2$ is to add constants to $T_n$ and remove $\mathbf{L}^2$ or $\mathbf{L}'^2$ from Eqs. (52–56).

An additional impact of the electronic orbital angular momentum $\mathbf{L}$ is that it makes the quantum number $\Lambda$ both an angular momentum quantum number and an electronic quantum number. Thus, basis states having different values $\Lambda$ also have different electronic-vibrational states.

The operator $B(r)$ is defined

$$B(r) = \frac{\hbar^2}{2\mu r^2} \tag{59}$$

allowing one to write the rotational kinetic energy of the nuclei as

$$\frac{\hbar^2}{2\mu r^2} \mathbf{R}^2(\theta \, \phi) = B(r) \, \mathbf{R}^2(\theta \, \phi). \tag{60}$$

Evaluation of this term does not require integration over the Euler angles because angular momentum theory gives the eigenvalues of angular momentum operators. However, in the evaluation of $\langle n|H|n \rangle$ the integration over the internuclear distance $r$ remains to be performed. The vibrational matrix element of $B(r)$,

$$\langle v|B(r)|v \rangle = \int_0^\infty \phi_v(r)\, B(r)\, \phi_v(r)\, dr \qquad (61)$$

$$= B_v \qquad (62)$$

is the rotational constant $B_v$. The vibrational eigenfunction $\phi_v(r) = \langle r|v \rangle$ is that from Eq. (40) or (42). One then defines the rotational Hamiltonian as

$$H_{\mathrm{rot}} = B_v\, \mathbf{R}^2(\theta\,\phi). \qquad (63)$$

### 3.2. *The fine structure Hamiltonian*

A widely used phenomenological fine structure Hamiltonian, $H_{\mathrm{fs}}(r\,\phi\,\theta\,\chi)$ of Eq. (50) expressed in terms of angular momentum operators, is written as

$$H_{\mathrm{fs}} = H_{\mathrm{SO}} + H_{\mathrm{SS}} + H_{\mathrm{SR}} \qquad (64)$$

where the spin-orbit interaction is

$$H_{\mathrm{SO}} = A_n(r)\, \mathbf{L} \cdot (S), \qquad (65)$$

the spin-spin interaction is

$$H_{\mathrm{SS}} = \frac{2}{3}\lambda_n(r)\, \left(3S_{z'}^2 - \mathbf{S}^2\right), \qquad (66)$$

and the spin-rotation interaction is

$$H_{\mathrm{SR}} = \gamma_n(r)\, \mathbf{N} \cdot \mathbf{S}. \qquad (67)$$

### 3.3. *Hamiltonian matrix elements in Hund's case (a)*

The matrix elements of

$$H = H_{\mathrm{rot}} + H_{\mathrm{SO}} + H_{\mathrm{SS}} + H_{\mathrm{SR}} \qquad (68)$$

can be evaluated using the eigenfunction given in Eq. (40). The rotation matrix elements are complete eigenfunctions that depend on the angular coordinates $\phi$, $\theta$, and $\chi$ and quantum numbers $J$, $M$, and $\Omega$, but they do not satisfy the mathematical definition of angular momentum. For example, an angular momentum state is specified by the angular momentum $\mathbf{J}$ and one component by convention chosen to be the $z$ component $J_z$. However, the rotation matrix element has two magnetic quantum numbers,

$$J_z D_{M\Omega}^{J^*}(\phi\,\theta\,\chi) = J_z\, e^{iM\phi} d_{M\Omega}^{J}(\theta)\, e^{i\Omega\chi} \tag{69}$$

$$= -i\frac{\partial}{\partial\phi}\, D_{M\Omega}^{J^*}(\phi\,\theta\,\chi) \tag{70}$$

$$= M\, D_{M\Omega}^{J^*}(\phi\,\theta\,\chi) \tag{71}$$

$$J_{z'} D_{M\Omega}^{J^*}(\phi\,\theta\,\chi) = J_{z'}\, e^{iM\phi} d_{M\Omega}^{J}(\theta)\, e^{i\Omega\chi} \tag{72}$$

$$= -i\frac{\partial}{\partial\chi}\, D_{M\Omega}^{J^*}(\phi\,\theta\,\chi) \tag{73}$$

$$= \Omega\, D_{M\Omega}^{J^*}(\phi\,\theta\,\chi) \tag{74}$$

one more than the mathematical definition of angular momentum admits. The above would indicate that $J_z$ and $J_{z'}$ commute and indeed they do commute when applied to a rotation matrix element. However, when the true commutator $J_z$ and $J_{z'}$ is evaluated (*i.e.*, when the commutator is applied to an angular momentum eigenfunction),

$$[J_z, J_{z'}] = i\,[\sin(\phi)\, J_x + \cos(\phi)\, J_y]\sin\theta, \tag{75}$$

one sees that the commutator vanishes only when $\theta = 0$, that is, when the $z$ and $z'$ axes coincide. One should recall that when we write a commutator such as $[J_z, J_{z'}]$ one really means $[J_z, J_{z'}]|JM\rangle$. That is, an angular momentum commutator must be applied to an angular momentum state. One cannot expect to obtain the expected commutator formula if the commutator is allowed to operate on some function that is not an angular momentum state, and the Wigner $D$-function is just such a function.

Application of the raising and lowering operators $J_\pm$,

$$J_\pm |JM\rangle = C_\pm(JM) |J, M \pm 1\rangle, \tag{76}$$

where

$$C_\pm(JM) = \sqrt{J(J+1) - M(M\pm 1)}, \tag{77}$$

$$= \sqrt{(J \mp M)(J \pm M + 1)}, \tag{78}$$

and $J_\pm'$,

$$J_\pm' |J\Omega\rangle = C_\pm(J\Omega) |J, \Omega \pm 1\rangle$$
$$= \sqrt{J(J+1) - \Omega(\Omega \pm 1)}\, |J, \Omega \pm 1\rangle, \tag{79}$$

to the rotation matrix elements again show that $D_{M\Omega}^J(\phi\,\theta\,\chi)$ and $D_{M\Omega}^{J*}(\phi\,\theta\,\chi)$ do not possess the mathematical properties of an angular momentum eigenfunction:

$$J_\pm D_{M\Omega}^J(\phi\,\theta\,\chi) = -\sqrt{J(J+1) - M(M \mp 1)}\, D_{M\mp1,\Omega}^J(\phi\,\theta\,\chi) \qquad (80)$$

$$J_\pm D_{M\Omega}^{J*}(\phi\,\theta\,\chi) = \sqrt{J(J+1) - M(M \pm 1)}\, D_{M\pm1,\Omega}^{J*}(\phi\,\theta\,\chi) \qquad (81)$$

$$J_\pm' D_{M\Omega}^J(\phi\,\theta\,\chi) = \sqrt{J(J+1) - \Omega(\Omega \pm 1)}\, D_{M,\Omega\pm1}^J(\phi\,\theta\,\chi) \qquad (82)$$

$$J_\pm' D_{M\Omega}^{J*}(\phi\,\theta\,\chi) = -\sqrt{J(J+1) - \Omega(\Omega \mp 1)}\, D_{M,\Omega\mp1}^{J*}(\phi\,\theta\,\chi) \qquad (83)$$

Instead of raising $M$ on $D_{M\Omega}^J(\phi\theta\chi)$, $J_+$ lowers $M$, Eq. (80), and inserts an unexpected minus sign before the square root. Equation (83) is the result of interest here. It is used to calculate the matrix elements of $H_{\text{rot}}$, $H_{\text{SO}}$, $H_{\text{SS}}$, and $H_{\text{SR}}$. Note that $J_+'$ acting on the complex conjugate of the rotation matrix element lowers $\Omega$ whereas $J_-'$ raises $\Omega$, and the unexpected (and frequently unreported) minus sign on the right side of Eq. (83).

Equations (76) and (79), the equation

$$S_\pm |SM_S\rangle = C_\pm(SM_S) |S, S \pm 1\rangle, \qquad (84)$$

and the equation

$$S_\pm' |S\Sigma\rangle = C_\pm(S\Sigma) |S, \Sigma \pm 1\rangle, \qquad (85)$$

are to be distinguished from Eqs. (80–83). The former give the results for application of angular momentum operators to angular momentum states whereas Eqs. (80–83) give the result for application of the differential operators $J_\pm$ and $J_\pm'$ to the mathematical functions $D_{M\Omega}^J(\phi\,\theta\,\chi)$ and $D_{M,\Omega}^{J*}(\phi\,\theta\,\chi)$.

Spin cannot be expressed in terms of spatial or angular coordinates. Thus, spin operators can act only on spin states. Application of spin operators to the case (b) basis (42) is straightforward because the spin states $|SM_S\rangle$ explicitly appear in the basis. Application of spin operators to the case (a) basis is more subtle because no spin state is seen directly in $D_{M,\Omega}^{J*}(\phi\,\theta\,\chi)$ or even in the inverse Clebsch-Gordan series (38) for $D_{M,\Omega}^{J*}(\phi\,\theta\,\chi)$. If the right side of the inverse Clebsch-Gordan series (38) is multiplied by $\langle S\Sigma'|S\Sigma\rangle$, and the series is rewritten as

$$D_{M,\Omega}^{J*}(\phi\,\theta\,\chi) = \sum_{M_N=-N}^{N} \sum_{M_S=-S}^{S} \langle NM_N SM_S|JM\rangle$$
$$\times \langle N\Lambda S\Sigma|J\Omega\rangle\, D_{M_N\Lambda}^{N*}(\phi\,\theta\,\chi)\, D_{M_S\Sigma}^{S*}(\phi\,\theta\,\chi)\langle S\Sigma|S\Sigma'\rangle, \quad (86)$$

now there is a spin state for the molecule-fixed spin operator $S'_\pm$ to act upon.

$$S'_\pm D^{J*}_{M,\Omega}(\phi\,\theta\,\chi) = C_\pm(S\Sigma') \sum_{M_N=-N}^{N} \sum_{M_S=-S}^{S} \langle NM_N SM_S | JM\rangle \tag{87}$$

$$\times \langle N\Lambda S\Sigma | J\Omega\rangle D^{N*}_{M_N\Lambda}(\phi\,\theta\,\chi) D^{S*}_{M_S\Sigma}(\phi\,\theta\,\chi) \langle S\Sigma | S, \Sigma' \pm 1\rangle$$

$$= C_\pm(S\Sigma') \sum_{M_N=-N}^{N} \sum_{M_S=-S}^{S} \langle NM_N SM_S | JM\rangle \tag{88}$$

$$\times \langle N\Lambda S, \Sigma' \pm 1 | J\Omega\rangle D^{N*}_{M_N\Lambda}(\phi\,\theta\,\chi) D^{S*}_{M_S,\Sigma'\pm 1}(\phi\,\theta\,\chi)$$

$$= C_\pm(S\Sigma) D^{J*}_{M,\Omega\pm 1} \tag{89}$$

The diagonal Hund's case (a) matrix elements of the Hamiltonian are

$$\langle nv\Lambda S\Sigma\Omega | H | nv\Lambda S\Sigma\Omega\rangle = T_v + B_v \left[ J(J+1) - \Omega^2 + S(S+1) - \Sigma^2 \right]$$

$$+ A_v\Lambda\Sigma + \frac{2}{3}\lambda_v \left[ 3\Sigma^2 - S(S+1) \right] + \gamma_v \left[ \Sigma\Omega - S(S+1) \right] \tag{90}$$

where, for example, $A_v$ is an electronic-vibrational matrix element

$$A_n(r) = \int\int \cdots \int \psi^*_n(\mathbf{r}'_1\mathbf{r}'_2 \cdots \mathbf{r}'_{N-1}\,\rho\,\zeta; r) A(\mathbf{r}'_1\mathbf{r}'_2 \cdots \mathbf{r}'_{N-1}\,r\,\rho\,\zeta)$$

$$\times \psi_{n'}(\mathbf{r}'_1\mathbf{r}'_2 \cdots \mathbf{r}'_{N-1}\,\rho\,\zeta; r)\, \mathbf{dr}'_1\mathbf{dr}'_2 \cdots \mathbf{dr}'_{N-1}\, d\rho\, d\zeta, \tag{91}$$

$$A_v = \langle v | A_n(r) | v\rangle. \tag{92}$$

We have followed tradition in dropping the $n$ quantum number from $A_{nv}$ in the above. The diagonal elements of the spin-spin and spin-rotation terms are handled similarly,

$$\lambda_v = \langle v | \lambda_n(r) | v\rangle \tag{93}$$

$$\gamma_v = \langle v | \gamma_n(r) | v\rangle. \tag{94}$$

Some off-diagonal elements are diagonal in the $|n\rangle|v\rangle$ electronic-vibrational states but off-diagonal in the $|J\Lambda\Sigma\Omega\rangle$ angular momentum basis states. For these matrix elements, parameters such as $B_v$ and $\gamma_v$ remain valid,

$$\langle nv\Lambda S\Sigma\Omega | H_{\text{rot}} + H_{\text{SS}} | n\Lambda S, \Sigma \pm 1, \Omega \pm 1\rangle =$$

$$- B_v \left[ J(J+1) - \Omega(\Omega \pm 1) \right]^{\frac{1}{2}} \left[ S(S+1) - \Sigma(\Sigma \pm 1) \right]^{1/2}$$

$$- \frac{\gamma_v}{2} \left[ J(J+1) - \Omega(\Omega \pm 1) \right]^{\frac{1}{2}} \left[ S(S+1) - \Sigma(\Sigma \pm 1) \right]^{1/2}. \tag{95}$$

The matrix elements of the same Hamiltonian terms can be off-diagonal in both the electronic-vibrational and angular momentum states.

$$\langle nv\Lambda S\Sigma\Omega|H_{\rm rot} + H_{\rm SO}|n'v', \Lambda\pm 1, S, \Sigma\mp 1, \Omega\rangle$$

$$= \langle nv|\left[2B(r) + A(r)\right]\left(L'_+ + L'_-\right)|n'v'\rangle\left[S(S+1) - \Sigma(\Sigma\mp 1)\right]^{1/2}$$

$$= \left(\langle BL^+\rangle_{nn'} + \frac{1}{2}\langle AL^+\rangle_{nn'}\right)\left[S(S+1) - \Sigma(\Sigma\mp 1)\right]^{1/2} \tag{96}$$

$$\langle nv\Lambda S\Sigma\Omega|H_{\rm rot}|n'v', \Lambda\pm 1, S\Sigma, \Omega\pm 1\rangle = \langle BL^+\rangle_{nn'}\left[J(J+1) - \Omega(\Omega\pm 1)\right]^{1/2} \tag{97}$$

where $\langle BL^+\rangle_{nn'}$ and $\langle AL^+\rangle_{nn'}$ are constants.

The matrix elements of two different terms in $H$ can have the same functional dependence on the angular momentum quantum numbers. For example, first order matrix elements of $H_{\rm SS}$ and the second order matrix elements of $H_{\rm SO}$ both contribute terms of the form

$$f(\mathbf{r}'_1\mathbf{r}'_2\cdots\mathbf{r}'_{N-1}\, r\,\rho\,\zeta)\,(S'_+S'_+ + S'_-S'_-). \tag{98}$$

A semi-empirical parameter $\alpha_v$ is created which includes both first order spin-spin and second order spin-orbit contributions.

$$\langle nv\Lambda S\Sigma\Omega|H|n'v', \Lambda\pm 2, S, \Sigma\mp 2, \Omega\rangle$$

$$= \frac{1}{2}\alpha_v\left\{\left[S(S+1) - \Sigma(\Sigma\mp 1)\right]\left[S(S+1) - (\Sigma\mp 1)(\Sigma\mp 2)\right]\right\}^{1/2}. \tag{99}$$

For $n' = n$ and $v' = v$ and $|\Lambda| = 1$, the above equation removes the degeneracy of $\Lambda = \pm 1$ states. Brown and Carriangton[7] and Lefebvre-Brion and Field[8] give detailed discussions of matrix elements in Hund's cases (a) and (b).

### 3.4. *Centrifugal corrections to molecular parameters*

The angular momentum terms in the vibrational Schrodinger Eq. (50) can be viewed as perturbations to the vibrational states. The first order correction produced by rotation of the nuclei are proportional to the matrix elements of $\mathbf{R}^2(\theta\,\phi)$, and higher corrections are proportional to matrix elements of higher powers of $\mathbf{R}^2(\theta\,\phi)$. The matrix elements of the rotational Hamiltonian are expressed as a power series of the matrix elements of $\mathbf{R}^2(\theta\,\phi)$,

$$\langle H_{\rm rot}\rangle = B_v\langle JM|\mathbf{R}^2|JM\rangle - D_v\langle JM|\mathbf{R}^4|JM\rangle + H_v\langle JM|\mathbf{R}^6|JM\rangle + \ldots \tag{100}$$

Terms in the fine structure Hamiltonian can also be centrifugally stretched,

$$\langle H_{\text{fs}} \rangle = (\text{first order } H_{\text{fs}}) + A_{D_v} \langle \mathbf{R}^2 \rangle + \frac{2}{3} \lambda_{D_v} \langle \mathbf{R}^2 \rangle + \gamma_{D_v} \langle \mathbf{R}^2 \rangle + \ldots \quad (101)$$

## 4. Finding the Molecular Parameters by Fitting a Measured Spectrum

In principle the parameters in the diatomic Hamiltonian can be found from *ab initio* calculation or from experiment. Only for $H_2$ can *ab inito* calculations achieve the accuracy with which the parameters can be found by fitting measured spectral line positions. The following algorithm has been implemented in FORTRAN.

1. Trial values for each of the molecular parameters are read from a data file, and experimental values of $J'$, $J''$, and the vacuum wavenumber $\tilde{\nu}_{ul}$ for each known line in a band are read from a second data file.

2. For given values of $J'$ and $J''$ from the experimental table, the upper and lower Hamiltonians are computed in the Hund's case (a) basis of length $\mathcal{N}$. The $\mathcal{N} \times \mathcal{N}$ Hamiltonian matrix is expressed as a sum of $\mathcal{N} \times \mathcal{N}$ submatrices each of which is multiplied by one of the molecular parameters. For example, the matrix of $\mathbf{R}^2$ is multiplied by $B_v$. Arrays of length $\mathcal{N}$ of the quantum numbers $n$ (electronic state), $\Lambda$, $\Sigma$, and $\Omega$ provide the information required to compute each matrix element.

3. The upper $H'$ and lower $H''$ Hamiltonian matrices are numerically diagonalized. The term values (*i.e.*, energy eigenvalues divided by $hc$) are the diagonal elements. The orthogonal (*i.e.*, real unitary) matrices $U'$ and $U''$ that diagonalize $H'$ and $H''$ are saved.

4. Each parameter such as $B_v$, $A_v$, $\lambda_v$ and its associated centrifugally stretched parameters such as $D_v$, $A_{D_v}$ and $\gamma_{D_v}$ contributes a submatrix to the Hamiltonian. Each submatrix is subjected to the same unitary transformation that diagonalized the total Hamiltonian. For example,

$$\mathcal{H}_{A_v} = U^{-1} H_{A_v} U. \quad (102)$$

5. The Hönl-London factors, $S(J', J)$ are computed from the case (a) basis and the matrices that diagonalized the upper and lower

Hamiltonians.

$$S(J', J) = (2J + 1) \left| \sum_{\Omega'} \sum_{\Omega} U'_{\Omega' J'} \langle J\Omega, n_p, \Omega' - \Omega | J'\Omega'_l \rangle U_{J\Omega} \right|^2$$
(103)

where $n_p = 1$ for one-photon transitions, $n_p = 2$ for second order transitions, *etc.* A transition for which the Hönl-London factor $S(J', J'')_{ij}$ vanishes is forbidden, a non-vanishing HLF indicates an allowed transition.

6 For each theoretically predicted spectral line, the experimental table is searched to find a line having the same wavenumber, $\tilde{\nu}_{ul}$, to within a specified tolerance (typically $0.1$ cm$^{-1}$), the same $J'$ and $J''$, and which has not already been identified with a previously computed theoretical spectral line.

7 A *derivative matrix* is computed. When completed, this matrix will be used to compute corrections to the trial values of the molecular parameters. Computation of the derivative matrix proceeds as follows: For each theoretical line which is accepted in step 6 the error in the theoretical wavenumber is computed,

$$\Delta\tilde{\nu}_j = \tilde{\nu}_{\text{theory}} - \tilde{\nu}_{\text{exper}},$$
(104)

and the contribution of each molecular parameter in the model to this error is computed,

$$\Delta\tilde{\nu}_j = \sum_p \frac{\partial\tilde{\nu}_j}{\partial c'_p} \Delta c'_p - \sum_q \frac{\partial\tilde{\nu}_j}{\partial c''_q} \Delta c''_q,$$
(105)

where the symbol $c'_p$ denotes one of the parameters for the upper state and $c''_q$ denotes one of the parameters for the lower state.

$$\frac{\partial\tilde{\nu}_j}{\partial c'_p} = \frac{\partial T_n(c'_p)}{\partial c'_p}$$

$$= \frac{\partial}{\partial c'_p} \sum_k \sum_l U^{-1}_{nk} H_{kl} U_{ln}$$
(106)

$$= \sum_k \sum_l \left[ \frac{\partial U^{-1}_{nk}}{\partial c'_p} H_{kl} U_{ln} + U^{-1}_{nk} \frac{\partial H_{kl}}{\partial c'_p} U_{ln} + U^{-1}_{nk} H_{kl} \frac{\partial U_{ln}}{\partial c'_p} \right].$$

The first and third terms cancel to give

$$\frac{\partial\tilde{\nu}_j}{\partial c'_p} = \sum_k \sum_l U^{-1}_{nk} \frac{\partial H_{kl}}{\partial c'_p} U_{ln}.$$
(107)

Since $H_{kl}(c'_p) = c'_p \, \mathcal{H}^{(p)}_{kl}$,

$$\frac{\partial \tilde{\nu}_j}{\partial c'_p} = \sum_k \sum_l U^\dagger_{nk} \mathcal{H}^{(p)}_{kl} \, U_{ln}. \tag{108}$$

This result is the reason for the computation described in step 4. For the lower states the corresponding result is

$$\frac{\partial \tilde{\nu}_j}{\partial c''_q} = -\sum_k \sum_l U^\dagger_{nk} \mathcal{H}^{(q)}_{kl} \, U_{ln}. \tag{109}$$

Notice the minus sign on the right side of this equation.

The derivative matrix has a column for each of the molecular parameters and a row for each experimental line for which a theoretical line has been found. For a given experimental line, the above two equations are used to fill out a row of the derivative matrix.

8 Steps 2–7 are repeated for all values of $J'$ and $J''$ for which there are experimentally measured line positions. At the conclusion of this step one has collected the information required to solve for $\mathbf{\Delta C}$ in the equation

$$\mathbf{\Delta \tilde{\nu}} = \mathbf{D} \, \mathbf{\Delta C} \tag{110}$$

where $\mathbf{\Delta \tilde{\nu}}$ is the column matrix composed of $\tilde{\nu}_{\text{theory}} - \tilde{\nu}_{\text{exper}}$, $\mathbf{D}$ is the derivative matrix described above, and $\mathbf{\Delta C}$ is the column matrix composed of corrections to the upper and lower molecular parameters. The method of least squares solution of this equation is as follows:

$$\mathbf{D}^T \, \mathbf{\Delta \tilde{\nu}} = \mathbf{D}^T \, \mathbf{D} \, \mathbf{\Delta C} \tag{111}$$

$$\left[\mathbf{D}^T \mathbf{D}\right]^{-1} \mathbf{D}^T \, \mathbf{\Delta \tilde{\nu}} = \mathbf{\Delta C} \tag{112}$$

where $\mathbf{D}^T$ is the transpose of the matrix $\mathbf{D}$. The variance in the line position is computed,

$$\sigma^2_{\tilde{\nu}} = \sum_j \frac{(\tilde{\nu}_{\text{theory}} - \tilde{\nu}_{\text{exper}})^2}{\mathcal{N}_f} \tag{113}$$

where the sum is carried over all experimental lines for which a matching theoretical line was found, and $\mathcal{N}_f$ is the number of molecular parameters being fitted. The variance-covariance matrix is computed,

$$V = \sigma^2_{\tilde{\nu}} \left[\mathbf{D}^T \mathbf{D}\right]^{-1}. \tag{114}$$

The diagonal elements of $V$ are variances (*i.e.*, error estimates) for the parameters. The off-diagonal elements are the covariances of the parameters.

9 The corrections $\Delta\mathbf{C}$ are made to the coefficients.

10 Steps 2–9 are repeated until the corrections $\Delta\mathbf{C}$ become negligibly small.

In a typical analysis of an experimental spectrum, not all of the parameters can be varied; some of them must be held fixed. Those parts of the program which allow some parameters to be varied while others held constant has not been described. Briefly, a logical array is created with each element corresponding to one of the parameters. When the array element is true the corresponding parameter is varied, when false the parameter is held fixed.

### 4.1. *Example of a spectrum fit*

The CH $A\,^2\Delta \leftrightarrow X\,^2\Pi$ system is frequently observed in flame spectra. The following shows some of submatrices from the upper and lower Hamiltonians that were fitted to line position measurements of Bernath, *et al.*[9] The matrix elements of the $A\,^2\Delta$ Hamiltonian were computed in a four term case (a) basis,

$$
\begin{aligned}
|A\,^2\Delta\rangle = &\ |\Lambda = -2, \Sigma = -0.5\rangle\ \langle\Lambda = -2, \Sigma = -0.5|A\,^2\Delta\rangle \\
&+ |\Lambda = -2, \Sigma = 0.5\rangle\ \langle\Lambda = -2, \Sigma = 0.5|A\,^2\Delta\rangle \\
&+ |\Lambda = 2, \Sigma = -0.5\rangle\ \langle\Lambda = 2, \Sigma = -0.5|A\,^2\Delta\rangle \\
&+ |\Lambda = 2, \Sigma = 0.5\rangle\ \langle\Lambda = 2, \Sigma = 0.5|A\,^2\Delta\rangle.
\end{aligned}
\tag{115}
$$

The $|\Lambda\Sigma\rangle$ basis states are the case (a) eigenfunctions function given in Eq. (40). Matrix elements such $\langle\Lambda\Sigma|A\,^2\Delta\rangle$ are from the matrix $U$ above that diagonalizes the Hamiltonian. A six term basis was chosen for the $X\,^2\Pi$ state,

$$
\begin{aligned}
|X\,^2\Pi\rangle = &\ |\Lambda = -1, \Sigma = -0.5\rangle\ \langle\Lambda = -1, \Sigma = -0.5|X\,^2\Pi\rangle \\
&+ |\Lambda = -1, \Sigma = 0.5\rangle\ \langle\Lambda = -1, \Sigma = 0.5|X\,^2\Pi\rangle \\
&+ |\Lambda = 1, \Sigma = -0.5\rangle\ \langle\Lambda = 1, \Sigma = -0.5|X\,^2\Pi\rangle \\
&+ |\Lambda = 1, \Sigma = 0.5\rangle\ \langle\Lambda = 1, \Sigma = 0.5|X\,^2\Pi\rangle \\
&+ |\Lambda = 0, \Sigma = -0.5\rangle\ \langle\Lambda = 0, \Sigma = -0.5|X\,^2\Pi\rangle \\
&+ |\Lambda = 0, \Sigma = 0.5\rangle\ \langle\Lambda = 0, \Sigma = 0.5|X\,^2\Pi\rangle
\end{aligned}
\tag{116}
$$

The last two terms in this basis are from a $^2\Sigma$ basis state (*i.e.* $\Lambda = 0$ state) and since $\Lambda$ is an electronic quantum number, the above basis is a mixture of two electronic states. Two electronic states are required for the $X\,^2\Pi$ state because the measurements of Bernath, *et al.*,[9] were performed with spectral resolution sufficient to resolve the $\Lambda$-doubling the $X$ state. However, $\Lambda$-doubling in the $A\,^2\Delta$ state is much smaller and was not resolved in the measurements. Thus, it was not necessary to model the $A$ state as a mixture of electronic states even though in principle it should be. Brazier and Brown[10] have observed $\Lambda$-doubling in the $A\,^2\Delta$ state of CH at very high resolution.

The spectrum fitting procedure used here departs from the standard practice in two ways. First, in standard practice the Hamiltonian matrix is subjected to Van Vleck transformations that reduce the dimensions of the Hamiltonian while at the same time including the influence many electronic and vibrational basis states and the numerous terms in the Hamiltonian yielding an "effective Hamiltonian." Second, this effective Hamiltonian is split into two submatrices of opposite parity. Term value differences are computed in both standard practice and the present procedure, but in the algorithm described above the Hönl-London factors are also computed for the purpose of determining which of the computed term value differences actually represent allowed spectra lines. This means that at the conclusion

Table 3. The molecular parameters for the $A\,^2\Delta$ and $X\,^2\Pi$ states of CH found from a fit to the (0,0) band line position measurements of Bernath, *et al.*[9] $\Lambda$-doubling in the $A\,^2\Delta$ is too small to be determined. $\Lambda$-doubling in the $X\,^2\Pi$ is accounted by inclusion of the $C\,^2\Sigma$ state. Of the 251 lines reported for the (0,0) band, 237 were fitted with a standard deviation of 0.011 cm$^{-1}$.

| Parameter | $A\,^2\Delta$ state | $X\,^2\Pi$ state | $C\,^2\Sigma$ |
|---|---|---|---|
| $B_v$ | 14.5661177(60) | 14.2081852(65) | 14.25597[a] |
| $10^3 \times D_v$ | 1.559677(12) | 1.46234(13) | 1.59621[a] |
| $10^8 \times H_v$ | 7.94944(23) | 9.5906(28) | 8.24[a] |
| $10^{12} \times L_v$ | −8.101(46) | 8.044(55)[a] | |
| $A_v$ | −1.0964(26) | 28.159(22) | |
| $10^2 \times \gamma_v$ | 4.179(15) | −3.814(15) | 3.712[a] |
| $10^6 \times \gamma_{D_v}$ | | 7.53(46) | 22.5[a] |
| $T_v$ | 1415.9633[a] | 24633.5045(95) | 33207.6126[a] |
| $< AL+ >$ | | 113.19(14) | |
| $< BL+ >$ | | 24.097(24) | |
| $10^3 \times < BL+ >_{D_v}$ | | −3.7047(58) | |

[a] Held fixed during the fitting computations.

Table 4. The Hamiltonian matrix of the CH $A^2\Delta$ state computed from the fitted molecular parameters given in Table 3 for $J = 10.5$. This Hamiltonian is composed of two independent submatrices because a second electronic state which could have mixed the $\Lambda = -2$ and $\Lambda = 2$ states was not included in the basis.

| $\Lambda$ | $\Sigma$ | $\Omega/\Omega$ | $\Lambda$<br>$\Sigma$<br>$-5/2$<br>$-2$<br>$-1/2$ | $-2$<br>$1/2$<br>$-3/2$ | $2$<br>$-1/2$<br>$3/2$ | $2$<br>$1/2$<br>$5/2$ |
|---|---|---|---|---|---|---|
| $-2$ | $-1/2$ | $-5/2$ | 2.629E+04 | $-153.$ | 0.00 | 0.00 |
| $-2$ | $1/2$ | $-3/2$ | $-153.$ | 2.635E+04 | 0.00 | 0.00 |
| $2$ | $-1/2$ | $3/2$ | 0.00 | 0.00 | 2.635E+04 | $-153.$ |
| $2$ | $1/2$ | $5/2$ | 0.00 | 0.00 | $-153.$ | 2.629E+04 |

Table 5. The orthogonal matrix that diagonalized the Hamiltonian matrix in Table 4. As with the Hamiltonian, the two submatrices here are independent. The elements are either approximately $\pm 1/\sqrt{2}$ or exactly zero. This indicates that the physical properties of this state are more nearly case (b) than case (a). Nevertheless, case (a) is used here as a basis, not a physical model, and the accuracy with which it physically represents the state is not relevant.

| $\Lambda$ | $\Sigma$ | $\Omega/\Omega$ | $\Lambda$<br>$\Sigma$<br>$-5/2$<br>$-2$<br>$-1/2$ | $-2$<br>$1/2$<br>$-3/2$ | $2$<br>$-1/2$<br>$3/2$ | $2$<br>$1/2$<br>$5/2$ |
|---|---|---|---|---|---|---|
| $-2$ | $-1/2$ | $-5/2$ | 0.77091166 | $-0.63694208$ | 0.00000000 | 0.00000000 |
| $-2$ | $1/2$ | $-3/2$ | 0.63694208 | 0.77091166 | 0.00000000 | 0.00000000 |
| $2$ | $-1/2$ | $3/2$ | 0.00000000 | 0.00000000 | 0.77091166 | 0.63694208 |
| $2$ | $1/2$ | $5/2$ | 0.00000000 | 0.00000000 | $-0.63694208$ | 0.77091166 |

of a successful spectrum fitting, one has perfectly collated tables of line positions and HLFs. Manual collation of separately computed line positions and rotational line strength factors (*i.e.*, HLF) is a messy, error prone task.

Although parity plays no part in the fitting process, parity and the $e/f$ parity are also computed for the purpose of comparing the fitted results with published parity designations.

Table 5 shows the orthogonal matrix that diagonalized the Hamiltonian matrix in Table 4 computed using the fitted parameters in Table 3.

Table 6 shows the Hamiltonian matrix of the CH $X^2\Pi$ state computed in the case (a) basis from the fitted molecular parameters given in Table 3 for $J = 10.5$. Inclusion of the $\Sigma$ state in the basis causes mixing of $\Lambda = -1$ and $\Lambda = 1$ submatrices of the $\Pi$ state. The parity matrix, Eq. (43), is given below the Hamiltonian. Next in the table is the orthogonal transformation matrix that diagonalized the Hamiltonian and which also diagonalizes the parity matrix. Finally, the energy and parity eigenvalues are given. The pairs of terms 3246.952–3251.937 and 2947.828–2943.948 are $\Lambda$ doublets.

Table 6. These tables show the Hamiltonian and parity matrices in the case (a) basis, the orthogonal transformation that diagonalized them, the term values (i.e., energy eigenvalues), the parity eigenvalues, and the e/f designation for the CH X $^2\Pi$ state. The unit of energy is the wavenumber, cm$^{-1}$.

**Hamiltonian**

| $\Lambda$ | $\Sigma$ | $\Omega/\Omega$ | $-1$ / $-1/2$ / $-3/2$ | $-1$ / $1/2$ / $-1/2$ | $1$ / $-1/2$ / $1/2$ | $1$ / $1/2$ / $3/2$ | $0$ / $-1/2$ / $-1/2$ | $0$ / $1/2$ / $1/2$ |
|---|---|---|---|---|---|---|---|---|
| $-1$ | $-1/2$ | $-3/2$ | 3100.098 | $-152.053$ | 0.000 | 0.000 | 259.289 | 0.457 |
| $-1$ | $1/2$ | $-1/2$ | $-152.053$ | 3099.704 | 0.000 | 0.000 | $-32.847$ | 260.326 |
| $1$ | $-1/2$ | $1/2$ | 0.000 | 0.000 | 3099.704 | $-152.053$ | 260.326 | $-32.847$ |
| $1$ | $1/2$ | $3/2$ | 0.000 | 0.000 | $-152.053$ | 3100.098 | 0.457 | 259.289 |
| $0$ | $-1/2$ | $-1/2$ | 259.289 | $-32.847$ | 260.326 | 0.457 | 33540.128 | $-160.854$ |
| $0$ | $1/2$ | $1/2$ | 0.457 | 260.326 | $-32.847$ | 259.289 | $-160.854$ | 33540.128 |

**Parity**

| $\Lambda$ | $\Sigma$ | $\Omega/\Omega$ | $-1$ / $-1/2$ / $-3/2$ | $-1$ / $1/2$ / $-1/2$ | $1$ / $-1/2$ / $1/2$ | $1$ / $1/2$ / $3/2$ | $0$ / $-1/2$ / $-1/2$ | $0$ / $1/2$ / $1/2$ |
|---|---|---|---|---|---|---|---|---|
| $-1$ | $-1/2$ | $-3/2$ | 0 | 0 | 0 | $-1$ | 0 | 0 |
| $-1$ | $1/2$ | $-1/2$ | 0 | 0 | $-1$ | 0 | 0 | 0 |
| $1$ | $-1/2$ | $1/2$ | 0 | $-1$ | 0 | 0 | 0 | 0 |
| $1$ | $1/2$ | $3/2$ | $-1$ | 0 | 0 | 0 | 0 | 0 |
| $0$ | $-1/2$ | $-1/2$ | 0 | 0 | 0 | 0 | 0 | $-1$ |
| $0$ | $1/2$ | $1/2$ | 0 | 0 | 0 | 0 | $-1$ | 0 |

**Transformation**

| $\Lambda$ | $\Sigma$ | $\Omega/\Omega$ | $-1$ / $-1/2$ / $-3/2$ | $-1$ / $1/2$ / $-1/2$ | $1$ / $-1/2$ / $1/2$ | $1$ / $1/2$ / $3/2$ | $0$ / $-1/2$ / $-1/2$ | $0$ / $1/2$ / $1/2$ |
|---|---|---|---|---|---|---|---|---|
| $-1$ | $-1/2$ | $-3/2$ | 0.50080826 | 0.49915421 | 0.50006687 | 0.49989666 | 0.00601326 | 0.00603814 |
| $-1$ | $1/2$ | $-1/2$ | $-0.49910819$ | 0.50084403 | 0.49986905 | $-0.50010304$ | $-0.00680257$ | 0.00528087 |
| $1$ | $-1/2$ | $1/2$ | 0.49910819 | $-0.50084403$ | 0.49986905 | $-0.50010304$ | 0.00680257 | 0.00528087 |
| $1$ | $1/2$ | $3/2$ | $-0.50080826$ | $-0.49915421$ | 0.50006687 | 0.49989666 | $-0.00601326$ | 0.00603814 |
| $0$ | $-1/2$ | $-1/2$ | $-0.00906120$ | 0.00057349 | $-0.00800386$ | $-0.00053385$ | 0.70704849 | 0.70706128 |
| $0$ | $1/2$ | $1/2$ | 0.00906120 | $-0.00057349$ | $-0.00800386$ | $-0.00053385$ | $-0.70704849$ | 0.70706128 |
| **Term value** | | | 3246.952 | 2947.828 | 2943.948 | 3251.937 | 33706.004 | 33383.191 |
| **parity - e/f** | | | $+e$ | $+e$ | $-f$ | $-f$ | $+e$ | $-f$ |

In the bottom parity row, the parity sign is followed by the $e/f$ designation. Elements of the $4 \times 4$ $\Lambda$-submatrix of the orthogonal transformation are approximately $\pm 1/2$ while those of the $2 \times 2$ $\Sigma$-submatrix are approximately $\pm 1/\sqrt{2}$. The elements connecting the $\Lambda$- and $\Sigma$-submatrices are much smaller in magnitude but are very important in obtaining molecular parameters that yield term value differences in good agreement with experimental line positions.

## 5. Diatomic Line Strengths in the Case (a) Basis

It was noted above that Eq. (7) is, with the appropriate change of coordinates, applicable to any quantum system. The equation holds for mathematical functions if they are expressed in spherical tensor form (see, *e.g.*, Rose,[11] Edmonds,[12] Brink and Satchler.[13]) For example, the components of spherical tensor $\mathbf{T}^{(1)}$ representing the vector $\mathbf{r}(xyz)$ are

$$T_1^{(1)} = -\frac{1}{\sqrt{2}}(x + iy), \tag{117}$$

$$T_0^{(1)} = z, \tag{118}$$

$$T_{-1}^{(1)} = \frac{1}{\sqrt{2}}(x - iy). \tag{119}$$

The spherical tensor $\mathbf{T}^{(q)}$ of rank $q$ behaves as if it were an angular momentum eigenfunction with $J = q$,

$$T_k^{(q)}(\mathbf{r}) = \sum_{\kappa=-q}^{q} T_\kappa^{(q)}(\mathbf{r}') D_{k\kappa}^{q*}(\phi\,\theta\,\chi). \tag{120}$$

For example, when the electric dipole operator $\mathbf{d}$ is expressed as a spherical tensor its matrix elements in the case (a) basis become

$$\langle n'v'J'M'|\mathbf{d}(\mathbf{r}_1\mathbf{r}_2\cdots\mathbf{r}_{N-1}\,r\,\rho\,\zeta)|nvJ'M\rangle = \langle n'v'|\mathbf{d}(\mathbf{r}_1'\mathbf{r}_2'\cdots\mathbf{r}_{N-1}'\,r\,\rho\,\zeta)|nv\rangle$$
$$\times \frac{\sqrt{(2J'+1)(2J+1)}}{8\pi^2} \tag{121}$$
$$\times \int_0^{2\pi}\int_0^{\pi}\int_0^{2\pi} D_{M'\Omega'}^{J'}(\phi\,\theta\,\chi)\, D_{k\kappa}^{1*}(\phi\,\theta\,\chi)\, D_{M\Omega}^{J*}(\phi\,\theta\,\chi)\, \sin\theta\, d\phi\, d\theta\, d\chi.$$

The electronic transition moment is defined,

$$\mathcal{R}(r) = \langle n'|\mathbf{d}(\mathbf{r}_1'\mathbf{r}_2'\cdots\mathbf{r}_{N-1}'\,r\,\rho\,\zeta)|n\rangle \tag{122}$$

and the triple integral is evaluated to give

$$\langle n'v'J'M'|\mathbf{d}|nvJM\rangle = \langle v'|(R(r)|v\rangle \sqrt{\frac{2J'+1}{2J+1}} \langle JM1k|J'M'\rangle \langle J\Omega1\kappa|J'M'\rangle.$$
(123)

With definition of the Franck-Condon factor,

$$q(v',v) = \langle v'|v\rangle^2,$$
(124)

expansion of $\mathcal{R}(r)$ in a Taylor series,

$$\mathcal{R}(r) = a_0 + a_1 r + a_2 r^2 \cdots,$$
(125)

and definition of the $r =$ centroids,

$$\bar{r}_i(v',v) = \frac{\langle v'|r^i|v\rangle}{\langle v'|v\rangle},$$
(126)

one can write the case (a) electric dipole transition matrix element as

$$\langle n'v'J'M'\Lambda'\Sigma'\Omega'|\mathbf{d}|nvJM\Lambda\Sigma\Omega\rangle$$
(127)

$$= [a_0 + a_1\bar{r}_1(v',v) + a_2\bar{r}_2(v',v)\cdots]\langle v'|v\rangle$$
(128)

$$\times \sqrt{\frac{2J'+1}{2J+1}} \langle JM1k|J'M'\rangle \langle J\Omega1\kappa|J'\Omega'\rangle\delta(\Sigma',\Sigma).$$

The transition moment is the sum over the upper and lower basis states of the case (a) transition matrix elements,

$$\langle n'v'J'M'|\mathbf{d}|nvJM\rangle = [a_0 + a_1\bar{r}_1(v',v) + a_2\bar{r}_2(v',v)\cdots]\langle v'|v\rangle$$
(129)

$$\times \sqrt{\frac{2J'+1}{2J+1}} \langle JM1k|J'M'\rangle \sum_{\Omega'}\sum_{\Omega} U'_{\Omega'J'}\langle J\Omega1\kappa|J'\Omega'\rangle U_{J\Omega}\,\delta(\Sigma',\Sigma).$$

The line strength is the square to the transition matrix summed over $M$ and $M'$,

$$S(n'v'J', nvJ) = [a_0 + a_1\bar{r}_1(v',v) + a_2\bar{r}_2(v',v)\cdots]^2\,\langle v'|v\rangle^2$$

$$\times (2J'+1)\left(\sum_{\Omega'}\sum_{\Omega} U'_{\Omega'J'}\langle J\Omega1\kappa|J'\Omega'\rangle U_{J\Omega}\,\delta(\Sigma',\Sigma)\right)^2$$

$$\times \frac{1}{2J+1}\sum_{M'=-J'}^{J'}\sum_{M=-J}^{J}\langle JM1k|J'M'\rangle^2.$$
(130)

This can be rewritten as

$$S(n'v'J', nvJ) = S_{ev}(n'v', nv)\, S(J', J) \qquad (131)$$

where

$$S_{ev} = [a_0 + a_1 \bar{r}_1(v', v) + a_2 \bar{r}_2(v', v) \cdots]^2 \langle v'|v\rangle^2 \qquad (132)$$

$$S(J', J) = (2J' + 1)\left( \sum_{\Omega'}\sum_{\Omega} U'_{\Omega' J'} \langle J\Omega 1\kappa | J'\Omega' \rangle U_{J\Omega}\; \delta(\Sigma', \Sigma) \right)^2. \qquad (133)$$

The diatomic line strength separates into the electronic-vibrational strength $S_{ev}$ and the rotational line strength factor $S(J', J)$. The rotational line strength factor $S(J', J)$ is also called the Hönl-London factor.

### 5.1. *RKR potentials and vibrational eigenfunctions*

The $V(r)$ potential function in Eq. (50) can be estimated by various semi-empirical methods. Given a $V(r)$ function or table of values versus $r$, one can numerically solve Eq. (50) to give the vibrational eigenfunctions $\psi_v(r) = \langle r|v\rangle$ and vibrational term values $G_v$. Cashion[14] gives a concise description of the Numerov solution of the one-dimensional Schrodinger equation. Tellinghuisen[15] describes a similar algorithm that gives accurate values for the rotational constants $B_v, D_v, H_v, \dots$. The reliability of a $V(r)$ potential is judged by how well the solutions agree with the experimental data used to calculate the $V(r)$ potential. The Rydberg-Klein-Rees method yields $V(r)$ from experimentally determined $G_v$ and $B_v$ values. Tellinghuisen's algorithm[16] is widely used. LeRoy's FORTRAN program LEVEL[17] combines several of these algorithms in very useable form.

### 5.2. *Computation of the diatomic line strength*

Solving for the vibrational eigenfunctions for the upper and lower electronic-vibrational states permits computation of the Franck-Condon factors and $r$-centroids. The electronic transition moment $\mathcal{R}(r)$ can be determined from measurements (*e.g.*, Luque and Crosley[18]) or from *ab initio* computations of the electronic eigenfunctions (*e.g.*, Arnold and Langhoff,[19] Chabalowski, *et al.*,[20] Cooper.[21]) Given a series expansion of the electron transition moment Eq. (125), Franck-Condon factors Eq. (124), $r$-centroids Eq. (126), and the Hönl-London factors Eq. (133), one can compute diatomic line strengths from Eqs. (130) and (131).

## 6. Example Applications of Line Strengths

Several example of computed spectra and some comparisons with recorded spectra are given below. Diagnostics applications include, for example, characterization of laser-induced optical breakdown (LIOB) phenomena using laser-induced breakdown spectroscopy (LIBS).[22] The prediction of a spectrum often requires much more than the line strength such as pressure, reaction and collision rates, electron density, *etc.*, but the line strength and temperature are the minimum information required. Computation of a synthetic spectrum begins with specification of the atoms and molecules of interest, the range of wavelengths to be covered, the spectral resolution (typically, the full line width at half-maximum intensity, FWHM), estimates of at least the relative population densities, and tables of line strengths. The basic algorithm consists of dividing the wavelength range from $\lambda_{\min}$ to $\lambda_{\max}$ into n equal wavelength bins, setting the value of the FWHM, and computation of the contribution of each line in the line strength tables to each wavelength bin.

### 6.1. *Free spontaneous emission*

The equation for free spontaneous emission of a spectral line is

$$I_{ul} = h\,\nu_{ul}\,A_{ul}\,N_u. \tag{134}$$

where $\nu_{ul}$ is the line frequency, $A_{ul}$ is the Einstein emission coefficient (transition probability per unit time), and $N_u$ is the population density of upper levels. For a thermal distribution of upper levels,

$$N_u = \frac{g_u\,N_0\,e^{-E_u/kT}}{Q} \tag{135}$$

where $E_u$ is the energy of the upper level, $g_u$ is the statistical weight (degeneracy) of the upper level, $N_0$ is the total population density, and $Q$ is the partition function. When the Einstein $A$ coefficient is expressed in terms of the line strength,

$$A_{ul} = \frac{64\pi^4\nu_{ul}^3}{3hc^3 g_u}\,S_{ul}, \tag{136}$$

the statistical weight is canceled in the emission equation,

$$I_{ul} = \frac{64\pi^4 N_0}{3hc^3 Q} \, \nu_{ul}^4 \, S_{ul} \, e^{-E_u/kT}. \tag{137}$$

This result shows one advantage of writing equations in terms of the line strength: The statistical weight is accounted for in a correctly computed line strength.

Our calling the quantity on the right side of Eq. (134) or Eq. (137) *intensity* is loose usage of the term. These equations predict the energy radiated isotropically per unit time per unit volume.[23] However, the energy falling on a detector is proportional to this quantity, and we will be dealing with relative measured intensities only (*i.e.*, intensity corrected for detector relative spectral sensitivity but not for absolute sensitivity.)

### 6.1.1. *An algorithm for computed emission spectra*

A synthetic emission spectrum is very similar to one recorded on an array detector. The spectrum from the minimum wavelength $\lambda_{\min}$ to maximum wavelength $\lambda_{\max}$ is broken in to a large number of wavelength bins (pixels). In the array detector, one or more spectral lines can contribute to the signal recorded on a wavelength pixel. A single line might contribute signal to more than one pixel. Equations (134) and (137) predict the total energy radiated from a spectral line but not how it distributed across the spectral line shape. If Eq. (137) is multiplied by a line shape function $g(\lambda_p, \lambda_k)$ where $\lambda_p$ is the center wavenumber of the $p$th pixel and $\lambda_k$ is the is the wavelength of the $k$th line in the table of line strengths, then a spectrum can be computed by scanning the line strength table and computing the contribution of each $k$th line to each $p$th pixel. We have found a simple Gaussian line shape is suitable in many cases,

$$g(\lambda_p, \lambda_k) = \text{const} \, e^{-4\ln(2)\,(\lambda_p - \lambda_k)^2 / \text{FWHM}^2} \tag{138}$$

where FWHM is the full line width at half of the maximum intensity. The following is a section of FORTRAN routine giving an example of how each line in a table of **kmax** line strengths contributes to each bin in a spectrum. A spectrum containing **npts** bins ranging from minimum wavenumber **wn_lo** to maximum wavenumber **wn_hi** is computed. Relative values of the quantity **peak(k)** are computed from Eq. (137) for a specified temperature.

```
      delwn = (wn_lo-wn_hi) / (npts-1)
      con = 2.0 * sqrt(alog(2.0))
      do 70 k=1, kmax
         FWHM_wn = FWHM * wn(k) / wnmax
         ndel = nint(2.5*FWHM_wn/delwn)
         if (wn(k).lt.wn_lo) go to 70
         if (wn(k).gt.wn_hi) go to 70
         n0 = nint((wn(k)-wnmin)/delwn) + 1
         nmin = n0 - ndel
         if (nmin.lt.1)  nmin = 1
         nmax = n0 + ndel
         if (nmax.gt.npts)  nmax = npts
         do n=nmin, nmax
            u = con * (wn(k)-x(n)) / FWHM_wn
            if (u.lt.9.21)  then
               y(n) = y(n) + peak(k) * exp(-u*u)
               if (y(n).gt.ymax)  ymax = y(n)
            end if
         end do
70    end do
```

where FWHM is the full width of the line at half of its maximum intensity.

Figure 1 shows hot C$_2$ Swan spectra at three resolutions. These synthetic spectra were computed from a table of line strengths prepared after the line position data of Phillips and Davis[24] and Tanabashi, *et al.*,[25] were fitted to Hund's case (a) matrix representations of the $d\,^3\Pi_g$ and $a\,^3\Pi_g$ states of C$_2$. The fit yielded a table of molecular parameters like Table 3 from which the upper and lower term values, parities, and HLF table were computed for the Swan system. Table 7 is example output from a FORTRAN program DHLF. This program computes the quantities shown in Table 7 for each each possible transition from the minimum values $J'$ and $J$ to the maximum values that are experimentally known. From the fitted values of $B_v$ and $G_v$, RKR $V(r)$ curves were computed for the $d\,^3\Pi_g$ and $a\,^3\Pi_g$ states. Numerical solutions of the upper and lower radial Schrodinger equations gave the vibrational eigenfunctions from which the Franck-Condon factors and $r$-centroids were computed. The electronic transition moment of Chabalowski, *et al.*,[20] was used. Finally, $R_e(r)$, $q(v',v)$, $\bar{r}_i(v',v)$, and $S(J',J)$ were combined to yield the line strength $S(n'v'J',nvJ)$.

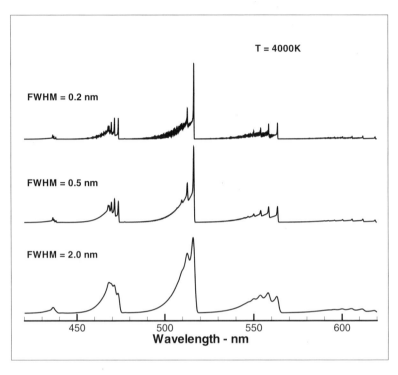

Fig. 1. $C_2$ Swan spectra at three different spectral resolutions. Spectra similar to these are seen in oxyacetylene flames, carbon LIBS, and carbon arcs.

One can compute synthetic spectra at conditions that might be difficult to obtain experimentally. For example, Fig. 2 shows a room temperature Swan spectrum. What appear to be fully resolved spectral lines at modest resolution are seen to be triplets at high resolution.

One can also use tables of diatomic line strengths to demonstrate such things as $\Lambda$-doubling and the influence of nuclear spin even when hyperfine structure is not resolved. Figure 3 shows a portion of the Swan (0,0) band for three isotopomers of diatomic carbon. When the two nuclei are identical the relative intensities of the two line in a $\Lambda$-doublet is strongly influenced by nuclear spin. When the nuclear spin of identical nuclei is zero, one of the lines in the $\Lambda$-doublet is missing. When the nuclear spin of identical nuclei is 0.5, the relative intensity of the doublet lines is 3:1. The line strength tables used to compute the spectra in Fig. 3 are based on the Fourier spectroscopy data of Amiot.[26]

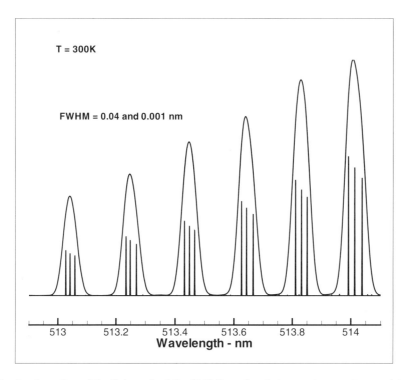

Fig. 2.  A portion of the R-branch of the (0,0) Swan band at room temperature and two different resolutions.

## 6.2.  *Using a measured spectrum to infer temperature*

Two methods for inferring temperature are discussed here. In the first method the Nelder-Mead algorithm starts with a trial temperature and continues to adjust the temperature estimate until the intensity differences between a measured spectrum and a synthetic spectrum are minimized. In the second method, which is a modification of the standard Boltzmann plot and is applicable to spectra in which individual lines are not resolved, Boltzmann plots of the intensity of unresolved lines are made for an assumed temperature and the process is iterated until the inferred temperature becomes constant.

### 6.2.1.  *Nelder-Mead emission spectrum fitting*

The Nelder-Mead[27] algorithm is an extremum algorithm. One provides the algorithm with a function which is typically the sum of the squared resid-

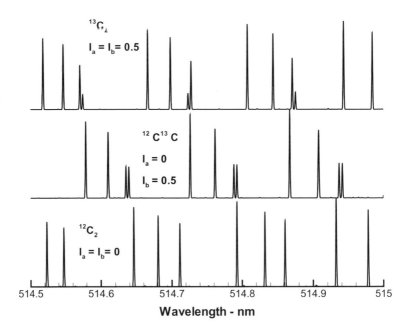

Fig. 3. The influence of nuclear spin on the Λ-doublets of homonuclear molecules. An electronic spin $S = 1$ produces the triplet structure of the Swan spectrum. Each line of the triplet is actually a Λ-doublet. Typically, the Λ-splitting can be resolved for only one of the triplets as seen in this figure. For the homonuclear $^{13}C_2$ molecule the nuclear spin $I = 0.5$ causes the lines of the resolved doublet have a 3:1 intensity ratio. For the heteronuclear $^{12}C^{13}C$ molecule the lines of the resolved Λ-doublet have equal intensities. For the terrestrially abundant $^{12}C_2$ molecule the nuclear spin $I = 0$ causes one line of each Λ-doublet to be missing.

uals between a computed quantity and a measured quantity. The Nelder-Mead algorithm adjusts the parameter values of the computed function until the sum of the squared residuals is minimized. The following describes a method for inferring temperature with a fitting function which is the sum of the squared residuals between the intensities of a measured spectrum and a synthetic spectrum. Temperature and optionally the line width (FWHM) are the fitting parameters.

The intensity of light cannot be negative, meaning that the sum of the residuals between measure and synthetic intensity would be a better fitting function than the sum of squared residuals. However, intensity baseline corrections to a recorded spectrum can yield small negative intensities, and the method of least squares is more familiar than so-called robust estimation. In our experience with applying the Nelder-Mead algorithm to

Table 7. Example output from program DHLF for the C$_2$ $d\,^3\Pi_g - a\,^3\Pi_u$ (0,0) band. The upper and lower Hamiltonians were computed in the Hund's cased (a) basis and numerically diagonalized. The HLFs are then computed for all possible transitions. Those having non-vanishing factors are allowed transitions. The same matrix that diagonalizes the Hamiltonian also diagonalizes the parity. In practice, lines having a HLF below a numerical cutoff factor, typically $(2J'+1)/1000$ are ignored. This factor should be lowered significantly if lines excited by laser induced fluorescence lines are of interest. The term value $F_{J'}$ and $F_J$, and their difference $\tilde{\nu}$ are expressed in wavenumbers, cm$^{-1}$.

| $J'$ | $J$ | | $p'$ | $p$ | $N'$ | $N$ | $F_{J'}$ | $F_J$ | $\tilde{\nu}$ | $S_{J'J}$ |
|---|---|---|---|---|---|---|---|---|---|---|
| 10.0 | 11.0 | P$_{33}$ | +e | −e | 11 | 12 | 21144.925 | 1789.844 | 19355.081 | 10.93358 |
| 10.0 | 11.0 | P$_{22}$ | +e | −e | 10 | 11 | 21103.565 | 1747.393 | 19356.172 | 10.86100 |
| 10.0 | 11.0 | P$_{11}$ | +e | −e | 9 | 10 | 21063.914 | 1706.583 | 19357.331 | 10.76667 |
| 10.0 | 10.0 | Q$_{32}$ | +e | −f | 11 | 10 | 21144.925 | 1711.267 | 19433.658 | .07993 |
| 10.0 | 10.0 | Q$_{33}$ | +e | −f | 11 | 11 | 21144.925 | 1750.556 | 19394.369 | .06569 |
| 10.0 | 10.0 | Q$_{21}$ | +e | −f | 10 | 9 | 21103.565 | 1673.605 | 19429.960 | .08602 |
| 10.0 | 10.0 | Q$_{22}$ | +e | −f | 10 | 10 | 21103.565 | 1711.267 | 19392.298 | .16883 |
| 10.0 | 10.0 | Q$_{23}$ | +e | −f | 10 | 11 | 21103.565 | 1750.556 | 19353.009 | .07206 |
| 10.0 | 10.0 | Q$_{11}$ | +e | −f | 9 | 9 | 21063.914 | 1673.605 | 19390.310 | .42475 |
| 10.0 | 10.0 | Q$_{12}$ | +e | −f | 9 | 10 | 21063.914 | 1711.267 | 19352.648 | .06485 |
| 10.0 | 9.0 | R$_{33}$ | +e | −e | 11 | 10 | 21144.925 | 1715.929 | 19428.995 | 9.89016 |
| 10.0 | 9.0 | R$_{32}$ | +e | −e | 11 | 9 | 21144.925 | 1679.439 | 19465.485 | .03381 |
| 10.0 | 9.0 | R$_{23}$ | +e | −e | 10 | 10 | 21103.565 | 1715.929 | 19387.636 | .04543 |
| 10.0 | 9.0 | R$_{22}$ | +e | −e | 10 | 9 | 21103.565 | 1679.439 | 19424.126 | 9.72374 |
| 10.0 | 9.0 | R$_{21}$ | +e | −e | 10 | 8 | 21103.565 | 1644.006 | 19459.559 | .04428 |
| 10.0 | 9.0 | R$_{12}$ | +e | −e | 9 | 9 | 21063.914 | 1679.439 | 19384.475 | .06858 |
| 10.0 | 9.0 | R$_{11}$ | +e | −e | 9 | 8 | 21063.914 | 1644.006 | 19419.908 | 9.67532 |
| 11.0 | 12.0 | P$_{11}$ | +f | −f | 10 | 11 | 21099.047 | 1742.501 | 19356.546 | 11.81470 |
| 11.0 | 12.0 | P$_{22}$ | +f | −f | 11 | 12 | 21141.354 | 1785.724 | 19355.629 | 11.85530 |
| 11.0 | 12.0 | P$_{33}$ | +f | −f | 12 | 13 | 21185.958 | 1831.094 | 19354.864 | 11.95901 |

Table 8. Part of a line strength table for the $C_2$ $d\,^3\Pi_g - a\,^3\Pi_u$ (0,0) band. This table also includes the experimental line wavenumbers, $\nu_{exp}$, to which Hamiltonian was fitted and, hence, the line strengths computed. The term value $F_{J'}$ and $F_J$, and their difference $\bar{\nu}$ are expressed in wavenumbers, cm$^{-1}$.

| $v'$ | $v$ | $J'$ | ′ | Br. | $p'$ | $p$ | $F_{J'}$ | $F_J$ | $\bar{\nu}$ | $\lambda$ | $S_{J'J}$ | $S_{n'v'J'nvJ}$ | $\bar{\nu}_{exp}$ | $\Delta\bar{\nu}$ |
|---|---|---|---|---|---|---|---|---|---|---|---|---|---|---|
| 0 | 0 | 11.0 | 12.0 | $P_{33}$ | $+f$ | $-f$ | 21186.48 | 1831.73 | 19354.76 | 516.5254 | 11.9400 | 18.08 | | |
| 0 | 0 | 11.0 | 12.0 | $P_{22}$ | $+f$ | $-f$ | 21141.86 | 1786.25 | 19355.61 | 516.5025 | 11.8700 | 17.98 | 19356.5170 | −0.0033 |
| 0 | 0 | 11.0 | 12.0 | $P_{11}$ | $+f$ | $-f$ | 21099.15 | 1742.64 | 19356.51 | 516.4785 | 11.7900 | 17.86 | | |
| 0 | 0 | 11.0 | 11.0 | $Q_{32}$ | $+f$ | $-e$ | 21186.48 | 1746.87 | 19439.62 | 514.2706 | 0.0753 | 0.1140 | | |
| 0 | 0 | 11.0 | 11.0 | $Q_{33}$ | $+f$ | $-e$ | 21186.48 | 1789.19 | 19397.29 | 515.3926 | 0.0666 | 0.1008 | | |
| 0 | 0 | 11.0 | 11.0 | $Q_{21}$ | $+f$ | $-e$ | 21141.86 | 1706.46 | 19435.40 | 514.3821 | 0.0808 | 0.1224 | | |
| 0 | 0 | 11.0 | 11.0 | $Q_{22}$ | $+f$ | $-e$ | 21141.86 | 1746.87 | 19395.00 | 515.4537 | 0.1563 | 0.2368 | | |
| 0 | 0 | 11.0 | 11.0 | $Q_{23}$ | $+f$ | $-e$ | 21141.86 | 1789.19 | 19352.68 | 516.5809 | 0.0680 | 0.1029 | | |
| 0 | 0 | 11.0 | 11.0 | $Q_{11}$ | $+f$ | $-e$ | 21099.15 | 1706.46 | 19392.69 | 515.5150 | 0.3702 | 0.5606 | | |
| 0 | 0 | 11.0 | 11.0 | $Q_{12}$ | $+f$ | $-e$ | 21099.15 | 1746.87 | 19352.28 | 516.5914 | 0.0625 | 9.4646E-02 | | |
| 0 | 0 | 11.0 | 10.0 | $R_{33}$ | $+f$ | $-f$ | 21186.48 | 1751.24 | 19435.24 | 514.3863 | 10.9000 | 16.50 | 19435.2510 | −0.0088 |
| 0 | 0 | 11.0 | 10.0 | $R_{32}$ | $+f$ | $-f$ | 21186.48 | 1711.79 | 19474.70 | 513.3442 | 0.0313 | 4.7341E-02 | | |
| 0 | 0 | 11.0 | 10.0 | $R_{23}$ | $+f$ | $-f$ | 21141.86 | 1751.24 | 19390.62 | 515.5700 | 0.0412 | 6.2386E-02 | | |
| 0 | 0 | 11.0 | 10.0 | $R_{22}$ | $+f$ | $-f$ | 21141.86 | 1711.79 | 19430.08 | 514.5231 | 10.7400 | 16.26 | 19430.0850 | −0.0088 |
| 0 | 0 | 11.0 | 10.0 | $R_{21}$ | $+f$ | $-f$ | 21141.86 | 1673.71 | 19468.15 | 513.5168 | 0.0413 | 6.2602E-02 | | |
| 0 | 0 | 11.0 | 10.0 | $R_{12}$ | $+f$ | $-f$ | 21099.15 | 1711.79 | 19387.36 | 515.6567 | 0.0618 | 9.3644E-02 | | |
| 0 | 0 | 11.0 | 10.0 | $R_{11}$ | $+f$ | $-f$ | 21099.15 | 1673.71 | 19425.44 | 514.6459 | 10.7100 | 16.22 | 19425.4180 | 0.0234 |
| 0 | 0 | 12.0 | 13.0 | $P_{11}$ | $+e$ | $-e$ | 21137.71 | 1781.73 | 19355.97 | 516.4929 | 12.8400 | 19.45 | 19355.9580 | 0.0166 |
| 0 | 0 | 12.0 | 13.0 | $P_{22}$ | $+e$ | $-e$ | 21183.14 | 1827.83 | 19355.31 | 516.5107 | 12.8600 | 19.48 | | |
| 0 | 0 | 12.0 | 13.0 | $P_{33}$ | $+e$ | $-e$ | 21231.06 | 1876.27 | 19354.79 | 516.5245 | 12.9700 | 19.64 | 19354.7860 | 0.0050 |
| 0 | 0 | 12.0 | 12.0 | $Q_{12}$ | $+e$ | $-f$ | 21137.71 | 1786.25 | 19351.46 | 516.6135 | 0.0590 | 8.9317E-02 | | |
| 0 | 0 | 12.0 | 12.0 | $Q_{11}$ | $+e$ | $-f$ | 21137.71 | 1742.64 | 19395.07 | 515.4518 | 0.3216 | 0.4871 | | |
| 0 | 0 | 12.0 | 12.0 | $Q_{23}$ | $+e$ | $-f$ | 21183.14 | 1831.73 | 19351.41 | 516.6146 | 0.0596 | 9.0219E-02 | | |
| 0 | 0 | 12.0 | 12.0 | $Q_{22}$ | $+e$ | $-f$ | 21183.14 | 1786.25 | 19396.89 | 515.4034 | 0.1488 | 0.2254 | | |
| 0 | 0 | 12.0 | 12.0 | $Q_{21}$ | $+e$ | $-f$ | 21183.14 | 1742.64 | 19440.50 | 514.2472 | 0.0800 | 0.1211 | | |
| 0 | 0 | 12.0 | 12.0 | $Q_{33}$ | $+e$ | $-f$ | 21231.06 | 1831.73 | 19399.33 | 515.3385 | 0.0638 | 9.6608E-02 | | |
| 0 | 0 | 12.0 | 12.0 | $Q_{32}$ | $+e$ | $-f$ | 21231.06 | 1786.25 | 19444.81 | 514.1333 | 0.0730 | 0.1105 | | |

emission spectrum fitting, failure of least squares and robust fitting to give the same results indicates a problem with the data. This can also be said when the same experimental spectrum and the same line strength table are subjected to the modified Boltzmann plot algorithm described below. Because all three algorithms use the same experimental data and the same line strengths, they should yield the same temperature within the statistical error estimates.

Our spectrum fitting begins with an experimental spectrum whose intensity is digitized at discrete wavelengths. Many spectrometers are now equipped with array detectors thus providing spectra in the desired form. We often find it necessary to make minor wavelength corrections to spectra recorded with an array detector. This is accomplished by computing a synthetic reference spectrum at high resolution and making small constant, linear, or even occasionally quadratic corrections to the wavelength calibration of the array detector. The wavelengths in good synthetic spectra should be essentially as accurate as the experimental accuracy of the experimental measurements from which the line strength table was created. Normally, this is much better than the wavelength accuracy achieved with an array detector spectrum.

A trial temperature and FWHM are used to compute a synthetic spectrum for the same wavelengths that appear in the experimental spectrum, the relative intensities of the synthetic spectrum are normalized to the experimental spectrum, and the sum of the residuals or squared residuals between the intensities of the measured spectrum and normalized synthetic spectrum is computed. This sum is the function minimized by the Nelder-Mead algorithm.

In addition to allowing temperature and optionally the FWHM to be adjusted in the fitting process, we also optionally allow a baseline intensity correction having a constant, linear, or quadratic variation with wavelength to be fitted. However, as all who have ever faced the challenge of subtracting a baseline offset from a recorded spectrum know, baseline correction is sometimes more an exercise of courage than competence.

Figure 4 shows a fit to an unpublished cavity ring-down CH spectrum of Nemes.[28] This spectrum was excited in a microwave plasma discharge.

### 6.2.2. *Modified Boltzmann plot*

The Boltzmann plot is based on the simple concept that the logarithm of the Boltzmann factor yields the point-slope form of a straight line. Isolation

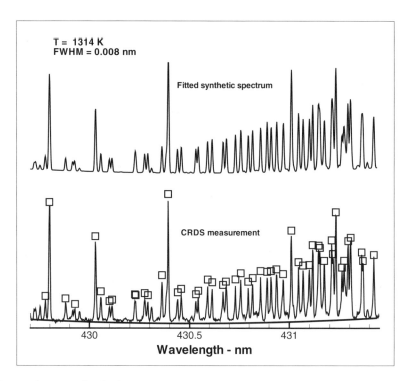

Fig. 4. An example of inferring temperature by fitting a synthetic spectrum to a measured CH ($A\,^2\Delta \leftrightarrow X\,^2\Pi$) spectrum. The Nelder-Mead algorithm was used to adjust the temperature of an assumed thermal distribution of states to make the synthetic spectrum best agree with the measured spectrum. A small baseline offset (show below the measured CRDS spectrum) having a quadratic variation with wavelength was also determined in the fit. The peaks below the squares are those used to prepare the modified Boltzmann plot shown in Fig. 8.

of the Boltzmann factor in Eq. (137) followed by taking the logarithm gives one,

$$\ln\left(\frac{3hc^3Q}{64\pi^4 N_0}\right) + \ln\left(\frac{I_{ul}}{\nu_{ul}^4 S_{ul}}\right) = -\frac{1}{kT}E_u. \tag{139}$$

The substitutions $E = hcF_u$ where $F_u$ is the upper term value and $\nu_{ul} = c\tilde{\nu}_{ul}$ where $\tilde{\nu}_{ul}$ is the line wavenumber and rearrangement gives,

$$\ln\left(\frac{I_{ul}}{\tilde{\nu}_{ul}^4 S_{ul}}\right) = -\frac{hc}{kT}F_u + \ln\left(\frac{3hQ}{64\pi^4 c N_0}\right). \tag{140}$$

A line strength table contains the quantities $F_u$, $\tilde{\nu}_{ul} = F_u - F_l$ and $S_{ul}$. A Boltzmann plot can be readily prepared from measured intensities $I_{ul}$ and a

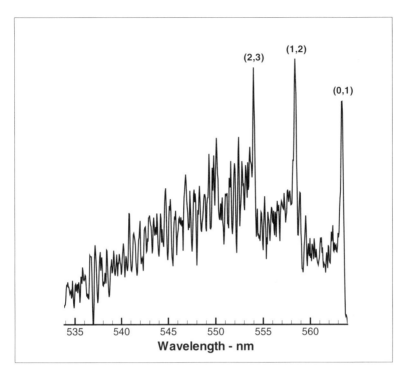

Fig. 5. An experimentally recorded spectrum of the $\Delta v = -1$ sequence of the $C_2$ Swan system.

table of line strengths. The measurement of spectral intensity requires that a relative spectral sensitivity calibration be performed, but the absolute intensity factor can be placed with the other constants collected on the right in Eq. (140). The Boltzmann plot described above assumes that the individual spectral lines are fully resolved.

Figure 5 shows a measured $\Delta v = -1$ sequence of the $C_2$ Swan system. Figure 6 compares the region from the (2,3) to (1,2) band heads with a synthetic spectrum at much higher resolution. Comparison of the measured spectrum with the high resolution synthetic spectrum demonstrates that individual lines are not resolved in the former. A line of wavenumber $\tilde{\nu}_i$ and total intensity $I_i$ contributes $g(\tilde{\nu}_i, \tilde{\nu}) I_i$ intensity at wavenumber $\nu$ where $g(\tilde{\nu}_i, \tilde{\nu})$ is the line shape function. The intensity at wavenumber $\tilde{\nu}$ has contributions from several unresolved spectral lines.

$$I(\tilde{\nu}) = \sum_i g(\tilde{\nu}_i, \tilde{\nu}) \, I_i(\tilde{\nu}_i)$$

$$= \text{const} \sum_i g(\tilde{\nu}_i, \tilde{\nu}) \, \tilde{\nu}_i^4 \, S_i \, e^{-hcF_i'/kT} \tag{141}$$

where $\tilde{\nu}_i = \tilde{\nu}_{ul}$ and $F_i' = F_u$ for the $i$th line. The summation is multiplied by unity in the form $\exp[-hc(F_0' - F_0')/kT]$ and the above rewritten as

$$I(\tilde{\nu}) = \text{const} \, e^{-hc\,F_0'/kT} \sum_i g(\tilde{\nu}_i, \tilde{\nu}) \, \tilde{\nu}_i^4 \, S_i \, e^{-hc\,(F_i'-F_0')/kT}$$

$$= \text{const} \, e^{-hc\,F_0'/kT} \, \Phi(\tilde{\nu}) \tag{142}$$

where $F_0'$ is the term value of an arbitrarily chosen upper state. Our choice for $F_0'$ is the upper term value of the line making the largest single contri-

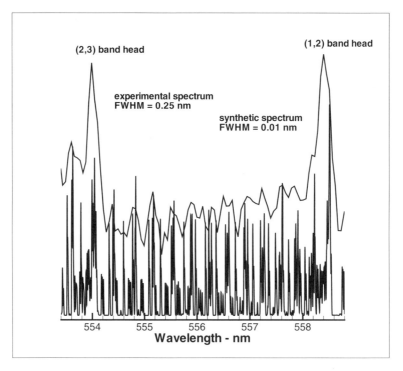

Fig. 6. The expanded region from the (2,3) band head to the (1,2) band head in Fig. 5 compared with a high resolution synthetic spectrum covering the same region. It is clear that the individual lines are not resolved in the experimental spectrum.

bution to the sum, and

$$\Phi(\tilde{\nu}) = \sum_i g(\tilde{\nu}_i, \tilde{\nu}) \, \tilde{\nu}_i^4 \, S_i \, e^{-hc \, (F_i' - F_0')/kT} \tag{143}$$

As with the standard Boltzmann plot, the exponential is isolated and the logarithm taken,

$$\ln\left[\frac{I(\tilde{\nu})}{\Phi(\tilde{\nu})}\right] = -\frac{hc}{kT} \, F_0' + \text{const}, \tag{144}$$

but now the modified Boltzmann plot is an iterative algorithm because computation of $\Phi(\tilde{\nu})$ requires that a trial value of temperature be assumed before the plot can be constructed. The temperature inferred from the plot is then used as the new trial temperature, and the process is repeated until the temperature given by the plot and the trial temperature converge. In our experience, convergence occurs in 3 to 5 iterations. An example modified Boltzmann plot is given in Fig. 7.

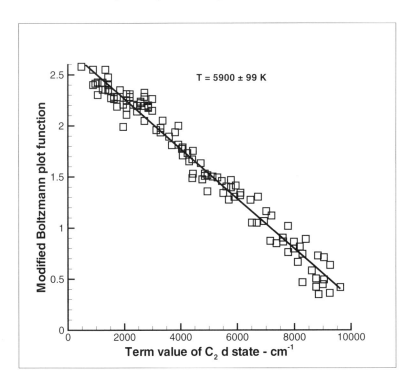

Fig. 7. A modified Boltzmann plot made from the spectrum in Fig. 5.

The statistical error estimate shown in Fig. 7 is taken from the standard deviation of the fitted straight line. Boltzmann plots, either standard or modified, typically yield unreliably optimistic statistical error estimates. A Nelder-Mead fit of a synthetic spectrum to the measured spectrum in Fig. 5 gave a temperature of 5865 K. A standard temperature measurement technology such as the type K thermocouple over the range of 70–1500 K is supported by thousands of man years of experience. A spectroscopic temperature measurement of 6000 K lacks such support. One is cautioned against making unrealistic claims for their accuracy, but the positive claim can be made that spectroscopy provides temperature measurements in environments where no other temperature probe would survive or could be applied.

A modified Boltzmann plot, Fig. 8 of the recorded CH spectrum in Fig. 4 yielded a temperature of $1328 \pm 28K$. Many of the lines in this spectrum are resolved individual lines while a few are mixtures of unresolved lines.

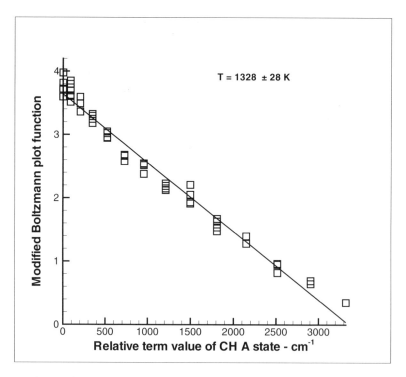

Fig. 8. A modified Boltzmann plot of the spectrum shown in Fig. 4. The baseline correction seen in Fig. 4 was applied before the Boltzmann plot was made.

### 6.2.3. *Synthetic absorption spectra*

The topic here is the spectral transmission of light from source passing through a diatomic gas,

$$\mathcal{T}_\lambda = e^{-\int_0^l k_\lambda \, dx} \tag{145}$$

where $k_\lambda$ is the absorption coefficient which is integrated over the optical path from 0 to $l$. The absorption coefficient is itself an integral, the integral of the absorption coefficient over the line shape of the absorbing spectral line.

The steps in an algorithm for computation of the absorption spectrum are:

1. Divide the spectrum from $\lambda_{\min}$ to $\lambda_{\max}$ into a large number of equally spaced wavenumber bins.

2. Select a line shape function, for example, the Gaussian of Doppler broadening or the Voigt function.

3. Specify the total population density $N_0$, absorption path length $l$, temperature $T$, and compute the partition function $Q$.

4. In a table of line strengths for each line of wavenumber $\lambda_0$ falling between $\lambda_{\min}$ and $\lambda_{\max}$, compute the integrated absorption coefficient $k_0(\lambda_0)$. For the $j$ line from the line list,

$$k_0(\lambda_j) = \frac{2\pi^2 N_0}{3\epsilon_0 h Q} \frac{1}{\lambda_j} S(n'v'J', nvJ) \, e^{-hcF_{nvJ}/kT} \tag{146}$$

5. Using the integrated absorption coefficient $k_0(\lambda_j)$ and the selected line shape function compute the contribution of each line to each wavelength bin. The bin whose wavelength is nearest the wavelength of a line selected from the line list is found and the contribution of the line to the spectral absorption coefficient for that bin is computed from the line shape function,

$$\Delta k = k_0(\lambda_j) \int g(\lambda - \lambda_j) \, d\lambda \tag{147}$$

Gauss-Legendre numerical integration suffices for evaluation of the above integrals if the bin width is chosen to be narrow enough that $g(\lambda - \lambda_j)$ is a slowly varying function across a bin. One then steps to the next bin of longer wavelength, again computes to contribution to the absorption coefficient, and continues stepping to longer wavelengths until the contribution of the $j$th line to the bin becomes negligible, Fig. 9. This process is repeated for the bins of lower wavelength.

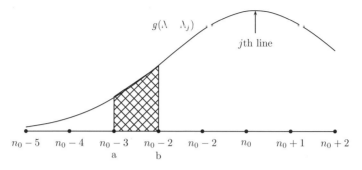

Fig. 9. Computation of an absorption spectrum. The spectrum of the absorption coefficient $k(\lambda)$ is divided into a large number of wavelength bins. The contribution of the $j$th line from a table of line strengths to each bin is computed. The integrated absorption coefficient $k_0(\lambda_j)$ is computed and Gauss-Legendre integration from $a$ to $b$ of the product of the line shape function $g(\lambda - \lambda_j)$ and $k_0(\lambda_j)$ gives the contribution of the $j$ line to the bin. The $j$th line is picked from the line list and its contribution to the $n_0$ bin is computed. Then the contribution of the $j$th line to the next bin to the right is computed, and the computations are marched to the next bin right until the contribution from the $j$ line becomes negligibly small. Finally, the computations are marched to left until the contribution of the $j$ line becomes negligibly small. Possible line shape functions $g(\lambda - \lambda_j)$ are that for Doppler broadening or the Voigt function.

6  The transmission is computed from the integrated spectral absorption coefficient for an assumed homogenous absorber and absorption path $l$

$$\mathcal{T}(\lambda) = e^{-k(\lambda)\,l}. \tag{148}$$

7  The last step is numerical integration of the spectral transmission $\mathcal{T}(\lambda)$ over the spectrometer bandpass function.

Figure 10 is an example synthetic absorption spectrum corresponding to the CRDS spectrum in Fig. 4.

### 6.2.4. Computing tunable laser LIF spectra

Laser induced fluorescence (LIF) poses an interesting application of line strengths because LIF can require two line lists, one for the excitation transitions and a second for the fluorescence transitions. Nonlinear excitation produced by depletion of the lower state of the excitation transition by an intense laser and collisional quenching of the excited state makes accurate prediction of LIF a challenge. The model described below is the simplest possible LIF model. The laser beam is assumed too weak to significantly deplete the lower excitation state and spontaneous emission from

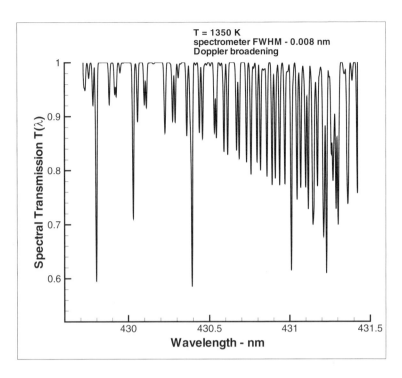

Fig. 10. A computed CH absorption spectrum.

the excited states is assumed to be the only decay process. With these assumptions the algorithm for computation of a synthetic LIF spectrum produced when the laser wavelength is scanned from $\lambda_{min}$ to $\lambda_{max}$ is much simpler.

1  Divide the spectrum into a large number of bins from $\lambda_{min}$ to $\lambda_{max}$. Each of these bins corresponds to a wavelength produced by the laser as it is scanned from $\lambda_{min}$ to $\lambda_{max}$.

2  Select spectral distribution functions for the laser beam and the absorption. In the examples shown below, it was assumed that when the tunable laser is set to the wavelength $\lambda_0$, it emits over a narrow Gaussian distribution of wavelengths $g_L(\lambda - \lambda_0)$ centered about $\lambda_0$. The wavelengths of laser photons absorbed by the initial molecular state are assumed to be controlled by Doppler broadening $g_D(\lambda - \lambda_j)$ at temperature $T$. Gauss-Hermite numerical integration was chosen for integration of the product of $g_L \, g_d$.

3 The rate of excitation from lower state $l$ to upper state $u$ is given by

$$\frac{dN_{l\to u}}{dt} = N_l B_{lu} I_L \int g_L(\lambda - \lambda_L) g_D(\lambda - \lambda_{lu}) \, d\lambda$$

$$= \frac{8\pi^3}{3h^2 g_l} I_L N_l S_{lu} \int g_L(\lambda - \lambda_L) g_D(\lambda - \lambda_{lu}) \, d\lambda \quad (149)$$

in which $N_l$ is the population density of the initial state, $B_{lu}$ is the Einstein absorption coefficient, and $g_l$ is the statistical weight of the lower state. An initial thermal distribution of lower states,

$$N_l = \frac{N_0 \, g_l}{Q} e^{-hcF_l/kT}, \quad (150)$$

is essentially unchanged by a very weak laser beam.

$$\frac{dN_{l\to u}}{dt} = \frac{8\pi^3 N_0 I_L S_{lu}}{3h^2 Q} e^{-hcF_l/kT} \int g_L(\lambda - \lambda_L) g_D(\lambda - \lambda_{lu}) \, d\lambda \quad (151)$$

4 Given the assumption that the only decay mechanism of the the excited states is spontaneous emission, the steady-state population of excited states is the solution of

$$\sum_l \frac{dN_{l\to u}}{dt} = \sum_f \frac{dN_{u\to f}}{dt}$$

$$= \sum_f A_{uf} = 1/\tau_u \quad (152)$$

where $A_{uf}$ is the Einstein emission coefficient and $\tau_u$ is the lifetime of the excited state.

$$N_u = \frac{8\pi^3 N_0 I_L \tau_u}{3h^2 Q} \sum_l S_{lu} e^{-hcF_l/kT} \int g_L(\lambda - \lambda_L) g_D(\lambda - \lambda_{lu}) \, d\lambda \quad (153)$$

5 The remaining computation is simple free spontaneous emission from $N_u$ excited states,

$$I_{uf} = \frac{64\pi^4 c}{3} \frac{A_{uf} N_u}{g_u (2J_u + 1) \lambda_{uf}^4}, \quad (154)$$

where is the $g_u$ is the electronic statistical weight of the excited state.

In summary, as the laser wavelength is scanned (*i.e.*, as one steps from one wavelength bin to the next) the excitation line strength table is scanned and the population density of excited states is computed from Eq. (153). After the full range from $\lambda_{min}$ to $\lambda_{max}$ has been covered, the the emission line strength table is scanned to compute the fluorescence intensity corresponding to each laser wavelength bin.

This LIF model is applicable only to low pressure gases probed by weak laser beams. However, in its region of applicability this model can yield useful results. Figure 11 compares CN violet system LIF and spontaneous emission spectra over a range of temperatures. The left side of the figure shows that LIF R-branch lines are more intense then the corresponding P-branch lines in agreement with the measurements shows in Fig. 3 of Sims, *et al.*[29]

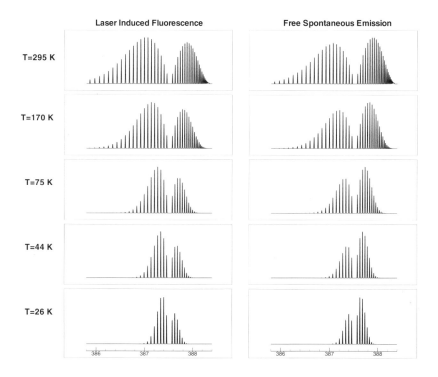

Fig. 11. A comparison of synthetic LIF spectra (on the left) and synthetic free spontaneous emission spectra. The P and R branches of these CN $B\,^2\Sigma^+ - X\,^2\Sigma^+$ (0,0) band spectra are well resolved, but the relative intensities of the two branches are different in LIF and spontaneous emission.

# References

1. R. C. Johnson (1927), "The Structure and Origin of the Swan Band Spectrum of Carbon," Philos. Trans. R. Soc. London, Ser. A, **226**, 157-230.
2. R. C. Johnson and R. K. Asundi(1929), "The Structure of the High Pressure Carbon Bands and the Swan System," Proc. R. Soc. London, Ser. A, **124**, 668-688.
3. E. U. Condon and G. H. Shortley (1935), *The Theory of Atomic Spectra*, University Press, Cambridge.
4. R. C. Hilborn (1981), "Einstein coefficients, cross sections, f values, dipole moments, and all that," Am. J. Phys. **50**, 982-986, Erratum **51**, 471 (1982), revision available at http://arxiv.org/abs/physics/0202029 (2002).
5. A. Thorne, U. Litzen, and S. Johansson (1999), *Spectrophysics: Principles and Applications*, Springer, Netherlands (1999).
6. J. M. Brown, J. T. Hougen, K.-P. Huber, J. W. C. Johns, I. Kopp, H. Lefebvre-Brion, A. J. Merer, D. A. Ramsay, J. Rostas, and R. N. Zare (1975), "The Labeling of Parity Doublet Levels in Linear Molecules," J. Mol. Spectrosc. **50**, 500 (1975).
7. J. M. Brown and A. Carrington (2003), *Rotational Spectroscopy of Diatomic Molecules*, University Press, Cambridge.
8. H. Lefebvre and R. W. Field (2004), *The Spectra and Dynamics of Diatomic Molecules*, Elsevier, Amsterdam.
9. P. F. Bernath, C. R. Brazier, T. Olsen, R. Haileu, and W. T. M. L. Fernando (1991), "Spectroscopy of the CH Free Radical," J. Mol. Spectrosc. **147**, 16-26.
10. C. R. Brazier and J. M. Brown (1984), "A measurement of the lambda-type doubling spectrum of the CH radical by microwave-optical double resonance: further characterization of the $A\,^2\Delta$ state," Can. J. Phys. **62**, 1563-1578.
11. M. E. Rose (1957), *Elementary Theory of Angular Momentum*, Wiley & Sons, New York.
12. A. R. Edmonds (1960), *Angular Momentum in Quantum Mechanics*, Princeton University Press, Princeton.
13. D. M. Brink and G. R. Satchler (1993), *Angular Momentum*, Clarendon, Oxford.
14. J. K. Cashion (1963), "Testing of Diatomic Potential-Energy Functions by Numerical Methods," J. Chem. Phys. **39**, 1872-1877.
15. J. Tellinghuisen (1987), "An Improved Method for the Direct Computation of Diatomic Centrifugal Distortion Constants," J. Mol. Spectrosc. **122**, 455-461.
16. J. Tellinghuisen (1972), "Fast, accurate RKR computations," J. Mol. Spectrosc. **44**, 194-196.
17. R. J. LeRoy (2007), "A Computer Program for Solving the Radial Schrodinger Equation for Bound and Quasibound Levels," University

of Waterloo Chemical Physics Research Report CP-663, available at http://leroy.uwaterloo.ca/programs/.

18. J. Luque and D. R. Crosley (1995), "Electronic transition moment and rotational transition probablities in CH. I. $A^2\Delta - X^2\Pi$ system," J. Chem. Phys. **104**, 2146-2156.

19. J. O. Arnold and S. R. Langhoff, "A theoretical study of the electronic transition moment for the $C_2$ Swan band system," J. Quant. Spectrosc. Radiat. Transfer **19**, 461-466.

20. C. F. Chabalowski, R. J. Buenker, S. D. Peyerimhoff (1981), "Theoretical study of the electronic transition moments for the $d^3\Pi_g - a^3\Pi_u$ (Swan) and $e^3\Pi_g - a^3\Pi_u$ (Fox-Herzberg) bands in $C_2$," Chem. Phys. Lett. **83**, 438-440.

21. D. M. Cooper (1981), "Theoretical electronic transition moments for the Ballik-Ramsay, Fox-Herzberg, and Swan systems of $C_2$," J. Quant. Spectrosc. Radiat. Transfer **26**, 113-118.

22. C. G. Parigger (2006), "Laser-induced breakdown in gases: experiments and simulation," in *Laser Induced Breakdown Spectroscopy*", eds. A.W. Miziolek, V. Palleschi, I. Schechter, (Cambridge University Press, New York), 171-193; and references therein.

23. E. E. Whiting, A. Schadee, J. B. Tatum, J. T. Hougen, and R. W. Nicholls (1980), "Recommended Conventions for Defining Transition Moments and Intensity Factors in Diatomic Molecular Spectra," J. Molec. Spectrosc. **80**, 249-256.

24. J. G. Phillips and S. P. Davis (1968), "The Berkeley Analysis of Molecular Spectra, Vol. 2, The Swan System of the $C_2$ Molecule and the Spectrum of the HgH molecule," Univ. California, Berkeley.

25. A. Tanabashi, T. Hirao, T. Amano, and P. F. Bernath (2007), "The Swan system of $C_2$: A global analysis of Fourier transform emission spectra," Astrophys. J. Suppl. Ser. **169**, 472-484.

26. C. Amiot (1983), "Fourier spectroscopy of the $^{12}C_2$, $^{13}C_2$, and $^{12}C^{13}C$ (0,0) Swan bands," Astrophys. J. Suppl. Ser. **52**, 329-340.

27. J. A. Nelder and R. Mead (1965), "A simplex method for function minimization," Computer J. **7**, 308-313.

28. L. Nemes (2008), private communication.

29. I. R. Sims, J. -L. Queffelec, A. Defrance, C. Rebrion-Rowe, D. Travers, P. Bocherel, B. R. Rowe, and I. W. M. Smith (1994), "Ultralow temperature kinetics of neutral-neutral reactions. The technique and results for the reactions $CN+O_2$ down to 13 K and $CN+NH_3$ down to 25 K," J. Chem. Phys. **100**, 4229-4241.

30. H. Ito, K. Oda, Y. Kawamura, and H. Saitoh (2006), "Measurements of density and sticking probability of $CN(X^2\Sigma^+)$ radicals by laser-induced fluorescence spectroscopy," Spectrochimica Acta A **67**, 39-47.

Chapter 5

# Optical Emission Spectroscopy of $C_2$ and $C_3$ Molecules in Laser Ablation Carbon Plasma

N. A. Savastenko[1,*] and N. V. Tarasenko[2]

[1]*INP-Greifswald, 17489 Greifswald, Felix-Hausdorff-Str. 2, Germany*
[2]*B. I. Stepanov Institute of Physics, National Academy of Sciences of Belarus, 220072 Minsk, Nezavisimosti Ave. 68, Belarus*
*E-mail address: savastenko@inp-greifswald.de*

Laser ablation of graphite is now routinely used to produce various materials, like carbon-based nanostructures and thin films. The characteristics of the synthesized nanostructures and deposited films depend crucially on the properties of the ablation plume, e.g., the abundance and velocity distribution of the various species contained within the plume and the small cluster formation process. This chapter provides an overview of some recent research concerning the formation and excitation of $C_2$ and $C_3$ molecules in plasma produced by single and double pulse laser ablation of graphite.

## 1. Introduction

Carbon-based new materials, like fullerenes, carbon nanostructures and nanotubes, and thin films of diamond-like carbon (DLC) or carbon nitride ($CN_x$) have been the focus of an increasing amount of the recent literature. The interest in research in this area is driven by many unexplored new properties of these materials and extensive technological applications. The ablation of graphite by using laser radiation is an established route to produce various carbon nanostructures [Kroto et al. 1985, Iijima et al. 1991, Puretzky et al. 2000(a)] and thin films [Chrisey,

& Hubler 1994]. The most important advantage of this technique is the ability to produce unique or metastable materials that cannot be made under thermal conditions. The unique kind of species generated in plasma controls the nucleation and growth processes of nanostructures and thin films. Therefore the characteristics of the synthesized nanostructures and deposited films depend crucially on the properties of the ablation plume, e.g., the abundance and velocity distribution of the various species contained within the plume and the small cluster formation process [Chrisey & Hubler 1994; Voevodin & Donley 1996; Riascos et al. 2004; Abdelli-Messaci et al. 2005; Cappelli et al. 2007]. It was reported that high quality DLC films are obtained at moderate laser irradiances where molecular $C_2$ formation is prominent [Dwivedi & Thareja 1995]. Nucleation and hence growth of thin films were found to depend on small carbon clusters ($C_2$ and $C_3$) behaviour [Ikegami et al. 2001]. Moreover, $C_2$ molecules are considered to be one of the main precursors of DLC films [Gong et al. 2007]. The amount of excited $C_2$ was found to correlate with the production rate of single-wall nanotubes (SWNTs) [Arepalli et al. 2000]. Therefore an improved knowledge of dynamics of the small cluster formation is important for the purpose of optimizing the conditions for depositing high quality films and growing SWNTs with high yield.

In order to understand the processes leading to cluster formation, several techniques can be applied. They include optical emission spectroscopy (OES) [Sasaki et al. 2002(a), 2002(b); Tasaka et al. 1995; Anisimov et al. 1995; Harilal 2001], optical absorption spectroscopy [Morrow et al. 1994; Puretzky et al. 2000 (a)], laser-induced fluorescence [Sasaki et al. 2002(a), 2002(b); Puretzki et al. 2000 (a, b); Ikegami et al. 2001; Goldsmith & Kearsley 1990], mass spectroscopy [Krajnovich 1995; Rohlfing 1990; Kokai et al. 1997; Koo et al. 2002; Torrisi et al. 2006] etc. Among these, time resolved OES can be considered as the most suitable method to study the cluster formation processes. The OES technique with time and space resolution can provide information about the processes involved in plume evolution. It allows to monitor the density distributions of mono-atomic, diatomic and triatomic carbon. Moreover, the method is applicable to the detection of growth process of heavier clusters by observation of continuum emission [Sasaki et al. 2002 (a)].

In this chapter, we would like to summarize our own studies concerning the formation and excitation of $C_2$ and $C_3$ molecules in plasma produced by single and double pulse laser ablation of graphite. Firstly, we shall give a brief overview of mechanisms for small cluster formation at nanosecond laser ablation. Thereafter we shall focus on researches performed by our group.

## 2. Main Sources of $C_2$ and $C_3$ Molecules in Laser Ablation Plume

In the pulsed laser ablation of a graphite target, carbon clusters, molecules, atoms and ions are formed by many concurrent and interrelated mechanisms like direct ablation, chemical reactions and non-reactive collisions. The $C_2$ and $C_3$ molecules can originate from three general mechanisms: direct laser ablation from the target, formation via dissociative or recombinative processes [Nemes et al. 2006, 2007; Harano et al. 1993; Iida & Yeung 1994; Harilal et al. 2001; Anselment et al. 1987]. The dominant route of formation is determined by the experimental parameters like laser wavelength, laser fluence, and background pressures.

Krajnovich reported that $C_2$ and $C_3$ molecules were emitted directly from the target surface [Krajnovich 1995]. He made the conclusion on the basis of the study of laser ablation at fluencies close to threshold, where collisional interactions were minimized.

The strip-off (dissociative) mechanism for the formation of carbon clusters was proposed by Harano et al. [Harano et al. 1993]. They found a discrepancy between the amount of carbon atoms involved in the particles and the highest possible amount of carbon atoms estimated by considering the laser energy used for ablation. On the basis of this discrepancy Harano et al. suggested, that the large particles were not formed from the gaseous species in the plume. They were supposed to be produced as direct strip-off fragments on the target surface.

Iida and Yeung suggested that small molecules and carbon atoms were produced in their experiments via dissociation of larger carbon particles [Iida & Yeung 1994]. The dissociative mechanism was supported by the identical dependences of vibrational temperatures on the laser power density with different pressure of background gas. Iida and Yeung explained the effect of laser power density on the size of the graphite particles in terms of dissociative model. They suggested that at low power density, large particles came out from the target surface by inter-layer dissociation. At higher power densities, the probability increased for intra-layer bond breakage. As a result, smaller clusters can be formed. Iida and Yeung supposed that molecules were formed via dissociation from the carbon particles not only in the ablation stage but also after the laser pulse was terminated. This suggestion was supported by the observed long duration of the $C_2$ band emission.

The role of a background gas is ambiguous in the cluster formation process. The increase in pressure of ambient gas resulted in creation of hotter and denser plasma. In turn, that caused the decrease in particle size by dissociation. On the other hand, hot and dense plasma is favourable for recombination.

Carbon clusters were found to form via fragmentation of larger clusters and carbon dust [Bloomfield et al. 1985; Cox et al. 1988; Kaizu et al. 1997; Moriwaki et al. 1997]. Anselment et al. reported that the dominant mechanism for production of excited $C_2$ molecules was low-energy electron collisions with carbon cluster $C_n$ followed by dissociation [Anselment et al. 1987].

Besides collision-induced dissociation, photofragmentation of clusters should be taken into account. It was shown that $C_3$ can arise through photodissociation of larger carbon cluster $C_n$ [Choi et al. 2000; Geusic et al. 1986; Cao et al. 2002; Fueno & Taniguchi 1999]:

$$C_n \rightarrow C_3 + C_{n-3}. \tag{1}$$

According to the recombinative model $C_2$ and $C_3$ molecules can be formed by three-body collisions [Mann 1978]:

$$C + C + M \rightarrow C_2 + M , \tag{2}$$

$$C_2 + C + M \rightarrow C_3 + M . \tag{3}$$

The third body (M) is responsible for removing excess energy into translation degrees of freedom.

The appearance of high pressure (HP) bands in $C_2$ spectrum can be considered as evidence supporting recombinative mechanism for $C_2$ formation [Little & Browne 1987]. Little & Browne showed that the population of the upper level of the high pressure bands $(d^3\Pi_g, v=6)$ occurred via the following sequence of reactions:

$$C + C + M \rightarrow C_2(^5\Pi_g, v) + M ,\tag{4}$$

$$C(^5\Pi_g, v) + M \rightarrow C_2(^5\Pi_g, v=0) + M ,\tag{5}$$

$$C(^5\Pi_g, v=0) + M \rightarrow C_2(d^3\Pi_g, v=6) + M .\tag{6}$$

The first step is the formation of metastable $^5\Pi_g$ state (4). Equation (5) reflects the vibrational and rotational relaxation. The population of the $d^3\Pi_g(v=6)$ state occurs via the potential curve crossing (6).

An alternative pathway of carbon dimer and trimer formation without collisional cooling was suggested by Monchicourt [Monchicourt 1991]. He proposed that excited carbon dimers ($C_2^*$) were produced by binary associative collisions between "fast" carbon atoms ($\tilde{C}$) and "slow" carbon atoms ($C$):

$$C + \tilde{C} \rightarrow C_2^*.\tag{7}$$

The excited carbon trimers $C_3^*$ were most probably formed via binary collisions of "fast" carbon atoms ($\tilde{C}$) with "slow" carbon dimers ($C_2^*$):

$$C_2^* + \tilde{C} \rightarrow C_3^*.\tag{8}$$

The recombinational (nucleation) mechanism lies in the basis of the models for the formation of the large carbon clusters [Kroto et al. 1985; Kroto et al. 1991; Iijima 1993]. Moreover, the gas-phase reaction between $C_2$ and $C_3$ molecules can be considered to be an initial step in carbon particle formation [Mann 1978].

## 3.  OES of $C_2$ and $C_3$ Molecules in Single Pulse Laser Ablation Plasma

### 3.1. *Dependence of plume characteristics on ablation parameters*

The spectroscopic studies of carbon plasma have been carried out using a variety of irradiation conditions (laser wavelengths, laser pulse duration, and laser energy density) and the ambient gas nature and pressure. An ArF laser was applied for laser ablation of graphite target by Claeyssens et al. [Claeyssens et al. 2003]. Optical emission spectroscopy diagnostic of carbon plasma created by an KrF excimer laser (248 nm) pulses was performed in [Abdelli-Messaci et al. 2005; Chen et al. 1991; Claeyssens et al. 2002; Shinozaki et al. 2002; Radhakrishnan et al. 2007;]. A XeCl$_2$ excimer laser was employed for ablation in [Acquaviva & De Giorgi 2002(a), 2002(b); Burakov et al. 2001; Savastenko 1998]. Laser ablation of graphite was carried out using 1064 nm and/or 532 nm radiation from Nd:YAG laser in [Abdelli-Messaci et al. 2002; Burakov et al. 2001, 2002; Harilal et al. 1996, 1997; Harilal 2001; Gong et al. 2007; Iida & Yeung 1994; Ikegami et al. 2004; Keszler & Nemes 2004; Monchicourt 1991; Nemes et al. 2005, 2006; Park et al. 2005; Riascos et al. 2004; Shinozaki et al. 2002; Torrisi et al. 2006; Varga & Nemes 1999].

The difference between the behaviour of plasma species, induced by laser beams with different wavelength, is due to the process by which the ablated species are produced and that by which the plasma interacts with the laser beam. The use of ultraviolet (UV) laser beams is more effective for penetration into the sample [Amiranoff et al. 1979; Geertsen et al. 1994]. On the other hand, the main advantage in the use of near infrared (NIR) radiation with low energy photons is that they are less likely to invoke photochemistry in the ablation phenomenon. The ablation threshold decreases at shorter wavelengths (UV) and increases at larger wavelengths (NIR). At UV wavelength, smaller carbon clusters are produced due to the photo- fragmentation of carbon clusters [Murray & Peeler 1994; Krajnovich 1995; Gaumet et al. 1993], while IR lasers with low energy photons produce larger clusters $C_n^+$ (n $\geq$ 3–24) and $C_2$, $C_3$ species [Torrisi et al. 2006; Cappelli et al. 2007; Gaumet et al. 1993]. At low photon energy, the species are characterized by low kinetic energy and low chance for photo-fragmentation.

Generally, decreasing the laser pulse duration leads to the increase of the mean thermal velocity and the plasma temperature [Claeyssens et al. 2002]. Highly charged ions are generated in the course of picosecond and femtosecond ablation. The differences in plume behavior due to different laser pulses may be attributable in part to the different time of plasma-laser interaction [Claeyssens et al. 2002]. In the case of nanosecond ablation, the plume characteristics depend on a number of factors, such as laser fluence and wavelength, and the background pressure. In contrast, some plasma parameters appear to be independent of the laser wavelength and the precise pulse duration, when ablating in vacuum with short laser pulse duration (< 100 ps) [Claeyssens et al. 2002].

At low laser fluence, characteristics of the species in the carbon plasma should be no different than in experiments on thermal graphite sublimation. At higher laser intensities, non-thermal effects start to dominate [Krajnovich 1995].

The environment in which ablation is carried out is one of the controlling parameters determining the cluster formation process [Miller 1994; Sasaki et al. 2002(b)]. The ambient atmosphere brings changes in plume expansion dynamic. In a background gas, the movement of the plasma is restricted. With an increase in the background gas pressure, it takes longer time for the plasma to expand. But even under the same pressure conditions, velocities of molecular species differ when plume propagates in Ar and He background gases [Park et al. 2005]. It is attributed to the differences in momentum-transfer and hydrodynamical effects. A buffer gas not only confines the plasma plume but also promotes the gas-phase reactions between target atoms. Since formation of larger clusters involves fusion of two smaller species, it requires third body collisions. The buffer gas provides the third body collisions required for stabilization.

## 3.2. *Temporal and spatial profiles of $C_2$ and $C_3$ molecules in laser produced plasma*

Typically, in order to identify the carbon plasma species, the time integrated plasma emission spectra are recorded and analyzed [Sasaki et al. 2002(a); Abdelli-Messaci et al. 2005; Burakov et al. 2001; Gong et al. 2007; Harilal et al. 1997; Ikegami et al. 2004; Keszler & Nemes 2004; Nemes et al. 2005, 2006; Savastenko 1998; Park et al. 2005; Riascos et al. 2004; Shinozaki et al. 2002; Varga & Nemes 1999]. Time-resolved OES is usually used to track the evolution of species in plasma [Burakov et al. 2002; Chen et al. 1991; Harilal et al. 1997, 2001; Iida & Yeung 1994; Monchicourt 1991; Park et al. 2005; Radhakrishnan et al. 2007].

The typical setup for spectroscopic studies of laser ablation plasma consists of a stainless steel vacuum chamber, a laser, and a photomultiplier tube coupled with a monochromator or a spectrograph equipped with a CCD camera. The chamber is equipped with several side arms, some of which are sealed by quartz windows. A target is mounted inside the chamber in such a way that the target surface could be irradiated with a laser beam. The target can be positioned at different angles relative to the direction of the incident laser. The emission spectrum from plasma is usually viewed normal to its expansion direction and imaged using appropriate focusing lenses onto the slit of a monochromator or spectrograph. For spatially resolved studies, different regions of the plasma plume are focused onto the monochromator/ spectrograph slit.

When taken in the neighbourhood of the target and immediately after the irradiation of the target, the spectra comprise overlapping continuum emission and the emission of exited species [Abdelli-Messaci et al. 2005; Gong et al. 2007]. The continuum emission can be due to Bremsstrahlung radiation and radiative recombination or to incandescence of hot carbon particles [Rolfing et al. 1984; Harilal et al. 1997]. The analyses of the emission spectra show the presence of C [Abdelli-Messaci et al. 2005; Burakov et al. 2001; Gong et al. 2007; Sasaki et al. 2002; Savastenko 1998], $C^+$ [Abdelli-Messaci et al. 2005; Burakov et al. 2001; Gong et al. 2007; Riascos et al. 2004; Sasaki et al. 2002; Savastenko 1998], $C_2$ [Abdelli-Messaci et al. 2005; Burakov et al.

2001; Sasaki et al. 2002; Savastenko 1998; Radhakrishnan et al. 2007; Riascos et al. 2004] and $C_3$ [Burakov et al. 2001; Radhakrishnan et al. 2007; Savastenko 1998] species. The occurrence of ionized species depends on the laser fluence. The emission from ionized mono-atomic carbon was observed at laser fluences above a threshold of 2.4 J cm$^{-2}$ [Sasaki et al. 2002]. Abdelli-Messaki et al. reported emission form $N^+$ and $N_2^+$ species in plasma generated in nitrogen [Abdelli-Messaci et al. 2005]. The emission from $N_2^+$ appeared at 1 J cm$^{-2}$, whereas NII emission appeared at 10 J cm$^{-2}$.

Table 1 presents the peak wavelengths and their assignments for the main emission lines and bands recorded under the following conditions: laser fluence, 7 J cm$^{-2}$; pressure, 100 Torr [Savastenko 1998; Burakov et al. 2001]. The ablation was performed in helium using a XeCl eximer

Table 1. Main emission lines and bands recorded during graphite ablation for 100 Torr He pressure.

| Molecular species | Wavelength (nm) | Vibrational band | Electronic transition |
|---|---|---|---|
| $C_2$ | 612.24 | 1-3 | $d^3\Pi_g - a^3\Pi_u$ |
| Swan system | 605.84 | 2-4 | |
| | 600.00 | 3-5 | |
| | 594.80 | 4-6 | |
| | 558.55 | 1-2 | |
| | 554.07 | 2-3 | |
| | 550.19 | 3-4 | |
| | 547.03 | 4-5 | |
| | 516.52 | 0-0 | |
| | 512.93 | 1-1 | |
| | 509.77 | 2-2 | |
| | 473.71 | 1-0 | |
| | 471.52 | 2-1 | |
| | 469.76 | 3-2 | |
| | 468.48 | 4-3 | |
| | 467.86 | 5-4 | |
| | 466.87 | 6-5 | |
| | 438.20 | 2-0 | |
| | 437.10 | 3-1 | |
| | 436.50 | 4-2 | |
| | 589.9 | 6-8 | |
| | 543.4 | 6-7 | |

Table 1. (Continued)

| $C_2$ | 385.22 | 0-0 | $C^1\Pi_g$ - $A^1\Pi_u$ |
|---|---|---|---|
| Deslandres– | 360.73 | 1-0 | |
| D'Azembuja | 359.29 | 2-1 | |
| system | 358.76 | 3-2 | |
| $C_3$ | Broad emission | 000-000 | $A^1\Pi_u$ - $X^1\Sigma^+_g$ |
| Comet Head | band with a | | |
| Sysytem | maximum at | | |
| | around | | |
| | 400 nm | | |

| Atomic and ionic species | | Transition | |
|---|---|---|---|
| CI | 247.86 | $3s^1P^0_1 \rightarrow 2p^{2\,1}S_0$ | |
| CII | 250.91 | $3p^{3\,2}D^0_{3/2} \rightarrow 2p^{2\,2}P_{1/2}$ | |
| | 251.21 | $3p^{3\,2}D^0_{3/2} \rightarrow 2p^{2\,2}P_{1/2}$ | |
| | 283.7 | $3p^{\,2}P^0_{3/2} \rightarrow 2p^{2\,2}S_{1/2}$ | |
| | 392.1 | $4s^{\,2}S_{1/2} \rightarrow 3p^{\,2}P^0_{3/2}$ | |
| | 396.8 | $4f^{1\,4}D_{1/2} \rightarrow 3d^{1\,4}D^0_{1/2}$ | |
| | 513.3 | $3p^{1\,4}P_{3/2} \rightarrow 3s^{1\,4}P^0_{1/2}$ | |

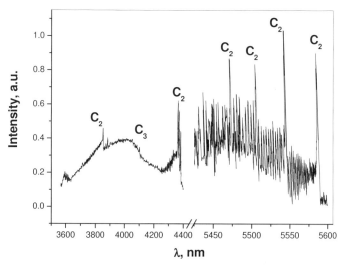

Fig. 1. Typical optical emission spectrum of the plume of graphite ablation in helium atmosphere at 100 Torr and 7.0 J cm$^{-2}$.

laser. Figure 1 represents the fragments of the time integrated spectrum recorded under the same conditions. This spectrum was taken at the proximity of the target (0.2 cm from the target surface).

Under these conditions, the spectrum was dominated by $C_2$ emission of the Swan system ($d^3\Pi_g$ - $a^3\Pi_u$) not only near the target, but also at the distance of 1.0 cm form the target,

The so-called high-pressure (HP) bands were also recorded including the (6–7) band at 543.4 nm and (6–8) band at 589.9 nm. The high-pressure bands are known to be part of the Swan system with $v'$ = 6 levels excited at relatively high pressures [Pearse & Gaydon 1976]. The appearance of the HP bands implies the formation of $C_2$ from the recombination of carbon atoms [Little & Browne 1987].

The emission of the $C_2$ Deslandres–D'Azembuja system was also observed. Both Swan and Deslandres–D'Azambuja bands were found to decrease in intensities with increasing distance from the target surface [Savastenko 1998]. On the other hand, Gong et al. observed that the emission of the $C_2$ Swan bands decreased and the $C_2$ Deslandres–D'Azambuja bands increased with increasing distance from the target surface [Gong et al. 2007]. Acquaviva and Giorgi, in contrast, reported that the Swan bands were predominant in the outer region of plume, whereas the $C_2$ Deslandres–D'Azambuja bands were more intensive near the target surface [Acquaviva & Giorgi 2002(b)]. No $C_2$ Deslandres–D'Azambuja bands were observed by Keszler and Nemes [Keszler & Nemes 2004]. Acquaviva and De Giorgi explained the spatial behavior of $C_2$ emissions by the different efficiencies in populating the electronic states [Acquaviva & Giorgi 2002(b)]. The electronic energy of the $d^3\Pi_g$ state (Swan system) is 2.48 eV, whereas that of $c^1\Pi_g$ state is 4.24 eV. At higher temperature, more levels can be populated via collisions. Thus, the efficiency in populating higher energy levels is higher near the target. This explanation can be also applicable to the results reported in [Savastenko 1998]. As the plasma temperature is lower in the outer region of the plasma, the efficiency in populating all levels by collisions decreases for greater distances. On the other hand, $C_2$ molecule in excited states can be formed as a result of atomic carbon recombination. The recombination processes started when the plasma temperature began to fall. In this case, an increase can be expected in the concentration of electronic excited molecules in outer region of the plume.

In some cases, the Mulliken band ($D^1\sum_u$ - $X_1\sum_g^+$) can be observed in a carbon plume [Dreyfus et al. 1987; Rohlfing 1988; Shinozaki et al. 2002].

A broad emission in the 360–440 nm region is attributable to electronically excited $C_3$ molecules. Though the spectroscopy of $C_3$ has a very long history, the detailed analysis of vibronic structure of its spectrum can be difficult. The main reasons for this complexity are the unusually large Renner-Teller effect and a very low ground state bending frequency. The $A^1\Pi_u$ - $X^1\Sigma^+_g$ emission spectrum of $C_3$ near 405 nm is often referred as a "quasi-continuum" or "pseudo-continuum" [Luque et al. 1997; Monchicourt 1991]. The literature on spectroscopy of $C_3$ molecules is provided in two excellent review papers by Van Orden and Saykally, and Weltner and van Zee as well as in some more recent works by Nemes [Weltner & van Zee 1989, Van Orden & Saykally, Nemes et al. 2006, 2007]. It should be stressed here that, though the emission around 400 nm was observed in carbon plasma by several authors, it was not always assigned to the $C_3$ molecules. Nemes et al. provided a detailed analysis of the possible origins of the continuum emission around 400 nm [Nemes et al. 2006].

The emission at around 400 nm was attributed to $C_3$ molecules by Shinozaki et al., Puretzky et al. and Monchicourt [Shinozaki et al. 2002; Monchicourt 1991; Puretzky et al. 2000(a)]. Anselment et al. observed a structureless continuous background in the 350–650 nm region with a peak at 400 nm [Anselment et al. 1987]. However, they did not ascribe it to any species. Rohlfing observed a continuum emission in the range from 300 to 800 nm [Rohlfing 1988]. He postulated that this emission was due to blackbody emission from hot carbon particles. Rohlfing also noted that blackbody model was unable to reproduce the magnitude and narrowness of the continuum maximum at 400 nm. The emission near 400 nm was observed by Arepalli et al. [Arepalli et al. 2000]. They reported that the emission with a maximum at about 400 nm may be due to radiation from $C_3$ ($A^1\Pi_u$ - $X^1\Sigma^+_g$), $C_2$ ($C^1\Pi_g$ - $A^1\Pi_u$) or to particulate continua [Arepalli & Scott 1999; Arepalli et al. 2000].

The emission presented in Fig. 1 is qualitatively similar to those observed by Shinozaki et al., Puretzky et al. and Monchicourt [Shinozaki et al. 2002; Monchicourt 1991; Puretzky et al. 2000(a)]. The red blackbody emission from large particles was not observed under conditions of Fig. 1. Thus, emissions near the 400 nm may be ascribed to electronically excited $C_3$ molecules rather than blackbody emission.

Not only $C_2$ emission intensity presented a maximum at the neighbourhood of the target but also intensity of the $C_3$ emission decreased with the distance from the target surface. Table 2 presents the changes in $C_2$ and $C_3$ normalized emission intensities with the distance from the target surface. The data were obtained by recording the emission at $\lambda = 436.8$ and 404.98 nm. The ablation was performed in He atmosphere at laser fluence of 7.0 J cm$^{-2}$ and pressure ranged from 0.1 Torr to 400 Torr [Savastenko 1998].

As it can be seen from the Table 2, the lower the pressure, the faster the intensities decreased with a distance. $C_2$ ($C_3$) emission intensity decreased by a factor of 1.3 (1.7) under pressure of 400 Torr, whereas it decreased by a factor 3.4 (6.2) in plasma generated under pressure of 0.1 Torr. Such behaviour of $C_2$ and $C_3$ species may be explained by the plasma confinement effect. At lower gas pressure, the effect of plasma confinement is smaller.

Table 2. Intensity variation with distance from the target surface (l) for 436.8 nm and 404.98 nm spectral emission from $C_2$ and $C_3$, respectively.

| $p$, Torr | $C_2$ | | | $C_3$ | | |
|---|---|---|---|---|---|---|
| | Intensity $(I_1)$, a.u., $l_1 = 0.2$ cm | Intensity $(I_2)$, a.u., $l_2 = 1.0$ cm | $I_1/I_2$ | Intensity $(I_1)$, a.u., $l = 0.2$ cm | Intensity $(I_2)$, a.u., $l = 1.0$ cm | $I_1/I_2$ |
| 400 | 1.00 | 0.77 | 1.3 | 1.00 | 0.60 | 1.7 |
| 100 | 0.74 | 0.41 | 1.8 | 0.41 | 0.19 | 2.2 |
| 0.1 | 0.67 | 0.20 | 3.4 | 0.56 | 0.09 | 6.2 |

As it is shown in Table 2, the emission of $C_2$ and $C_3$ molecules was clearly observed within the 10 mm from the target. Abdelli-Messaci et al. observed no $C_2$ emission beyond a distance of 7 mm from the target surface [Abdelli-Messaci et al. 2005]. Differences between data in Table

2 and those reported by Abdelli-Messaci et al. may be explicable in terms of differences in the ablation conditions. In the latter case, the experiments were performed in nitrogen atmosphere. Therefore the reaction between $C_2$ and $N_2$ or (C and N) affected $C_2$ spatial distribution. Besides, $C_2$ species can be dissociated under conditions reported by Abdelli-Messaci et al. [Abdelli-Messaci et al. 2005].

$C_2$ and $C_3$ species can arise in the plume as a result of collisional processes. The addition of helium increases the probability of collisions. On the other hand, a background gas confines the plasma. Hence, it is expected that the intensity of $C_2$ and $C_3$ emission becomes weaker with decreasing gas pressure (see Table 2). A similar observation was made by Iida and Yeung [Iida & Yeung 1994]. On the other hand, the molecular emission intensity observed in ambient He gas was higher than that in vacuum at the same laser fluence [Sasaki et al. 2002]. Claeyssens et al. observed the weak emission of $C_2$ bands very close to the target at pressure up to $5 \ 10^{-7}$ Torr [Claeyssens et al. 2002]. They also reported broad emission in the range 360–450 nm, which can be attributed to $C_3$ molecules.

The temporal distributions of emitting species represent a convolution of different factors that govern the temporal history of the molecules e.g. their formation mechanism, velocity and radiative and/or collisional decay rates. Thus, study the temporal profiles of emission from different species allows us to follow their respective production and destruction processes.

Typical temporal profiles of $C_2$ species in electronic excited state ($d^3\Pi_g$) are given in Fig. 2. The distributions were obtained by monitoring the spectral emission from $C_2$ (at $\lambda = 558.55$ nm and $\lambda = 543.4$ nm corresponding to the (1 2) and (6 7) transitions $d^3\Pi_g$ - $a^3\Pi_u$ of $C_2$ Swan system) recorded for different irradiation conditions. Laser ablation was performed using a Nd:YAG laser with emission at the fundamental and second harmonic lines (1064 nm and 532 nm). The spectra were recorded at a distance 1 mm from the target surface. The experiments are described in detail in [Burakov et al. 2002]. It is clearly seen in Fig. 2 that the (1 2) and (6 7) Swan band emission intensities exhibited similar temporal variation. Emission from $C_2$ excited species reached its maximum 300 ns after the incidence of laser pulse, and lasted a few µs.

In plasma induced by 1064 nm irradiation, the temporal profile of the emission of $C_2$ molecules was characterized by a spike emission, followed by a broad emission (Fig. 2a). In contrast, only the broad peak appeared in plume produced by 532 nm (Fig. 2b). This difference can be attributed to different laser fluences in these two cases.

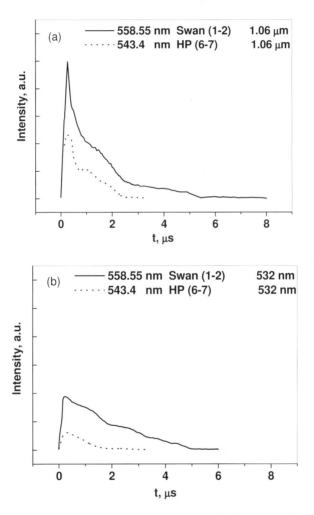

Fig. 2. Intensity variation with time for 558.55 nm and 543.4 nm spectral emission from $C_2$ observed at 1 mm away from the target for different irradiation conditions: 1064 nm, 6 J cm$^{-2}$ (a) and 532 nm, 3 J cm$^{-2}$ (b). The pressure was kept at 600 Torr.

Harilal et al. observed similar dependence of emission temporal profiles on the laser fluence [Harilal 2001; Harilal et al. 1996]. They reported the appearance of twin peak structure in the temporal pattern for some threshold laser intensity. Authors attributed this twin peak distribution to the formation of fast and slow components in plasma above threshold laser irradiance. The fast component was considered to appear due to recombination processes. The slow component was supposed to result from the dissociation of larger carbon clusters.

Figure 3 presents the temporal distributions of $C_3$ molecules in electronic excited state $(A^1\Pi_u)$. The distributions were obtained by monitoring the spectral emission from $C_3$ (at $\lambda = 404.98$ nm recorded for different irradiation conditions: 1064 nm, 6 J cm$^{-2}$ (1) and 532 nm, 3 J cm$^{-2}$ (2). The experimental conditions were the same as those in Fig. 2.

As is evident from comparison of Figs. 1 and 2, the emission lines from different species possess similar temporal profiles. Both $C_3$ and $C_2$ emission were characterized by narrow peak followed by a plateau at higher laser fluence and by a broad emission at lower laser fluence. This fact implies that $C_3$ molecules were likely formed via recombination reaction (8).

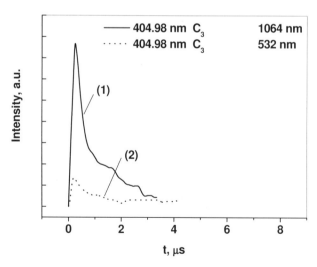

Fig. 3. Intensity variation with time for 404.9 nm spectral emission from $C_3$ observed at 1 mm away from the target for different irradiation conditions: 1064 nm, 6 J cm$^{-2}$ (1) and 532 nm, 3 J cm$^{-2}$ (2). The pressure was kept at 600 Torr.

The study of temporal profiles can suggest a scenario for the evolution of the species in the plume. Though the natural excited-states lifetimes of molecules are of a few tens of nanosecond, the electronically excited $C_2$ and $C_3$ molecules are observable for several hundred microseconds. This can be considered as evidence that excited species are formed via gas-phase reactions.

### 3.3. *Temperature determination*

Emission spectroscopy is often used to determine the molecular vibrational and rotational temperatures. Plasma expansion into a vacuum is accompanied by the non-equilibrium distributions of vibrational and rotational temperature. By expansion in a relatively high pressure ambient, a thermal equilibrium can be established between the vibrational and rotational level populations [Radhakrishnan et al. 2007].

The vibrational and rotational temperatures of electronically excited $C_2$ molecules are usually obtained by simulation of the experimental spectra [Park et al. 2005]. The vibrational temperature can be also determined from the intensity ratio of the band heads of the $C_2$ Swan system [Iida & Yeung 1994; Chen et al. 1991]:

$$T_{vib} = \frac{hc}{k} \frac{G(v')-G(w')}{\ln\left( \dfrac{I_{v'v''}}{I_{w'w''}} \dfrac{\lambda_{v'v''}}{\lambda_{w'w''}} \dfrac{A_{w'w''}}{A_{v'v''}} \right)} . \tag{9}$$

Here $\dfrac{I_{v'v''}}{I_{w'w''}}$ is the intensity ratio of two transitions $(v'v'')$ and $(w'w'')$, $\lambda_{v'v''}$ and $\lambda_{w'w''}$ are the wavelengths corresponding to the transitions $(v'v'')$ and $(w'w'')$, $A_{v'v''}$ and $A_{w'w''}$ are the Einstein spontaneous emission coefficients, $G(v')$ and $G(w')$ are the term values of the vibrational levels $v'$ an $w'$, $k$ is the Boltzmann constant and $h$ is the Planck constant.

Table 3. The vibrational temperatures calculated from $C_2$ Swan bands at different distances from the target surface (l) in plasma produced under different pressure (p) of helium.

| $p$, Torr | $T_{vib}$, K | |
|---|---|---|
| | $l_1 = 0.2$ cm | $l_2 = 1.0$ cm |
| 400 | 16300 | 12200 |
| 100 | 18100 | 14500 |
| 0.1 | 19500 | 17800 |

Table 3 presents the spatially resolved measurements of vibrational temperatures of $C_2$ molecules in plasma generated in He atmosphere at different pressure. The laser fluence was 7.0 J cm$^{-2}$ [Burakov et al. 2001; Savastenko 1998].

The vibrational temperatures of $C_2$ molecules were much higher than the graphite melting point. This suggests that molecules were most likely formed in the laser plume after ablation stage. If $C_2$ were ablated directly from the surface, $C_2$ molecules would exhibit a temperature of target surface. The high values of vibrational temperatures imply that the most likely mechanism for the $C_2$ species formation is the recombination of carbon atoms rather than decomposition of ablated particles. Additional support for this mechanism is provided by appearance of high-pressure band in emission spectra of $C_2$.

As it can be seen from Table 3, the increase of helium pressure decreased the vibrational temperature and enhanced the $C_2$ emission (compare Tables 2 and 3). This indicates that addition of helium is favoured for cooling the molecular species and increasing the recombination rate. Similar results were obtained by Harilal et al. [Harilal et al. 1997].

Park et al. performed time-resolved measurements of vibrational and rotational temperatures of $C_2$ molecules as a function of pressure in different background gases [Park et al. 2005]. At low background gas pressure (0.1 Torr), no gas dependence was found on vibrational and rotational temperatures. At pressure above 0.5 Torr, however, the temperatures were gas dependent. The temperatures were found to increase with an increase of molecular weight of background gas for a pressure range 0.5–10 Torr. The highest vibrational and rotational temperatures were found to be 15000 K and approximately 9000 K at 200 ns after the laser ablation, respectively. The temperatures decreased rapidly within 800 ns. The rotational and vibrational temperatures remained constant (~5000 K) for nearly 20 µs, when ablation was performed in $O_2$ [Radhakrishnan et al. 2007]. The authors explained this result concerning the influence of exothermic chemical reactions between carbon species and oxygen on the vibrational-rotational distribution of $C_2$.

In contrast to data, reported by Park et al., Ikegami et al. observed that the vibrational temperature of $C_2$ was higher in nitrogen gas than in argon atmosphere [Ikegami et al. 2001]. Though in both cases the experiments were carried out at similar laser fluences of 2.8 and 2.6 J cm$^{-2}$, the pressures differed significantly (0.1–10 Torr and 600 Torr). Moreover, the data were obtained for different distances from the target surface. Thus, the differences in vibrational temperatures reported by the two research groups result from the fact that excited state $C_2$ formation is controlled by many concurrent and interrelated processes like collisional excitation and ionisation, chemical reactions, and dissociation of larger clusters. These processes are strongly dependent on ablation conditions and may vary greatly with the spatial separation from the target [Harilal et al. 1997].

The variation in the vibrational temperature of $C_2$ species was also studied as function of the laser irradiance. The vibrational temperature was found to increase with increasing laser irradiance until a certain value and then saturated. Harilal et al. explained the saturation effect by excitation to higher level followed by dissociation and ionisation. The other reasons can be quenching and plasma shielding [Harilal et al. 1997; Iida & Yeung 1994].

## 4. OES of $C_2$ and $C_3$ Molecules in Double Pulse Laser Ablation Plasma

### 4.1. *Double pulse ablation plasma*

The experiments on double pulse laser ablation plasma were first reported by Piepmeiier and Malmstadt in 1969 [Piepmeier & Malmstadt

1969] Since that time many efforts have been devoted to investigate the processes involved in generation and evolution of such a plasma [Scott & Strasheim 1970; Petukh, et al. 2000; Babushok et al. 2006; De Giacomo et al. 2007].

Experiments on double pulse ablation can be performed in two configurations: by focusing the two laser beams onto a target (collinear beam laser ablation) and onto two perpendicular targets (cross-beam laser ablation). Several authors used cross-beam configuration without focussing the second laser beam onto another target [Uebbing et al. 1991; Sdorra et al. 1992; Shinozaki et al. 2002; Stratis et al. 2000]. The interaction of the two plasmas or second laser beam and plasma produces different effects such as changes in the emission intensity [Piepmeier & Malmstadt 1969; Scott & Strasheim 1970; Corsi et al. 2004; St-Onge et al. 1998; Cristoforetti et al. 2006] and kinetic energy of the species [Capms et al. 2002; Aké et al. 2004]. The double pulsed approach can lead under certain conditions to the increase in the degree of ionization and plasma temperature [St-Onge et al. 1998, 2002; Colao et al. 2002; Shinozaki et al. 2002; Stratis et al. 2000; Burakov et al. 2003].

St-Onge et al. explained the temperature and emission intensity enhancement by that fact that significant amount of laser energy is absorbed by ablated matter rather than at the target surface in dual-laser plasma [St-Onge et al. 2002]. Recent studies also showed that effects in dual-pulse ablation plasma are not simply the result of added energy from another laser pulse [Mao et al. 2001; Sanginés et al. 2007; Windom et al. 2006]. In dual-pulse system, complex physical phenomena occur such as enhanced ablation and shock-wave induced plasma rarefaction (density reduction). The reduced density behind the shock wave of the first plasma enhances the second laser's interaction with the sample. That leads to a greater quantity of energy coupled into the resulting dual-pulse plasma. By coupling more energy into the resulting dual-pulse plasma, the temporal temperature profile of the plasma is altered, thereby altering the efficiency of collisional processes.

The double pulsed approach is often used in the laser induced breakdown spectroscopy (LIBS) technique, as it allows increasing its sensitivity through a better coupling of laser energy to the ablated matter [Piepmeier & Malmstadt 1969; Scott & Strasheim 1970; Corsi et al.

2004; St-Onge et al. 1998, 2002; Colao et al. 2006; Noll et al. 2001, 2004; Burakov et al. 2008]. Ablation with two lasers is also applied for thin film deposition and nanoparticles formation [Witanachchi et al. 1995; Tselev et al. 1999; Lambert et al. 1999; Camps et al. 2002; Lillich et al. 1995; Burakov et al. 2005]. A double-pulse approach based on collinear beams was also used to produce single-wall carbon nanotubes (SWNTs) [Guo et al. 1995; Arepalli &. Scott 1999; Arepalli et al. 2000].

As the formation of thin films and nanostructures are supposed to depend on abundance of small clusters in plasma, the study of evolution of $C_2$ and $C_3$ species is of great interest. In double-pulse configuration, the production rate of SWNTs was much greater than the sum of the single-pulse production rates [Guo et al. 1995; Arepalli &. Scott 1999]. The same kind of synergism was found in emission intensity of $C_2$ and $C_3$ molecules [Arepalli &. Scott 1999; Burakov et al. 2002].

## 4.2. Time resolved emission from $C_2$ and $C_3$ molecules in double pulse ablated plasma

In order to compare the behaviour of $C_2$ and $C_3$ species in a single-laser ablated plume with that in dual-pulse ablated plasma, time-resolved optical emission spectroscopy was performed by Burakov et al. [Burakov et al. 2002]. The collinear beam configuration was used. In this configuration, a laser (1064 nm, Laser 1) was focused onto a target generating plasma, subsequently, an additional laser (532 nm, Laser 2) produced a second plasma from the same target. The delay between lasers was varied from 0 to 700 µs. As an example, the time evolution of the excited $C_2$ and $C_3$ species are presented in Fig. 4, as acquired using the dual-pulse configuration with laser-laser delay time of 1 µs , and for comparison, using the single-pulse (Laser 1 only or Laser 2 only) configuration. Several features were observed. First, there was a clear increment of the emission intensity form $C_2$ and $C_3$ in the double-pulse beam case. It is worth to notice that the effect is more than just additive. This synergy effect is due to the second plasma expanding through the remnant particle region of the first plasma. The particles of the first plasma cause collisions that result in increasing of $C_2$ and $C_3$ formation rate. Besides, the collisions may excite the non-emitting molecules. Another interesting effect is the narrowing of the temporal profiles of excited molecules in the dual-pulse plasma.

Since the phenomena in dual-pulse plasma are due to the interaction between the two non-stationary plasmas, it may be expected that they are dependent on laser-laser delay time.

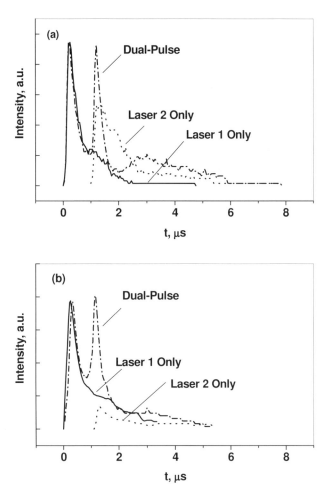

Fig. 4. Intensity variation with time for 473.7 nm spectral emission from $C_2$ (a) and 404.9 nm spectral emission from $C_3$ (b) observed for two different single pulses (Laser 1 Only or Laser 2 Only) and dual-pulse configuration with a fixed laser-laser delay $\tau = 1$ μs.

### 4.3. *Influence of pulse separation on emission intensity enhancement*

As discussed above, the dual-pulse emission enhancement can be explained in the context of the efficiency of collisional processes. It, in turn, depends at least on two parameters: the particle density and the expansion length of the first plasma. Both are directly related to the time delay between lasers. Based on the literature data, it seems reasonable to expect that the optimal pulse separation should be different for different plasma species [St-Onge et al. 2002; Burakov et al. 2008]. The dual-pulse enhancement in molecular emission intensities is shown vs. laser-laser delay time in Fig. 5.

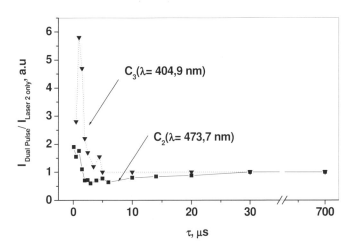

Fig. 5. Ratio of peak intensities of emission from $C_2$ and $C_3$ acquired for single pulse (Laser 2 Only) and dual-pulse configurations vs. laser-laser delay time.

The enhancement is expressed as a ratio of maximal intensity in dual-pulse configuration and Laser 2 only. For the relatively short laser–laser delay times ($\tau < 2$ µs), the enhancement in emission intensity of $C_2$ was observed ($I_{\text{dual-Pulse}}/I_{\text{Laser 2 Only}} > 1$). The maximum enhancement occurred when the first and second plasmas were generated simultaneously and corresponded to an approximately 2 fold increase in $C_2$ emission intensity enhancement. At longer delay times ($\tau > 30$ µs), no effect of first plasma on the second one was observed. At this time, the first plasma has undergone substantial decay, and the dual-pulse plasma is once again approaching conditions corresponding to a single-pulse environment. For the delay time ranging from 2 μs to 30 μs), the dual-peak configuration yielded a decrease in $C_2$ emission intensity as compared to single pulse experiments ($I_{\text{dual-Pulse}}/I_{\text{Laser 2 Only}} < 1$). This result appears at first glance to contrast with much of the literature data on dual-pulse ablation, which shows emission enhancement with dual-pulse configuration. To explain decrease in emission, it should be taken into account that the first plasma influences the molecular emission intensity in the second plasma indirectly. The exact mechanism of intensity enhancement is complex due to the combinations of changes in excitation and dissociation of plasma species as well as in dynamics of plasma. The first laser pulse generates a chain of complicated processes in the second plume, which can change the density of reagents in one of the reactions (1)–(8). It, in turn, can lead to increase in density of products ($C_2$ or $C_3$ species). It should be stressed here that decrease in $C_2$ emission enhancement coincided with the enhancement of $C_3$ emission intensity. A comparison of the dual-pulse to single-pulse $C_3$ emission intensity showed an increase by factor up to 6.0 with the dual-pulse scheme. At laser-laser delay time longer than 5 μs, no enhancement in $C_3$ emission intensity occurred, while decrease in $C_2$ intensity was still observed. It implies that the formation of carbon clusters $C_n$, $n > 3$ was more favourable at this stage of plasma decay. For laser-laser delay time larger that 30 μs, the density of the remnant particles in the first plasma was not sufficient to affect the processes in the second plume.

### 4.4. Influence of laser-laser delay time on the temporal profiles of excited $C_2$ and $C_3$ species in dual-pulse plasma

Both $C_2$ and $C_3$ temporal profiles were found to be narrowed in the dual-pulse plasma as compared to those in single-pulse configuration. Figure 6 illustrates the changes in the value of HWHM ($\Delta$, the half width at half maximum) in dual-pulse plasma as a function of laser-laser delay time. The influence of the first plasma is expressed in terms of relation between values of HWHM of the species temporal profiles in dual-pulse and single-pulse configurations ($\Delta_{\text{Dual-Pulse}} / \Delta_{\text{laser 2 Only}}$).

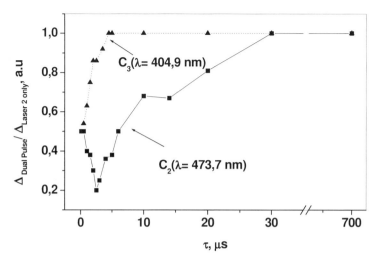

Fig. 6. Ratio of HWHM of $C_2$ and $C_3$ temporal profiles obtained by single pulse (Laser 2 Only) and dual-pulse configurations vs. laser-laser delay time.

It turns out that first plasma influences the emission intensity of $C_2$ and $C_3$ species and their temporal profiles in the second plume within the same laser-laser delay time intervals (up to 30 μs for $C_2$ and up to 5 μs for $C_3$). Though the exact mechanism of profile narrowing in dual-pulse configuration is not known, it may be assumed that the decrease in value of HWHM is related to changes in plasma dynamics and temporal temperature profiles in the second plasma plume.

## 5. Summary Remarks

Optical emission spectroscopy data are useful in identification of $C_2$ and $C_3$ molecules and monitoring their evolution in the plasma. The temporal distribution of plasma species is governed by different factors e.g. their formation mechanism, velocity and collisional decay rates. Thus, study the temporal profiles of emission from different species allows us to follow their respective production and destruction processes. That, in turn, can be used in building a model for nanoparticles or thin films formation.

# References

Abdelli-Messaci, S., Kerdja, T., Bendib, A. and Malek, S. (2002). Emission study of $C_2$ and CN in laser-created carbon plasma under nitrogen environment, *J. Phys. D*, 35, pp. 2772-2778.

Abdelli-Messaci, S., Kerdja, T., Bendib, A., Aberkane, S.M., Lafane, S. and Malek, S. (2005). Investigation of carbon plasma species emission at relatively high KrF laser fluences in nitrogen ambient, *Appl. Surf. Sci.*, 252, pp. 2012-2020.

Acquaviva, S. and De Giorgi, M. L. (2002). High-resolution investigations of $C_2$ and CN optical emissions in laser-induced plasmas during graphite ablation, *J. Phys. B*, 35, pp. 795-809 (a).

Acquaviva, S. and De Giorgi, M. L. (2002). Study of kinetics of atomic carbon during laser ablation of graphite in nitrogen by time- and space-resolved emission spectroscopy, *Appl. Surf. Sci.*, 197-198, pp. 21-26 (b).

Aké, S. C., Sobral, H., Sterling, E. and Villagrán-Muniz, M. (2004). Time of flight of dual-laser ablation carbon plasmas by optical emission spectroscopy, *Appl. Phys. A*, 79, pp. 1345-1347.

Amiranoff, F., Fabbro R., Fabre, E., Garban C., Virmont, J. and Weinfeld, M. (1979). Experimental Transport Studies in Laser-Produced Plasmas at 1.06 and 0.53 μm, *Phys. Rev. Lett.*, 43, pp. 522-525.

Anisimov, V. N., Baranov, V. Y., Grishina, V. G., Delkash, O. N., Serbrant, A. Y. and Stepanova, M. A. (1995). Interaction of Laser-Ablation Plasma Plume with Grid Screens, *Appl. Phys. Lett.*, 67, pp. 2923-2924.

Anselment, M., Smith, R. S., Daykin, E. and Dimauro, L. F. (1987). Optical-Emission Studies on Graphite in a Laser Vaporization Supersonic Jet Cluster Source, *Chem. Phys. Lett.*, 134, pp. 444 – 449.

Arepalli, S. and Scott, C.D. (1999). Spectral measurements in production of single-wall carbon nanotubes by laser ablation, *Chem. Phys. Lett.*, 302, pp. 139-145.

Arepalli, S., Nikolaev, P., Holmes, W. and Scott, C. D. (2000). Diagnostics of laser-produced plume under carbon nanotube growth conditions, *Appl. Phys. A*, 70, pp. 125 – 133.

Babushok, V. I., DeLucia F. C., Jr., Gottfried, J. L., Munson, C. A. and Miziolek, A. W. (2006). Double pulse laser ablation and plasma: Laser induced breakdown spectroscopy signal enhancement, *Spectrochim Acta B.*, 61, pp. 999-1014.

Bloomfield, L. A., Geusic, M. E., Freeman, R. R. and Brown, W. L. (1985). Negative and Positive Cluster Ions of Carbon and Silicon, *Chem. Phys. Lett.*, 121, pp. 33-37.

Burakov, V. S., Bokhonov, A. F., Nedel'ko, M. I., Savastenko, N. A. and Tarasenko, N. V. (2002). Dynamics of the emission of light by $C_2$ and $C_3$ molecules in a laser plasma produced by two-pulse irradiation of the target, *J. Appl. Spectr.*, 69, pp. 907 – 912.

Burakov, V.S., Tarasenko, N.V. and Savastenko, N.A. (2001). Plasma chemistry in laser ablation processes, *Spectrochim. Acta B*, 56, pp. 961-971.

Burakov, V. S., Bokhonov, A. F., Nedel'ko M. I. and Tarasenko, N. V. (2003). Change in the ionization state of a surface laser-produced aluminum plasma in double-pulse ablation modes, *Quantum Electronics*, 33, pp. 1065-1071.

Burakov, V. S., Tarasenko, N. V., Butsen, A. V., Rozantsev V. A. and Nedel'ko, M. I. (2005). Formation of nanoparticles during double-pulse laser ablation of metals in liquids, *EPJAP*, 30, pp. 107–113.

Burakov, V. S., Tarasenko, N. V., Nedel'ko M. I. and Isakov S. N., (2008). Time-resolved spectroscopic characterization of laser-produced plasma for quantitative analysis of glass samples, *Spectrochim. Acta B*, 63, pp. 19-26.

Camps, E., Escobar-Alarcón, L., Haro-Poniatowski, E. and Fernández-Guasti, M. (2002). Spectroscopic studies of two perpendicularly interacting carbon plasmas generated by laser ablation, *Appl. Surf. Sci.*, 197–198, pp. 239-245.

Cao, Z., Mühlhäuser, M., Hanrath, M. and Peyerimhoff, S. D. (2002). Study of possible photodissociation channels in linear carbon clusters C-n (n=4-6), *Chem. Phys. Lett.*, 351, pp. 327–334.

Cappelli, E., Orlando, S., Servidori, M. and Scilletta, C. (2007). Nano-graphene structures deposited by N-IR pulsed laser ablation of graphite on Si, *Appl. Surf. Science*, 254, pp. 1273-1278.

Chen, X., Mazumber, J. and Purohot, A. (1991). Optical-Emission Diagnostics of Laser-Induced Plasma for Diamond-Like Film Deposition, *Appl. Phys. A*, 52, pp. 328-334.

Choi H., Bise, R. T., Hoops, A. A., Mordaunt, D. H. and Neumark, D. M. (2000). Photodissociation of linear carbon clusters C-n (n=4-6), *J. Phys. Chem. A*, 104, pp. 2025-2032.

Chrisey, D.B. and Hubler, G.K. Pulsed laser deposition of thin films, Wiley, New York, 1994.

Claeyssens, F., Henley, S. J. and Ashfold, M. N. R. (2003). Comparison of the ablation plumes arising from ArF laser ablation of graphite, silicon, copper, and aluminium in vacuum, *J. Appl Phys.*, 94, pp. 2203-2211.

Claeyssens, F., Ashfold, M. N. R., Sofoulakis, E., Ristoscu, C. G., Anglos, D. and Fotakis, C. (2002). Plume emissions accompanying 248 nm laser ablation of graphite in vacuum: Effects of pulse duration, *J. Appl Phys.*, 92, pp. 6162-6172.

Colao, F., Lazic, V., Fantoni, R. and Pershin, S. (2002). A comparison of single and double pulse laser-induced breakdown spectroscopy of aluminium samples, *Spectrochim. Acta, Part B*, 57, pp. 1167-1179.

Corsi, M., Cristoforetti, G., Giuffrida, M., Hidalgo, M., Legnaioli, S., Palleshi, V., Salvetti, A., Tognoni, E. and Vallebona, C. (2004). Three-dimensional analysis of laser induced plasmas in single and double pulse configuration, *Spectrochim. Acta, Part B*, 59, pp. 723-735.

Cox, D. M., Reichmann, K. C. and Kaldor, A. (1988). Carbon clusters revisited: The "special" behavior of $C_{60}$ and large carbon clusters, *J. Chem. Phys.*, 88, pp. 1588-1597.

Cristoforett,i G, Legnaioli S, Pardini L, Palleschi V, Salvetti A. and Tognoni, E. (2006). Spectroscopic and shadowgraphic analysis of laser induced plasmas in the orthogonal double pulse pre-ablation configuration, *Spectrochim. Acta B*, 61, pp. 340-350.

De Giacomo, A., Dell'Aglio, M., De Pascale, O. and Capitelli, M. (2007). From single pulse to double pulse ns-Laser Induced Breakdown Spectroscopy under water: Elemental analysis of aqueous solutions and submerged solid samples, *Spectrochim Acta B*, 62, pp. 721-738.

Dreyfus, R.W., Kelly, R., and Walkup, R.E. (1987). Laser-induced fluorescence: Study of laser sputtering of graphite, *Nucl. Instrum. Meth. Res. B*, 23, pp. 557-561.

Dwivedi, R.K. and Thareja, R.K. (1995). Optical-emission diagnostics of $C_{60}$-containing laser-ablated plumes for carbon-film deposition, *Phys. Rev. B*, 51, pp. 7160-7167.

Fueno, H. and Taniguchi, Y. (1999). Ab-initio molecular orbital study of the isomerization reaction surfaces of $C_3$ and $C^-_3$, *Chem. Phys. Lett.*, 312, pp. 65–70.

Gaumet, J.J., Wakisaka, A., Shimizu, Y. and Tamori Y. (1993). Energetics for carbon clusters produced directly by laser vaporization of graphite - dependence on laser power and wavelength, *J. Chem. Soc. Faraday Trans.*, 89, pp. 1667-1670.

Geertsen C., Briand, A., Chartier, F., Lacour, J.-L., Mauchien, P., Mermet, J.-M. and Sjöström, S. (1994). Comparison between infrared and ultraviolet-laser ablation at atmospheric-pressure - implications for solid sampling inductively-coupled plasma spectrometry, *J. Anal. At. Spectrom.*, 9, pp. 17–22.

Geusic, M. E., McIlrath, T. J., Jarrold, M. F., Bloomfield, L. A. and Freeman, R. R. (1986). Photofragmentation of mass-resolved carbon cluster ions - observation of a magic neutral fragment, *J. Chem. Phys.*, 84, pp. 2421–2422.

Goldsmith, J. E. H. and Kearsley, D. T. B. (1990). $C_2$ creation, emission, and laser-induced fluorescence in flames and cold gases, *Appl. Phys. B*, 50, pp. 371-379.

Gong, Z.S., Sun, J., Xu, N., Ying, Z.F., Lu, Y.F., Yu, D. and Wu, J.D. (2007). Spectroscopic study on the evolution of graphite ablation plume in ECR argon plasma during the deposition of diamond-like carbon films, *Diamond and Related Materials*, 16, pp. 124–130.

Guo, T., Nikolaev, P., Thess, A., Colbert, D.T. and Smalley, R.E. (1995). Catalytic Growth of Single-Walled Nanotubes by Laser Vaporization, *Chem. Phys. Lett.*, 243, pp. 49-54.

Harano A., Kinoshita, J., Itou, K., Kitamori, T., Sawada, T. and Koda, S.(1993). An application of the laser-induced breakdown method to measurement of carbon particles produced from the laser ablation of a graphite rod, *J. Spectrosc. Soc. Japan*, 42, pp. 94- 101.

Harilal, S.S. (2001). Expansion dynamics of laser ablated carbon plasma plume in helium ambient, *Appl. Surf. Sci.*,172, pp. 103-109.

Harilal, S.S., Issac, R.C., Bindhu, C.V., Nampoori, V.P.N. and Valabhan C.P.G. (1997). Optical emission studies of $C_2$ species in laser-produced plasma from carbon, *J. Phys. D.*, 30, pp. 1703-1709.

Harilal, S.S., Riju, C., Issac, R.C, Bindhu, C. V., Nampoori, V. P. N. and Vallabhan, C. P. G. (1996). Temporal and spatial evolution of $C_2$ in laser induced plasma from graphite target, *J. Appl. Phys.*, 80, pp. 3561- 3565.

Iida, Y. and Yeung, E. S. (1994). Optical monitoring of laser-induced plasma derived from graphite and characterization of the deposited carbon film, *Appl. Spectr.*, 48, pp. 945-950.

Iijima, S. (1991). Helical microtubules of graphitic carbon, *Nature*, 354, pp. 56-58.

Ikegami, T., Nakanishi, F., Uchiyama, M. and Ebihara, K. (2004). Optical measurement in carbon nanotubes formation by pulsed laser ablation, *Thin Solid Films*, 457, pp. 7–11.

Ikegami, T., Nakao, M., Ohsima, T., Ebihara, K. and Aoqui, S. (2001). Impression of high voltage pulses on substrate in pulsed laser deposition, *J. Vac. Sci. Technol. A.*, 19, pp. 2737-2740.

Kaizu, K., Kohno, M., Suzuki, S., Shiromaru, H., Moriwaki, T. and Achiba Y. (1997). Neutral carbon cluster distribution upon laser vaporization, *J. Chem. Phys.* 106, pp. 9954-9956.

Keszler, A. and Nemes L. (2004). Time averaged emission spectra of Nd:YAG laser induced carbon plasmas, *J. Mol. Struct.*, *695–696*, pp. 211–218.

Kokai, F. and Koga, Y. (1997). Time-of-flight mass spectrometric studies on the plume dynamics of laser ablation of graphite, *Nucl. Instr. Meth. Phys. Res. B*, 121, pp. 387-391.

Koo, Y.-M., Choi, Y.-K., Lee, K. H. and Jung, K.-W. (2002). Mass spectrometric study of carbon cluster formation in laser ablation of graphite at 355 nm, *Bull. Korean Chem Soc.*, 23, pp. 309-314.

Krajnovich, D. J. (1995). Laser sputtering of highly oriented pyrolytic graphite at 248 nm, *J. Chem. Phys.*, 102, pp. 726 – 743.

Kroto, H. W., Allaf, A. W. and Balm, S. P. (1991). $C_{60}$: Buckminsterfullerene, *Chem. Rev.*, 91, pp. 1213-1233.

Kroto, H. W., Heath, J. R., O'Brien, S. C. O., Curl, R.F. and Smalley, R.E. (1985). $C_{60}$: Buckminsterfullerene, *Nature*, 318, pp. 162-163.

Lambert, L., Grangeon, F. and Autric, M. (1999). Crossed beam pulsed laser deposition of cryolite thin films, *Appl. Surf. Sci.*, 138–139, pp. 574-580.

Lillich, H., Aleandri, L.E., Jones, D.J., Rozière, J., Albers, P., Seibold, K., Freund, A., Zumbach, V. and Wolfrum, J. (1995). Production and characterization of noble metal clusters by laser ablation, *J. Phys. Chem.*, 99, pp. 12413-12421.

Little, C. E. and Browne, P. G. (1987). Origin of the high-pressure bands of $C_2$. *Chem. Phys. Lett.*, 134, pp. 560-564.

Luque, J., Juchmann, W. and Jeffries, J. B. (1997). Spatial density distributions of $C_2$, $C_3$, and CH radicals by laser-induced fluorescence in a diamond depositing dc-arcjet, *J. Appl. Phys.*, 82, pp. 2072 – 2081.

Mann, D. M. (1978). A possible first step in carbon particle formation: $C_2+C_3$, *J. Appl. Phys.*, 49, pp. 3485 – 3489.

Mao, S. S., Mao, X., Greif, R. and Russo, R. E. (2001). Influence of preformed shock wave on the development of picosecond laser ablation plasma, *J. Appl. Phys.*, 89, pp. 4096- 4098.

Miller, J. C. Laser Ablation. Principles and Application. Springer-Verlag, Berlin, Heidelberg, New York, 1994, pp. 141, 167.

Monchicourt, P. (1991). Onset of carbon cluster formation inferred from light emission in a laser-induced expansion, *Phys. Rev. Lett.*, 66, pp. 1430-1433.

Moriwaki, T., Kobayashi, K., Osaka, M., Ohara, M., Shiromaru, H. and Achiba, Y. (1997.) Dual pathway of carbon cluster formation in the laser vaporization, *J. Chem. Phys.*, 107, pp. 8927-8932.

Morrow, T., Sakeek, H. F., Astal, A. El., Graham, W. G. and Walmsley, D. G. (1994). Absorption and emission spectra of the YBCO laser plume J. Supercond. 7, pp. 823-827.

Murray, P.T. and Peeler, D.T. (1994). Pulsed laser deposition of carbon films: Dependence of film properties on laser wavelength, *J. Electron. Mater.*, 23, pp. 855-859.

Nemes, L., Keszler, A. M., Parigger, C. G., Hornkohl, J. O., Michelsen, H.A. and Stakhursky, V. (2006). The $C_3$ Puzzle: Formation of and spontaneous emission from the $C_3$ radical in carbon plasma, *Internet Electronic Journal of Molecular Design*, 5, pp. 150–167.

Nemes, L., Keszler, A.M., Hornkohl, J.O. and Parigger, C. G. (2005). Laser-induced carbon plasma emission spectroscopic measurements on solid targets and in gas-phase optical breakdown, *Appl. Opt.*, 44, pp. 3661–3667.

Nemes, L., Keszler, A.M., Parigger, C.G., Hornkohl, J.O., Michelsen, H. A. and Stakhursky, V. (2007). Spontaneous emission from the $C_3$ radical in carbon plasma, *Appl. Opt.*, 46, pp. 4032- 4040.

Noll, R., Bette, H., Brysch, A., Kraushaar, M., Monch, I., Peter, L. and Sturm, V. (2001). Laser-induced breakdown spectrometry — applications for production control and quality assurance in the steel industry, *Spectrochim. Acta Part B*, 56, pp. 637 –649.

Noll, R., Sattmann, R., Sturm, V. and Winkelmann, S. (2004). Space- and time-resolved dynamics of plasmas generated by laser double pulses interacting with metallic samples *J. Anal. At. Spectrom.*, 19, pp. 419-428.

Park, H.S., Nam, S.H. and Park, S.M. (2005). Time-resolved optical emission studies on the laser ablation of a graphite target: The effects of ambient gases, *J. Appl. Phys.*, 97, pp. 113103-1 - 113103-5.

Pearse, R. and Gaydon, W. B. The Identification of Molecular Spectra, 4th ed. Wiley, London, Chagman and Hall, New York, 1976, p. 87.

Petukh, M. L., Rozantsev, V. A., Shirokanov, A. D. and Yankovskii, A. A. (2000). The spectral intensity of the plasma of single and double laser pulses, *J. Appl. Spectrosc.*, 67, pp. 1097-1101.

Piepmeier, E.H. and Malmstadt, H.V. (1969). Q-Switched laser energy absorption in plume of an aluminum alloy, *Anal. Chem.*, 41, pp. 700–707.

Puretzki, A. A., Geohegan, D. B., Fan, X. and Pennycook, S. J. (2000). In situ imaging and spectroscopy of single-wall carbon nanotube synthesis by laser vaporization, *Appl. Phys. Lett.*, 76, pp. 182-184 (b).

Puretzky, A. A., Geohegan, D. B., Fan, X. and Pennycook, S. J. (2000). Dynamics of single-wall carbon nanotube synthesis by laser vaporization, *Appl. Phys. A*, 70, pp. 153–160 (a).

Radhakrishnan, G., Adams, P.M. and Bernstein, L.S. (2007). Plasma characterization and room temperature growth of carbon nanotubes and nano-onions by excimer laser ablation, *Appl. Serf. Sci.*, 253, pp. 7651-7655.

Riascos, H., Zambrano, G. and Prieto, P. (2004). Plasma characterization of pulsed-laser ablation process used for fullerene-like CN$_x$ thin film deposition, *Brazilian J. Physics*, 34, pp. 1583-1586.

Rohlfing, E. A. (1990). High-resolution time-of-flight mass-spectrometry of carbon and carbonaceous clusters, *J. Chem. Phys.*, 93, pp. 7851- 7862.

Rohlfing, E. A., Cox, D. M. and Kaldor, A. (1984). Production and characterization of supersonic carbon cluster beams, *J. Chem. Phys.*, 81, pp. 3322-3330.

Rohlfing, E.A. (1988). Optical-emission studies of atomic, molecular, and particulate carbon produced from a laser vaporization cluster source, *J. Chem. Phys.*, 89, pp. 6103-6111.

Sanginés, R., Aké, C. S., Sobral, H. and Villagrán-Muniz, M. (2007). Time resolved optical emission spectroscopy of cross-beam pulsed laser ablation on graphite targets, *Phys. Lett. A*, 367, pp. 351–355.

Sasaki, K., Wakasaki, T. and Kadota, K. (2002). Observation of continuum optical emission from laser-ablation carbon plumes, *Appl. Surf. Sci.*, 197-198, pp. 197-202 (a).

Sasaki, K., Wakasaki, T., Matsui, S. and Kadota, K.(2002). Distributions of C$_2$ and C$_3$ radical densities in laser-ablation carbon plumes measured by laser-induced fluorescence imaging spectroscopy, *J. Appl. Phys.*, 91, pp. 4033-4039 (b).

Savastenko, N. A. (1999). Formation of C$_2$ molecules in laser-produced plasma, *Izvestiya*, 1 pp. 90-93 (in Russian).

Scott, R.H. and Strasheim, A. (1970). Study of medium voltage and high-voltage controlled waveform spark sources for spectrometric analysis of aluminum-alloys, *Spectrochim. Acta Part B*, 25, pp. 311–332.

Sdorra, W., Brust, J. and Niemax, K. (1992). Basic investigations for laser microanalysis. 4.The dependence on the laser wavelength in laser ablation, *Mikrochim. Acta*, 108, pp. 1–10.

Shinozaki, T., Ooie, T., Yano, T., Zhao, J. P., Chen, Zh. Y. and Yoneda, M. (2002). Laser-induced optical emission of carbon plume by excimer and Nd:YAG laser irradiation, *Appl. Surf. Sci.*, 197-198, pp. 263-267.

St-Onge, L., Detalle, V. and Sabsabi, M. (2002). Enhanced laser-induced breakdown spectroscopy using the combination of fourth-harmonic and fundamental Nd:YAG laser pulses, *Spectrochim. Acta Part B*, 57, pp. 121–135.

St-Onge, L., Sabsabi, M. and Cielo, P. (1998). Analysis of solids using laser-induced plasma spectroscopy in double-pulse mode, *Spectrochim. Acta Part B*, 53, pp. 407-415.

Stratis, D.N., Eland, K.L. and Angel, S.M. (2000). Dual-pulse LIBS using a pre-ablation spark for enhanced ablation and emission, *Appl. Spectrosc.*, 54, pp. 1270–1274.

Tasaka, Y., Tahako, M. and Usami, S. (1995). Optical emission analysis of triple-fold plume formed at pulsed IR laser ablation of graphite, *Jpn. J. Appl. Phys.*, 34, pp. 1673-1680.

Torrisi, L., Caridi, F., Margarone, D., Picciotto, A., Mangione, A. and Beltrano, J.J. (2006). Carbon-plasma produced in vacuum by 532 nm - 3 ns laser pulses ablation, *Appl. Surf. Sci.*, 252, pp. 6383-6389.

Tselev, A. E., Gorbunov, A. and Pompe, W. (1999). Features of the film-growth conditions by cross-beam pulsed-laser deposition, *Appl. Phys. A*, 69, pp. 353-358.

Uebbing, J., Brust, J., Sdorra, W., Leis, F. and Niemax, K. (1991). Reheating of a laser-produced plasma by a $2^{nd}$ pulse laser, *Appl. Spectrosc.*, 45, pp. 1419–1423.

Van Orden, A. and Saykally, R. J. (1998). Small carbon clusters: Spectroscopy, structure, and energetics, *Chem. Rev.*, 98, pp. 2313 - 2357.

Varga, F. and Nemes, L. (1999). Emission spectroscopic studies of $CO_2$-laser induced graphite plasmas, *J. Mol. Struct.*, 480- 481, pp. 273 – 279.

Voevodin, A.A. and Donley, M.S. (1996). Preparation of amorphous diamond-like carbon by pulsed laser deposition: A critical review, *Surf. Coat. Technol.*, 82, pp. 199-213.

Weltner, W. and van Zee, R. J. (1989). Carbon molecules, ions, and clusters, *Chem. Rev.*, 89 pp. 1713-1747.

Windom, B.C., Diwakar, P.K. and Hahn, D.W. (2006). Dual-pulse laser induced breakdown spectroscopy for analysis of gaseous and aerosol systems: Plasma-analyte interactions, *Spectrochim. Acta Part B*, 61, pp. 788–796.

Witanachchi, S., Ahmed, K., Sakthivel, P. and Mukherjee, P. (1995). Dual-laser ablation for particulate-free film growth, *Appl. Phys. Lett.*, 66, pp. 1469-1471.

Chapter 6

# Intra-Cavity Laser Spectroscopy of Carbon Clusters

S. Raikov[1,*] and L. Boufendi[2]

[1]*Institute of Physics, 220072 Minsk, 68 Nezavisimosti Ave, Belarus*
[2]*Ecole Polytechnique de l'Université d'Orléans,
45067 Orleans Cedex 2, France*
*E-mail address: raikov@imaph.bas-net.by*

Techniques of ultra-sensitive intra-cavity laser spectroscopy and cavity ring-down spectroscopy are applied for measurements of carbon cluster formation. Electronic spectra of carbon clusters $C_n$ with $5 \leq n \leq 10$ are reported for the first time. Applications of ultra-sensitive spectroscopy extend to optical diagnostics of vapors and/or transient plasma.

## 1. Introduction

Carbon cluster formation and fragmentation show interesting dynamics in transient plasma, for example in laser ablation plasma. These dynamics may be monitored with in-situ laser spectroscopy techniques. Fullerene growth peculiarities require detailed experimental studies to explore initial and intermediate stages of the formation processes. Theoretical predictions indicate that buckminsterfullerene, $C_{60}$, and other closed structures are formed from precursor clusters $C_n$ with $n \leq 10$. These precursor clusters are typically present in carbon vapor. One can assume [Alekseev & Dyujev 2001 (a, b); 2002] that the evolution of carbon particles occurs according to typical gas-cooling processes: (i) chains are formed for cluster sizes of less than 10 atoms; (ii) rings are formed with increasing number of atoms; and (iii) more complex carbon

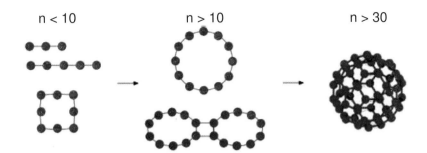

Fig. 1. Probable equilibrium geometries of the low-energy structural isomers for small ($n \leq 10$), intermediate ($n > 10$) and large ($n > 30$) carbon clusters.

structures are formed. Figure 1 illustrates these processes. Notice that the complex carbon structures are direct precursors of fullerenes.

Application of quantum chemistry [Alekseev & Dyujev 2001 (a)] shows that the most probable precursors for fullerenes are two-ring clusters bonded between rings. The algorithm for these predictions is designed to evaluate all probable transformations from multi-ring clusters into a fullerene. Subsequent computations may focus on kinetics of carbon vapor transformations. To accomplish this, the concentration should be defined for two-ring clusters that evolve into a fullerene [Alekseev & Dyujev 2002].

The formation processes described above are common for nucleation and aggregation of any cluster or nano-particle. Measurements of electronic spectra of carbon clusters and their ions are of current research interest in nano-materials and microelectronic plasma processing, as well as in astrophysical investigations of comets and interstellar clouds.

Carbon clusters possibly show diffuse interstellar bands in the uv-visible region [Douglas 1977; Smith *et al.* 1977]. This particular hypothesis can be tested in the laboratory pending availability of carbon clusters for measurement of these gas-phase electronic carbon-cluster spectra. Gas-phase electronic spectra of only three polyatomic bare carbon chains $C_3$, $C_4$ and $C_5$ can be found in the literature and databases. First measurements were reported [Maier *et al.* 2002, 2004] in an attempt to detect the origin bands of $C_4$ (378.9 nm) and $C_5$ (510.9 nm) in diffuse clouds.

## 2. Cluster Formation in Laser Ablation Plume

The earliest measurements of bare carbon clusters have been reported over fifty years ago [Honig 1954]. These experiments used a heated substrate as a cluster source. In later experiments [Rohlfing *et al.* 1984; Bloomfield *et al.* 1985] laser ablation was utilized. The laser ablation method allows us to investigate and model cluster properties in the laboratory. For smallest ($n \leq 10$) clusters, the multiply-bonded chain is the structure that contains the smallest number of dangling bonds [Bernholc & Phillips 1986]. Bending of a chain into a ring causes removal of dangling bonds at the ends of the chain. The formation of a ring is favorable only for comparatively large clusters. Clusters in the $C_{10}$–$C_{20}$ size range are predicted to exist primarily as mono-cyclic rings. For this size-range a minute angle strain is associated with closing the ring. Additional stability arises from forming an extra C-C bond [Van Orden & Saykally 1998]. For large aggregating clusters one finds that rings and three-dimensional structures become reasonable possibilities, provided a low energy path exists from the chains to form these structures. While it is well known that fullerenes are the most stable isomers for even-numbered clusters larger than $C_{32}$, it is unknown experimentally what cluster sizes are thresholds for transitions to occur from chain to ring structures and to fullerenes.

Products of laser ablation typically include atoms, ions and electrons. Clusters and larger liquid and solid particles can be formed as well, either directly by a laser-solid interaction and further dissociation, or by condensation in the expanding ablation plume.

A variety of models describe laser ablation; however, not many models consider formation of clusters [Bogaerts *et al.* 2003]. Some models describe cluster and larger particle formation and growth during expansion of the ablated material, based on condensation and nucleation theories [Kar & Mazumder 1994; Callies *et al.* 1998] or effusion theory [Schittenhelm *et al.* 1996]. Using simplified modeling, information is obtained about cluster size, distribution and rate of cluster growth. The cluster formation efficiency is determined by the rate of collisions between clusters, vapor atoms and buffer gas [Wood *et al.* 1998; Han *et al.* 2002]. Efficient cluster formation is expected when the buffer gas pressure is sufficiently high, causing increased rate of collisions of the ablated species. Condensation may occur at low pressure and/or vacuum conditions [Zhigilei 2003].

Different mechanisms can play a role for direct cluster ablation from the irradiated target. The dynamics of ablation plume formation during early stages and the mechanisms of cluster ejection were theoretically investigated using large-scale molecular dynamics simulations [Zhigilei 2003; Zhigilei *et al.* 2003]. In accordance with these simulations the cluster composition of the ablation plume strongly depends on the irradiation conditions and on the interplay of various processes during evolution of the ablation plume. Spatially resolved simulations of plume formation dynamics reveal segregation effects of different cluster sizes in the expanding plume. A relatively low density of small/medium clusters is observed in the region adjacent to the surface, where large ($n > 30$) clusters are formed. Medium-size clusters ($10 < n < 30$) dominate in the middle of the plume and only small ($n \leq 10$) clusters and monomers are observed near the front of the expanding plume.

## 3. Electronic Spectroscopy of Carbon Clusters

The geometric structure of small and/or intermediate cluster ions can be seen in some detail when using ion chromatography [Van Orden & Saykally 1998]. However, corresponding data for neutral carbon species are limited, even though progress has been made in identifying the electronic absorption spectra of some linear clusters. Table 1 shows a summary of recent observations of electronic spectra. The electronic transitions of the carbon rings could not yet be identified other than in work reported in a neon matrix [Fulara *et al.* 2004] where electronic and infrared absorption spectra are reported for linear and probably cyclic cluster cation $C_6^+$. Moderate amounts of small and especially intermediate size clusters can only be produced in the gas phase. Direct spectroscopic measurement of these species is challenging.

Experimental observations both in absorption and in emission of the electronic transitions of $C_n$ ($n > 3$) and other similar clusters are rather complex. Spectroscopic information on neutral carbon clusters is

Table 1. Electronic spectra observation from neutral carbon chains, $C_n$ ($n > 3$).

| $C_n$ | Neon or Argon Matrix | The Gas Phase |
|---|---|---|
| $C_4$ | Freivogel *et al.* 1996; 1997 (a) | Linnartz *et al.* 2000 |
| $C_5$ | Forney *et al.* 1996; Freivogel *et al.* 1997 (c) | Motylewski *et al.* 1999; this work |
| $C_6$ | Forney *et al.* 1995 | This work |
| $C_7$ | Forney *et al.* 1996; Freivogel *et al.* 1997 (c) | |
| $C_8$ | Zhao *et al.* 1996 | This work |
| $C_9$ | Forney *et al.* 1996 | |
| $C_{10}$ | Freivogel *et al.* 1997 (b) | This work |
| $C_{11}$ | Forney *et al.* 1996 | |
| $C_{12}$ | Freivogel *et al.* 1997 (b) | |
| $C_{13}$ | Forney *et al.* 1996 | |
| $C_{14}$ | Freivogel *et al.* 1997 (b) | |
| $C_{15}$ | Forney *et al.* 1996 | |

primarily obtained by measuring IR-vibrational frequencies. Yet only a few electronic bands have been recorded in inert gas matrices.

Experimental results of electronic spectra from carbon species larger than $C_2$ and $C_3$ were only reported in studies of carbon vapor condensed in argon or neon matrixes [Maier 1998]. Absorption spectra were observed in an inert environment at approximately 4–5 Kelvin, using an approach that combines mass selection with matrix isolation spectroscopy. In this experimental technique, mass-selected carbon cluster anions are deposited from the gas phase into a low-temperature matrix. Electronic absorption spectra can then be recorded for the anion and for the corresponding neutral cluster after uv-irradiation of stable species in the matrix. Mass selected concentrations of typically $10^{15}$–$10^{16}$ cm$^{-3}$ are accumulated in matrix experiments. These concentrations of transient molecules are much higher than in gas-phase experiments. Availability of the matrix spectra allows us to better design the gas-phase experiments due to the fact that the wavelength range for the transitions has been determined. The expected gas-matrix shifts can be estimated from both sets of data available for a number of species. These shifts can be in the 100–200 cm$^{-1}$ range for the longer chain species [Maier 1998]. The uncertainty is acceptable and promising for a search of spectroscopic signatures in the gas phase. A few electronic transitions of carbon anion chains, $C_n^-$, have been studied previously using resonant photo-detachment spectroscopy [Van Orden & Saykally 1998; Maier 1998]. Experimental observation and identification of electronic transitions of $C_n$ in the gas phase will help to find the steadiest variants of stereo-chemical structure important for cluster formation. Moreover, it will allow us to *in-situ* monitor intermediate products in plasma-chemical reactors.

As mentioned above, gas-phase electronic spectra were recorded experimentally for only three polyatomic carbon clusters, namely for linear chains $C_3$, $C_4$ and $C_5$. Amongst these, $C_3$ is now well-characterized in its ground and several excited electronic states [Van Orden & Saykally 1998; Northrup *et al.* 1991; Mladenovich *et al.* 1994]. In addition, efforts include interstellar exploration of this carbon cluster. The identification of $C_3$ through its origin band at 405.2 nm [Maier *et al.* 2001] led to the several further studies in diffuse and translucent interstellar clouds [Roueff *et al.* 2002], including correlations with well-characterized $C_2$ [Oka *et al.* 2003; Adamkovics *et al.* 2003]. The electronic spectra of larger neutral pure carbon chains are of special interest in astrophysics, because electronic bands of $C_2$ and $C_3$ have been identified in diffuse and translucent interstellar clouds [Haffner & Meyer 1995]. For larger, neutral clusters limited experimental information is available in the gas phase. Bands of small carbon chains, $C_4$ and $C_5$, have been previously recorded using ultra-high sensitive cavity ring-down spectroscopy [Motylewski *et al.* 1999; Linnartz *et al.* 2000].

Once again, we emphasize that despite progress in investigations of small carbon clusters, more efforts are necessary for detailed understanding of growth mechanisms. An experimental approach is required for *in situ* measurements with sufficient sensitivity to resolve growth mechanisms. Promising candidates for the proposed experimental work are direct absorption spectroscopy techniques, namely cavity ring-down spectroscopy and intra-cavity laser spectroscopy. Figures 2 and 3 illustrate typical experimental arrangements.

Cavity ring-down spectroscopy [O'Keefe & Deacon 1988; Scherer *et al.* 1997] has been proposed recently for detection of gas-phase electronic spectra of $C_n$ [Motylewski *et al.* 1999; Linnartz *et al.* 2000].

The absorbent is placed inside a passive cavity that is illuminated by narrow-band short-pulsed laser radiation. Measurements of the pulse decay-time allows us to determine absorption profiles for weakly absorbing species. An effective absorption length up to 100 km can be achieved using cavity ring-down spectroscopy.

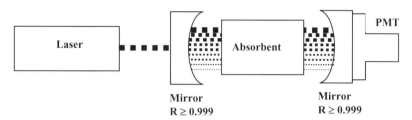

Fig. 2. Schematic arrangement for cavity ring-down spectroscopy. Multiple pass is indicated, a "PMT" photomultiplier records transmitted radiation.

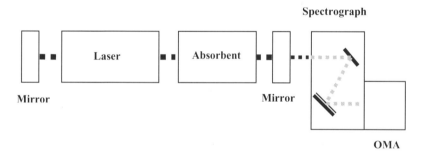

Fig. 3. Schematic arrangement for intra-cavity laser spectroscopy. "OMA" is an Optical Multichannel Analyzer.

An electronic spectrum of gas-phase carbon clusters for $n > 3$ was recorded for linear $C_5$ (near 511-nm) for the first time in 1999 [Motylewski *et al.* 1999]. The band origin was measured for the $C_5$ electronic $^1\Pi_u \leftarrow X\,^1\Sigma_g^+$ transition. The band origin at 19566.3 cm$^{-1}$ was found to be red-shifted by 29 cm$^{-1}$ from the 5 Kelvin value in a neon matrix. The gas-phase $C_5$ clusters for the measurements were created using two different methods. The first method was developed for cavity ring-down spectroscopy of different molecules and clusters [Motylewski & Linnartz 1999; Birza *et al.* 2002]. It is based on generating planar plasma by a discharge in a supersonic gas pulse using a $3 \, cm \times 100 \, \mu m$ slit nozzle and the following composition: 3% HCCH and 1% $CO_2$ in Ne or 0.2% DCCD in He [Motylewski et al. 1999]. The second method uses conventional cluster generation in laser ablation by focusing coherent radiation onto a graphite target. Pulsed laser ablation has been successfully used previously in the study of ir-spectra of carbon chains. Primary advantages of laser ablation include generation of pure carbon clusters at high density in a comparatively small volume. A combination of laser ablation with cavity ring-down spectroscopy shows some complexity and disadvantages. For detailed measurements the first method was initially selected.

Linear $C_4$ clusters in the gas phase were observed using cavity ring-down spectroscopy through supersonic planar plasma [Linnartz et al. 2000] by recording the absorption spectrum near 380 nm. The band origin of the $^3\Sigma_u^- \leftarrow X \, ^3\Sigma_g^-$ electronic transition of $C_4$ was found at 26384.9 $cm^{-1}$ (or 379-nm). Several ab initio calculations show a predicted band origin between 408 nm and 455 nm. Similar methods can be applied for recording of electronic spectra of combined clusters such as $HC_nN$ or $NC_n$ [Gottlieb et al. 2000; Linnartz et al. 2001; Linnartz et al. 2002].

Considering the success of cavity ring-down spectroscopy, application of intra-cavity laser spectroscopy is recommended for measurement of gas-phase carbon clusters that are generated during laser ablation. This experimental approach is elaborated in the next Section.

## 4. Intra-cavity Laser Spectroscopy

Measurement of electronic spectra of gas-phase clusters is effective with the ultra-high sensitive absorption method of intra-cavity laser spectroscopy (ICLS) (see reviews [Burakov et al. 1997; Baev et al. 1999; Burakov & Raikov 2002]). Most importantly, studies of cluster dynamics can be initiated to understand mechanisms of formation and fragmentation in transient plasma due to the inherent temporal resolution of ICLS. Intra-cavity spectroscopy can be nicely combined with pulsed laser ablation processes [Burakov et al. 1995]. When compared with cavity ring-down spectroscopy ICLS allows us to record simultaneously many spectral band-profiles and lines using a single broadband continuum light source. Intra-cavity spectroscopy is valuable for studies over large spectral regions: Spectral multiplexing can be used resulting in an increased data acquisition rate. ICLS also compares favorably with laser-induced fluorescence (LIF) technique. LIF is widely applied for plasma diagnostics when investigating plasma phenomena. It has been shown that spontaneous emission of the laser gain medium limits the minimum detectable absorption. This minimum can be as low as $10^{-10}$ cm$^{-1}$ [Baev *et al.* 1999]. Also, absorption spectroscopy allows us to directly measure absolute species concentration in plasmas.

The laser in ICLS is equivalent to an un-damped cavity with extremely high finesse in the order of $10^5$ to $10^{10}$. The broadband emission spectrum of the laser/optics segment (see Fig. 3) is very sensitive to narrow-line absorption in the cavity due to the enormous effective absorption length (arising from multiple pass in the high-finesse cavity). The highest value established with a continuous-wave (cw) dye laser is about 100,000 km. However, pulsed broadband tunable lasers are most suitable for transient plasma and/or vapor diagnostics.

With high amplification of a probe laser all non-selective (broadband) cavity losses are compensated and not affected by band absorption. Cavity losses include for example continuous absorption (which is comparatively high in ablation plasma), light scattering due to the presence in a plume of solid and liquid micro- and nanoparticles, diffraction losses, and mirror transmissions. The high spectral intensity of the probe laser source (or the small divergence of the laser radiation) allows us to employ a compact diffraction device with adequate spectral resolution for broadband absorption measurements. A compact diffraction device would be for example an echelle grating that operates in high spectral orders and/or double dispersion. Noteworthy is that the scattered light in the spectral device as well as continuous emission of high luminous plasma plume (which reduces the apparent absorption) can be neglected in an ICLS optical arrangement. The high sensitivity of ICLS and its large dynamic range (when both the amplitude in a band center and the total absorption are used) allows us to process both strong and weak absorption bands recorded in a single spectrum.

The basics of ICLS of plasma are well documented in the literature [e.g., see Burakov & Raikov 2002]. For ICLS the instantaneous laser intensity within the absorption band (line) profile $I(v,t)$ can be expressed with the help of an equation similar to Beer-Lambert's relation:

$$I(v,t) = I_0(v,t)\exp[-K(v)L*(t)],\qquad\qquad(1)$$

where $I_0(v,t) = I_0(v,0)f(t)$ is identical to $I(v,t)$ when no further losses occur in the cavity. Usually this intensity is recorded at the un-suppressed frequencies close to the spectral bands. Here, $f(t)$ indicates the temporal pulse shape in the spectral range of the observed absorption band; $K(v)$ is the frequency dependent absorption coefficient; $L*(t) = ctl/L$ is the effective optical path length of laser radiation through an absorbent in the cavity; $c$ is the velocity of light; $L$ is the cavity length; and $l$ is the absorbing layer length.

The measured value corresponds to the integrated intensity during the laser pulse: $H(v,\tau) = \int_0^\tau I(v,t)dt$, where $\tau$ is the pulse duration. The relative laser intensity (transmission profile) can be written as follows:

$$H(v,\tau)/H_0(v,\tau) \sim \frac{\int\limits_0^\tau f(t)\exp[-K(v)L*(t)]dt}{\int\limits_0^\tau f(t)dt}.\qquad\qquad(2)$$

For the majority of laser pulses Equation (2) can be analytically integrated [Burakov & Raikov 2002]; however, we numerically calculate the frequency-dependent absorption coefficient $K(v)$ from the experimentally measured relative transmission profile $H(v,\tau)/H_0(v,\tau)$, and the measured temporal profile $f(t)$ of the probe laser.

For an increase of the dynamic range and increase of precision of the quantitative absorption measurements (for example, in spectrochemical analysis) the total absorption is typically used in application of the intra-cavity technique [Burakov et al. 2000]. The total absorption can be computed as follows

$$A = \int_{-\infty}^{\infty} \left[ 1 - \frac{\int_{0}^{\tau} I(v,t)\,dt}{\int_{0}^{\tau} I_0(v,t)\,dt} \right] dv. \tag{3}$$

Nonlinear competition of intra-cavity modes of the broadband (multimode) radiation does not noticeably modify the spectral bands. This agrees with previously recorded experimental data, and it is further substantiated by comparisons with modeled intra-cavity absorption spectral bands (lines) using *a priori* known broadening mechanisms [Burakov *et al.* 2000]. Additional agreement can be established using a conventional approach, namely by introducing an interferometer of known transmission characteristics to simulate a selectively absorbing medium in a non-selective cavity with pulsed, multimode dye laser radiation.

Specific to ICLS, small frequency shift may be of concern: Spectral band shifts can cause mismatched profile centers of dye gain ($v_0^L$) and of investigated absorption band ($v_0$). For laser radiation and homogeneously broadened profiles the shift does not exceed $10^{-3}$ nm. The shift may however lead to saturated absorption. Saturation induces additional broadening which is modeled with Lorentzian line shapes. The amplitude of a spectral profile is typically diminished at the center $K(v_0)$. Due to various broadening mechanisms the aggregate width of a line equals $\Delta v(\sqrt{1+S})$. The amplitude amounts to $K(v_0)/(1+S)$, where $S$ is the saturation parameter. Yet saturation effects can be neglected for relatively weak absorption bands recorded in comparatively low cluster density.

## 5. Experimental Results

The general experimental apparatus and associated procedures for ICLS with laser sampler for ablation has been developed previously for ablation plasma diagnostics [Burakov *et al.* 1995] and trace level spectrochemical analyses of solid samples [Burakov *et al.* 1996]. For the present work some of the equipment has been customized for ICLS.

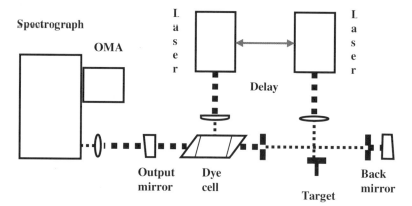

Fig. 4. Schematic experimental arrangement setup illustrating major components. "OMA" is an Optical Multichannel Analyzer attached to the spectrograph. The arrow indicates variable delay between the laser pulses.

Figure 4 shows the schematic experimental setup that includes two synchronized lasers and a high-resolution spectrograph. A tunable pulsed multimode dye laser generates smooth (close to structureless within our spectral resolution) broadband spectra in the uv to ir range (375–1000 nm) with variable pulse duration ($\tau$) of 25 to 200 ns. The spectral width of the laser radiation depends on the specific dye but is usually in the 20 to 30 nm range. The cavity length ($L$) is about 15 cm. The laser-pumped probe laser can be replaced by a flash-lamp pumped dye laser with pulse duration of 1–10 µs in order to increase the effective absorption length. In the latter case, however, the time resolution does not allow us to carry out precise dynamical measurements. A flash-lamp pumped laser is favored for measurements of ultra-trace contents of species in the pulsed ablation plume, for example, in the longer afterglow region.

Ablation plasma in the cavity of the probe dye-laser is usually generated by focusing ruby (or Nd:glass) radiation onto a plane surface of a high-purity graphite target (fragment of graphite rod for arc spectrochemical analysis). The ruby (or Nd:glass) laser is operated at the fundamental wavelength in low or high intensity mode: (i) low intensity mode with maximum irradiance of about 10 MW/cm$^2$ (up to 300 µs pulse width), and (ii) high intensity mode (ns pulse duration) with irradiance of about 10 GW/cm$^2$ (25 ns). Both lasers are synchronized using a controllable delay. Single pulse operation is elected for intra-cavity broadband qualitative observations of cluster absorption. Nd:YAG or eximer laser radiation at high pulse repetition rate is advantageous for statistical quantitative measurements of species densities, especially in spectrochemical analysis of solid materials — see recent work in laser-induced breakdown spectroscopy [Burakov *et al.* 2004; Bel'kov *et al.* 2005].

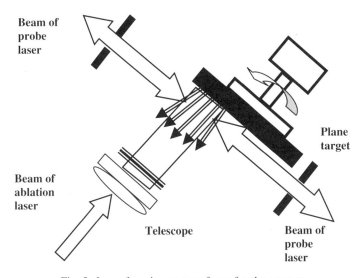

Fig. 5. Laser focusing on a surface of a plane target.

Figure 5 illustrates the focusing arrangement. Usually the laser beam is focused on the sample surface by means of a 50 mm focal length plano-convex quartz (glass) lens. The sphero-cylindrical telescope can also be applied to increase the dimension (absorbing layer length — *l*) of the ablation plume along the optical axis of the probe dye laser.

The telescope can also be used for heterogeneous target material. A sample-holding positioner is used to sustain the ablation plume in the dye laser cavity with a controllable distance (usually 0.5-mm steps) of the probing beam from the target surface, and to preselect the irradiated surface zone. The device can be located within a hermetic chamber in which a buffer gas (usually air or argon at normal pressure) creates the desired environment. An ohmically heated graphite tubular furnace of ultrahigh purity was utilized in some experiments instead of the planar graphite target. This allowed us to increase the initial temperature (up to approximately 3000 Kelvin) of the graphite surface for the low intensity mode of laser ablation. Figure 6 shows the low intensity focusing variant. Such device is typically used in spectro-chemical analysis with a graphite furnace electro-thermal atomizer.

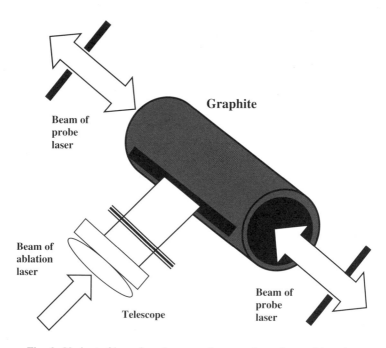

Fig. 6. Variant of laser focusing on an inner surface of a graphite tube.

The dye laser spectra including absorption bands and lines of interest are dispersed by a high-resolution diffraction spectrograph (theoretical resolution is about 0.001 nm) and recorded with an optical multichannel analyzer (OMA) with one or four inter-connected linear CCD array detectors. In addition, necessary plasma parameters for the radiating expanding plume are determined using conventional optical emission spectroscopy and bi-dimensional imaging of the ablation plume.

## 6. Intra-cavity Absorption Spectra of Carbon Clusters

Experimental conditions for carbon (and any other) cluster production including characteristics of an ablation laser should be optimized to increase the efficiency of cluster formation. Optimization is primarily needed due to comparatively low densities of such specific particles and their short lifetime, excluding, of course, well-known and stable species $C_2$ and $C_3$. In general, the laser ablation phenomenon can be divided into two separate processes: (i) evaporation of the solid target and formation of the laser plasma; (ii) expansion of the ablated material into a background gas. Depending on the laser characteristics, laser ablation of solids comprises complex processes of heating, melting, vaporization, ejection of atoms, ions, molecules, clusters and larger solid and liquid particles, dissociation, shock waves, plasma initiation and plasma expansion, ablation plume expansion in the afterglow, and condensation. Laser-induced plasma characterization will be essential for determining the optimal conditions for cluster formation and growth. A multitude of experimental diagnostics should be used simultaneously to extract accurate information regarding physical and chemical processes that occur.

Intra-cavity Laser Spectroscopy (ICLS) has been applied previously [Burakov *et al.* 1979] in the study of dynamics of $C_2$ and CN. Also, ICLS has been used for characterization of the pre-breakdown [Burakov *et al.* 1990], plasma and afterglow [Burakov *et al.* 1992] stages of ablation plume. However, further work is required to understand the underlying physics and chemistry of the complex processes of carbon (or any other) cluster formation or fragmentation. In this context some preliminary experiments have been initiated with the aim of optimizing the laser ablation procedure. These preliminary/subsidiary experiments predominantly include bi-dimensional imaging (with a CCD photocamera) geometrical shape and size of: (i) the crater formed in the target material and (ii) the radiating ablation plume. Figures 7 and 8 display recorded images.

Figures 9–12 show the band origins of electronic transitions of four carbon chains, $C_5$, $C_6$, $C_8$ and $C_{10}$ in the gas phase. All intra-cavity spectra presented below have been recorded in argon at pressure close to normal. The spectra for the last three species were recorded to our knowledge for the first time. Only matrix spectra for these clusters were available previously. As for $C_5$, the same band origin has been recorded for the electronic transition $^1\Pi_u \leftarrow X \ ^1\Sigma_g^+$ using cavity ring-down spectroscopy in laser ablation plume [Motylewski *et al.* 1999].

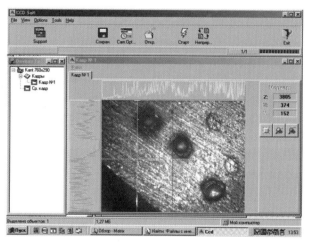

Fig. 7. Typical craters formed in the target material in high intensity mode of laser ablation. The average diameter of a crater is about 100 µm. [Photo courtesy of V. Rosantsev who was our co-worker in our lab, but is now retired.]

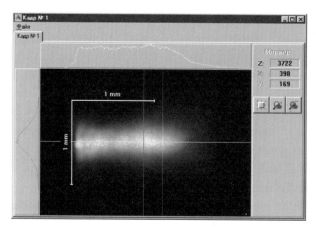

Fig. 8. Radiating plume recorded in high-intensity mode laser-ablation of a target in air at normal pressure.

Comparisons of both $C_5$ spectra (as well as with spectra recorded in supersonic slit/nozzle plasma) obtained by different absorption techniques in similar cluster production conditions (laser ablation) show overall similarity including major details and red satellites near the wing

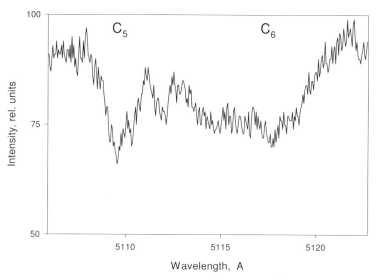

Fig. 9. The gas-phase electronic spectra of $C_5$ ($^1\Pi_u \leftarrow X\,^1\Sigma_g^+$, $0_0^0$) and $C_6$ ($^3\Sigma_u^- \leftarrow X\,^3\Sigma_g^-$, $0_0^0$) measured in high intensity mode of laser ablation.

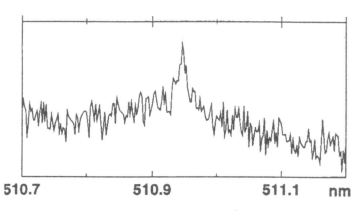

Fig. 10. Electronic spectrum of $C_5$ ($^1\Pi_u \leftarrow X\,^1\Sigma_g^+$, $0_0^0$) in laser ablation experiment presented in [Motylewski *et al.* 1999].

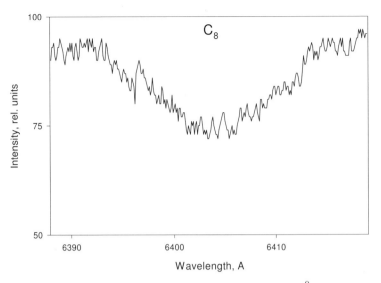

Fig. 11. The gas-phase electronic spectrum of $C_8$ ($^3\Sigma_u^- \leftarrow X\,^3\Sigma_g^-$, $0\,^0_0$) measured in high intensity mode of laser ablation.

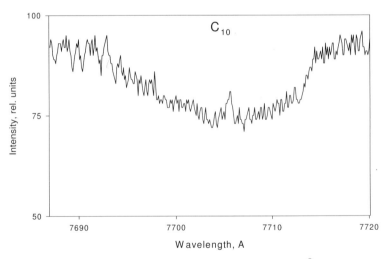

Fig. 12. The gas-phase electronic spectrum of $C_{10}$ ($^3\Sigma_u^- \leftarrow X\,^3\Sigma_g^-$, $0\,^0_0$) measured in high intensity mode of laser ablation.

of the band. However, the blue satellite located very close to the band center and more clearly discernible in supersonic spectra (DCCD in He) and very weak (on a level of the visibility limit) in laser ablation and supersonic (HCCH + $CO_2$ in Ne) spectra [Motylewski *et al.* 1999] is not clearly visible in intra-cavity spectra. This difference may be caused by spectral interference from other species in cavity ring-down experiments — or it may be caused by specific characteristics of the selected instruments for the measurements.

As for the larger investigated carbon clusters such comparison can be attempted only for matrix spectra with rather low resolution, thus, the band profiles cannot be directly compared.

For conditions of the rather "soft" mode of laser-target interaction at red and ir wavelengths of 0.69 μm or 1.06 μm, respectively, a thermal target ablation mechanism dominates (due to low photon energy). As graphite displays large differences between inter-layer and intra-layer bond strengths, layer by layer evaporation of graphite is accompanied by formation of large carbon complexes for small laser irradiance. The carbon complexes can create various carbon clusters and molecules, including the smallest ones, $C_3$ and $C_2$.

Equally, for conditions of high irradiance thermal ablation mechanisms prevail, as energy of radiation quanta (photons) is insufficient for direct solid-vapor transitions. Irrespective, only carbon species of the least sizes (mainly ions) will be produced in an ablation plume in the initial stages. In our experimental conditions consecutive processes can occur in $C_n$ formation, owing to condensation of carbon species in an ablation plume during its expansion and cooling. For example, carbon atoms may attach to and/or combine with rather steady $C_3$ and $C_2$ molecules (recombination mechanism). The best signal to noise ratio occurs in the afterglow of the ablation plume for all recorded cluster bands (see Figs. 9–12).

In both modes of laser ablation qualitatively similar electronic spectra of $C_5$, $C_6$, $C_8$ and $C_{10}$ have been recorded. This indicates that both possible mechanisms occur as carbon clusters evolve. In the "soft" ablation mode all recorded bands are notably weaker. For all case-studies further optimization of experimental conditions will be necessary to achieve more reproducible absorption spectra in transient plasma.

Further processing of measured broadband intra-cavity spectra of $C_n$ as well as available matrix spectra of these species is indicated in the search for cyclic carbon isomers. Unfortunately, the only reliable experimental work utilizes a neon matrix [Fulara *et al.* 2004] to address measurement of electronic absorption spectra of the cyclic cluster cation $C_6^+$. As for other cyclic carbon clusters, a possible approach is comparison of unassigned cluster fragments in recently recorded broadband intra-cavity spectra with new theoretical predictions.

## 7. Conclusions

The following summarizes major points of this work. Firstly, electronic gas-phase spectra of neutral carbon chains larger than $C_5$ have been recorded for the first time. Secondly, $C_n$ clusters ($3 < n \leq 10$) could be formed due to dissociation processes when using low-irradiance ablation laser radiation. Thirdly, high-irradiance radiation caused carbon particles ($3 < n \leq 10$) growth from recombination processes. Fourthly, use of the discussed measurement techniques allowed us to record gas-phase electronic spectra for odd-numbered ($C_7$, $C_9$), cyclic and larger linear carbon clusters ($10 < n < 15$).

Experimental study of small size carbon cluster ($n \geq 5$) will lead to a fundamental understanding of their electronic spectra and their composition. In particular, the results will be useful in building a chemical model for carbon cluster formation during laser ablation of graphite or in plasma processing reactors. In an applied direction, this opens the possibility of monitoring reactive intermediates, in turn, monitoring of changing conditions in, for example, synthesis of fullerenes. And last but not least, these results point to certain conclusions about the relevance of such species for interpretation of astrophysical observations. Astrophysical observations and diffuse interstellar band measurements can help indicate which sort and what sizes of carbon chains should be considered for further gas-phase investigations.

## Acknowledgement

The editors would like to thank Professor Christian G. Parigger at the University of Tennessee Space Institute for careful reading of this chapter and for his help in the revision and correction of the text for slight errors in the usage of scientific English.

## References

M. Adamkovics, G.A. Blake, B.J. McCall. Astrophys. J., 2003, **595**, 235.

N.I. Alekseev, G.A. Dyujev. J. Techn. Phys., 2001, **71**, 67.

N.I. Alekseev, G.A. Dyujev. J. Techn. Phys., 2001, **71**, 71.

N.I. Alekseev, G.A. Dyujev. J. Techn. Phys., 2002, **72**, 121.

V.M. Baev, T. Latz, P.E. Toschek. Appl. Phys. B, 1999, **69**, 171.

M.V. Bel'kov, V.S. Burakov, V.V. Kiris, N.M. Kozhukh, S.N. Raikov. J. Appl. Spectrosc., 2005, **72**, 352.

J. Bernholc, J.C. Phillips. J. Chem. Phys., 1986, **85**, 3258.

P. Birza, T. Motylewski, D. Khoroshev, A. Chirokolava, H. Linnartz, J.P. Maier. Chem. Phys., 2002, **283**, 119.

L.A. Bloomfield, M.E. Geusic, R.R. Freeman, W.L. Brown. Chem. Phys. Lett, 1985, **121**, 33.

Bogaerts, Z. Chen, R. Gijbels, A. Vertes. Spectrochim. Acta B, 2003, **58**, 1867.

V.S. Burakov, P.Ya. Misakov, S.V.Nechaev, S.N. Raikov. J. Appl. Spectrosc., 1979, **30**, 625.

V.S. Burakov, P.A. Naumenkov, S.N. Raikov. Opt. Commun., 1990, **80**, 26.

V.S. Burakov, P.A. Naumenkov, S.V.Nechaev, S.N. Raikov. J. High Temp. Chem. Proc., 1992, **1**, 471.

V.S. Burakov, S.N. Raikov, N.A. Savastenko, N.V. Tarasenko. J. Mol. Struct., 1995, **349**, 281.

V.S. Burakov, P.Ya. Misakov, S.N. Raikov. Fresenius J. Anal. Chem., 1996, **355**, 361.

V.S. Burakov, S.N. Raikov, N.V. Tarasenko. J. Appl. Spectrosc., 1997, **64**, 293.

V.S. Burakov, A.V. Isaevich, S.N. Raikov. J. Appl. Spectrosc., 2000, **67**, 449.

V.S. Burakov, S.N. Raikov. J. Appl. Spectrosc., 2002, **69**, 425.

V.S. Burakov, V.V. Kiris, P.A. Naumenkov, S.N. Raikov. J. Appl. Spectrosc., 2004, **71**, 676.

G. Callies, H. Schittenhelm, P. Berger, H. Hügel. Appl. Surf. Sci., 1998, **127-129**, 134.

A.E. Douglas. Nature, 1977, **269**, 130.

D. Forney, J. Fulara, P. Freivogel, M. Jakobi, D. Lessen, J.P. Maier. J. Chem Phys., 1995, **103**, 48.

D. Forney, P. Freivogel, M. Grutter, J.P. Maier. J. Chem Phys., 1996, **104**, 4954.

P. Freivogel, M. Grutter, D. Forney, J.P. Maier. Chem. Phys. Lett., 1996, **249**, 191.

P. Freivogel, M. Grutter, D. Forney, J.P. Maier. J. Chem Phys., 1997, **107**, 22.

P. Freivogel, M. Grutter, D. Forney, J.P. Maier. J. Chem Phys., 1997, **107**, 4468.

P. Freivogel, M. Grutter, D. Forney, J.P. Maier. J. Chem Phys., 1997, **107**, 5356.

J. Fulara, E. Riaplov, A. Batalov, I. Shnitko, J.P. Maier. J. Chem. Phys., 2004, **120**, 7520.

C.A. Gottlieb, A.J. Apponi, M.C. McCarthy, P. Thaddeus, H. Linnartz. J. Chem. Phys., 2000, **113**, 1910.

L.M. Haffner, D.M. Meyer. Astrophys. J., 1995, 4**53**, 450.

M. Han, Y. Gong, J. Zhou, C. Yin, F. Song, N. Muto, T. Takiya, Y. Iwata. Phys. Lett. A, 2002, **302**, 182.

R.E. Honig. J. Chem. Phys., 1954, **22**, 126.

Kar, J. Mazumder. Phys. Rev. E, 1994, **49**, 410.

H. Linnartz, O. Vaizert, T. Motylewski, J.P. Maier. J. Chem Phys., 2000, **112,** 9777.

H. Linnartz, O. Vaizert, P. Cias, L. Grüter, J.P. Maier. Chem. Phys. Lett., 2001, **345**, 89.

H. Linnartz, D. Pfluger, O. Vaizert, P. Cias, P. Birza, D. Khoroshev, J.P.

Maier. J. Chem. Phys., 2002, **116**, 924.

J.P. Maier. J. Phys. Chem., 1998, **102**, 3462.

J.P. Maier, N.M. Lakin, G.A.H. Walker, D.A. Bohlender. Astrophys. J., 2001, **553**, 267.

J.P. Maier, G.A.H. Walker, D.A. Bohlender. Astrophys. J., 2002, **566**, 332.

J.P. Maier, G.A.H. Walker, D.A. Bohlender. Astrophys. J., 2004, **602**, 286.

M. Mladenovich, S. Schmatz, P.J. Botschwina. J. Chem. Phys., 1994, **101**, 5891.

T. Motylewski, H. Linnartz. Rev. Sci. Instrum., 1999, **70**, 1305.

T. Motylewski, O. Vaizert, T.F. Giessen, H. Linnartz, J.P. Maier. J. Chem. Phys., 1999, **111**, 6161.

F.J. Northrup, T.J. Sears, E.A. Rolhfing. J. Mol. Spectrosc., 1991, **145**, 74.

T. Oka, J.A. Thorburn, B.J. McCall, S.D. Friedman, L.M. Hobbs, P. Sonnentrucker, D.E. Welty, D.G. York. Astrophys. J., 2003, **582**, 823.

O'Keefe, D.A.G. Deacon. Rev. Sci. Instrum., 1988, **59**, 2544.

E.A. Rohlfing, D.M. Cox, A. Kaldor. J. Chem. Phys., 1984, **81**, 3322.

E. Roueff, P. Felenbok, J.H. Black, C. Gry. Astronomy and Astrophys., 2002, **384**, 629.

J.J. Scherer, J.B. Paul, A. O'Keefe, R.J. Saykally. Chem. Rev., 1997, **97**, 25.

H. Schittenhelm, G. Callies, P. Berger, H. Hügel. J. Phys. D: Appl. Phys., 1996, **29**, 1564.

W.H. Smith, T.P. Snow, D.G. York. Astrophys. J., 1977, **218**, 124.

Van Orden, R.J. Saykally. Chem Rev., 1998, **98**, 2313.

R.F. Wood, J.N. Lebouef, K.R. Chen, D.B. Geohegan, A.A. Puretzky. Appl. Surf. Sci., 1998, **127-129**, 151.

Y. Zhao, E. De Beer, C. Xu, T. Taylor, D. Neumark. J. Chem Phys., 1996, **105**, 4905.

L.V. Zhigilei. Appl. Phys. A, 2003, **76**, 339.

L.V. Zhigilei, E. Leveugle, B.J. Garrison, Y.G. Yingling, M.I. Zeifman, Chem Rev., 2003, **103**, 321.

Chapter 7

# Dynamics of Laser-Ablated Carbon Plasma for Thin Film Deposition: Spectroscopic and Imaging Approach

Raj K. Thareja[1,*] and Ashwini K. Sharma[2,†]

[1]*Department of Physics and Centre for Laser Technology, Indian Institute of Technology Kanpur, Kanpur – 208 016, Uttar Pradesh, India*
*E-mail address: thareja@iitk.ac.in*
[2]*Department of Physics, Indian Institute of Technology Guwahati, North Guwahati – 781 039, Assam, India*
†*E-mail address: aksharma@iitg.ernet.in*

Pulsed laser ablation has grown rapidly over the last few years. It has shown tremendous promise in the growth of nanoparticles, heterostructures, quantum dots, superlattices, nanotubes, nanowires, and nanorods, to name a few. Formation of these structures when deposited on a substrate depends upon various experimental parameters like laser radiation wavelength, fluence, ambient gas which in turn dictates as to what kind of distribution of the emitted species within the plasma will give rise to their formation. In the present chapter we discuss the formation of $C_2$ and CN molecules using pulsed laser ablation of graphite in nitrogen ambient. We correlate the space- and time-resolved optical emission spectroscopic studies with the plasma dynamics of the expanding carbon plasma using fast photography. A one-to-one correlation of our results suggests the presence of CN and $C_2$ molecules in different regions of the expanding plasma.

## 1. Introduction

Carbon based nanomaterials have attracted great scientific and technological attention, since the discovery of fullerenes in 1985 (Kroto *et al.*, 1985). A wide range of nanostructures such as graphene, carbon nanotubes, nanowires, onions, nanocages and nanobuds have been observed (Rao and Dresselhaus, 2001). Most of the electronic and optical properties of nanomaterials differ significantly from the bulk materials (Edelstein and Cammarata, 1998). Utilization of these newly found properties of nanomaterials in numerous potential applications has led to a thrust towards nanomaterial research. Carbon nanomaterials have often been discussed as the material of the future (Bogdanov *et al.*, 2000). Carbon nanoparticles have generated much interest as they exist over a diverse range of systems, from lightning discharge in atmosphere and interstellar dust to arc generated soot and vacuum deposited thin films (Lopinski *et al.*, 1998). Recently, carbon nanoparticles have found use in many industrial applications like electronic devices, field emission devices, supercapacitors, microsensor preconcentrators, non-linear optical devices, dye in paint, filler in rubber and protective coatings, thus leading to intensive research in the field (Amaratunga *et al.*, 1996; Bezryadin *et al.*, 1999; Yu *et al.*, 2001; Diederich *et al.*, 1999; Siegal *et al.*, 2002; Li *et al.*, 2002; Vincent *et al.*, 2002; Donnet and Voet, 1976; Chen *et al.*, 2004; Harilal *et al.*, 1996).

Various methods like carbon arc technique (Amaratunga *et al.*, 1996), microwave-plasma chemical vapour deposition (Yu *et al.*, 2001), supersonic cluster beam deposition (Diederich *et al.*, 1999), and pulsed laser ablation (Chen *et al.*, 2004) have been used to produce carbon nanoparticles. In the present work, the method of pulsed laser ablation has been used for carbon nanoparticle production. A Nd:YAG laser pulse typically lasting $\sim 10$ ns, produces sufficient power density to ablate a graphite target (Harilal *et al.*, 1996). The occurrence of atomic, ionic and molecular carbon in the plasma plume depends on the laser intensity. It has been observed that at higher laser intensities ($\sim 10^9$–$10^{12}$ W/cm$^2$) the emission spectrum is dominated by atomic and ionic species (CI to CV) but at later times and large distances from the target surface molecular species resulting from recombination of ionic species are observed. At lower laser intensities ($\sim 10^8 - 10^9$ W/cm$^2$) emission from molecular $C_2$ is dominant even at earlier times and close to the target (Thareja and Abhilasha, 1994). The blue-green Swan band emissions from $C_2$ form one of the oldest observed and assigned molecular electronic spectra (Shea, 1927). The $C_2$ radical has been found important in astrophysics and flame spectroscopy (Gaydon, 1974; Herzberg, 1950), (Goldsmith and Kearsley, 1990). The number densities and energy states of various carbon species produced during ablation affect the characteristics of thin films and various carbonaceous nanomaterials prepared (Abhilasha *et al.*, 1993; Thareja *et al.*, 1997; Misra *et al.*, 1999). $C_2$ radical being the dominant species at low laser energies, attains great importance for the production of carbon nanoparticles.

There are several diagnostic techniques for characterization of laser ablated plasma and these include optical emission spectroscopy (OES) (Misra and Thareja, 1999; Dwivedi *et al.*, 1998), optical absorption spectroscopy (Morrow *et al.*, 1994; Geohegan, 1991), laser induced fluorescence (LIF) (Goldsmith and Kearsley, 1990; Nakata *et al.*, 1995; Ikegami *et al.*, 2001), time resolved spectroscopy (Harilal *et al.*, 1994), fast photography (Misra and Thareja, 1999; Ikegami *et al.*, 2001), and electrostatic probes (Mayo *et al.*, 1999). Among these time resolved spectroscopy is the most suitable method to study the dynamics of laser produced plasma (Harilal *et al.*, 1996), however, it does not give much information on non-emissive species in lower electronic states. The LIF spectroscopy and optical absorption spectroscopy techniques are used for analysis of such non-emissive species. LIF is a very sensitive technique and provides an additional advantage over optical absorption of being offering a possibility of spatially resolved. In the present studies, we have used optical emission spectroscopy and dynamic fast photography to understand the dynamics and evolution of carbon plasma and nanoparticles. The technique of laser-induced fluorescence spectroscopy is employed to measure the ground state number density of $C_2$ molecules in plasma plumes generated by pulsed-laser ablation of graphite.

## 2. Experimental Details

The experimental configuration used for the optical emission diagnostics (space- or time- resolved) of laser-ablated graphite target is shown in Fig. 1. A Q-switched Nd:YAG laser (DCR-4G, Spectra Physics) with a pulse width of 8 ns at full width half maximum (FWHM) creates plasma in vacuum and in presence of ambient nitrogen. It delivers energy of 1 J/pulse (maximum) at a repetition rate of 10 Hz, operating in the fundamental mode ($\lambda = 1.064\ \mu m$). The laser has a Gaussian limited mode structure with beam divergence less than 0.5 mrad and beam diameter of 8 mm. The laser energy is monitored on a calibrated power meter (Model 30 A, Ophir) by placing it in the path of the laser beam.

The laser beam from Nd:YAG laser is focused using a lens of focal length of 45 cm on the solid target to a spot size of ~400 $\mu m$ inside the vacuum chamber. The target is continuously rotated and translated with an external motor so that each laser pulse falls on a fresh target surface. The vacuum chamber is evacuated to a pressure of around $10^{-5}$ mbar using a rotary pump backed turbo molecular pump. The chamber is purged with the ambient nitrogen several times before introducing it in a controlled manner. The plasma radiation emitted is imaged on an array of optical fiber cable with a lens of focal length 16 cm so as to have one-to-

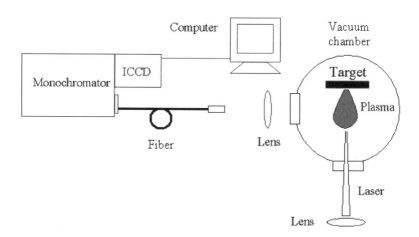

Fig. 1. Experimental setup for optical emission spectroscopy and imaging.

one correspondence with the plasma and its image. The other end of the fiber is coupled with the entrance slit of a Monochromator (Jobin Yvon, HRS 2/HR 320). The output is detected with a gated intensified charge-coupled device (ICCD) (DH 720, Andor Technology) attached at the exit slit of the monochromator. Fast gate pulses (> 2 ns) acting as a shutter for the ICCD are supplied at variable delay time with respect to the ablating pulse. The gate width is kept at 10 or 20 ns while recording the spectra. ICCD is triggered by Nd:YAG laser used for ablation.

In order to record the two-dimensional side-on images of the overall visible emission from the expanding plasma in vacuum and in ambient nitrogen, the ICCD is removed from the monochromator and is placed perpendicular to the direction of the laser radiation (in place where fiber is kept, cf Fig. 1). A zoom lens (Sigma) is attached at its front. The focal length of the lens is adjusted to 69 cm. The aperture of the lens is kept at minimum to avoid any excess radiation entering the ICCD. The calibration of each pixel correspond to 57.2 $\mu$m. The images are recorded and used to estimate plume front position, plume velocity, plume length, vapor pressure and temperature, and shock temperature, respectively.

In order to study the LIF from graphite target, the ablated carbon plasma is probed with a dye laser (PDL-3, Spectra Physics) operating at a wavelength $\lambda = 516.5$ nm. A solution of Coumarin 500 (Mol. Wt. 257.21, *Exciton-USA*) dye in spectroscopic grade methanol was used as the laser dye. Tuning range of dye solution is between 485–546 nm. The dye solution was pumped by a Q-switched Nd:YAG laser (Quanta Ray INDI-40-10, Spectra Physics) operating in the third harmonic ($\lambda = 355$ nm, $E = 80$ mJ), the dye laser beam is referred to as probe beam. The time delay between the ablating and probe laser pulse is adjusted using a delay generator (SRS, DG535). Fluorescence from the plume is collected from the top window of the plasma chamber by a lens and imaged onto an entrance slit of a monochromator using an optical fiber bundle. The output from the monochromator is detected using a gated ICCD connected to the computer. Time-resolved LIF signal is monitored using a digitizing oscilloscope (Agilent 54615 B, 500 MHz). The two dimensional images of the fluorescence were recorded using ICCD in place of monochromator.

## 3. Laser-Ablated Plasma in Presence of an Ambient Gas

Deposition of thin films using pulsed-laser ablation involves optimization of various parameters viz. laser wavelength and incident energy on a target, choice of a proper substrate and its temperature, ambient gas and its pressure, and target-substrate distance. However, at the same time it is important to understand the transport mechanism of ablated material and its interaction with the ambient atmosphere prior to subsequent deposition on a suitable substrate. During the expansion of the plasma in ambient gases it is observed that there exists a high density and chemically active region at the plume front where vigorous reaction may lead to the formation of compounds. This region is also characterized by high temperature as compared to that in the plasma. Knowing the location of such a region helps in optimizing the target-substrate distance for the deposition of thin films. This can be achieved by taking two-dimensional images of the expanding plume using gated ICCD. The emission from the plume contains information about the spatial and temporal distribution of various species (atoms, ions, molecules) and can be studied selectively. ICCD images of the plume have been used to calculate the position of the plume front, its velocity, and plume length, respectively.

### 3.1. *Optical emission spectroscopy*

In order to understand the formation of carbon nitride (CN), a graphite target was ablated in nitrogen ambient and OES was used, both spatially and temporally. For the formation of CN we can write the reaction as follows $C + N_2 = CN + N - 2$ eV. The emission spectra of CN bands were recorded in ambient nitrogen pressure ranging from $10^{-3}$ Torr to 1 Torr. The fluence was fixed at 12 Jcm$^{-2}$. The following CN bands (violet $B^2 \Sigma^+ - X^2 \Sigma^+$ band system of the sequence $\Delta v = -1$) were observed in the emission spectrum: (0–1) at 421.6 nm, (1–2) at 419.7 nm, (2–3) at 418.1 nm, (3–4) at 416.8 nm, (4–5) at 415.6 nm, and (5–6) at 415.2 nm, respectively (Dwivedi, 1997). No CN band heads were observed at $10^{-3}$ Torr.

At 0.01 Torr, CN bands were observed at and beyond 5 mm from the target surface. Increasing the distance from 5 to 9 mm, the intensity of the band head shifted to larger delay time, increased gradually, and was maximum at 9 mm, as shown in Fig. 2 for the strongest band head at 415.2 nm. All other band heads showed similar behavior. Increasing the pressure to 0.1 Torr, all band heads were observed at and beyond 4 mm and followed the similar spatial and temporal intensity behavior as at 0.01 Torr.

At 1 Torr all band heads were observed at and beyond 3 mm. However, the intensity first increased to a maximum at 5 mm and then decreased both spatially and temporally, the decrease in intensity was more prominent for band heads at 415.6 and 415.2 nm, respectively. Figures 3(a) and 3(b) shows such a comparison for band heads at 421.6 and 415.2 nm. The fact that increasing the ambient nitrogen pressure decreased the distance at which CN was formed suggests that pressure has a very important role in controlling the CN formation. Comparing Fig. 2 with Fig. 3(b) we observe that at a given distance the intensity maximum shifted to a larger delay time with increasing pressure. With

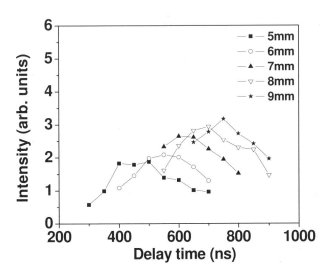

Fig. 2. Intensity variation of CN band head at 415.2 nm with delay time at various distances from the target surface at 0.01 Torr ambient nitrogen.

this information the velocity of the band head at 415.2 nm was found to be around $2 \times 10^6$, $5 \times 10^5$, and $4 \times 10^5$ cm s$^{-1}$ at 0.01, 0.1, and 1 Torr, respectively. Decrease in velocity with pressure is due to plume confinement.

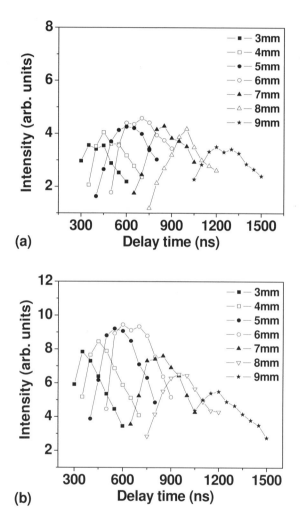

(a)

(b)

Fig. 3. (a) Temporal variation of CN band head at 421.6 nm with delay time at various distances away from the target surface at 1 Torr ambient nitrogen. (b) Temporal variation of CN band head at 415.2 nm with delay time at various distances away from the target surface at 1 Torr ambient nitrogen.

A time integrated spectrum of CN band heads at various ambient pressures at 9 mm away from the target surface is shown in Fig. 4. CN band formation started at 300 ns delay time with respect to the laser pulse in the pressure range studied.

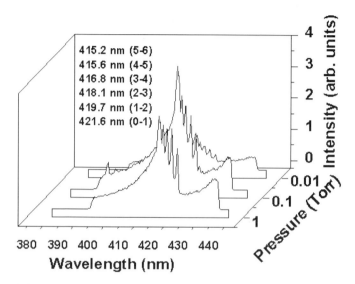

415.2 nm (5-6)
415.6 nm (4-5)
416.8 nm (3-4)
418.1 nm (2-3)
419.7 nm (1-2)
421.6 nm (0-1)

Intensity (arb. units)

4
3
2
1
0

0.01
0.1
1

Pressure (Torr)

380 390 400 410 420 430 440

**Wavelength (nm)**

Fig. 4. Variation of CN band head emission intensity with ambient nitrogen pressure at 9 mm away from the target surface.

We also studied the spatial and temporal dependence of $C_2$ Swan band (d $^3\Pi_g$ – a $^3\Pi_u$) corresponding to the sequence $\Delta v = 0$ with band heads (0–0) at 516.5, (1–1) at 512.9, (2–2) at 509.7, (3–3) at 507.0, and (4–4) at 505.6 nm, and $\Delta v = -1$ with band heads (0–1) at 563.5, (1–2) at 558.5, (2–3) at 554.0, (3–4) at 550.1, and (4–5) at 547.0 nm, respectively (Sharma, 2005). A typical time integrated spectrum showing these bands at 1 Torr nitrogen ambient is shown in Fig. 5.

Consider the band head sequence corresponding to $\Delta v = 0$. In vacuum, the intensity of the band heads attained a maximum at 1 mm and then decreased with delay time. With increasing distance from the target surface the intensity maximum shifted to larger delay times. Figure 6(a) shows the temporal variation of emission intensity for the band head at 516.5 nm at various distances from the target surface.

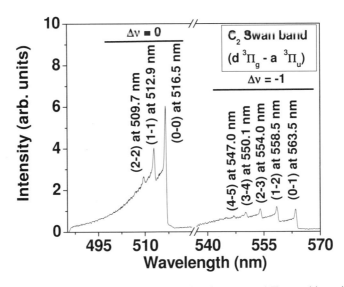

Fig. 5. Time integrated spectrum of $C_2$ Swan band system at 1 Torr ambient nitrogen.

Band heads at 512.9 and 509.7 nm also showed similar behavior. Band heads at 507.0 and 505.6 nm were weak and were not considered for the analysis. The bands were observed at and beyond 100 ns delay time with respect to the laser pulse. At 0.01 and 0.1 Torr, all band heads showed similar behavior as in vacuum, however, beyond 6 mm, the intensity of the band heads again increased, as shown in Fig. 6(b) at 0.1 Torr. Increasing the pressure to 1 Torr showed the similar intensity variation with time and distance from the target surface as in vacuum, Fig. 6(c). The velocity of the band head at 516.5 nm was found to be $8.3 \times 10^5$, $9 \times 10^5$, $1 \times 10^6$, and $7.5 \times 10^5$ cm s$^{-1}$ in vacuum and at 0.01, 0.1, and 1 Torr, respectively.

For the band head sequence corresponding to $\Delta v = -1$, all band heads showed similar behavior as that observed for the band head sequence corresponding to $\Delta v = 0$ at the corresponding pressures with the only difference that at 0.01 Torr, intensity maxima was observed at 0 mm (at the vicinity of the surface).

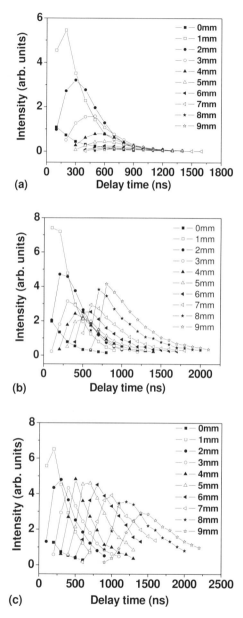

Fig. 6. Temporal variation of $C_2$ Swan band head at 516.5 nm at various distances from the target surface (a) in vacuum and (b) 0.1 Torr ambient nitrogen. (c) Temporal variation of $C_2$ Swan band head at 516.5 nm at various distances from the target surface at 1 Torr ambient nitrogen.

The velocity of the band head at 563.5 nm was found to be $2.5 \times 10^5$, $1 \times 10^6$, $8.3 \times 10^5$, and $8.3 \times 10^5$ cm s$^{-1}$ in vacuum and at 0.01, 0.1, and 1 Torr, respectively. Comparison of the velocities of the two band heads of $C_2$ shows that band head at 516.5 nm is slightly more energetic than at 563.5 nm.

The forgoing analysis shows that CN band attained maximum intensity at large distances from the target surface indicating the formation of CN by gas phase reaction between the ablated graphite and the ambient nitrogen. On the other hand, $C_2$ bands attained maximum intensity very close to the target surface. Thareja (Thareja *et al.*, 2002) used very high fluence ($\sim 95$ Jcm$^{-2}$) at laser wavelength of 1.064 μm and observed carbon atomic and ionic species in the emission spectrum of the plasma. Therefore, they attributed the formation of $C_2$ to the atomic recombination of carbon species in presence of ambient nitrogen. However, with much lower fluence (12 Jcm$^{-2}$) no carbon atomic lines were observed in the emission spectra. Therefore, the formation of $C_2$ cannot be only due to the atomic recombination of carbon atoms or ions produced from the target and three-body recombination (Dwivedi, 1997) in ambient gas. Formation of $C_2$, however, may also result directly from the dissociation of higher clusters. Figure 7 shows the spatial variation of

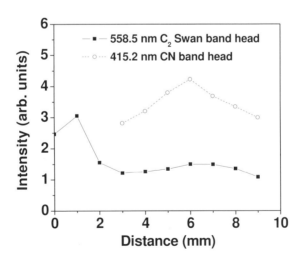

Fig. 7. Comparison of CN and $C_2$ band head intensities at 1 Torr ambient nitrogen.

CN band head at 415.2 nm and $C_2$ band head at 558.5 nm in nitrogen ambient at 1 Torr. This indicates that beyond 4 mm $C_2$ dissociates into atomic C and reacts with $N_2$ as per the reaction mentioned above. We observed a contrasting feature in the temporal profile of $C_2$ and CN band at a given distance, the profile being broader for $C_2$ as compared to CN (compare Figs. 3 and 6). The $C_2$ molecule dissociates into two C atoms and one of the atoms react with $N_2$ and forms CN. Remaining C atom reacts with another C atom (obtained from the dissociation of higher clusters) in presence of gas atom N (atomic recombination) and forms $C_2$ hence observed to last for a longer time. Thareja (Thareja *et al.*, 2002) studied the formation of CN by comparing the intensity variation of $C_2$ Swan band (c $^1\Pi_g$ – a $^3\Pi_u$) corresponding to the sequence $\Delta v = -1$ and CN band (violet $B^2 \Sigma^+ - X^2 \Sigma^+$ band) corresponding to the sequence $\Delta v = 0$.

Experiments were also carried out at varying laser fluences in order to understand the formation of neutral atoms and ions of carbon, $C_2$ and CN molecules in the plasma plume (Kushwaha and Thareja, 2008). It is observed that the fluence plays a major role in the formation of molecules in vapor phase (Kushwaha and Thareja, 2008; Sharma, 2005). At an early stage of plasma formation, the collisional process dominates the recombination process whereas at a later stage the recombination dominates. Therefore, we expect the formation of atoms and ions at an early stage of ablation. At later stages, the ions recombine to form neutral atoms which may further combine to form molecules. These molecules have longer lifetime since the collisions are less and the possibility of dissociation into subsequent atoms is reduced. At high laser fluence $> 60$ $Jcm^{-2}$, atomic and ionic emissions of carbon dominated the spectrum. Spatially-resolved data show that the collisions are reduced gradually as one moves away from the target surface since the plasma continues to expand thereby increasing the mean free path between the electrons, atoms, molecules etc.

At moderate laser fluence, the plasma emission is dominated by $C_2$ Swan bands ($d^3\Pi_g - a^3\Pi_u$) of the sequence ($\Delta v = -1, 0, +1$) and CN violet ($B^2\Sigma^+ - X^2\Sigma^+$) bands of the sequence ($\Delta v = -1$) over the ionized carbon species which contributes only at early stages ($< 100$ ns) of emission from the expanding plasma very close to target surface.

Thus, above analysis suggests an important role that both the laser fluence and the ambient pressure play in the formation of molecules, here CN.

A detailed investigation of laser ablated carbon plasma was carried out to understand the dynamics of nanoparticle formation. The spectra are dominated by continuum close to the target, whereas discrete lines of atomic and ionic species of carbon are observed away from the target, however at later stages of expansion, the atomic and ionic lines disappear and the Swan band system of molecular $C_2$ appears at 270 ns after laser ablation. Figure 8 shows the temporally resolved spectrum of carbon at a laser fluence of 20 J cm$^2$ recorded at a distance of 3 mm from the target in the wavelength range 440–580 nm, the spectrum is dominated by $C_2$ species. It is observed that $C_2$ emission increases with time initially, attains a maximum value at 210 ns and thereafter decays and is not observed after 600 ns. $C_2$ emission from $\Delta v = -1, 0, +1$ sequences of the Swan band $(d^3\Pi_g - a^3\Pi_u)$ (where $\Delta v = v' - v''$ is the difference of vibrational quantum number between the upper ($v'$) and lower ($v''$) states of transitions) is observed to dominate the spectra. The vibrational temperature of $C_2$ was calculated from respective band head intensities measured experimentally from the recorded optical emission spectrum.

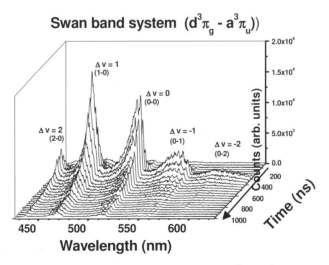

Fig. 8. Temporal evolution of $C_2$ Swan bands $(d^3\Pi_g - a^3\Pi_u)$.

The relative population in each vibrational level $N_{v'}$ can be obtained using theoretical Franck-Condon factors (Spindler, 1965) $F_{v'v''}$ and the band head intensity $I_{em}^{v'v''}$ observed at frequency $v$, and is given by the relation

$$N_{v'} = \frac{C I_{em}^{v'v''}}{D_v F_{v'v''} v^4} \qquad (1)$$

where C is a constant, $D_v$ represents a correction factor for the detection system, and $v'$ and $v'$ are the vibrational levels of the upper and lower electronic states, respectively. In thermal equilibrium the population $N_{v'}$ of the initial state is proportional to $\left[ \dfrac{-G(v')hc}{k_B T_{vib}} \right]$, therefore we have

$$\frac{N_{v'}}{N_{v''}} = \exp\left[ -\frac{G(v') - G(v' = 0)}{k_B T_{vib}} hc \right] \qquad (2)$$

where $G(v')$ is the term value for the upper vibrational level $v'$, $k_B$ is the Boltzmann constant, $T_{vib}$ is the molecular vibrational temperature, $h$ is the Planck's constant, and $c$ is the speed of light. The vibrational temperature 300 ns after plasma formation at a distance of 3 mm away from the target was found to be 7300 K for $\Delta v = +1$, and 9200 K for $\Delta v = -1$.

Figure 9 shows the temporal variation of radial plasma expansion parallel and perpendicular to target surface. Figure 9(a) shows the splitting of the plume at 30 ns comprising of an almost stationary component *a* close to the target surface and a component *b* moving away from the target surface. The stationary part may be the result of collisions between the ejected particles of the plume and the ambient gas in the high pressure region of the early expansion which results in the formation of Knudsen layer with stopped and/or backward moving material (Kelly, 1990), whereas the moving part consists of molecular carbon, predominantly $C_2$. At 270 ns, the moving component again bifurcates into two parts, the fast and the slow moving components as shown in the Fig. 9(b). The intensity of faster species *b* decreases after re-splitting and finally dies out ~ 600 ns whereas the formation of the

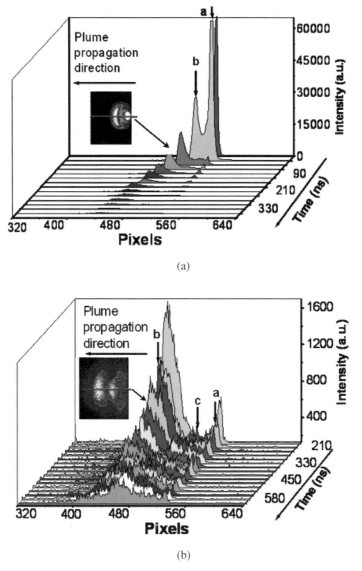

(a)

(b)

Fig. 9. Time-resolved optical emission spectra of laser ablated carbon plume. (a) Plume splitting at early delay time, (b) re-splitting of the plume at later delay time. Inset shows the images of the expanding plume.

slower species $c$ starts at $\sim 270$ ns and gradually increases. The slower species can be observed even several μs after the plasma formation.

The re-splitting of the plasma plume into fast and slow moving component is attributed to the fact that at later times various ionic and molecular species near the target surface recombine to form longer carbon chains and hence leading to nanoparticle (NP) formation. These NPs being heavier and bulkier get decelerated due to opposing ambient gas pressure whereas molecular $C_2$ easily penetrate through the ambient gas.

## 3.2. Fast photography

In vacuum the ablated material expands with its thermal kinetic energy. However, when ambient gas is introduced, the ablated material suffers collisions with the ambient gas atoms and tends to loose energy thereby following different dynamics of motion depending upon the ambient conditions (Sharma and Thareja, 2004).

In vacuum plasma expands freely and therefore a linear variation is expected between the delay time and plume front position. However, in presence of an ambient gas the collisions between the ambient gas and the ablated material from the target result in the formation of shock waves as well as deceleration of expanding plume. An interesting aspect associated with such interaction is the stopping of the plume which has been shown to be a useful way to optimize the target-substrate distance for the thin film deposition (Misra and Thareja, 1999). Fast photography has been used extensively to visualize the propagation of laser-ablated plasmas in vacuum and in ambient atmosphere (Geohegan, 1992; Huddlestone, 1965). At background gas pressures $< 10^{-2}$ mbar, a large fraction of ions stream through the background gas loosing very little of their energy whereas at pressures $> 10^{-2}$ mbar, essentially all the kinetic energy of the ions is lost in collisions with the background gas particles that occur within a very small volume near the target resulting in shock/blast waves (Liberman and Velikovich, 1986; Zeldovich and Raizer, 1967). The interaction of laser plasma with the ambient gas therefore results in recombination interactions, collisional and collision-less interactions, shock/blast wave interactions, micro-instabilities in the laser plasma and turbulent interactions. The shock/blast wave and drag models have been used to understand the plasma dynamics (Sharma and Thareja, 2004; 2005). Gas phase nanoparticle synthesis reported in the literature (Geohegan *et al.*, 1998) provides a novel way to understand the transportation and subsequent deposition of nanoparticles on a substrate, and control on the particle size.

Interaction of laser with the target results in the formation of plasma which is a strong source of UV radiation. In presence of an ambient gas this radiation interacts with the ambient and results in an increase in the density in a very narrow region which propagates in the ambient atmosphere with speed more than that of the local ion sound speed given by $c_s = (Zk_BT_e/m_i)^{1/2}$ and results in the formation of shock wave. $Z$ is the average ion charge, $k_B$ is the Boltzmann constant, $T_e$ is the electron temperature, and $m_i$ is the ion mass. Here mass of the ejected vapor is less than that of the ambient gas in motion.

In order to understand the plume dynamics it is essential to identify the following boundaries, namely; the contact front or the expansion front, compressed front and the shock front. In the present context the plasma pushes and compresses the ambient gas in front of it. It acts as a piston and almost all the ejected particles are concentrated near the boundary called the contact front or the expansion front (this front is dynamic). A thin region in front of the contact surface is compressed between the piston and the ambient atmosphere and is termed as compressed front. The region between the ejected material (driven gas) and the ambient gas is the shock front. A shock wave represents a discontinuity formed near the leading edge of the plume when the surrounding ambient gas is pressed by the ablated material. Figure 10 shows the schematic of the expanding plasma in an ambient atmosphere showing various regions of interest.

According to the Taylor-Sedov (T-S) theory of spherical blast waves emanating from strong point explosion, the shock position (Freiwald and Axford, 1975) is defined by

$$R = \xi_o \, (E_o/\rho_o)^{1/5} t^{2/5} \tag{3}$$

where $t$ is the delay time, $\rho_o$ is the ambient gas density, $E_o$ is the amount of energy released during the explosion and $\xi_o$ is a constant given by

$$\xi_o = [(75/16\pi)(\gamma_v - 1)(\gamma_v + 1)^2/(3\gamma_v - 1)]^{1/5} \qquad (4)$$

where $\gamma_v$ is the ratio of the specific heat for the plume. This model is restricted between two extremes i.e., in a range of $R$ beyond which the model breaks down (Dyer *et al.*, 1990). The first limit requires that the mass of the gas encompassed by the shock wave be much greater than the initial explosion mass ($M_o$) i.e.,

$$R >> R_1 = (3M_o/2\pi\rho_o)^{1/3} \qquad (5)$$

The second limit requires that the pressure driving the front ($P_1$) be much greater than that ahead of the propagation ($P_o$) of the explosion front i.e.,

$$R_2 = (E_o/P_o)^{1/3} \qquad (6)$$

Thus the model is expected to hold for $R_1 << R << R_2$ or

$$(3M_o/2\pi\rho_o)^{1/3} << R << (E/P_o)^{1/3} \qquad (7)$$

A far more general expression suitable for modeling the plasma dynamics in pulsed laser ablated plasma can be written from Eq. (3) as

$$R = at^n \qquad (8)$$

where, $a$ is a constant, $n$ is an exponent and has a value of 0.4 for an ideal shock condition and 1 for free expansion (Sharma and Thareja, 2004).

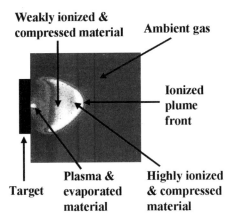

Fig. 10. Various regions in an expanding laser-ablated aluminum plume at 0.1 Torr nitrogen ambient and at 140 ns delay time with respect to ablating pulse.

The plasma parameters, namely, pressure, specific volume and density $(P_1, V_1, \rho_1)$ in the plasma region can be evaluated just behind the plasma front in terms of known parameters of undisturbed gas $(P_o, V_o, \rho_o)$ using the laws of conservation of mass, momentum and energy, respectively, (Zeldovich and Raizer, 1967). It is shown that

$$P_1/P_0 = [\{(\gamma_v + 1)V_0 - (\gamma_v - 1)V_1\}/\{(\gamma_v + 1)V_1 - (\gamma_v - 1)V_0\}] \quad (9)$$

$$V_1/V_0 = [\{(\gamma_v - 1)P_1 + (\gamma_v + 1)P_0\}/\{(\gamma_v + 1)P_1 + (\gamma_v - 1)P_0\}] \quad (10)$$

$$T_1/T_0 = P_1V_1/P_0V_0 \quad (11)$$

Also, the velocity in the undisturbed gas and in the plasma region is given by

$$u_0^2 = V_0[(\gamma_v + 1)P_1 + (\gamma_v - 1)P_0]/2 \quad (12)$$

$$u_1^2 = V_0[\{(\gamma_v - 1)P_1 + (\gamma_v + 1)P_0\}/\{(\gamma_v + 1)P_1 + (\gamma_v - 1)P_0\}]/2 \quad (13)$$

For the case of a strong shock wave we have $P_1/P_0 \gg 1$. The limiting density is therefore

$$\rho_0/\rho_1 = V_1/V_0 \approx (\gamma_v - 1)/(\gamma_v + 1) \quad (14)$$

In the limit $P_1/P_0 \to \infty$,

$$P_1 = 2\rho_0 u_0^2/(1 + \gamma_v) \quad (15)$$

$$T_1 = P_1 M/(R_g \rho_1) \quad (16)$$

where $R_g$ is the gas constant. $P_1$ and $T_1$ are the vapor temperature and vapor pressure just behind the shock front.

Assume the ablated particles to be an ensemble moving along $X$ direction. The equation of motion

$$nm(\partial \mathbf{v}/\partial t) + (\mathbf{v} \cdot \nabla)\mathbf{v} = -\nabla P_e \quad (17)$$

where $n$ is the number density, $\mathbf{v}$ the velocity, $m$ the mass of the particle, and $P_e$ the plasma pressure, when used together with a drag force $\mathbf{F} = -\beta \mathbf{v}$ and the Maxwell-Boltzmann distribution of the density of the particles, give

$$\partial v_x/\partial t = -\beta v_x \quad (18)$$

Applying suitable boundary conditions we get

$$x = X_{max}[1 - \exp(-\beta t)] \qquad (19)$$

where $X_{max} = v_{x0}/\beta$, and is referred to as the maximum distance to which a plume can expand in an ambient atmosphere. The above equation is expressed in the following form

$$R(t) = R_0(1 - e^{-\beta t}) \qquad (20)$$

where $R = x$, and $R_0 = X_{max}$, respectively.

Equation (18) shows that the plasma plume progressively slows down and eventually comes to rest due to the drag force (at a place where the ambient pressure equals the plume pressure). The distance from the target at which the propagation of the plume ceases is called the stopping distance or plume length.

The time-resolved ICCD images due to $C_2$ Swan band and CN violet band recorded using a tunable optical filter tuned at 516.5 nm for $C_2$ ($d^3\Pi_g - a^3\Pi_u$) and 415.2 nm for CN ($B^2\Sigma^+ - X^2\Sigma^+$) transitions in nitrogen ambient at pressures of 1.2, 12 and 120 mbar at 100, 350 and 600 ns delays with respect to ablating pulse and at a fluence of 32 Jcm$^{-2}$, respectively, are shown in Fig. 11. At low pressure ($\sim 10^{-3}$ mbar, vacuum) the emission is observed till 500 ns, however, in ambient nitrogen the plasma survived for a longer time, $\sim 5000$ ns, at 1.2 mbar. The early stage of expansion ($< 100$ ns) shows the atomic and ionic carbon near the target surface, but as plasma evolved, the electronically excited molecules ($C_2$ and CN) start to form due to collisions and chemical reactions at larger delays.

The ambient gas has a different influence on the propagation dynamics and plume shape of the molecular $C_2$ and carbon nanoparticles (NPs). The molecular $C_2$ plume develops into a hemispherical expansion as predicted by the shock wave model, whereas the NPs plume expansion remains almost forward directed as a consequence of the high mass ratio between the NPs and the ambient gas molecules. The plume front dynamics for the molecular $C_2$ and NPs can be explained by using a simple physical approach based on the balance between plume linear momentum variation and the external pressure force (Predteceensky and Mayorov, 1993; Amoruso *et al.*, 2008).

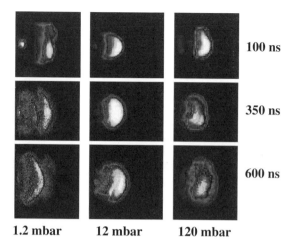

Fig. 11. ICCD images of the expanding plume at 1.2, 12, 120 mbar ambient nitrogen pressure (Kushwaha and Thareja, 2008).

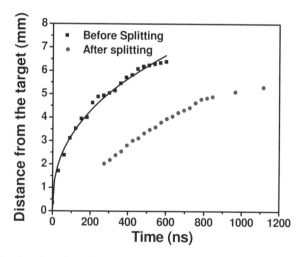

Fig. 12. R-T plot for the plasma front expansion of molecular $C_2$ and carbon nanoparticles (Yadav, D., Gupta, V. and Thareja, R. K. (2009), Evolution and imaging of nanoparticles observed in laser ablated carbon plume, *J. Appl. Phys.*, 106, pp. 064903-1-064903-7).

Figure 12 shows the R-T plots for the molecular $C_2$ and carbon NPs. By fitting the experimental points with the general equations of plasma expansion models, we get the equations for the expansion of plasma front. The evolution of molecular $C_2$ follows shock-wave model, $R = 0.515t^{0.4}$ and the expansion of NPs formed after re-splitting of plasma plume is well described by the drag model, $R = 9.102 (1 - \exp^{-0.00094\,t})$. Here $R$ is taken in mm and $t$ is taken in ns.

In the case of the molecular $C_2$ plume, the model considers plume and background gas boundary as a hemispherical thin layer of radius $R$ moving at velocity $u_b$ and experiencing the force due to the background gas pressure $p_g$. Hence, the equations of motion for $R$ and $u_b$ is,

$$\frac{1}{4\pi}\frac{d}{dt}\left[\left(2M_b + \frac{4}{3}\pi R^3\rho_g\right)u_b\right] = -R^2 P_g \tag{21}$$

where $u_b(t) = dR/dt$, $M_b$ is the confined molecular $C_2$ plume mass and $\rho_g$ is the ambient gas density. $2/3\pi R^3\rho_g$ represents the mass of the gas swept away by the expanding plume at a distance $R$ at time $t$.

The initial conditions are $R(t=0)=0$ and $u_b(t=0)=u_{b,0}$. Solving this equation numerically by obtaining fits to the experimental data using an iterative minimization procedure and taking the value of $u_{b,0}$ from the R-T plot in Fig. 12, we obtain the value of molecular $C_2$ mass as $M_b \approx 3\times10^{-11}$ kg.

In the case of NPs plume, we consider a one-dimensional case by considering the plume confined in a front layer acting as a piston of mass $M_c$ and section $S$ on the ambient gas, and sweeping it away during its motion. Hence, the equations of motion for the NPs plume front $x$ and its velocity $u_c$ becomes

$$\frac{d}{dt}\left[\left(\frac{M_s}{S} + \rho_g x\right)u_c\right] = -P_g \tag{22}$$

where $u_c(t) = dR/dt$. The initial conditions are $x(t=0)=0$ and $u_c(t=0)=u_{c,0}$. Solving this equation numerically by obtaining fits to the experimental data using an iterative minimization procedure and taking the value of $u_{c,0}$ from the R-t plot in Fig. 12, we obtain a value of $M_c/S \approx 3.4\times10^{-6}$ kg/m$^2$. By considering the full width at half maximum of the NP plume in Fig. 12, the average transverse size of NP plume is found to be 4 mm, which leads to $S = 1.25\times10^{-5}$ m$^2$, which in turn leads to an estimated value of the NPs plume mass $M_c \approx 4.25\times10^{-11}$ kg.

Two immiscible fluids with different densities when accelerated in a direction perpendicular to an interface gives rise to some kind of perturbations at the interface. If the acceleration is from a heavier to a lighter fluid, the interface is unstable whereas if the acceleration is from a lighter to a heavier fluid, the interface is stable. Such an instability is called Rayleigh-Taylor (RT) instability. Depending upon the interaction (extent of diffusion) of the ablated material with the ambient gas, plasma-gas boundary may show peculiar but interesting features like plume splitting and RT instability during the course of plasma expansion (Sharma and Thareja, 2005; Harilal *et al.*, 2003). These features cannot be realized easily using diagnostic techniques other than the ICCD imaging. A typical image of the expanding plasma in ambient nitrogen atmosphere is shown in Fig. 13. The perturbations can be clearly seen at the plasma-gas interface in all three sides of the expanding plasma. A detailed account of RT instability is given in the references (Chandrasekhar, 1981; Taylor, 1950).

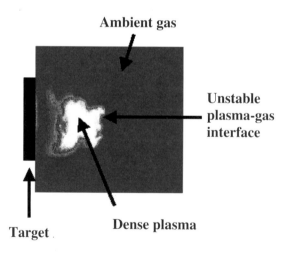

Fig. 13. Laser ablated plume showing unstable front.

The effect of the curvature of magnetic lines of force and RT instability on an expanding carbon plasma in presence of an external magnetic field is studied and reported in the literature (Neogi, 1999). In the absence of an external magnetic field, the occurrence of RT instability is reported based on snowplow model (Baranov, 1993). The images of the plume along with emission spectroscopy can be used effectively to quantify the instability as RT instability and to estimate the growth time of instability (Sharma and Thareja, 2005).

The plume penetrating the ambient gas gets split into a fast and a slow moving component, the slower component being the one that is decelerated because of the difference in the density of propagating plasma and the ambient gas which exerts a resistive force. The stratification of plume front in slow and fast components is reported in the literature (Wood *et al.*, 1997; Harilal *et al.*, 2003, Sharma and Thareja, 2005; Geohegan, 1992). However, the faster component is observed to undergo re-splitting only at a definite pressure regime (here ~ 1.2 mbar) (Harilal *et al.*, 2003). This critical pressure (~ 1.2 mbar) at which re-splitting occurs comes within the transition from non-collisional to collisional interaction among ablated species and the ambient gas. The nature of re-splitting is found to depend on the ambient gas pressure; re-splitting in the faster moving part decreases as the pressure increases and is almost absent at 120 mbar. The increase in pressure results in the confinement of the plume and the mutual diffusion of the carbon plasma species into $N_2$ gas or vice-versa decreases, which results in turbulence near plasma-gas interface. The instability, referred to as RT instability, is manifested by the distortion in the plume front boundary. The behavior is mainly observed in the decelerated part of the plume. The exponential growth of instability is given by ~ $e^{\eta t}$ (Harilal, 2003),

$$\eta^2 = Ka(\rho_p - \rho_g)/(\rho_p + \rho_g) \tag{23}$$

where $\rho_p$ and $\rho_g$ are the plasma and background gas densities, a is the acceleration of plasma front and K is the unstable mode wave vector. The plasma front is stable when $\eta^2 > 0\,(\rho_p > \rho_g)$ and unstable when $\eta^2 < 0\,(\rho_p < \rho_g)$. No such instability is observed at pressure of 1.2 mbar. As the pressure increases, at the penetration region the density of plasma becomes lesser than that of the ambient gas giving rise to instability on the plasma-gas interface, Fig. 11.

### 3.3. *Lased-induced fluorescence*

Laser-induced fluorescence (LIF) refers to selective excitation of a molecule from a lower energy state (ground state) to a higher state followed by emission either at the excitation wavelength or at a different wavelength. The selective excitation helps in calculating the density of the lowest energy state (ground state) of the system under consideration. The technique can be used effectively to obtain information about the molecular states involved in an emission process, collisional cross-sections, and distribution of population densities of various species participating in a chemical reaction (Demtröder, 2003).

In this section we discuss the use of LIF technique to measure the ground state number density of $C_2$ molecules. The expanding plasma is probed with a laser sheet from a Nd:YAG pumped dye laser tuned at $C_2$ transition $a^3\Pi_u - d^3\Pi_g$ at 516.5 nm (0, 0) and LIF from transition $d^3\Pi_g-a^3\Pi_u$ at 563.5 nm (0, 1). The laser beam from the dye laser is sent to probe the expanding plasma when the spontaneous emission from the $C_2$ molecules has subsided and most of the molecules have returned back to the ground state. Since $a^3\Pi_u$ is $C_2$ ground state, the LIF intensity from $d^3\Pi_g$ level may be taken as proportional to number density.

When the laser ablated carbon plasma is produced with pump laser fluence of 20 J/cm$^2$, the most intense LIF signal at 563.5 nm is obtained when delay between the pump and probe laser is set at 600 ns, as spontaneous emission from $C_2$ subsides by that time. Figure 14 shows such a scheme where the laser ablated carbon plasma is probed with dye laser ($\lambda = 516.5$ nm, E = 4 mJ) at a distance of $d = 3$ mm from the target surface.

The variation of LIF intensity with delay between the pump and probe laser showed LIF peak intensity decreases and shifts temporally to higher value as we increase the pump energy and can be attributed to decrease as well as slow formation of molecular $C_2$ at higher laser. The LIF images obtained at various delays between the pump and the probe beams have been used to measure the ground state number density of $C_2$ molecules. LIF images are calibrated with complementary absorption measurements in terms of absolute number densities of ground state species.

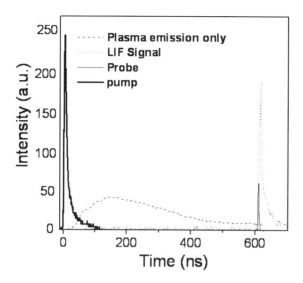

Fig. 14. LIF scheme that is used to ascertain the delay time of the probe laser with respect to the pump laser at which there is no more spontaneous emission from the carbon plasma (Yadav, D., Gupta, V. and Thareja, R. K. (2009), Evolution and imaging of nanoparticles observed in laser ablated carbon plume, *J. Appl. Phys.*, 106, pp. 064903-1-064903-7).

Assuming linear dependence between ground state density and LIF intensity, the calibration factor relating the ground state number density to the measured LIF signal $f_0 = n_0/S_{LIF}$ is determined from

$$\int n_0(x)dx = f_0 \int S_{LIF}(x)dx \qquad (24)$$

where the integral on the right-hand side of the equation is obtained from the corresponding LIF image (Dutouquet and Hermann, 2001). $S_{LIF}(x)$ is the pixel-wise LIF intensity along the direction of probe beam. The integration variable $x$ is the position following the probe laser beam axis. From the absorption experiment the value of $\int n_0(x)dx = 1.2 \times 10^{16}$ m$^{-2}$. This gives $f_0 = 5 \times 10^{15}$ m$^{-3}$ which can then be applied to calibrate LIF images.

Figure 15 shows the ground state number density of $C_2$ molecules at various delays. The maximum value of ground state density of $C_2$ molecules, $n_0 = 1.12 \times 10^{22}$ m$^{-3}$ is observed at a delay of 600 ns. The ground state number density of $C_2$ molecules obtained by LIF

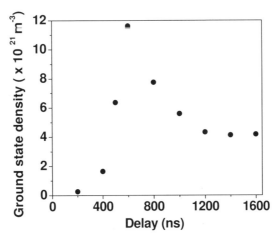

Fig. 15. Ground state number density of $C_2$ molecule at various delay times (Yadav, D., Gupta, V. and Thareja, R. K. (2009), Evolution and imaging of nanoparticles observed in laser ablated carbon plume, *J. Appl. Phys.*, 106, pp. 064903-1-064903-7).

investigation is in good agreement with the mass of molecular part of plume obtained in the previous analysis based on the hemispherical model.

## 4. Conclusions

In this chapter, we discussed the formation of $C_2$ and CN molecules in the laser-ablated carbon plasma at different ambient gas pressures and laser intensities. We studied the formation of CN band (violet $B^2 \Sigma^+ - X^2 \Sigma^+$ band system corresponding to the sequence $\Delta v = -1$) when graphite was ablated in nitrogen ambient using OES. Emission spectrum showed strong spatial as well as pressure dependence. In order to study the formation of CN band we also investigated the $C_2$ Swan band (d $^3\Pi_g$ – a $^3\Pi_u$) sequence corresponding to $\Delta v = 0$ and $-1$, respectively, at various ambient nitrogen pressures at the same fluence. The plume dynamics is discussed by imaging the expanding plasma using an ICCD camera. The images also showed splitting of plasma plume into stationary and moving components in a specific pressure regime attributed to the formation of nanoparticles (NPs) as a result of recombination of various ionic and molecular species near the target surface at later times. The mass of molecular $C_2$ and NPs in the plasma is estimated by considering the balance between plume linear momentum variation and the external pressure force, with the assumption that molecular $C_2$ follows hemispherical expansion whereas NPs follow 1-D expansion as inferred from images. LIF investigations are performed for the $C_2$ transition $d^3\Pi_g - a^3\Pi_u$ to measure the ground state number density of $C_2$ molecules.

## Acknowledgement

Work is supported by Department of Science and Technology (New Delhi, India).

## References

Abhilasha, Prasad, P. S. R. and Thareja, R. K. (1993). Laser-produced carbon plasma in an ambient gas, *Phys. Rev. E*, 48, pp. 2929-2933.

Amaratunga, G. A. J., Chhowalla, M., Kiely, C. J., Alexandrou, I., Aharonov, R. and Devenish, R. M. (1996). Hard elastic carbon thin films from linking of carbon nanoparticles, *Nature*, 383, pp. 321-323.

Amoruso, S., Bruzzese, R., Wang, X. and Xia., J. (2008). Propagation of a femtosecond pulsed laser ablation plume into a background atmosphere, *Appl. Phys. Lett.*, 92, pp. 041503-1-041503-3.

Baranov, V. Yu., Derkach, O. N., Grishina, V. G., Kanevskii, M. F. and Sebrant, A. Yu. (1993). Dynamics and stability of an expanding laser-induced plasma in a low-density gas, *Phys. Rev. E*, 48, pp. 1324-1330.

Bezryadin, A., Westervelt, R. M. and Tinkham, M. (1999). Self-assembled chains of graphitized carbon nanoparticles, *Appl. Phys. Lett.*, 74, pp. 2699-2701.

Bogdanov, A. A., Deininger, D. and Dyuzhev, G. A. (2000). Development prospects of the commercial production of fullerenes, *Tech. Phys.*, 45, pp. 521-527.

Chandrasekhar, S. (1981). *Hydrodynamic and Hydromagnetic Stability* (Oxford University Press, London).

Chen, G. X., Hong, M. H., Chong, T. C., Elim, H. I., Ma, G. H. and Ji. W. (2004). Preparation of carbon nanoparticles with strong optical limiting properties by laser ablation in water, *J. Appl. Phys.*, 95, pp. 1455-1459.

Demtröder, W. (2003). *Laser Spectroscopy* (Springer-Verlag, Berlin).

Diederich, L., Barborini, E., Piseri, P., Podestà, A., Milani, P., Schneuwly, A. and Gallay R. (1999). Supercapacitors based on nanostructured carbon electrodes grown by cluster-beam deposition, *Appl. Phys. Lett.*, 75, pp. 2662-2664.

Donnet, J. B. and Voet, A. (1976). *Carbon Black Physics, Chemistry and Elastomer Reinforcement* (Dekker, New York).

Dutouquet, C. and Hermann, J. (2001). Laser-induced fluorescence probing during pulsed-laser ablation for three-dimensional number density mapping of plasma species, *J. Phys. D: Appl. Phys.*, 34, pp. 3356-3363.

Dwivedi, R. K. (1997). *Laser ablated plumes for thin carbon film deposition*, Ph. D. thesis, IIT Kanpur, India.

Dwivedi, R. K., Singh, S. P. and Thareja, R. K. (1998). Ion probe diagnostics of laser ablated plumes for thin carbon films deposition, *Int. J. Mod. Phys. B*, 12, pp. 2619-2633.

Dyer, P. E., Issa, A. and Key, P. H. (1990). Dynamics of excimer laser ablation of superconductors in an oxygen environment, *Appl. Phys. Lett.*, 57, pp. 186-188.

Edelstein, A. S. and Cammarata, R. C., Nanomaterials: Synthesis, Properties and applications (CRC Press, 1998).

Freiwald, D. A. and Axford, R. A. (1975). Approximate spherical blast theory including source mass, *J. Appl. Phys.*, 46, pp. 1171-1175.

Gaydon, A.G. (1974). *The Spectroscopy of Flames* (Chapman and Hall, London).

Geohegan, D. B. (1991). *Laser Ablation: Mechanisms and Applications* (Springer, Heidelberg).

Geohegan, D. B. (1992). Fast intensified-CCD photography of $YBa_2Cu_3O_{7-x}$ laser ablation in vacuum and ambient oxygen, *Appl. Phys. Lett.*, 60, pp. 2732-2734.

Geohegan, D. B., Puretzky, A. A., Duscher, G. and Pennycook, S. J. (1998). Time-resolved imaging of gas phase nanoparticle synthesis by laser ablation, *Appl. Phys. Lett.*, 72, pp. 2987-2989.

Goldsmith, J. E. M. and Kearsley, D. T. B. (1990). $C_2$ creation, emission, and laser-induced fluorescence in flames and cold gases, *Appl. Phys. B*, 50, pp. 371-379.

Harilal, S. S., Radhakrishnan, P., Nampoori, V. P. N. and Vallabhan, C. P. G. (1994). Temporal and spatial evolution of laser ablated plasma from $YBa_2Cu_3O_7$, *Appl. Phys. Lett.*, 64, pp. 3377-3379.

Harilal, S. S., Issac, R. C., Bindhu, C. V., Nampoori, V. P. N. and Vallabhan, C. P. G. (1996). Temporal and spatial evolution of $C_2$ in laser induced plasma from graphite target, *J. Appl. Phys.*, 80, pp. 3561-3565.

Harilal, S. S., Bindhu, C. V., Tillack, M. S., Najmabadi, F. and Gaeris, A. C. (2003). Internal structure and expansion dynamics of laser ablation plumes into ambient gases, *J. Appl. Phys.*, 93, pp. 2380-2388.

Herzberg, G. (1950). *Molecular Spectra and Molecular Structure. I. Spectra of Diatomic Molecules* (Van Nostrand Reinhold, New York).

Huddlestone, R. H. and Leonard, S. L. (eds.) (1965). *Plasma Diagnostic Techniques* (Academic Press, New York).

Ikegami, T., Ishibashi, S., Yamagata, Y., Ebihara, K., Thareja, R. K. and Narayan, J. (2001). Spatial distribution of carbon species in laser ablation of graphite target, *J. Vac. Sci. Technol. A*, 19, pp. 1304-1307.

Kelly, R. (1990). On the dual role of the Knudsen layer and unsteady, adiabatic expansion in pulse sputtering phenomena, *J. Chem. Phys.*, 92, pp. 5047-5056.

Kroto, H. W., Heath, J. R., O'Brien, S. C., Curl, R. F. and Smalley, R. E. (1985). $C_{60}$: Buckminsterfullerene, *Nature*, 318, pp. 162–163.

Kushwaha, A. and Thareja, R. K. (2008). Dynamics of laser-ablated carbon plasma: formation of $C_2$ and CN, *Appl. Optics*, 47, pp. G65-G71.

Liberman, M. A. and Velikovich, A. L. (1986). *Physics of Shock Waves in Gases and Plasmas* (Springer-Verlag, Berlin).

Li, D., Liu, Y., Yang, H. and Qian, S. (2002). Femtosecond nonlinear optical properties of carbon nanoparticles, *Appl. Phys. Lett.*, 81, pp. 2088-2090.

Lopinski, G. P., Merkulov, V. I., and Lannin, J. S. (1998). Semimetal to semiconductor transition in carbon nanoparticles, *Phys. Rev. Lett.*, 80, pp. 4241-4244.

Mayo, M., Newman, J. W., Sharma, A., Yamagat, Y. and Narayan, J. (1999). Electrostatic measurement of plasma plume characteristics in pulsed laser evaporated carbon, *J. Appl. Phys.*, 86, pp. 2865-2871.

Misra, A. and Thareja, R. K. (1999). Laser-ablated plasma for deposition of aluminum oxide films, *Appl. Surf. Sci.*, 143, pp. 56-66.

Misra, A., Mitra, A. and Thareja, R. K. (1999). Diagnostics of laser ablated plasmas using fast photography, *Appl. Phys. Lett.*, 74, pp. 929-931.

Morrow, T., Sakeek, H. F., Astal, A. El, Graham, W. G. and Walmsley, D. G. (1994). Absorption and emission spectra of the YBCO laser plume, *J. Supercond.*, 7, pp. 823-828.

Neogi, A. (1999). *Temporal and spatial evolution of laser ablated carbon plasma in ambient gas and magnetic field,* Ph. D. thesis, IIT Kanpur, India.

Nakata, Y., Kumuduni, W. K. A., Okada, T. and Meada, M. (1995). Two-dimentional laser-induced fluorescence imaging of non-emissive species in pulsed laser deposition process of $YBa_2Cu_3O_{7-x}$, *Appl. Phys. Lett.*, 66, pp. 3206-3208.

Predteceensky, M. R. and Mayorov, A. P. (1993). Expansion of laser plasma in oxygen at laser deposition of HTSC films: theoretical model, *Appl. Supercond.*, 1, pp. 2011-2017.

Rao, A. M. and Dresselhaus, M. S. (2001). *Nanostructured Carbon for Advanced Applications*, G. Benedek, P. Milani, and V. G. Ralchenko (eds.) (Kluwer Academic, The Netherlands).

Sharma, A. K. (2005). *Formation and characterization of nitrides and oxides during reactive pulsed laser ablation*, Ph. D. thesis, IIT Kanpur, India.

Sharma, A. K. and Thareja, R. K. (2004). Characterization of laser-produced aluminum plasma in ambient atmosphere of nitrogen using fast photography, *Appl. Phys. Lett.*, 84, pp. 4490-4492.

Sharma, A. K. and Thareja, R. K. (2005). Plume dynamics of laser-produced aluminum plasma in ambient nitrogen, *Appl. Surf. Sci.*, 243, pp. 68-75.

Shea, J. D. (1927). The structure of the Swan bands, *Phys. Rev.*, 30, pp. 825-843.

Siegal, M. P., Overmyer, D. L., Kottenstette, R. J., Tallant, D. R. and Yelton, W. G. (2002). Nanoporous-carbon films for microsensor preconcentrators, *Appl. Phys. Lett.*, 80, pp. 3940-3942.

Spindler, R. J. (1965). Franck-Condon factors based on RKR potentials with applications to radiative absorption coefficients, *J. Quant. Spectrosc. Radiat. Transfer*, 5, pp. 165-204.

Yu, J., Wang, E. G. and Bai, X. D. (2001). Electron field emission from carbon nanoparticles prepared by microwave-plasma chemical-vapor deposition, *Appl. Phys. Lett.*, 78, pp. 2226-2228.

Taylor, G. (1950). The Instability of Liquid Surfaces when Accelerated in a Direction Perpendicular to their Planes. I, *Proc. R. Soc. Lond. A*, 201, pp. 192-196.

Thareja, R. K. and Abhilasha (1994). $C_2$ from laser produced carbon plasma, *J. Chem. Phys.*, 100, pp. 4019-4020.

Thareja, R. K., Dwivedi, R. K. and Abhilasha (1997). Role of ambient gas on laser-ablated plumes for thin carbon film deposition, *Phys. Rev. B*, 55, pp. 2600-2605.

Thareja, R. K., Dwivedi, R. K., and Ebihara, K. (2002). Interaction of ambient nitrogen gas and laser ablated carbon plume: Formation of CN, *Nucl. Instr. And Meth. In Phys. Res. B*, 192, pp. 301-310.

Vincent, D., Petit, S. and Chin, S. L. (2002). Optical limiting studies in a carbon-black suspension for subnanosecond and subpicosecond laser pulses, *Appl. Opt.*, 41, pp. 2944-2946.

Wood, R. F., Chen, K. R., Leboeuf, J. N., Puretzky, A. A. and Geohegan, D. B. (1997). Dynamics of plume propagation and splitting during pulsed-laser ablation, *Phys. Rev. Lett.*, 79, pp. 1571-1574.

Yadav, D., Gupta, V. and Thareja, R. K. (2009), Evolution and imaging of nanoparticles observed in laser ablated carbon plume, *J. Appl. Phys.*, 106, pp. 064903-1-064903-7.

Zeldovich, Y. B. and Raizer, Y. P. (1967). *Physics of Shock Waves and High-Temperature Hydrodynamic Phenomena* Vol. I (Academic, New York).

Chapter 8

# Laser Spectroscopy of Transient Carbon Species in the Context of Soot Formation

Václav Nevrlý,[1] Michal Střižík,[1] Petr Bitala,[1] Zdeněk Zelinger[2]

[1]*Technical University of Ostrava, Faculty of Safety Engineering,
Lumírova 13, 700 30 Ostrava-Vyškovice, Czech Republic*
[2]*J. Heyrovský Institute of Physical Chemistry, v.v.i., Academy of Sciences
of the Czech Republic, Dolejškova 3, 182 23 Prague 8, Czech Republic*

Proper understanding of soot formation chemistry is strongly dependent on experimental studies of transient carbon species. For more than two decades, experimental approaches based on advanced laser spectroscopic methods are successfully used when dealing with non-invasive detection and monitoring of transient carbon-containing species in gas phase relevant to astronomical applications, plasma processing and combustion research. This chapter aims at summarizing the state of the art in the field of laser spectroscopy of transient carbon species important for soot and polyaromatic hydrocarbons (PAH) formation process. After a brief overview of soot formation theories, plasma generation methods together with laser spectroscopic techniques employed for optical diagnosis of laboratory plasma are discussed. Main emphasis is laid upon a description of tunable diode laser absorption spectroscopy applied in investigation of carbon-containing radicals and ions.

## 1. Introduction

The spectra of carbon-containing transient species have been studied extensively for more than a century. Carbon chain radicals, their charged analogues (ions) and slightly hydrogenated carbon clusters are of exceptional interest in astronomy as they are suggested to be possible carriers of several of mysterious diffuse interstellar bands (DIBs)[1,2]. In terrestrial conditions, small carbon clusters are known to be present in hydrocarbon flames[3] and together with hydrocarbon radicals and ions are investigated as important combustion intermediates and precursors of carbonaceous soot particles. Finally, the above-mentioned reactive (transient) carbon species are involved in plasma assisted processes like chemical vapor deposition (CVD) and synthesis of fullerenes, carbon nano-tubes and nano-crystaline diamonds (NCD), during which soot formation is often responsible for deposit contamination and process instabilities[4].

A large number of candidate molecules like reactive free radicals ($C_2$, $C_3$, $C_2H$, $C_3H_3$, $C_6H_5$), hydrocarbon cations ($C_3H_3^+$), polyaromatic hydrocarbons (PAH) and many other species have been investigated in the context of soot formation[5]. Nevertheless, the clear identification of direct soot precursor still seems to be unforeseeable without further improvements in diagnostic capabilities. From the practical point of view, the development of measurement techniques and strategies for *in situ* diagnostics is crucial to the effective control of phenomenon of soot formation not only under laboratory conditions (e.g. in a harsh environment of flame and hydrocarbon plasma).

In the following text, the term *transient species* will stand for short-lived atoms, molecules and ions in gas phase under standard laboratory conditions. A detailed discussion about terminology characterizing the above-mentioned types of species is stated e.g. by Herzberg[6] and Hirota and Endo[7]. It is highly reactive free radicals and ions that are usually called transient species. Free radicals can be defined as open-shell molecular systems, i.e. systems with one or more unpaired electrons, and thus nonzero spin ($S \neq 0$). Generally, it can be a case of unstable neutral species and some ions. However, this definition also includes chemically stable molecules, such as triplet-state $O_2$ and $C_6H_6$, and on the contrary, it excludes some short-lived highly reactive species like singlet-state $C_2$, $C_3$ and $CH_2$, when $S = 0$. In addition, only in some ions, i.e. molecular and atomic systems carrying positive or negative charges, an unpaired electron simultaneously occurs as with e.g. a molecule of $C_2H^-$ anion.

Therefore, in the case of hydrocarbon ions what is often meant are not free radicals in the true sense of the word[8].

Transient carbon species are present in hydrocarbon flames and other soot-forming systems[9-11]. These species, especially simple carbon clusters and hydrogenated carbon chains are spontaneously formed in the course of processes in carbon and hydrocarbon plasmas. Simultaneously, it has been proved that they occur in atmospheres around carbon stars and in interstellar clouds irradiated with intensive UV (ultraviolet) radiation. In spite of the very low concentrations in which transient carbon species occur in these environments, it is a case of entirely decisive intermediates in the chemistry of gas phase carbon. Thus the study of transient carbon species is a necessary precondition for understanding these, from the chemical point of view, very complex systems. The great significance of various forms of carbon bonds is observed increasingly in areas ranging from astrophysics and combustion chemistry to molecular physics; spectroscopic studies with a view to determine the structure of carbon clusters[3,12] play an important role here.

## 2. Soot Formation Chemistry

Soot formation chemistry in its full comprehensiveness has not been examined thoroughly yet. That is why it is often designated as one of unresolved fundamental problems in the area of combustion research.

Purely in a phenomenological way, the mechanism of soot formation can be divided into the pyrolysis of hydrocarbons, the formation of soot nuclei, the polymerization of simple hydrocarbons and the subsequent condensation of aromatic hydrocarbons leading to PAH formation, the heterogeneous growth of particle surface (due to reactions between gas molecules and the particle surface), the coagulation of soot particles and finally the agglomeration, oxidation and fragmentation reactions of soot in interaction with gas phase species. When considering the formation and the growth of soot particles as a process taking place in a certain time, the mentioned phases cannot be differentiated strictly.

What remains to be the cardinal issue is especially the detailed chemical mechanism of forming the first and subsequently the second aromatic ring. The historical ambiguity of an answer to this issue gave rise to two basic schools of opinion: although at present rather a radical mechanism of aromatic ring formation is already generally accepted, an alternative ionic mechanism will be mentioned briefly in this chapter as well.

## 2.1. *Radical mechanism of soot formation*

The radical mechanism presupposes the production of free radicals by the initiation reaction, which can be hydrocarbon pyrolysis. Produced hydrocarbon radicals can, in a flame or in a plasma, participate in a series of consecutive and/or competitive chemical reactions leading to the formation of larger molecular fragments and stable intermediates (such as molecular hydrogen or carbon monoxide). In addition to the propagation reactions of pyrolytically-generated radicals with other molecules of hydrocarbons, the growth of soot particles is mainly influenced by the branching of the growing chain as a result of reaction between the radicals and the reactive intermediates being produced, and also by recombination reactions. To reactions supporting the formation and growth of soot particles, oxidation processes taking place in the high-temperature zone, where transported soot particles react with strong oxidizing agents, such as molecular oxygen or OH radicals, are competitive. These oxidation processes lead to a loss of mass and changes in the physical structure of soot particles; the extent of occurring oxidation reactions depends heavily on the distribution of OH radicals and the structure of temperature field.

Bockhorn et al.[13], on the basis of studies of flat premixed low pressure flames of acetylene, propane and benzene, proposed a scheme of the production of the first aromatic ring with the initial reaction of diacetylene with the $C_2H$ radical to form a branched hydrocarbon radical attacked subsequently by acetylene. By closing the ring, a phenylacetylene radical was then produced. Cole et al.[14] investigated mechanisms of formation of aromatic species in flames of aliphatic hydrocarbons and arrived at the conclusion that the key intermediate for the formation of benzene by the radical mechanism including the closing of the ring was the 1,3-butadienyl radical. The study of vinylacetylene pyrolysis[15] showed that at temperatures more than 1600 K, the recombination reaction of $n$-$C_4H_3$ radical with $C_2H_2$ was the dominant pathway for phenyl radical production. Westmoreland *et al.*[16] identified the reactions of $n$-$C_4H_5$ and $n$-$C_4H_3$ radicals with acetylene as key pathway for benzene formation. In order to explain the formation of the first aromatic ring, Miller and Melius[17] distinguished the role of $C_3H_3$ radical when considering a formation of benzene in acetylene flames. A potential role of C5 species in benzene formation was also investigated[18,19].

Nevertheless, benzene is not taken as a single possible precursor in the process of further growth of soot particles. Besides the above-mentioned phenylacetylene, other candidates are styrene[15] and a cyclopentadiene radical[20,21]. One of dominant reaction paths of the subsequent PAHs condensation is represented by a sequence of hydrogen-abstraction / acetylene-addition (HACA) steps[22-24]. Because this monograph is devoted above all to small carbon clusters, readers who require more information on the subsequent heterogeneous growth of surface of soot particles, the coagulation of them and agglomeration processes are kindly invited to refer to reviews dealing with these problems[25,26].

## 2.2. *Ionic mechanism of soot formation*

The ionic theory of soot particle formation, advocated by H.F. Calcote, D.G. Keil[27] and A.B. Fialkov[28] among others, presupposes the participation of chemiions of the $C_3H_3^+$ type in the earliest phases of the process of soot formation. An ion serves as nucleus, on which the addition of acetylene, diacetylene and substances of similar types occurs at carbon chain lengthening. Supporters of the ionic mechanism of first aromatic ring formation consider the radical mechanism to be too slow to account for rapid soot formation rates in flames. One can assume that straight-chain ions can undergo, in the environment of a flame or plasma, cyclization reactions much faster than electro-neutral hydrocarbon radicals.

According to the ionic theory, one of pathway for the formation of $C_3H_3^+$ in the reaction zone is the chemiionization reaction between the CH radical and $C_2H_2$ (in the ground or electronically excited states), when an electron is abstracted[29]. The $C_3H_3^+$ chemiion can also be formed by transferring a proton from another charged species occurring naturally in a flame or plasma (e.g. $CHO^+$) to the $C_3H_2$ radical. The repeated and relatively rapid exothermic addition of electro-neutral molecules of the $C_2H_2$ type to the linear $C_3H_3^+$ isomer (propargyl) at first and subsequently to the formed ion with the lengthened carbon chain then leads to the growth of the ion and in the end to its cyclization to form $PAH^+$ (e.g. $C_{19}H_{11}^+$). With lengthening the carbon chain ion, the value of its electron recombination coefficient, meaning the probability of reaction between the ion and the electron (released by the initiation chemiionization reaction), also increases. Due to recombination with electrons or dissociation of growing ions, neutral species are produced. They can further participate in the process of soot growth described in the introductory paragraph of subchapter 2. Free electrons present in the environment of a plasma or flame can be, in some cases, retained by larger electro-neutral molecules to form carbanions. Dissociation reactions of ions simultaneously generate small positively charged carbon clusters that can serve as starters for forming new longer-chain hydrocarbon ions. The cycle of PAH formation controlled by the ionic mechanism is thus closed.

Some observations provided results supporting the hypothesis of controlling the process of soot *nucleogenesis* by the ionic mechanism[30]. On the other hand, a number of experiments that contest this hypothesis have been carried out[31]. And just modern spectroscopic techniques could become one of tools for discovering a definite answer to the issue of the mechanisms by means of which the earliest phases of soot particle forming take place.

### 2.3. *Some interesting transient carbon species*

In addition to commonly known hydrocarbon radicals (methyl-, vinyl-, propargyl-, allyl-, phenyl-radical) and ions, to which importance to soot formation has already been attached for many years, the role of diacetylene as well as longer carbon chain molecules has been investigated lately.

Kaiser *et al.*[32] studied, using the approach combining *ab initio* calculations with mass spectroscopy, in cross-linked molecular beams, the production of so-called "hydrogen deficient" carbon molecules in the framework of combustion chemistry and soot formation process. Their results have proved a possible presence of these molecules in reactions taking place in hydrocarbon flames, where slightly hydrogenated carbon chains (species of the $C_4H$ or $C_5H$ type) can be formed from unsaturated hydrocarbons by their reactions with small carbon clusters ($C_2$ or $C_3$). Above all $C_2$ addition was designated as one of probable pathways for carbon chain lengthening in the environment of a flame, hydrocarbon plasma and also in the universe.

Carbon clusters are fascinating examples of the richness and diversity of carbon chemistry. Thanks to the great binding flexibility of carbon (capability to form single, double and triple bonds), carbon chains occur in an extensive range of structural forms. A very interesting feature of small linear carbon clusters ($n \leq 11$, where $n$ is odd) is the unusually low vibration frequency of bending modes in relation to the central carbon atom[12,33-36]. Just these low-frequency large-amplitude bending modes can be a cause of isomerization of carbon chains into carbon rings.

For carbon clusters of size less than 20 carbon atoms, *ab initio* calculations predict the existence of two most stable structures: linear structure for small clusters and cyclic one for larger clusters[37]. Results of the first theoretical calculations of carbon cluster structure were published in[38,39]. They predict the linear form for $C_n$ carbon clusters with $n \leq 10$ and the ring form for clusters with $n > 10$. Odd-numbered linear carbon clusters have $^1\Sigma$ singlet ground states, whereas even-numbered linear carbon clusters have $^3\Sigma$ triplet ground states.

## 3. General Experimental Approaches and Instrumentation

The combination of infrared (IR) and visible (VIS) spectroscopy of mass-selected clusters (see Chapter 3.2.4) isolated in noble-gas matrices[40] and high resolution infrared absorption spectroscopy of gas phase[3,41] contributed substantially to the characterization of the structure and dynamics of small carbon clusters. *Ab initio* calculations applied to this group of molecules provide the better-quality prediction of their chemical and physical properties (see e.g.[42,43]) including their rovibrational spectra as well. Three basic limitations on spectral high resolution observation of gas phase unstable molecular clusters exist, namely:

(1) production of molecular clusters in sufficient concentrations,
(2) availability of detection methods of sufficient sensitivity and spectral resolution for recording rovibrational spectra,
(3) high-quality prediction of energy levels of molecular systems that will enable the accurate prediction of spectral band positions and dipole moment sizes.

Transient carbon and hydrocarbon species have in general short lifetimes (up to a few µs) and are present in very low concentration levels (often less than $10^9$ molecules/cm$^3$), even if the sample volume is generated in laboratory conditions.

For the above-mentioned reasons, several basic factors should be considered in the preparation of high resolution spectroscopic experiments. Above all, influences causing the broadening of spectral lines are to be eliminated, especially in a case of considerably dense rotationally resolved spectra of species having more complicated molecular structures.

## 3.1. Experimental set-up

Spectroscopic measurements are usually performed at reduced pressure, which diminishes the effects of so-called pressure (collision) broadening. To achieve the required spectral resolution (usually in the order of several thousandths cm$^{-1}$), the measured volume is closed in a vacuum chamber at the pressure < 20 Torr. In these conditions, the width of lines is limited by Doppler broadening. The effect of this type of spectral line broadening can be avoided by: (a) applying "Doppler-free" methods of laser spectroscopy; (b) using supersonic jet (molecular beam) techniques.

### 3.1.1. *Spectroscopy in supersonic jets*

Lately, supersonic jet techniques have become an integral part of the prevailing majority of spectroscopic studies of transient species. The first important aspect of these techniques is the already-mentioned decrease in the Doppler width of spectral lines in the molecular beam.

In a molecular flow characterized by a uniform direction of vector of translation motion of molecules, chemical reactions do not take place due to the absence of molecular collisions. Another aspect of no less importance in connection with the high resolution spectroscopy of transient species is the strong cooling of molecules ($T_{rot} \approx 10$ K) in the course of supersonic expansion taking place in almost adiabatic mode. In these conditions, only the lowest vibrational rotation energy levels are thermally populated, which results in a considerable simplification of the structure of rotationally resolved spectra, limitation on spectral line overlaps and elimination of occurrence of hot and combination bands[44].

### 3.1.2. *Spectroscopy in absorption cells*

Experimental set-ups based on the preparation of transient species in static and flow-through absorption cells represent a traditional approach in high resolution molecular spectroscopy.

Techniques, by which transient species are produced in cells, are usually various types of electric discharge, photolysis and also pyrolysis. A special case is the preparation of radicals on the basis of chemical reactions initiated by IR laser radiation. Wallaart *et al.*[45] produced the $C_2$ radical in a cell (filled with a CO/Ar/He mixture) by the CO-laser vibrational excitation of carbon monoxide.

### 3.1.3. *Multipass arrangement*

To increase sensitivity in the course of measurements of very low concentrations of transient species, methods of laser absorption spectroscopy using the arrangement that ensures a multiple passing of a laser beam through the measured volume, and thus an optical path extension, are applied. Curl and Tittel[46] state three basic types of multireflection cells that are used in measurements in cells and also supersonic jets. They are as follows: (a) White cells; (b) Herriot cells; (c) astigmatic mirror cells.

In the case of the first two types of cells the maximum achievable optical path is limited by a multiple of distance between two opposite mirrors and the number of possible reflections without the occurrence of overlapping points (usually realized approximately 25 reflections). Cells of astigmatic mirrors, which are a modified variant of Herriot cell, enable mutual overlaps of spots formed by individual passes on both the mirrors. In this way, it is possible to achieve even more than a hundred passes through the measured volume.

### 3.2. Production of transient carbon species

Before the spectroscopic measurement in gas phase, the careful selection and optimization of techniques, by means of which a sufficient amount (detectable concentrations) of transient species can be produced, must be carried out. Several well-established techniques for the production of transient species for spectroscopic investigation are briefly summarized in the following subchapters. The correct selection of a precursor molecule is, in this phase of experiment preparation, wholly crucial. Generally, by using a suitable procedure it is possible to prepare transient species with molecular structure much more complicated than the structure of precursor itself. For example, in a discharge plasma containing acetylene ($C_2H_2$), transient species with a long carbon chain are produced. On the contrary, techniques of photolysis and pyrolysis are based on the principle of dissociation of precursor molecule. Moreover, the selection of a carrier gas is of similar importance. For this purpose, noble gases are used almost exclusively, of which Ar, He and Ne are used most frequently. When applying preparation techniques based on electric discharges in gases, Ar is preferred, because discharges in it are usually more stable. A helium discharge is used in the case of application of Penning ionization with helium atoms.

## 3.2.1. *Electric discharge*

A low-temperature plasma containing transient carbon species can be prepared by the electric discharge excited by high voltage (application of electric field to the system of electrodes).

All the following types of electric field have proved themselves in the production of free radicals and ions for the purpose of spectroscopic studies[47]: continuous (d.c.) discharges; low-frequency (a.c.) discharges; radio-frequency (RF) discharges and microwave (MW) discharges.

Whereas the nature of glow discharge in gases is quite well known, the neutral chemistry of discharges controlled rather by dissociative electron impact is not clarified in detail. On the contrary, the production of ions in the environment of electric discharge plasma has been examined much better in the course of tens years devoted to the investigation of ion-molecular reactions[46].

Herzberg[6] described a so-called "afterglow" discharge technique; as for this technique, the continuous discharge takes place in a lateral tube entering the absorption cell, which limits undesirable effects caused by the electric discharge in the absorption spectrum. This modification is however suitable only in a case of investigating the species with a longer lifetime that is sufficient for preventing the species from extinction in the measured part of absorption cell.

The electric discharge can be generated inside a hollow cathode. In this case, it is suitable to cool the cathode with liquid nitrogen or by a water flow system. Simultaneously with the hollow cathode, the concentration modulation technique, sometimes called population modulation or on-off modulation, can be applied conveniently. The advantage of this procedure consists in the utilization of a modulated a.c. (order of tens to thousands Hz) discharge; a signal from the detector being processed by means of the phase-sensitive (lock-in) amplifier that evaluates merely the 1f harmonic component of the signal, i.e. at the identical frequency of discharge modulation. In this way, a marked increase in signal-to-noise ratio (SNR) is achieved. The experimental set-up for concentration modulation was described by e.g. Davies[48] and by Civis *et al.*[49]

From the experimental point of view, it is often desirable to differentiate neutral radicals from ions. Although the above-described method of concentration modulation does not differentiate directly these two types of species from each other, with regard to a substantial difference in characteristic lifetime it is however possible to adapt the modulation frequency to the detection of required type of species. The modulation frequency corresponds to the lifetime of the species detected. For relative longer-lived neutrals, frequencies less than about 1 kHz are usually used, whereas for very short-lived ions it is necessary to apply higher modulation frequencies of about 100 kHz.

A method of direct differentiation between charged and neutral species in electric discharges is so-called velocity modulation[50-52]. In this technique, the polarity of electrodes within the discharge cell is periodically inverted, which affects the direction of motion of generated electrically-charged species. Thus not only ions can be distinguished from neutral species, but also cations from anions.

## 3.2.2. *Flash photolysis*

Although photolytic techniques for the preparation of hydrocarbon radicals became widely used no sooner than the period of introduction of lasers as high-energy monochromatic radiation sources, Norrish and Porter[53] had developed, already many years before, the technique of flash photolysis used up to the present.

One of early applications of this method was the production of $CH_2$ radical[6] by the photolytic dissociation of diazomethane ($CH_2N_2$) or ketene ($C_2H_2O$) for the purpose of interpretation of a spectral band of comets in the 405 nm region. By coincidence, the $C_3$ radical was identified by Douglas[54] on the basis of isotopic studies in the course of investigation into just this spectral band in the year 1951.

Nowadays a focused beam of a pulsed excimer laser enabling periodically repeated measurements in the range of 10–100 Hz is most commonly used for photolysis; the required radical being usually produced by releasing a bromine or iodine atom substituted in a molecule of chosen precursor (halide). Thus it is possible to obtain, in a very selective way, hydrocarbon radicals (e.g. $C_2H$, $C_2H_3$, $C_3H_3$, $C_3H_5$)

in concentrations sufficient for the application of high resolution spectroscopic methods[7].

### 3.2.3. *Pyrolysis*

In contrast to photolytic dissociation based on the initiation of a sequence of chemical reactions by photon absorption, pyrolysis is connected with reaction mechanisms induced by violent intermolecular collisions. It is a case of heat-induced dissociation that is usually carried out by heating a sample to a temperature sufficient for the dissociation of the weakest bond in a precursor molecule.

As follows from the given principle, it is the selection of a suitable precursor (and a respective way of its chemical synthesis), that is pivotal in the sufficiently selective generation of transient species by the process of pyrolysis. In the course of spectroscopic experiments in absorption cells, where there is a need to ensure as uniform as possible distribution (concentration) of the studied species in the whole measured volume of the cell, the applicability of pyrolysis is limited to relatively longer-lived transient species.

In spite of the above-mentioned disadvantages, pyrolysis is used successfully in the production of hydrocarbon radicals, especially in combination with supersonic jet techniques. The increased pre-expansion temperature is, however, associated in this case with a somewhat higher rotation temperature in the region of supersonic jet beam of molecules ($T_{rot} \approx 40$ K) in comparison with other techniques of preparation (electric discharge or flash photolysis). Liu *et al.*[55] produced the methyl ($CH_3$) radical by the pulsed pyrolysis of *tert*-butyl nitrite at the temperature of about 1500 K. Hirota mentions the preparation of allyl ($C_3H_5$) radical by the pyrolysis of 1,5-hexadiene, allyl bromide or allyl iodide at approximately 800°C.

### 3.2.4. *Laser ablation*

A well-tried manner of the generation of carbon clusters, radicals and ions of carbon chains is laser ablation as well. This technique is based on a beam of pulsed (mostly UV excimer) laser focused on the surface of a rotating or translating target coated with graphite or another material being examined.

As a result of local heating to several thousand Kelvin, the material evaporates and a so-called laser plume is produced. By combination of laser ablation and electric discharge, in the region of the plume being generated, required carbonaceous radicals and ions can be produced on the basis of suitably selected supersonic jet beam and parameters of the electric discharge.

Wide variety of transient carbon species (C, $C^+$, $C_2^+$, $C^{++}$, $C_3^+$) was produced in[56-59] using the combined technique of pulsed KrF excimer laser ablation and modulated RF discharge, see Fig. 1. In a similar experimental set-up, Bratescu et al.[60] applied the method of laser absorption spectroscopy in time-resolved measurements (kinetic studies) of atomic carbon.

The most frequently used method of production of rather large carbon clusters is the laser gasification of a carbon target with the subsequent supersonic expansion in an inert carrier gas, usually He or Ar.

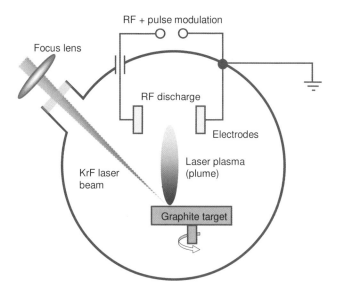

Fig. 1. Experimental set-up for combined technique of pulsed laser ablation and modulated RF discharge.

Neutral, anion- and cation-carbon clusters are produced directly in a laser-initiated plasma in the range from one to hundreds carbon atoms. For the first time, this technique was used by Smalley and his collaborators. Subsequently, it was applied by Kroto *et al.*[61] at the discovery of fullerenes. Carbon atoms are ablated and carried by helium pulses of the pressure of 10 bar. Small carbon clusters are generated in dense plasma in the extent of slits of the source before their adiabatic expanding into a vacuum chamber. By each laser pulse, clusters of various sizes are produced. In addition, laser ablation can be suitably supplemented by "time-of-flight" (TOF) measurement for mass-selected species[62]. This method makes it possible to define more exactly the species studied; in this way, carbon clusters are periodically (in pulsed mode) separated in a buffer gas on the basis of various masses. The adjustment of a time window in a box-car integrator enables the selection of species based on flight time determined by its mass.

### 3.2.5. *Flame*

Quite a number of carbon-containing transient species are produced via natural reaction mechanisms taking place in the course of hydrocarbon combustion[63]. Hydrocarbon flames are characterized by the intensive spectra of $C_2$ and CH radicals the presence of which in the combustion zones causes a characteristic bluish color of the flame. Some other hydrocarbon radicals have been successfully detected by the laser spectroscopic methods in a harsh environment of the flame in the framework of combustion process investigation[64].

The ionic character of the flame has been known and studied for already more than a hundred years. In spite of this, no ion (as well as many radicals decisive to the process of soot particles formation, such as $C_3H_3$, $C_3H_5$ and others) has been detected by *in situ* non-invasive optical methods in a flame environment[28].

For the purposes of spectroscopic investigation into transient species at reduced pressure, standard laboratory burners generating laminar premixed flames (e.g. McKenna burner or calibration burner published by Hartung *et al.*[65]) are utilized. Even provided the laboratory conditions are defined as well as that, the spectroscopic detection of transient species in a flame is complicated, however, by several factors: (a) the complex structure of high-temperature spectra; (b) the presence of a multicomponent mixture of reactants, intermediates and stable products of combustion, including solid soot particles; (c) higher total pressure.

## 4. Methods of Laser Spectroscopy

The unambiguous detection of individual transient species requires employing methods of a high degree of sensitivity, accuracy and sufficiently fast response time. The introduction of lasers into spectroscopy caused a revolution in molecular dynamics. Methods of high resolution laser spectroscopy enable non-invasive studies of transient species based on specific interactions between electromagnetic radiation (light) and matter. The methods of laser spectroscopy as well as suitable types of instrumentation (lasers, optical elements, detectors, etc.) are described in detail in excellent teaching texts by Hollas[66] and Demtröder[67], and also in reviews edited by Andrews and Demidov[68] and most recently by Lackner[69]. The purpose of this chapter thus is not to replace these bibliographies; the above-mentioned books are highly recommended to inquiring readers.

This chapter summarizes very shortly the selected methods of laser spectroscopy that play a key role in the studies of transient carbon species, including hydrocarbon radicals being of importance to the formation of soot particles. Furthermore, an overview of specific applications of particular methods of laser spectroscopy in the framework of investigation into the physical and chemical properties of these species is provided. In the publications referred to in this chapter, additional information on practical aspects of implementation of the given methods, their limitations and also present-day trends in their applications can be found.

### 4.1. Laser absorption spectroscopy

In addition to the oldest methods of emission spectroscopy, the experimental observation of absorption spectra represents historically one of fundamental tools for the study of transient species. At present, the potential of direct absorption methods in the field of rotational spectroscopy in gas phase is increased by: (a) progress in the area of development of widely tunable laser sources with a narrow (sub-Doppler) line-width; (b) increased availability of infrared technologies.

In many spectroscopic applications, especially in the middle infrared region, traditional tunable semiconductor diode lasers (TDLs) are replaced by difference frequency generation (DFG), optical parametric oscillators (OPO), quantum cascade (QC) and color center (CC) lasers.

The effective observation of very low concentrations of transient species by the methods of laser absorption spectroscopy requires, besides the multipass arrangement, the application of modulation techniques and phase-sensitive detection in connection with a possibility of averaging the periodically repeated measurements in pulsed mode (using lock-in amplifiers, and also box-car integrators).

Another possibility of increasing the sensitivity of laser absorption methods is the placing of the sample being studied into an optical cavity formed by a pair of highly reflective mirrors. Two variants of these cavity-enhanced methods are intracavity laser absorption spectroscopy (ICLAS)[70,71] and cavity ringdown spectroscopy (CRDS)[72]. The latter method, designated sometimes as cavity ringdown laser absorption spectroscopy (CRLAS)[73-76], has become, especially in connection with supersonic jet techniques, an important tool for spectroscopic investigations into carbon chain radicals and ions.

### 4.1.1. *Infrared tunable diode laser absorption spectroscopy (TDLAS)*

Direct infrared absorption spectroscopy provides a tool for observing the fine rotational structure that accompanies vibrational transitions, and furnishes thus the most direct experimental characterization of a molecular structure, bonds and vibration dynamics[77].

Rotationally resolved spectra of several small hydrocarbon radicals (CH, $CH_2$, $CH_3$, $C_2H$, $C_2H_3$, $C_2H_5$, $C_3H_3$, $C_3H_5$) were successfully measured by TDLAS applied in absorption cells and supersonic jets, see review of Hirota[78] and references herein. The infrared diode laser

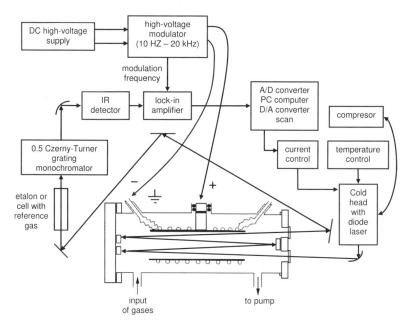

Fig. 2. Experimental set-up for IR laser diode spectroscopy of transient species produced in hollow cathode.

spectrum was investigated during the search for the rovibrational spectra of the $\nu_2$ bending mode of $C_3$ radical in the $\tilde{a}^3\Pi_u$ electronic state[49]. The concentration modulation technique by applying a 9.3 kHz a.c. high voltage to a hollow cathode was employed (see Fig. 2).

Series of relatively strong lines were observed within the 700 cm$^{-1}$ region. The spectrum was recorded phase-sensitively in the discharge plasma of acetylene diluted in helium. The carrier seemed to be a shortly living ion or radical, such as $C_2H_3^+$ and $C_2H$, judging from the response to the high (~ 10 kHz) frequency modulation. However, when the frequency was reduced to 1 kHz, the line intensity increased about 5 times, indicating that the carrier lifetime in the discharge was about 1 ms. Series of acetylene hot band $(\nu_2 + \nu_5 - \nu_2)$ lines were conclusively assigned to the observed IR spectra.

The combination of infrared laser diode spectroscopy and laser ablation as production method enables to record the rotationally resolved spectra corresponding to modes of asymmetric vibration of small carbon

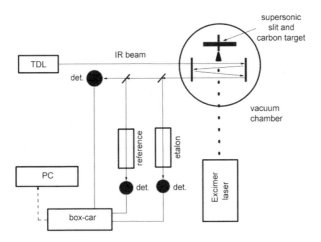

Fig. 3. Experimental set-up for IR laser diode spectroscopy of carbon clusters produced by UV laser ablation.

clusters. The experimental set-up illustrated in Fig. 3 utilizes an infrared beam of tunable diode laser for probing carbon clusters produced in the supersonic jet.

A substantial increase in sensitivity can be achieved by a multipass optical set-up. After multiple passing through a cell, the infrared beam is focused on the MCT (Mercury cadmium telluride) detector. The fast a.c. amplifier of the detector will enable the time-resolved detection of weak absorption signals that will be then transmitted by means of the box-car integrator and collected in a PC. Part of the infrared beam is used for recording the spectra of a reference gas and the etalon fringes (traditional part of laser diode spectroscopy).

The optimal tool for carbon cluster identification would be MW spectroscopy, but owing to the zero permanent dipole moment, pure rotation spectra of carbon clusters are not measurable. Nevertheless, they can be studied by rotationally resolved infrared spectroscopy that enables a detailed analysis of the fine molecular structure. Triplet splitting of $C_4$ in gas phase has shown to be unresolvable for detection by means of IR spectroscopy on the level of Doppler broadening[79-81]. However, for the case of $C_6$, the infrared spectrum shows a partially resolved triplet splitting for the four lowest P- and R- transitions[82].

In the work of Weltner and Van Zee[12] a rule that the amount of knowledge of carbon clusters decreases monotonously with the size of them, i.e. the number of carbons, was expressed. At present, this rule is still valid, although during last years the level of knowledge of carbon clusters has been raised substantially. On the basis of high resolution laser spectroscopy[37], the measurement of infrared absorption of carbon clusters in cryogenic matrices[40,83], photoelectron spectroscopy[84,85] and *ab initio* calculations[42,86], accurate values of vibration energy for all linear $C_2 - C_9$ carbon clusters and several vibration frequencies for linear $C_{10} - C_{13}$ clusters are known at present.

The majority of experimental information on $C_{11}$ carbon clusters and larger ones is based on studies of gas ion chromatography of carbon ion clusters. Very little experimental data on neutral carbon clusters exists so far. Clear spectroscopic information on a carbon cluster of this size is known only for linear $C_{13}$. Giesen *et al.*[37] measured 76 rovibrational absorption transitions using infrared diode laser spectroscopy and a supersonic beam. Because only even-numbered rotation states were observed and only P- and R-branch transitions were present, the spectrum received was characteristic of a centre-symmetric linear carbon cluster in non-degenerated vibration state. These days, the $C_3$ radical is very well characterized in its ground state and several excited electronic states[87,88]. Very detailed information was acquired concerning ground electronic states of $C_4 - C_7$ linear clusters, $C_9$ and $C_{13}$ just by means of gas phase infrared laser spectroscopy[89-92]. Similar information on analogous cyclic isomers would be exceedingly valuable.

### 4.2. *Laser-magnetic resonance (LMR)*

In contrast to laser absorption spectroscopy based on variability in wavenumber of tunable laser line, the principle of LMR[93] consists simply in "tuning" the wavenumber of a molecular transition line to a line of a fixed frequency laser. In this case, the transition frequency shift is caused by the so-called Zeeman effect connected with the application of a magnetic field. With reference to the fact that the Zeeman shift of molecules in $^1\Sigma$ states is very small, this method is primarily suitable for the spectroscopy of radicals, where an unpaired electron spin produces a significant magnetic moment. A lot of transient species including CH, $CH_2$, $C_2H$ radicals have been detected by LMR spectroscopy[66].

The basis of instrumentation required for LMR purposes consists of sufficiently intensive, discretely tunable gas phase (CO, $CO_2$, HF, $N_2O$, HCN and $H_2O$) lasers emitting the radiation of wavelengths from the middle to the far IR spectral region coinciding with a large number of vibration bands of transient species, including hydrocarbon radicals. As early as 1971, Evenson *et al.*[94] published the application of LMR to the detection of CH radical in an oxygen-acetylene flame placed in a $H_2O$ laser cavity.

### 4.3. *Laser-induced fluorescence (LIF)*

In the case of LIF, a laser line is accommodated to the centre of vibration or vibronic transition and then is left constant. In this regard, LIF differs markedly from both the above-mentioned methods. LIF is inherently a background-free method, and that is why it has a wide range of applications in the spectroscopy of transient species and the diagnostics of chemically reactive flows, including a hydrocarbon flame[95,96]. In these environments, it is often applied with spatial resolution. The undesirable spectral coincidence with chemiluminescent emission lines of radicals (above all $C_2$ and CH) in a flame can be, in this case, avoided by an increase in intensity of a short laser pulse (typically 3-10 ns)[64].

The range of applications of this method to the study of transient species is, however, restricted to those species that show sufficient quantum yields. Moreover, it cannot be applied to the study of the systems that have unstable (dissociative or pre-dissociative) excited electronic states.

### 4.4. *Nonlinear laser spectroscopy*

In comparison with LIF and other methods based on the incoherent process, in the course of application of coherent optical techniques such signal is emitted that resembles in character a beam of laser radiation. The required phase coherence in the direction of the signal is usually ensured by means of nonlinear optical processes. Methods of laser spectroscopy based on the process of four-wave mixing provide considerable advantages when investigating transient species. It is especially the case of a so-called "Doppler-free" technique, enabling the simultaneous effective discrimination of background signal. The implementation of these methods brings, on the other hand, higher requirements for experimental set-ups (e.g. beam alignment).

The method of coherent anti-Stokes Raman spectroscopy (CARS)[97] and its experimental version (resonant CARS, BOXCARS) have been used first of all in combustion process diagnostics. Attal $et$ $al.$[98] applied resonance enhanced CARS in $C_2$ concentration measurement.

What represents a promising tool for the field of spectroscopy of transient carbon species is the method of degenerate four-wave mixing (DFWM)[99]. By means of DFWM, e.g. $C_2$ and $C_3$ radicals have been detected[100-103]. Mazzotti $et$ $al.$[104] tested, in the course of investigation into electronic spectra of a $C_2$ molecule, the resonance-enhanced variant two-color resonant four-wave mixing (TC-RFWM).

The relative simplicity of polarization spectroscopy (PS) in comparison with DFWM increases the range of applications of this method in the area of diagnostics of processes in plasma and also in a flame. Using the method PS in the visible region, concentrations of $C_2$ radical have been mapped in the environment of a flame[105]. Li $et$ $al.$[106] applied this method to the infrared region. In this way they proved the potential of this method as a tool for the rotation-vibration spectroscopy.

### 4.5. Multi-photon ionization spectroscopy

The resonant multi-photon ionization (REMPI) method employs a multi-photon process. A population of resonant intermediate state is pumped by one- or more- ($n$-) photon absorption process. Consequently one or more ($m$-) laser photons need to be absorbed to cause the ionization process.

The use of the REMPI method is limited to neutral molecules; however in combination with the methods of mass spectroscopy, e.g. the above-mentioned (TOF) method, it enables the unambiguous assignment of various transient species. Various carbon containing radicals ($C_2$, $C_3$, CH, $CH_2$, $CH_3$, $C_2H$, $C_2H_5$, $C_3H_5$) were detected by this method[107].

Schmidt *et al.*[108] reported optical detection of $C_nH_3$ ($n = 9,11,13$) by two-color two-photon ionization.

These days, combined REMPI-TOF studies are one of the fundamental tools for investigations into the chemical kinetics of soot formation in hydrocarbon flames[109]. Using this approach, primarily the process of formation of the first and the second aromatic ring is examined. In comparison with the other mentioned methods, it is the partially invasive character that is a disadvantage of REMPI. In spite of advanced techniques of sampling, this process is still connected with a certain degree of uncertainty.

A number of other resonant multi-photon methods have been successfully used in the spectroscopy of transient carbon species. For instance, the method of resonance enhanced multi-photon electron detachment (REMPED)[110] and its variant resonant two-color photodetachment (RM2PD) variant have proved themselves in the investigation into electronic spectra of anions of linear carbon chains[111].

## 5. Overview and Perspectives

A tremendous progress has so far been performed in the field of studying small hydrocarbon molecules and carbon clusters. Nevertheless we are still too far from a detailed understanding of mechanisms leading from small carbon chains and rings in gas phase to solid phase products of carbon chemistry (highly organized structures, such as fullerenes and carbon nanomaterials, or amorphous soot particles, respectively). Decomposition of gas phase precursors and formation of one- to three-ring aromatic hydrocarbons represent critical steps of this route.

Thanks to its outstanding advantage enabling *in situ* monitoring of individual transient species occurring in plasmas or flames, high resolution spectroscopy can serve as a very helpful tool for characterizing carbon clusters, investigating the kinetics of chemical pathways, and disclosure of mechanisms taking place in such environments. To accept a challenge like that, however, novel experimental approaches employing highly sensitive techniques for detection of short-lived radicals and ions will be necessary. The development in this area still remains a task for both theoretical and experimental chemists, physicists and spectroscopists.

## Acknowledgments

The authors thank Ms. Radmila Jelínková for her assistance in language proofreading of the chapter.

## References

1. P. Thaddeus and M. C. McCarthy, *Spectrochim. Acta A* **57**, 757–774 (2001).
2. J. Fulara and J. Krełowski, *New Astron. Rev.* **44**, 581–597 (2000).
3. A. Van Orden and R. J. Saykally, *Chem. Rev.* **98**, 2313–2357 (1998).
4. N. Aggadi, C. Arnas, F. Bénédic, C. Dominique, X. Duten, F. Silva, K. Hassouni and D. M. Gruen, *Diam. Rel. Mater.* **15**, 908–912 (2006).
5. A. Hamins, In *Environmental Implications of Combustion Processes*, I. K. Puri (Ed.). Boca Raton: CRC Press, 1993, 71–95.
6. G. Herzberg, *The Spectra and Structures of Simple Free Radicals*. Mineola, New York: Dover Publications. 2003.
7. E. Hirota and Y. Endo, In *Vibration-Rotational Spectroscopy and Molecular Dynamics*, D. Papousek (Ed.). Singapore: World Scientific, 1997, 1–55.
8. P. B. Davies, *Annu. Rep. Prog. Chem., Sect. C*, **84**, 43–64 (1987).
9. Q. L. Zhang, S. C. O'Brien, J. R. Heath, Y. Liu, R. F. Curl, H. W. Kroto and R. E. Smalley, *J. Phys. Chem.* **90**, 525–528 (1986).
10. P. Gerhardt, S. Loffler and K. H. Homann, *Chem. Phys. Lett.* **137**, 306–310 (1987).
11. H. W. Kroto and K. McKay, *Nature* **331**, 328–331 (1988).
12. W. Weltner and R. J. Van Zee, *Chem. Rev.* **89**, 1713–1747 (1989).
13. H. Bockhorn, F. Fetting and H. W. Wenz, *Ber. Bunsenges. Phys. Chem.* **87**, 1067–1073 (1983).
14. J. A. Cole, J. D. Bittner, J. P. Longwell and J. B. Howard, *Combust. Flame* **56**, 51–70 (1984).
15. M. B. Colket, *Symposium (International) on Combustion* **21**, 851–864 (1988).
16. P. R. Westmoreland, A. M. Dean, J. B. Howard and J. P. Longwell, *J. Phys.Chem.* **93**, 8171–8180 (1989).
17. J. A. Miller and C. F. Melius, *Combust. Flame* **91**, 21–39 (1992).
18. C. F. Melius, M. E. Colvin, N. M. Marinov, W. J. Pitz and S. M. Senkan, In *Twenty-sixth Symposium (International) on Combustion*, A. R. Burgess and F. L. Dryer (Eds.). Pittsburgh: Combustion Institute, 1996, 685–692.
19. L. V. Moskaleva, A. M. Mebel and M. C. Lin, In *Twenty-sixth Symposium (International) on Combustion*, A. R. Burgess and F. L. Dryer (Eds.). Pittsburgh: Combustion Institute, 1996, 521–526.
20. N. M. Marinov, W. J. Pitz, C. K. Westbrook, M. J. Castaldi and S. M. Senkan, *Combust. Sci. Technol.* **116**, 211–287 (1996).
21. M. J. Castaldi, N. M. Marinov, C. F. Melius, J. Huang, S. M. Senkan, W. J. Pitz and C. K. Westbrook, In *Twenty-sixth Symposium (International) on Combustion*,

A. R. Burgess and F. L. Dryer (Eds.). Pittsburgh: Combustion Institute, 1996, 693–702.

22. M. Frenklach, D. W. Clary, W. C. Gardiner Jr. and S. E. Stein, *Proc. Combust. Inst.* **20**, 887–901 (1984).
23. J. D. Bittner and J. B. Howard, *Symposium (International) on Combustion* **18**, 1105–1116 (1981).
24. M. Frenklach, D. W. Clary, W. C. Gardiner Jr. and S. E. Stein, *Symposium (International) on Combustion* **21**, 1067–1076 (1988).
25. Z. A. Mansurov, *Combust. Explos.* **41**, 727–744 (2005).
26. B. S. Haynes and H. G. Wagner, *Prog. Energy Combust. Sci.* **7**, 229–273 (1981).
27. H. F. Calcote, D. B. Olson and D. G. Keil, *Energy Fuels* **2**, 494–504 (1988).
28. A. B. Fialkov, *Prog. Energy Combust. Sci.* **23**, 399–528 (1997).
29. A. N. Hayhurst and H. R. N. Jones, *Nature* **296**, 61–63 (1982).
30. H. F. Calcote and D. G. Keil, *Pure Appl. Chem.* **62**, 815–824 (1990).
31. V. J. Hall-Roberts, A. N. Hayhurst, D. E. Knight and S. G. Taylor, *Combust. Flame* **120**, 578–584 (2000).
32. R. I. Kaiser, T. N. Le, T. L. Nguyen, A. M. Mebel, N. Balucani, Y. T. Lee, F. Stahl, P. v. R. Schleyer and H. F. Schaefer, *Faraday Discuss.* **119**, 51–66 (2001).
33. K. Raghavachari and J. S. Binkley, *J. Chem. Phys.* **87**, 2191–2197 (1987).
34. J. M. L. Martin, J. P. Francois and R. Gijbels, *J. Chem. Phys.* **93**, 8850–8861 (1990).
35. J. M. L. Martin, J. P. Francois and R. Gijbels, *J. Comput. Chem.* **12**, 52–70 (1991).
36. J. Kurtz and L. Adamowicz, *Astrophys. J.* **370**, 784–790 (1991).
37. T. F. Giesen, U. Berndt, K. M. T. Yamada, G. Fuchs, R. Schieder, G. Winnewisser, R. A. Provencal, F. N. Keutsch, A. Van Orden and R. J. Saykally, *ChemPhysChem.* **4**, 242–247 (2001).
38. K. S. Pitzer and E. Clementi, *J. Am. Chem. Soc.* **81**, 4477–4485 (1959).
39. R. Hoffmann, *Tetrahedron* **22**, 539–545 (1966).
40. P. Freivogel, M. Grutter, D. Forney and J. P. Maier, *Chem. Phys.* **216**, 401–406 (1997).
41. T. F. Giesen, A. Van Orden, H. J. Hwang, R. S. Fellers, R. A. Provencal and R. J. Saykally, *Science* **265**, 756–759 (1994).
42. J. M. L. Martin and P. R. Taylor, *J. Phys. Chem.* **100**, 6047–6056 (1996).
43. R. O. Jones, *J. Chem. Phys.* **110**, 5189–5200 (1999).
44. W. Demtröder, *Molecular Physics. Theoretical Principles and Experimental Methods.* Weinheim: WILEY-VCH, 2005.
45. H. L. Wallaart, B. Piar, M. Y. Perrin and J. P. Martin, *Chem. Phys. Lett.* **246**, 587–593 (1995).
46. R. F. Curl and F. K. Tittel, *Annu. Rep. Prog. Chem., Sect. C* **98**, 219–272 (2002).
47. P. F. Bernath, *Annu. Rev. Phys. Chem.* **41**, 91–122 (1990).
48. P. B. Davies and P. A. Martin, *Chem. Phys. Lett.* **136**, 527–530 (1987).
49. S. Civis, Z. Zelinger and K. Tanaka, *J. Mol. Spectrosc.* **187**, 82–88 (1998).
50. C. S. Gudeman, M. H. Begemann, J. Pfaff and R. J. Saykally, *Phys. Rev. Lett.* **50**, 727–731 (1983).
51. Z. Zelinger, A. Bersch, M. Petri, W. Urban and S. Civis, *J. Mol. Spectrosc.* **171**, 579–582 (1995).
52. Z. Zelinger, S. Civis, P. Kubat and P. Engst, *Infrared Phys. and Techn.* **36**, 537–543 (1995).

53. R. G. W. Norish and G. Porter, *Nature* **164**, 658–658 (1949).

54. A. E. Douglas, *Astrophys. J.* **114**, 466–468 (1951).

55. Z. A. Liu, R. J. Livingstone and P. B. Davies, *Chem. Phys. Lett.* **291**, 480–486 (1998).

56. Z. Zelinger, M. Novotny, J. Bulir, J. Lancok, P. Kubat and M. Jelinek, *Contrib. Plasm. Phys.* **43**, 426–432 (2003).

57. J. Bulir, M. Novotny, M. Jelınek, J. Lancok, Z. Zelinger and M. Trchova, *Diam. Rel. Mater.* **11**, 1223–1226 (2002).

58. M. Jelinek, J. Lancok, R. Tomov and Z. Zelinger, *Spectrochim. Acta A* **58**, 1513–1521 (2002).

59. J. Bulir, M. Novotny, M. Jelinek, L. Jastrabik and Z. Zelinger, *Surf. Coat. Technol.* **173-174**, 968–972 (2003).

60. M. A. Bratescu, N. Sakura, D. Yamaoka, Y. Sakai, H. Sugawara and Y. Suda, *Appl. Phys. A-Mater. Sci. Process.* **79**, 1083–1088 (2004).

61. H. W. Kroto, J. R. Heath, S. C. O'Brien, R. F. Curl and R. E. Smalley, *Nature* **318**, 162–163 (1985).

62. P. Neubauer-Guenther, T. F. Giesen, U. Berndt, G. Fuchs and G. Winnewisser, *Spectrochim. Acta A* **59**, 431–441 (2002).

63. K. K. Kuo, *Principles of Combustion (2$^{nd}$ Edition)*. New York: J. Wiley & Sons, 2005.

64. K. C. Smyth and D. R. Crosley, In *Applied Combustion Diagnostics*, K. Kohse-Höinghaus and J. B. Jeffries (Eds.). New York: Taylor & Francis, 2002.

65. G. Hartung, J. Hult and C. F. Kaminski, *Meas. Sci. Technol.* **17**, 2485–2493 (2006).

66. J. M. Hollas, *High-Resolution Spectroscopy. (2$^{nd}$ Edition)*. Chichester: John Wiley & Sons, 1998.

67. W. Demtröder, *Laser Spectroscopy (3$^{rd}$ Edition)*. Berlin, Heidelberg: Springer, 2003.

68. L. Andrews and A. A. Demidov (Eds.) *An Introduction to Laser Spectroscopy. (2$^{nd}$ Edition)*. New York: Kluwer Academic/Plenum Publishers, 2002.

69. M. Lackner (Ed.) *Lasers in Chemistry*. Weinheim: WILEY-VCH, 2008.

70. A. Campargue, F. Stoeckel and M. Chenevier, *Spectrochim. Acta Rev.* **13**, 69–88 (1990).

71. A. Kachanov, A. Charvat and F. J. Stoeckel, *J. Opt. Soc. Am. B* **11**, 2412–2421 (1994).

72. T. Motylewski and H. Linnartz, *Rev. Sci. Instrum.* **70**, 1305–1012 (1999).

73. J. J. Scherer, J. B. Paul, A. O'Keefe and R. J. Saykally, *Chem. Rev.* **97**, 25–52 (1997).

74. J. J. Scherer, D. Voelkel, D. J. Rakestraw, J. B. Paul, C. P. Collier, R. J. Saykally and A. O'Keefe, *Chem. Phys. Lett.* **245**, 273–280 (1995).

75. J. B. Paul and R. J. Saykally, *Anal. Chem.* **69**, A287–A292 (1997).

76. R. A. Provencal, J. B. Paul, E. Michael and R. J. Saykally, *Photon. Spect.* **32**, 159–166 (1998).

77. G. Winnewisser, T. Drascher, T. Giesen, I. Pak, F. Schmulling and R. Schieder *Spectrochim. Acta A* **55**, 2121–2142 (1999).

78. E. Hirota, *Annu. Rep. Prog. Chem., Sect. C* **96**, 95–138 (2000).

79. W. R. M. Graham, K. I. Dismuke and W. Weltner Jr., *Astrophys. J.* 204, 301–310 (1976).

80. J. R. Heath and R. J. Saykally, *J. Chem. Phys.* **94**, 3271–3272 (1991).

81. N. Moazzen-Ahmadi, J. J. Thong and A. R. McKellar, *J. Chem. Phys.* **100**, 4033–4038 (1994).

82. H. J. Hwang, A. Van Orden, K. Tanaka, E. W. Kuo, J. R. Heath and R. J. Saykally, *Mol. Phys.* **79**, 769–776 (1993).

83. J. Szczepanski, S. Ekern, C. Chapo and M. Vala, *Chem. Phys.* **211**, 359–366 (1996).

84. D. W. Arnold, S. E. Bradforth, T. N. Kitsopoulos and D. M. Neumark, *J. Chem. Phys.* **95**, 8753–8764 (1991).

85. C. S. Xu, G. R. Burton, T. R. Taylor and D. M. Neumark, *J. Chem. Phys.* **107**, 3428–3436 (1997).

86. J. M. L. Martin, J. El-Yazal and J. P. Francois, *Chem. Phys. Lett.* **242**, 570–579 (1995).

87. F. J. Northrup, T. J. Sears and E. A. Rohlfing, *J. Mol. Spectrosc.* **145**, 74–88 (1991).

88. M. Mladenovic, S. Schmatz and P. Botschwina, *J. Chem. Phys.* **101**, 5891–5899 (1994).

89. J. R. Heath and R. J. Saykally, In *On Clusters and Clustering: from Atoms to Fractals*, P. J. Reynolds (Ed.). New York: Elsevier Science, 1993, 7–21.

90. T. F. Giesen, A. Van Orden, H. J. Hwang, R. S. Fellers, R. A. Provencal and R. J. Saykally, *Science* **265**, 756–759 (1994).

91. N. Moazzen-Ahmadi and J. J. Thong, *Chem. Phys. Lett.* **233**, 471–476 (1995).

92. A. Van Orden, R. A. Provencal, F. N. Keutsch and R. J. Saykally, *J. Chem. Phys.* **105**, 6111–6116 (1996).

93. P. B. Davies, *J. Phys. Chem.* **85**, 2599–2607 (1981).

94. K. M. Evenson, H. E. Radford and M. M. Moran Jr., *Appl. Phys. Lett.*, **18**, 426–429 (1971).

95. K. Kohse-Höinghaus, *Prog. Energy Combust. Sci.* **20**, 203–279 (1994).

96. J.W. Daily, *Prog. Energy Combust. Sci.* **23**, 133–199 (1997).

97. P. R. Régnier and J. P. E. Taran, *Appl. Phys. Lett.* **23**, 240–242 (1973)

98. B. Attal, D. Débarre, K. Müller-Dethlefs and J. P. E. Taran, *Rev. Phys. Appl.* **18**, 39–50 (1983).

99. R. L. Farrow and D. J. Rakestraw, *Science*, **257**, 1894–1900 (1992).

100. K. Nyholm, M. Kaivola and C. G. Aminoff, *Opt. Commun.* **107**, 406–410 (1994).

101. C. F. Kaminski, I. G. Hughes and P. Ewart, *J. Chem. Phys.* **106**, 5324–5332 (1997).

102. S. Williams, J. D. Tobiason, J. R. Dunlop and E. A. Rohlfing, *J. Chem. Phys.* **102**, 8342–8358 (1995).

103. D. S. Green, T. G. Owano, S. Williams, D. G. Goodwin, R. N. Zare and C. H. Kruger, *Science* **259**, 1726–1729 (1993).

104. F. J. Mazzoti, E. Achkasova, R. Chauhan, M. Tulej, P. P. Radi and J. P. Maier, *Phys. Chem. Chem. Phys.* **10**, 136–141 (2008).

105. K. Nyholm, M. Kaivola and C. G. Aminoff, *Appl. Phys. B* **60**, 5–10 (1995).

106. Z. S. Li, M. Rupinski, J. Zetterberg and M. Alden, *Proc. Comb. Inst.* **30**, 1629–1636 (2005).

107. M. N. R. Ashfold, S. G. Clement, J. D. Howe and C. M. Western, *J. Chem. Soc., Faraday Trans.* **89**, 1153–1172 (1993).

108. T. W. Schmidt, A. E. Boguslavskiy, T. Pino, H. Ding and J. P. Maier, *Int. J. Mass Spectrom.* **228**, 647–654 (2003).

109. C. S. McEnally, L. D. Pfefferle, B. Atakan and K. Kohse-Höinghaus, *Prog. Energy Combust. Sci.* **32**, 247–294 (2006).
110. M. Ohara, H. Shiromaru and Y. Achiba, *J. Chem. Phys.* **106**, 9992–9995 (1997).
111. M. Tulej, D. A. Kirkwood, G. Maccaferri, O. Dopfer and J. P. Maier, *Chem. Phys.* **228**, 293–299 (1998).

Chapter 9

# Developing New Production and Observation Methods for Various Sized Carbon Nanomaterials from Clusters to Nanotubes

Toshiki Sugai

*Toho University,*
*Miyama 2-2-1, Funabashi, 274-8510 Japan*
*E-mail: sugai@chem.sci.toho-u.ac.jp*

Carbon nano-materials have been attracting significant interest because of their unique structures and properties which would be useful for many practical applications. It is however hard to control these unique structures and properties which are strongly dependent on their size. We have been working with these carbon nano-materials, ranging from carbon clusters and fullerenes to carbon nanotubes, and have been developing new production, purification, and observation methods. In this chapter these methods are presented along with realized novel structures and properties.

## 1. Introduction

Nano-carbon materials, such as fullerenes and carbon nanotubes (CNTs), have been attracting significant recent attention and are considered to be principal materials in the fields of nano-science and nano-technology.[1-5] These materials have various structures and properties owing to the co-valent bonds of the carbon atoms, and their bonding patterns vary from linear sp, planar $sp^2$ and tetrahedral $sp^3$ forming rigid and stable structures. These special properties of carbon atoms have contributed to the opening of a new era for nano science and technology in terms of realizing novel properties with mass produced stable materials. The size distribution ranges from nm of the diameters of fullerenes and CNTs to $\mu$m and mm of the diameters of capsules and the lengths of carbon nanotubes. Especially, CNTs have two typical sizes of nm as the diameter and of mm as the length simultaneously, suggesting that we can bring useful quantum effects from the molecular scale diameters to our world through the macroscopic length by attaching electrodes or other macroscopic devices. Using the unique mechanical, electronic, and chemical properties of these structures, these nanocarbon materials are now beginning to be applied in various devices, including electronic devices,[6–11] nano mechanical devices,[12,13] field emitters,[14] and energy-related devices.[15–19] Since their properties are strongly dependent on the nanoscale structures, controlling their structures and purity is crucial for applications as well as fundamental research. Not only whole structures such as the diameters and numbers of layers, but also detailed structures such as defects determine the overall properties and performance of devices.

A range of controllable production methods to minimize defects have been developed for nanocarbon materials. For example, fullerenes and SWNTs have been selectively produced using a steady arc discharge and a laser furnace technique, by changing the buffer gas, temperature, and metal catalyst.[4,20–22] Chemical vapor deposition techniques for CNTs are also drawing attention for their potential for control of structure.[23–25] In spite of extensive development of these techniques, difficulties in controlling nanocarbon structures still persist. Within our group, we have developed a novel production method, the so-called high-temperature pulsed-arc discharge (HTPAD) technique, a combination techniques between the laser furnace and the steady arc discharge.[26–31] Ambient temperatures are controlled by a furnace like the laser furnace production method and currents are also controlled by a power supply like the steady arc techniques. Both parameters are crucial for the control of nano structures. In the HTPAD technique, the duration of the pulsed arc discharge is controlled and varied to manipulate the structure of the product. As a result of wide controllability with catalyst manipulation, it is feasible to produce novel materials, such as high-quality double-walled carbon nanotubes with narrow diameters, small diameter distributions, and less defects.[29,30] This pulsed arc discharge method was originally developed and applied to produce gas phase clusters.[32] A high energy density of $10^{17}$ W/m$^3$ can vaporize any material having a high boiling temperature, and the vaporized products can be annealed into stable structures for a relatively longer duration compared to laser vaporization. These advantages allow for the investigation of the growth mechanisms of various nanocarbon materials as well as to control their structures. The nanocarbon structures are determined at the early stages of their growth on the time scale of milliseconds,[33–35] similar to the pulse width of HTPAD. For example, "cap structures" of SWNTs, which are the half spheres of fullerenes and which determine the whole structures of SWNTs, have been reported to be produced within this time scale.[33–37] Thus, HTPAD has the potential for a high degree of controllability in the production of nanocarbon materials.

These bulk production methods are key technologies for this field. However, almost all of these production methods rely on self-assembly processes, so that production processes must be followed by purification and identification. For the purification for fullerenes, chromatography is the main tool and has been applied to various fullerene samples since the discovery of mass production of fullerenes.[38,39] The method is powerful enough to isolate structural isomers and to clarify property dependence on molecular structures.[39] These achievements have been realized due to the soluble properties of fullerenes. However, isolation of larger fullerenes, the so-called higher fullerenes, has been difficult because of their low solubilities.[39] For other nanocarbon materials, which are typically insoluble in any solvent, purification has been a big problem and a major issue. A number of methods have been developed to purify CNTs utilizing burning, filtration, chromatography, electrophoresis, centrifuging etc.[40–45] Burning assisted purification has been widely used for separating SWNTs from amorphous carbon,[40] however low selectivity and defects through purification have been serious problems. On the other hand, purification methods utilizing dispersion techniques have much higher selectivity. For example, several groups have realized partial chirality isolation using this dispersion technique together with column or electrophoresis separations.[42,45] However, the dispersion results in very dilute CNT samples and low throughput of the purification processes.

Our DWNTs, using HTPAD, are not produced selectively, rather they are produced together with SWNTs and amorphous carbon.[29–31] To purify DWNTs, we developed a "burning method" which utilizes much higher chemical durability of DWNTs against oxidation compared to that of SWNTs. This durability may come from layer interactions, which do not exist in SWNTs.[29,30] This method has been applied to various DWNTs from CVD, proving its universality and usefulness.[46,47] Furthermore, we have developed a dispersion-assisted burning method to realize high efficiency and high yield.[31] The purified DWNTs have other distinct properties in addition to chemical durability. We applied the DWNTs as high-performance

AFM tips, and demonstrated their mechanical as well as chemical durability.[48] Sophisticated TEM observations of DWNTs also clarified the strong interactions between layers,[49] which should contribute to their chemical and mechanical durability. The electronic properties of DWNTs were also measured by utilizing them as channels in field effect transistors.[50]

For identification of nanocarbon materials, several methods, including NMR,[38] X-ray diffraction,[51–53] ultra-high vacuum scanning tunnel microscopy (UHV-STM),[54–56] and TEM[3,57] have been developed and modified. For example, the structures of endohedral metallofullerenes have been extensively studied, since their unique electronic properties are dominantly determined by their structures.[39] They have varying cage sizes, numbers of metal atoms, and metal species, all of which affect their properties. The hollow spaces of fullerenes can hold molecules within, for example, tri-metallic nitride of $Sc_3N@C_{80}$,[58] and metal carbide of $Sc_2C_2C_{84}$.[59] These species need to be purified before identification by conventional methods, requiring large amounts ($\sim$ mg) of purified ($> 99.9\%$) samples and long preparation times ($\sim$ months). It is thus necessary and crucial to develop new methods to study the stability and the structures of such novel nanocarbon materials.

Gas phase ion mobility spectrometry combined with mass spectrometry (IMS/MS) is one of the candidates for structural identification. This method has been applied to various species including clusters,[60–62] fullerene-related materials,[63] and poly-peptides. The method has now been utilized for CNTs.[64–66] High-sensitivity and high-selectivity originating from ion detection allow us to use small amounts of nanocarbon mixtures for which conventional identification methods cannot be applied. Furthermore, high-speed ($\sim$ ms) gas-phase detection yield structural information on intermediate species and reactions. This method utilizes the fact that ion mobility or the ion drift velocity in a gas phase are inversely proportional to the cross-section of the ion and buffer gas used for the measurement. Since the cross-section directly corresponds to the ion structure, we can deduce the structures of nanocarbon materials from the mobility. In this chapter, we present high-resolution ion gas mobility measurements on Sc-metallofullerenes. The results show that all metallofullerenes have carbide structures as well as usual endohedral structures. These two structures seems to be converted by laser irradiation, wherein carbon cages shrink when exposed to high desorption/ionization laser fluence.

This IMS/MS method is powerful and sensitive enough to explore novel nanocarbon materials, however, it requires ionization and dispersion of measured materials in gas phase. This is not easy for most condensed nanocarbon materials since they have a tendency to aggregate strongly via van der Waals interactions through their large surface area. Usually, the production of ionized and dispersed nanocarbon materials is accomplished through laser irradiation[67] and laser ablation of a cluster source.[1,68] The former, however, could change the structure of the measured materials, for example, from CNTs to fullerenes,[66,69] and the latter cannot produce sufficient amounts of larger cluster ions more than $C_{1000}$ to be detected even by sensitive mass spectrometry.[1] For exploring novel nanocarbon materials by IMS/MS, we have developed high-pressure laser vaporization which produces a wide range of carbon clusters up to $C_{2000}$ as dispersed ions to be detected by IMS/MS.[70] These novel nanocarbon materials, the so-called giant carbon clusters or ultra-small carbon materials, range from $C_{100}$ to $C_{100000}$ in terms of the number of carbon atoms, and there are no standard identification and purification methods for them. Nanocarbon materials less than $C_{100}$ and greater than $C_{100000}$ have been well studied as fullerenes and CNTs. Novel structures of nano- diamonds,[71] onions,[72] and graphite balls[73] have been reported with new properties in this size regime. These giant carbon clusters will significantly expand the possibilities of nanocarbon materials.

As mentioned above, the development of new production, purification, and structural analysis methods are described in this chapter.

## 2. Experimental

### 2.1. *Pulsed Arc Discharge of and Property Measurements on DWNTs*

The HTPAD apparatus consists of a furnace, a quartz tube, electrical feedthroughs, pure carbon or catalytic metal-doped carbon electrodes, a water-cooled trap, and a homemade pulsed HV power supply (Fig. 1). A buffer gas, Ar or He, was allowed to pass through the quartz tube ($\phi$25 mm), where the temperature was controlled between 25 and 1400°C using the furnace. The flow and pressure of the buffer gas were respectively regulated at 300 cm$^3$/s and 500–800 Torr. The power supply delivered a breakdown pulsed HV voltage of 1 kV for 10 $\mu$s followed by a long and low voltage pulse of 50 $\mu$s – 100 ms and 40 V. Pulsed arc discharges between the electrodes were produced. The current increased from 0 to 100 A in 50 $\mu$s keeping the current constant up to 300 ms. The actual voltage difference between the

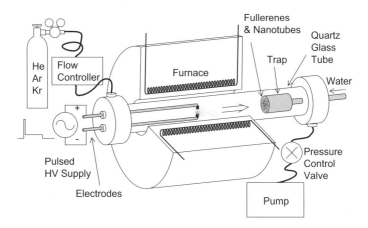

Fig. 1.  Schematic diagram of the high-temperature pulsed arc discharge apparatus.

electrodes was 1 kV at the beginning and decreased to 20 V, as determined by the ionization potential of the buffer gas.[22,74] The power consumption between the electrodes was almost constant after 50 $\mu$s when the current reached a constant value. The repetition rate was controlled to maintain a duty factor at 3%. The products were collected in a water-cooled trap. The dependence of production efficiencies of DWNTs was studied as a function of the catalytic metals used (Ni, Co, Y, La and their mixtures) and temperature. Other experimental conditions like the pulse duration, the buffer gas and the pressure were investigated and were set up to be 600 $\mu$s Ar, and 1 atm as a local optimum at least. For the dependence on the catalyst metal, electrodes were prepared from carbon and metal powder with different atomic ratios.[30] The trapped as-produced DWNTs with metal catalyst, SWNTs, fullerenes, and amorphous carbon were then subjected to purification. The fullerenes were washed out with carbon disulfide and the catalytic metal was removed by sonication in concentrated HCl. This prevents degradation of DWNTs induced by the subsequent oxidation.[31] The sonicated DWNTs were rinsed with distilled water and oxidized in air at 200 $\sim$ 400°C for a day. The sample was again sonicated in concentrated HCl and rinsed with distilled water. Finally, oxidation in air at around 500°C for one or two hours eliminated SWNTs and residual amorphous carbon from the sample.

For the further selective purification of DWNTs, we also developed a dispersion-assisted purification technique. The as-produced purified samples, obtained from the as-produced soot by heating in air at 360°C, were dispersed into 100 ml of 1% of sodium dodecyl sulfate (SDS) solution. The solution was centrifuged at 4800 G and decanted to remove the metals and nanocapsules residue. The supernatant, which is the solution of CNTs wrapped with SDS, was mixed with 2 g of fumed silica (particle size: 7 nm, Aldrich). The fumed silica was used to prevent the bundling of CNTs in the liquid and solid phases even after losing the surfactant through oxidation. This mixture was dried to a powder, and was then oxidized by hot air at 360°C and hydrogen peroxide to remove residual SWNTs and amorphous carbon. This dispersion significantly enhanced the selectivity to remove SWNTs.[31]

The products of each process were characterized by scanning electron microscopy (SEM: Hitachi S-900), high-resolution transmission electron microscopy (TEM: JEOL JEM 2010), and Raman spectroscopy (Jobin Yvon HR-800) to detect CNTs. Raman measurements were carried out in a micro-Raman mode wherein the excitation lasers ($Ar^+$: 488.0 and 514.5 nm, HeNe: 632.8 nm) were focused to a tight spot ($\phi \sim \mu$m).

The purified DWNTs were applied to AFM tips like SWNTs and MWNTs.[75,76] Our DWNTs have narrow diameters compared with MWNTs and have the layer interaction contributing to mechanical rigidity, which cannot be found in SWNTs. Bundles of the material were aligned on a knife-edge,[77] transferred and then attached to a conventional silicon tip (OMCL-AC240TS-C2) in an SEM, as described in a previous report.[76] The tips have the DWNT bundle sticking out about several hundred nm from the top of the tip. To quantify the effective resolution of the present DWNT tips as compared to MWNT-tips (Daikenkagaku Co. Ltd.) and standard Si-tips (RTESP (MPP-11100), Veeco Probes), individual SWNTs (CNI, HiPco) were imaged by AFM with the three different tips, DWNT-, MWNT- and Si-tips, under similar experimental conditions. Samples of individual SWNTs were prepared by spin-coating 1,2-dichloroethane dispersion solutions of SWNTs on a silicon oxide substrate. AFM images were obtained using tapping-mode AFM (Nanoscope IV, Dimension 3100, Veeco Digital Instruments) in air.

The transport properties of the CNTs were also measured by constructing field effect transistors (FET) to characterize the CNTs, particularly to know the difference between SWNTs and DWNTs.[50] A heavily doped Si substrate (525 $\mu$m) was used as a back gate with a Ti/Au gate electrode (100/400 nm). An $SiO_2$ insulating layer (100 nm) was grown on top of the substrate by thermal oxidation. Electron beam lithography and photolithography were used for fabrication of the source and drain electrodes (Ti/Au; 3/15 nm). The source-drain gap was 400 nm. DWNTs were dispersed in N,N-dimethylformamide solutions and dropped onto the substrate. This method provides us with a suitable tube density on the substrate. A semiconductor parameter analyzer (Agilent, 4156C) and a low temperature probe station (Nagase, BCT-21MRF) were used to measure the drain current as a function of gate bias or drain bias voltages.[50]

## 2.2. *Ion Mobility and Mass Spectrometry*

Sc metallofullerene samples are prepared in a large-scale DC arc discharge apparatus using Sc graphite composite rods as anodes (Sc 1.6 at. %, Toyotanso Inc.). A discharge current of 500 A was used along with He buffer gas at 40 Torr.[39,78] The metallofullerenes were extracted from the soot by $CS_2$.

A schematic of the IMS/MS measurement system is shown in Fig. 2, which was developed by Prof. M. F. Jarrold's group at Indiana University.[79] This high-resolution ion-mobility apparatus has been previously described in detail.[79] The apparatus consists of a source region coupled directly to a 63 cm long drift tube. Metallofullerene extracts are drop-coated onto a copper rod and then desorbed/ionized by a focused pulsed excimer laser (XeCl 308 nm). The resulting ions are introduced into the drift tube, which

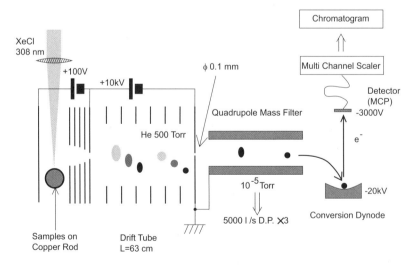

Fig. 2. Schematic diagram of the ion mobility spectrometer/mass spectrometer (IMS/MS) system developed by Prof. M. F. Jarrold's group at Indiana University.

contains helium buffer gas at 500 Torr and has a drift voltage of 10 kV. After travelling through the drift tube, the fullerene ions were separated according to their mobility (cf. Fig. 2): larger ions moved slowly (longer drift times) while smaller ions moved faster (shorter drift times). These ions are then mass analyzed using a quadrupole mass spectrometer (Extrel Inc.) and drift-time distributions are recorded with a multichannel analyzer (Tennelec/Nucleus Inc. MCS-II). These simultaneous mobility and mass measurements enable us to investigate their structures.

### 2.3. *High-pressure Laser Vaporization*

Large carbon clusters were produced using the high-pressure laser-vaporization cluster source and were analyzed by a reflectron TOF mass spectrometer.[70,80–82] The source consists of a preload valve, a main valve, an evaporation room with an x stage and holes for a laser and a cluster beam, and a rectangular graphite plate (5×10×1 mm) mounted on a y-z stage (see Fig. 3). The sample was periodically translated in a zigzag motion by the y-z stage so that the laser vaporization occurred on a fresh surface of the sample plate. A General Valve and a Jordan Valve were used as the preload valve and the main valve, respectively. Three stepping motors controlled the x, y, z positions of the each stage.

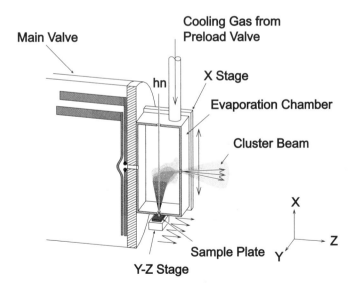

Fig. 3.  Schematic diagram of a high-pressure laser vaporization cluster source.

Operation of the cluster source was executed as follows. First, a cooling (buffer) gas, typically He, was introduced into the evaporation chamber from the preload valve. The pressure inside the evaporation chamber was estimated to be $1 \sim 10$ Torr, realizing a quasi-fluxless condition, yielding giant clusters.[83] After 700 $\mu$s from the cooling gas injection, a 532 nm light pulse from a YAG laser was introduced and used to evaporate the sample. Vapor from the sample collided with the introduced cooling gas and aggregated into large clusters. After 1000 $\mu$s, the clusters formed were introduced into the TOF mass spectrometer by supersonic jets of He from the main valve, and then their mass was analyzed.

## 3. Results and Discussion

### 3.1. Production and Characterization of DWNTs by High-temperature Pulsed Arc Discharge

Figure 4 shows TEM and Raman spectra of the products of HTPAD at different temperatures. The DWNTs were produced in the high-temperature region over 1200°C. The TEM images in Fig. 4(a) show that DWNTs are produced above 1200°C with the Y/Ni catalyst (1.0/4.2 at. %), which are known to only produce SWNTs. It is important to note that the DWNTs were produced at temperatures over 1200°C while maintaining other conditions. This temperature dependence was also confirmed by Raman spectroscopy with excitation at 632.8 nm (cf. Fig. 4(b)), wherein the peaks at 214 and 136 cm$^{-1}$ correspond to the inner and outer diameters of DWNTs of 1.15 and 1.83 nm, respectively. The diameter evaluation was carried out using the relation $d = 248/\omega$, where $d$ is the diameter of the tube in nm and $\omega$ is the frequency of the radial breathing modes (RBMs) in cm$^{-1}$.[84] The results show that the DWNT synthesis can be carried out under conditions that are almost the same as those of SWNTs except for the temperature. Also, the present DWNT synthesis does not require an additional sulfur catalyst and a hydrogen buffer gas, which are required for the reported steady arc synthesis of DWNTs.[85,86] A special catalyst has been thought to be necessary to produce DWNTs before this finding.

The RBM Raman spectra (cf. Fig. 4(b)) reveal that the average diameter of SWNTs increased with temperature below 1200°C. The spectral feature changed suddenly at 1200°C, suggesting that DWNTs are synthesized when the temperature is sufficiently high to produce thick SWNTs with some critical tube diameters. The same temperature threshold for DWNTs was

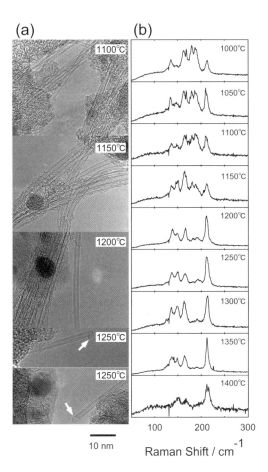

Fig. 4. (a) TEM images, and (b) Raman spectra of raw soot produced by the high-temperature pulsed arc discharge at various furnace temperatures.

also observed in the Raman spectra obtained with excitation at 488 and 514.5 nm. Such temperature thresholds for DWNTs were also observed in CCVD synthesis of DWNTs.[23] Although the DWNTs are produced in the temperature range between 1200 and 1350°C, the yields of DWNTs together with those of SWNTs are reduced at 1400°C. This temperature dependence is similar to that of SWNTs produced by the laser furnace method.[87] These results strongly suggest that the growth of the present DWNTs is closely related to that of SWNTs.

TEM images show that the ends of the DWNTs have doubly capped structures at 1250°C (cf. Fig. 4). These structures have also been observed

in CNT samples obtained by other methods.[23,85,86] Such capped structures
are known as potential candidates for precursors of SWNTs;[36] thus, these
doubly capped structures should be the precursors of DWNTs as well. On
the basis of the temperature and catalyst dependence on the DWNTs synthesis, the outer layers and outer caps are considered to play crucial roles
in the early growth of DWNTs.

The inner and outer diameters estimated by Raman spectra
(cf. Fig. 4(b)) were confirmed by TEM statistical observations (cf. Fig. 5).
The diameter distribution of DWNTs shows that the inner and outer diameters are 0.8–1.2 nm and 1.6–2.0 nm with peaks at 1.0 and 1.8 nm,
respectively. The diameters of SWNTs, however, are distributed from 1.2
to 1.6 nm with the most abundant diameter of 1.4 nm (cf. Fig. 5), which
falls between 1.0 and 1.8 nm of the inner and outer diameters of DWNTs,
respectively. For the production of DWNTs, large outer diameters greater
than 1.6 nm seem to be crucial. Without thicker outer tubes, the inner
tubes become too narrow to be formed. Narrow carbon nanotubes have
tightly rolled graphene-sheet and highly strained structures.[5] The typical
diameters of SWNTs are around 1 to 2 nm regardless of the production

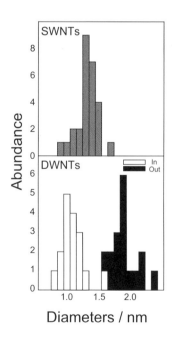

Fig. 5.  Diameter distribution of SWNTs and DWNTs of raw soot.

method used.[5] Since the diameter difference between the inner and outer tubes is around 0.7 nm in DWNTs,[86] the diameters of the outer tubes must be more than 1.6 nm to keep the diameters of the inner tubes within the favorable range of 1 to 2 nm. The temperature of 1200°C seems to be the critical temperature.

TEM and Raman observations show that the yield of DWNTs is strongly dependent on the concentration of Y and La,[30,31] and optimum conditions are achieved at 2.5 and 0.5 at. %, respectively. Under this optimum condition, the concentration of DWNTs among all CNTs is estimated to be 30%. In the absence of lanthanides, no DWNTs, but only SWNTs, are produced. Excess lanthanide doping decreases the production efficiency of both SWNTs and DWNTs, and converts almost all the carbonaceous materials into amorphous carbon. This lanthanide dependence was confirmed by the RBMs of Raman spectra,[84] which show enhancement of DWNT production together with that of thick SWNTs. A similar enhancement of the average diameter of SWNTs with Y has also been observed for steady arc discharge.[88] Thick SWNTs are essential for the effective synthesis of DWNTs with steady arc discharge and CVD.[23,85,86] These results are closely related to the temperature dependence of the production of DWNTs as discussed in the previous paragraph.

This lanthanide effect may be attributed to the strong interaction between lanthanide atoms and carbonaceous materials, as observed for various carbon materials, such as carbon-metal clusters[89,90] and metallofullerenes,[39,52] where electrons transfer from the lanthanide to the carbonaceous material. This interaction actually enlarges the size distribution of the carbon cage of metallofullerenes compared to that of pure carbon fullerenes,[39] which is similar to the enlargement of the diameter of the observed SWNTs. This effect can be understood in terms of a lanthanide atom curving a graphene sheet around it with this interaction. Some yttrium carbon clusters have a curved graphene sheet surrounding a center Y atom,[89,90] suggesting that Y is destructive for the $sp^2$ graphene structures and effective for producing metallofullerenes and amorphous carbon. Actually, only amorphous carbon is produced with excess of Y of 7.5 and La of 5 at. % as described above. Considering the two effects of the production of larger nanotubes and amorphous carbon, an optimum concentration clearly exists. These temperature and catalyst effects suggest that the production of DWNTs is mainly dominated by the diameter, which can be utilized for the selective production of DWNTs by CVD through control of the diameters of the metal catalyst particles.

### 3.2. *Purification of DWNTs, and their properties*

Figure 6 shows the changes observed in TEM images and Raman spectra of as-produced DWNTs through the purification process by hot air oxidation at 500°C. The TEM images in Fig. 6(a) show that the SWNTs are degraded together with amorphous carbon through the process. This is also supported by Raman spectra obtained with an excitation laser of 633 nm (cf. Fig. 6(b)) showing that the DWNTs peaks at 214 and 136 cm$^{-1}$ are particularly enhanced as the purification proceeds while the SWNTs peaks around 150 cm$^{-1}$ are diminished. Since these Raman spectra cor-

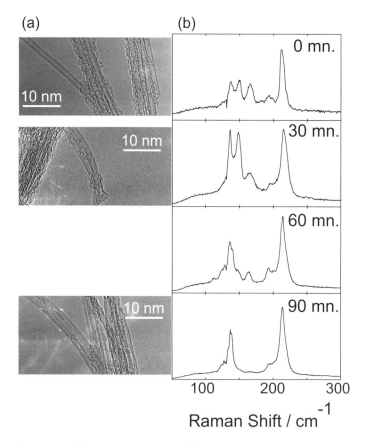

Fig. 6. Changes in (a) TEM images, and (b) radial breathing Raman spectra of as-produced DWNTs during purification by hot air oxidation at 500°C. The peaks located around 150 cm$^{-1}$ disappear as the longer oxidation are employed, showing the elimination of the contaminated SWNTs.

respond to radial breathing modes (RBMs) of CNTs and their frequency in cm$^{-1}$ is inversely proportional to the diameter of CNTs,[84] we can identify the peaks at 214 and 136 cm$^{-1}$ are corresponds to those of the inner and the outer layers of DWNTs, respectively while the peaks around 150 cm$^{-1}$ come from SWNTs. These results clearly show the higher chemical durability of DWNTs against oxidation compared to SWNTs, originating in the layer interaction of DWNTs.[91–94] This method has been applied to various DWNTs obtained using CVD, thereby proving its universality and usefulness.[46,47] The chemical durability of DWNTs from CVD has been determined to be several hundred times that for SWNTs.[47]

In these successful cases, the contaminant SWNTs are much narrower than the DWNTs and they are not bundles but rather are completely dispersed. Since narrower SWNTs are reported to be unstable[95] compared to thicker ones, we need to make a fair comparison between DWNTs and SWNTs having the same diameter to deduce the true effect of the layer interaction and to achieve selective and efficient purification. In addition to the diameter effect, CNTs bundling effects should also play a crucial role in the purification of DWNTs, since the layer and bundle effects come from similar layered graphene structures. If these two effects contribute the chemical durability of CNTs against oxidation similarly, SWNTs in a thick bundle could be burned slower than DWNTs in a thin bundle or isolated DWNTs. Our recent results of dispersion-assisted purification show that the diameter and bundle effects have significant influence on purification, however DWNTs are twice as durable as SWNTs for the same diameter of 1.6 nm, which is due to the layer interaction. Using this dispersion-assisted purification, we succeeded in purifying DWNTs up to 95% (cf. Fig. 7). These unique properties of DWNTs allow us to purify the present DWNTs, and may be one of the main causes for the much longer lifetimes observed compared to SWNTs, for field electron emitters, as mentioned earlier.[96] These DWNTs are still the smallest diameters and narrowest diameter distributions achieved so far.[23,85,97–100]

The layer interaction was observed not only for chemical durability but also within the structure itself. Sophisticated TEM observations were carried out by Dr. Suenaga and Prof. Iijima's group to detect structural correlations between the inner and outer layers of the DWNTs.[49] In this method, interference in TEM images originating in the carbon chiral network and chiral angles is detected by manipulating the tilt angles of the specimen holder. The results showed that 75% of the chiral angles of our DWNTs are correlated with the layer interaction. If there was no interaction, only

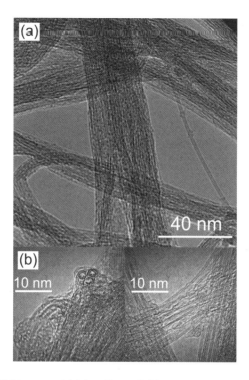

Fig. 7.   TEM images of (a) bundles and (b) individual purified DWNTs.

50% would be correlated. Those properties originating in the layer interaction have been utilized to construct devices as further described below.

### 3.3.   DWNT AFM tips

Table 1 shows the results of performance analyses of AFM tips made of Si, MWNTs and DWNTs, together with image parameters of isolated HiPco SWNTs. The lateral resolution is defined as the ratio between the full width and the height. Since the cross-section of an isolated SWNT used here should be circular, a better tip should show lateral resolution close to unity. The estimated tip radius ($R$) is defined as $R = h^2/(8v)$, where $h$ and $v$ are widths and heights of the SWNT images, respectively.[48] All performance parameters of the DWNT tips exceed those of the other types of tips. SEM observations show the diameter of the DWNT tips range from 5 to 10 nm suggesting that the attached DWNTs are bundles. The obtained high resolution show that only the very top of the bundle consists

Table 1. Performance parameters of AFM tips made of Si, MWNTs, and DWNTs, along with image parameters of HiPco SWNTs.

| Tip type | Full width/nm | Height/nm | Resolution | Tip radius/nm |
|----------|---------------|-----------|------------|---------------|
| Si | 33.0 | 0.95 | 35 | 143 |
| MWNTs | 27.4 | 0.84 | 33 | 112 |
| DWNTs | 3.91~6.64 | 0.36~1.13 | 4.9~11 | 3.3~5.3 |

of an individual DWNT with a narrow diameter less than 2 nm. Since the performance of the DWNT tip is comparable to that reported for SWNT tips,[75] the individual DWNTs at the very top of the tip should contribute to the high-resolution images.

This tip performance is due to the thinness and robustness of DWNTs, which would come from the layer interaction. While we have not prepared SWNT tips with the current manipulation techniques, the reported performance of SWNT tips is very similar to that of the DWNT tips.[75] Furthermore, the resolution of the SWNT tip is significantly affected by the tip morphology, where the length of the isolated SWNTs or a bundle sticking out must be less than 100 nm.[75] The lengths of the current DWNTs are not always shorter than 100 nm but are around several hundred nm, suggesting that the DWNTs could be more suitable for high-resolution AFM tips. A molecular mechanics calculation[101] indicated that the extent of cross-sectional deformation of an isolated SWNT increases dramatically as the tube diameter increases. If an inner tube is added to the SWNT to form DWNTs, this trend can be reversed. Moreover, the presence of defects mostly affects the bending rigidity of SWNTs as the number of defects increases,[102] whereas in DWNTs, the number of defects in the inner tubes is much less than that of SWNTs, leading to larger bending rigidity. These mechanical properties of DWNTs ensure stable scans for at least three hours on hard silicon oxide surfaces without any tip wear or degradation in lateral resolution.

## 3.4. *DWNT Field Effect Transistor*

Transport and conductivity measurements of DWNTs were carried out by varying the gate voltage ($V_{GS}$), while keeping drain-source voltage ($V_{DS}$) at 51 mV and the temperature at 23 K.[50] All the devices showed metallic or semiconducting behavior. In particular, semiconducting devices exhibited both p- and n-type actions, the so-called ambipolar characteristics with a

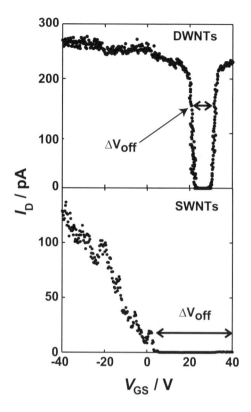

Fig. 8. Typical FET transfer characteristics of semiconducting DWNTs and SWNTs.

highly symmetric curve. Figure 8 shows typical drain current-gate voltage characteristics ($I_D$-$V_{GS}$) of the semiconducting DWNT devices. All the ambipolar DWNTs had off current region in the scanned gate voltage ($\Delta V_{off}$) which are also shown in the figure. The voltage width of the off-state region, $\Delta V_{off}$, ranged from 8 to 17 V with an average of 13 V in terms of DWNTs. The transfer characteristics of SWNT ($d = 1.3$–$1.6$ nm) devices were observed under the same conditions, for comparison. The SWNT devices showed either metallic or only p-type semiconducting behavior (cf. Fig. 8), suggesting a much larger $\Delta V_{off}$. The voltage width of the off-state region has been reported to depend on the band gap of the nanotubes employed.[103,104] A similar observation has also been reported for the ambipolar characteristics in large-diameter SWNTs.[105] Since the band gap of carbon nanotubes is inversely proportional to their diameters,[5] the width of the off-state region of DWNTs is primarily dominated by their diameters. In fact, the diameter distributions of SWNTs and DWNTs for the outer layers are 1.3–1.6 nm and 1.6–2.0 nm, respectively.

To further investigate FET operation of DWNTs and SWNTs, a comparison of the subthreshold swing ($S$) between the two types of CNTs, which is commonly used to characterize metal oxide semiconductor FETs, was performed.[106] The S factor is defined by the equation, $S \equiv dV_{GS}/d\log_{10} I_D$.[107] The obtained $S$ factors are 0.5–2 V per decade and 5–10 V per decade for DWNTs and SWNTs, respectively. The DWNTs show much higher $S$ factors than the SWNTs. Obviously, FETs employing DWNTs as channels provide higher performance than SWNTs-FETs for fundamental FET operation. DWNT channels can be a potential candidate for high-performance CNT-FETs. Recently electronic layer interaction has been reported,[93,94] which would make it possible to develop new types of electronic devices utilizing these layer interactions.

### 3.5. *Ion Mobility and Mass Spectrometry*

Figure 9 shows drift time distributions for positive ions of empty fullerenes and metallofullerenes obtained from solvent extractable $C_{80}$, $C_{82}$ and $Sc_n@C_{82}$ ($n = 1 \sim 3$) using the IMS/MS apparatus. $C_{80}^+$ and $C_{82}^+$ have a single peak, and the drift time of $C_{80}^+$ is clearly shorter than that of $C_{82}^+$. The obtained drift time is inversely proportional to the drift velocity, and proportional to the cross section of ions with the buffer He gas. Simple analysis of this cross section show that the spherical structures exactly correspond to those obtained by X-ray diffraction[39,108] and computational calculations.[109] This results indicate that the fullerene structures are not disturbed by the laser desorption and ionization, and the structural resolution of the apparatus is good enough to identify the fullerene size but is not sufficient to resolve structural isomers.[109] The ion of $Sc@C_{82}^+$ also shows a single peak, which corresponds exactly to that of $C_{82}^+$ and well annealed $C_{82}^+$ clusters.[62,110] These results are consistent with the endohedral structures determined by X-ray diffraction.[108] On the other hand, the di-metallofullerene $Sc_2C_{82}^+$ shows two distinct peaks. The main peak with a larger drift time and larger cross section corresponds to the peaks of $C_{82}^+$ and $Sc@C_{82}^+$. The other peak has a substantially smaller drift time, which is very similar to that of $C_{80}^+$ revealing that some of the $Sc_2C_{82}^+$ has the $C_{80}$ cage. The tri-metallofullerene $Sc_3@C_{82}^+$ shows only one peak close to $C_{80}^+$. The high-resolution ion mobility measurements enable us to detect these two structures for the first time.

*T. Sugai*

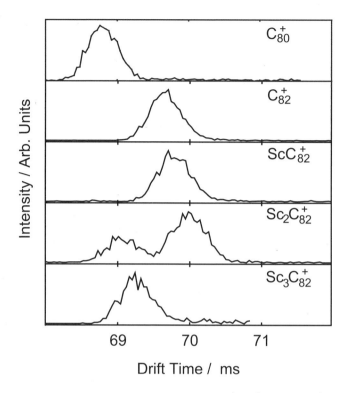

Fig. 9. High-resolution ion mobility distributions of $C_{80}^+$, $C_{82}^+$ and $Sc_nC_{82}^+$ ($n = 1 \sim 3$) produced from laser desorption/ionization of solvent extractable fullerenes. $Sc_2C_{82}^+$ shows two distinct peaks (see text for details).

Investigation of the whole species revealed that Sc mono-metallofullerenes have almost the same cross section as that of the empty fullerenes indicating that Sc metallofullerenes have endohedral structures in the size region less than $C_{100}$. In contrast, most of the $Sc_2@C_n^+$ ($n \leq 86$) have two cage sizes, one corresponds to that of $C_n^+$ and the other corresponds to that of $C_{n-2}^+$. These results strongly suggest that $Sc_2@C_n^+$ ions have a "carbide" structure, similar to $(Sc_2C_2)@C_{n-2}^+$, as well as $Sc_2@C_n^+$, the simple endohedral structure. Larger Sc di-metallofullerenes $Sc_2@C_n^+$ ($n \geq 88$) and Sc tri-metallofullerenes $Sc_3@C_n^+$ only have the smaller cage. These Sc tri-metallofullerenes may also possess carbide $(Sc_3C_2)@C_{n-2}^+$ structures. The drift times of the metallofullerenes are slightly larger than those of corresponding empty fullerenes (cf. Fig. 9). These differences are probably associated with the electronic properties of the metallofullerenes.

Electrons are transferred from the metal atoms to the cage, which result in, for example, $Sc^{+2}@C_{82}^{-2}$.[108] Since the ion mobility is mainly dominated by close interaction between the fullerene cage and He, $(Sc^{+2}@C_{82}^{-2})^+$ and $C_{82}^+$ are expected to have slightly different mobilities. Parallel differences have been observed for the mobilities of $In_n^-$ and $In_n^+$ clusters.[111]

Since this IMS/MS measurement requires laser desorption and ionization, carbide structures may be generated through the processes. Figure 10 shows the laser power dependence of the drift time distributions of $Sc_2@C_{82}^+$. The peak intensities for the shorter drift time increase as the laser power increases. These results clearly suggest that metallofullerene ions with the smaller (carbide) cages are produced through the desorption/ionization pro-

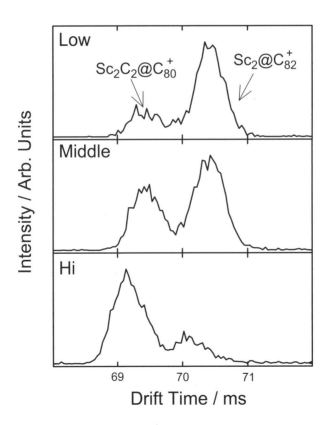

Fig. 10. Drift time distribution of $Sc_2@C_{82}^+$ produced under (a) low, (b) middle, and (c) high laser fluence. The peak with a shorter drift time becomes prominent as the laser fluence increases.

cesses. Fullerenes have an even number of carbon atoms to eliminate dangling bonds.[112] It is, therefore, natural that two carbon atoms (rather than one) should be encapsulated inside the cage during shrinkage. Although these new carbide structures are observed in the gas phase upon desorption/ionization, we have produced and identified $(Sc_2C_2)@C_{84}$ in macroscopic quantities.[59] Another group recently identified the carbide structures of$(Sc_3C_2)@C_{80}$ for Sc tri-metallofullerenes.[113] The results presented here show that the carbide encapsulated structure is universal. The results also clarified that novel structures can be identified by a series of IMS/MS measurements utilizing its high sensitivity and high throughput, which is available for mixtures and a wide range of materials. IMS/MS measurements are becoming popular, especially in biochemical research areas, and some of the measurement systems have been commercialized.[114] We are planning to develop a new IMS/MS system with much higher resolution and higher upper mass limit, and use it to explore new nanocarbon materials.

### 3.6. *Production of Large Carbon Clusters*

Figure 11 shows the mass spectrum of large positive carbon clusters up to $C_{2000}^+$ produced by the high-pressure laser vaporization cluster source. The broad peak of the mass spectrum shows resolved cluster signals around $C_{250}^+$ (cf. Fig. 11 inset), where the mass spacing was $C_2$. The mass resolution is not sufficient to observe individual peaks in the higher mass region, but they are very similar to fullerene peaks.[1,60] This similarity suggests that the giant clusters are not amorphous carbon but have structural similarities to fullerenes or multi-layered fullerenes of onions.[72]

Through the measurements of effects of the preload gas, cluster growth in the high-pressure laser-vaporization cluster source is expected to be as follows. A calm, stagnant, and quasi-fluxless condition is achieved by the preload valve. After the laser shot, the sample vapor is effectively and mildly cooled by the cooling gas. Expanding in the cooling gas, the cooled sample vapor gradually grows into large clusters. The fluxless condition has been found to be very important to produce various ultrafine particles,[83] which can hardly be realized for the usual laser vaporization cluster source.[1,68] These flux effects have been also observed in the production of CNTs by the laser furnace method,[34] where the diameters of SWNTs are strongly dependent on the gas flow speed.

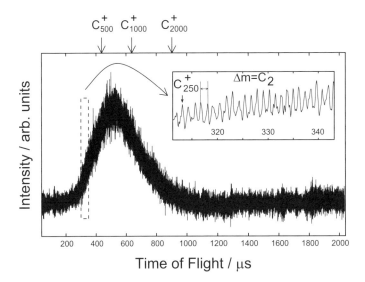

Fig. 11. Mass spectra of large carbon clusters up to $C_{2000}^+$ produced by a high-pressure laser-vaporization cluster source. The inset shows an expansion around $C_{250}^+$.

In the present cluster source, a combination of the so-called gas evaporation[83] and supersonic expansion[68] enables us to produce the large clusters. This system allow us to not only study the growth processes of nanocarbon materials and to produce novel materials, but also to investigate their structures since the giant clusters produced are well ionized and dispersed in the gas phase, suitable for IMS/MS. These kinds of well isolated and ionized sample preparation have been rarely realized in terms of nanocarbon samples since nanocarbon materials tend to stick to each other strongly. This new sample preparation and production methods would bring new dimensions to nanocarbon studies when it is combined with IMS/MS.

## 4. Summary

Control of nanocarbon material structure is still a difficult problem to overcome. The main reason for this originates in the carbon bonds which are strong and oriented. High-temperature is required to modify and rebuilt carbon networks, however, under such conditions, not only precise control but even direct measurements are difficult. Oriented carbon bonds are directly connected to the higher activation energy of bond reorganization and defects caused by misalignment. These difficulties are mirror images of the benefits of nanocarbon materials, such as high mechanical strength and chemical stability. We need to overcome these difficulties by developing new strategies. High-temperature pulsed arc discharge and ion mobility measurements, which have time control sequences, provide some options for future developments of nanocarbon materials.

## Acknowledgments

The author thanks Professor H. Shinohara, Dr. H. Yoshida, and Mr. H. Omote (Nagoya University) for the development of HTPAD and purification of DWNTs. The author also thanks Professor S. Iijima (Meijo University, NEC, AIST) and Dr. K. Suenaga for the state-of-the-art TEM observations, Professor T. Mizutani, Professors Y. Ohno and Dr. T. Shimada (Nagoya University) for measurements of FET characteristics. The author expresses his appreciation of Professor Y. Nakayama's and Professor S. Akita's (Osaka Pref. University) support for constructing DWNT AFM tips, and Mr. Kuwahara's (Nagoya University) support for evaluating tip performance. The author acknowledge the sincere support of Professor M. F. Jarrold (Indiana University) for the IMS/MS measurements on the metallofullerenes. This work has been supported by the JSPS Future Program on New Carbon Nanomaterials, Foundation Advanced Technology Institute, Shorai Foundation for Science and Technology, The Ogasawara Foundation for the Promotion of Science & Engineering, and Grants-in-Aid for Scientific Research (C)(2)(16510074), (C)(18510087), and (C)(20510095) of the Ministry of Education, Culture, Sports, Science and Technology of Japan.

## References

1. H. W. Kroto, J. R. Heath, S. C. O'Brien, R. F. Curl, and R. E. Smalley, $C_{60}$: Buckminsterfullerene, *Nature.* **318**, 162, (1985).
2. W. Krätschmer, L. D. Lamb, K. Fostiropoulos, and D. R. Huffman, Solid $C_{60}$: a new form of carbon, *Nature.* **347**, 354–358, (1990).
3. S. Iijima and T. Ichihashi, Helical microtubules of graphitic carbon, *Nature.* **354**, 56–58, (1991).
4. S. Iijima and T. Ichihashi, Single-shell carbon nanotubes of 1-nm diameter, *Nature.* **363**, 603–605, (1993).
5. R. Saito, G. Dresselhaus, and M. S. Dresselhaus, *Physical Properties of Carbon Nanotubes.* (Imperial College Press, London, 1998).

6. A. F. Hebard, M. J. Rosseinskuy, R. C. Haddon, D. W. Murphy, S. H. Glarum, T. T. M. Palstra, A. P. Ramirez, and A. R. Kortan, Superconductivity at 18 K in potassium-doped $C_{60}$, *Nature.* **350**, 600–601, (1991).

7. K. Tanigaki, T. W. Ebbesen, S. Saito, J. Mizuki, J. S. Tsai, Y. Kubo, and S. Kuroshima, Superconductivity at 33 K in $Cs_xRb_yC_{60}$, *Nature.* **352**, 222–223, (1991).

8. S. J. Tans, M. H. Devoret, H. Dai, A. Thess, R. E. Smalley, L. J. Geerlings, and C. Dekker, Individual single-wall carbon nanotubes as quantum wires, *Nature.* **386**, 474–477, (1997).

9. Z. Yao, H. W. C. Posta, L. Balents, and C. Dekker, Carbon nanotube intramolecular junctions, *Nature.* **402**, 217–328, (1999).

10. A. Bachtold, P. Hadley, T. Nakanishi, and C. Dekker, Logic circuits with carbon nanotube transistors, *Science.* **294**, 1317–1320, (2001).

11. P. G. Collins, M. S. Arnold, and P. Avouris, Engineering carbon nanotubes and nanotube circuits using electrical breakdown, *Science.* **292**, 706–709, (2001).

12. C. Cornell and L. Wille, *Solid State Commun.* **101**, 555, (1997).

13. M. Zhang, K. R. Atkinson, and R. H. Baughman, Multifunctional carbon nanotube yarns by downsizing an ancient technology, *Science.* **306**, 1358–1361, (2004).

14. Y. Saito, S. Uemura, and K. Hamaguchi, Cathode ray tube lighting elements with carbon nanotube field emitters, *Jpn. J. Appl. Phys.* **37**, L346–L352, (1998).

15. M. Endo, Y. A. Kim, T. Hayashi, K. Nishimura, T. Matushita, K. Miyashita, and M. S. Dresselhaus, Vapor-grown carbon fibers (VGCFs): Basic properties and their battery appplications, *Carbon.* **39**, 1287–1297, (2001).

16. O. Kimizuka, O. Tanaike, J. Yamashita, T. Hiraoka, D. N. Futaba, K. Hata, K. Machida, S. Suematsu, K. Tamamitsu, S. Saeki, Y. Yamada, and H. Hatori, Electrochemical doping of pure single-walled carbon nanotubes used as supercapacitor electrodes, *Carbon.* **46**, 1999–2001, (2008).

17. S. chun Mu, H. lin Tang, S. hao Qian, M. Pan, and R. zhang Yuan, Hydrogen storage in carbon nanotubes modified by microwave plasma etching and pd decoration, *Carbon.* **44**, 762–767, (2006).

18. J. W. Lee, H. S. Kim, J. Y. Lee, and J. K. Kang, Hydrogen storage and desorption properties of ni-dispersed carbon nanotubes, *Appl. Phys. Lett.* **88**, 143126, (2006).

19. J. Peet, A. B. Tamayo, X.-D. Dang, J. H. Seo, and T.-Q. Nguyena, Small molecule sensitizers for near-infrared absorption in polymer bulk heterojunction solar cells, *Appl. Phys. Lett.* **93**, 163306, (2008).

20. R. E. Haufler, Y. Chai, L. P. F. Chibante, J. Conceicao, C. Jin, L. S. Wang, S. Maruyama, and R. E. Smalley, *Mat. Res. Soc. Symp. Proc.* **206**, 627, (1991).

21. T. Guo, P. Nikolaev, A. Thess, D. T. Colbert, and R. E. Smalley, Catalytic growth of single-walled manotubes by laser vaporization, *Chem. Phys. Lett.* **243**, 49–54, (1995).

22. Y. Saito, M. Inagaki, H. Nagashima, M. Ohkohchi, and Y. Ando, Yield of fullerenes generated by contact arc method under He and Ar: dependence on gas pressure, *Chem. Phys. Lett.* **200**, 643–648, (1992).

23. R. R. Bacsa, C. Laurent, A. Peigney, W. S. Bacsa, T. Vaugien, and A. Rousset, Double-walled carbon nanotubes fabricated by a hydrogen arc discharge method, *Chem. Phys. Lett.* **323**, 566–571, (2000).

24. S. Maruyama, R. Kojima, Y. Miyauchi, S. Chiashi, and M. Kohno, Water-assisted highly efficient synthesis of impurity-free single-walled carbon nanotubes, *Chem. Phys. Lett.* **360**, 229, (2002).

25. K. Hata, D. N. Futaba, K. Mizuno, T. Namai, M. Yumura, and S. Iijima, Water-assisted highly efficient synthesis of impurity-free single-walled carbon nanotubes, *Science.* **306**, 1362–1364, (2004).

26. T. Sugai, H. Omote, and H. Shinohara, Production of fullerenes by high-temperature pulsed arc discharge, *Eur. J. Phys. D.* **9**, 369, (1999).

27. T. Sugai, H. Omote, S. Bandow, N. Tanaka, and H. Shinohara, Production of fullerenes by high-temperature pulsed arc discharge, *Jpn. J. Appl. Phys.* **38**, L477 – L479, (1999).

28. T. Sugai, H. Omote, S. Bandow, N. Tanaka, and H. Shinohara, Production of fullerenes by high-temperature pulsed arc discharge, *J. Chem. Phys.* **112**, 6000, (2000).

29. T. Sugai, H. Yoshida, T. Shimada, T. Okazaki, and H. Shinohara, New synthesis of high-quality double-walled carbon nanotubes by high-temperature pulsed arc discharge, *Nano Lett.* **3**, 769–773, (2003).

30. T. Sugai, T. Okazaki, H. Yoshida, and H. Shinohara, Syntheses of single- and double-wall carbon nanotubes by the HTPAD and HFCVD methods, *New J. Phys.* **6**, 21, (2004).

31. H. Yoshida, T. Sugai, and H. Shinohara, Fabrication, purification and characterization of double-wall carbon nanotubes via pulsed-arc discharge, *J. Phys. Chem. C.* **112**, 19908–19915, (2008).

32. G. Ganteför and H. R. Siekmann, Pure metal and metal-doped rare-gas clusters grown in a pulsed arc cluster ion source, *Chem. Phys. Lett.* **165**, 293–296, (1990).

33. T. Ishigaki, S. Suzuki, H. Kataura, W. Krätschmer, and Y. Achiba, Characterization of fullerenes and carbon nanoparticles generated with a laser-furnace technique, *Appl. Phys. A.* **70**, 121–124, (2000).

34. R. Sen, Y. Ohtsuka, T. Ishigaki, D. Kasuya, S. Suzuki, H. Kataura, and Y. Achiba, Time period for the growth of single-wall carbon nanotubes in the laser ablation process: evidence from gas dynamic studies and time resolved imaging, *Chem. Phys. Lett.* **332**, 467–473, (2000).

35. R. Sen, S. Suzuki, H. Kataura, and Y. Achiba, Growth of single-walled carbon nanotubes from condensed phase, *Chem. Phys. Lett.* **349**, 383–389, (2001).

36. D. B. Geohegan, H. Schittenhelm, X. Fan, S. J. Pennycook, A. A. Puretzky, M. A. Guillorn, D. A. Blom, and D. C. Joy, Condensed phase growth of single-wall carbon nanotubes from laser annealed nanoparticlulates, *Appl. Phys. Lett.* **78**, 3307–3309, (2001).

37. F. Kokai, K. Takahashi, M. Yudasaka, and S. Iijima, Laser ablation of grphite–co/ni and growth of single-wall carbon nanotubes in vortexes formed in an ar atmosphere, *J. Phys. Chem. B.* **104**, 6777–6784, (2000).

38. R. Taylor, J. P. Hare, A. K. Abdul-Sada, and H. W. Kroto, Isolation, separation and characterisation of the fullerenes $C_{60}$ and $C_{70}$: The third form of carbon, *J. Chem. Soc. Chem. Commun.* **20**, 1423–1425, (1990).

39. H. Shinohara, Endohedral metallofullerenes, *Rep. Prog. Phys.* **63**, 843–892, (2000).

40. I. W. Chiang, B. E. Brinson, A. Y. Huang, P. A. Willis, M. J. Bronikowski, J. L. Margrave, R. E. Smalley, and R. H. Hauge, Purification and characterization of single-wall carbon nanotubes (SWNTs) obtained from the gas-phase decomposition of CO (HiPco process), *J. Phys. Chem. B.* **105**, 8297–8301, (2001).

41. S. Bandow, A. M. Rao, K. A. Williams, A. Thess, R. E. Smalley, and P. C. Eklund, Purification of single-wall carbon nanotubes by microfiltration, *J. Phys. Chem. B.* **101**, 8839, (1997).

42. M. Zheng, A. Jagota, M. S. Strano, A. P. Santos, P. Barone, S. G. Chou, B. A. Diner, M. S. Dresselhaus, R. S. Mclean, G. B. Onoa, G. G. Samsonidze, E. D. Semke, M. Usrey, and D. J. Walls, Structure-based carbon nanotube sorting by sequence-dependent DNA assembly, *Science.* **302**, 1545–1548, (2003).

43. R. Krupke, F. Hennrich, H. v. Löhneysen, and M. M. Kappes, Separation of metallic from semiconducting single-walled carbon nanotubes, *Science.* **301**, 344–347, (2003).

44. M. S. Arnold, A. A. Green, J. F. Hulvat, S. I. Stupp, and M. C. Hersam, Sorting carbon nanotubes by electronic structure using density differentiation, *Nat. Nanotechnol.* **1**, 60–65, (2006).

45. K. Yanagi, Y. Miyata, and H. Kataura, Optical and conductive characteristics of metallic single-wall carbon nanotubes with three basic colors; cyan, magenta, and yellow, *Appl. Phys. Exp.* **1**, 034003, (2008).

46. Y. A. Kim, H. Muramatsu, T. Hayashi, M. Endo, M. Terrones, and M. S. Dresselhaus, Fabrication of high-purity, double-walled carbon nanotube buckypaper, *Chem. Vap. Deposition.* **12**, 327–330, (2006).

47. N. Kishi, T. Hiraoka, P. Ramesh, J. Kimura, K. Sato, Y. Ozeki, M. Yoshikawa, T. Sugai, and H. Shinohara, Enrichment of small-diameter

double-wall carbon nanotubes synthesized by catalyst-supported chemical vapor deposition using zeolite supports, *Jpn. J. Appl. Phys.* **46**, 1797–1802, (2007).

48. S. Kuwahara, S. Akita, M. Shirakihara, T. Sugai, Y. Nakayama, and H. Shinohara, Fabrication and characterization of high-resolution afm tips with high-quality double-wall carbon nanotubes, *Chem. Phys. Lett.* **429**, 581–585, (2006).

49. Z. Liu, K. Suenaga, H. Yoshida, T. Sugai, H. Shinohara, and S. Iijima, Determination of optical isomers for left-handed or right-handed chiral double-wall carbon nanotubes, *Phys. Rev. Lett.* **95**, 187406, (2005).

50. T. Shimada, T. Sugai, Y. Ohno, S. Kishimoto, T. Mizutani, H. Yoshida, T. Okazaki, and H. Shinohara, Double-wall carbon nanotube field-effect transistors: Ambipolar transport characteristics, *Appl. Phys. Lett.* **84**, 2412, (2004).

51. J. M. Hawins, A. Meyer, T. A. Lewis, S. Loren, and F. J. Hollander, Crystal structure of osmylated $C_{60}$: Confirmation of the soccer ball framework, *Science.* **252**, 312–313, (1991).

52. M. Takata, B. Umeda, E. Nishibori, M. Sakata, Y. Saito, M. Ohno, and H. Shinohara, Confirmation by x-ray diffraction of the endohedral nature of the metallofullerene $Y@C_{82}$, *Nature.* **377**, 46–49, (1995).

53. Y. Maniwa, H. Kataura, M. Abe, A. Udaka, S. Suzuki, Y. Achiba, H. Kira, K. Matsuda, H. Kadowaki, and Y. Okabe, Ordered water inside carbon nanotubes: formation of pentagonal to octagonal ice-nanotubes, *Chem. Phys. Lett.* **401**, 534–538, (2005).

54. T. Hashizume, X. D. Wang, Y. Nishina, H. Shinohara, Y. Saito, and T. Sakurai, Field ion-scanning tunneling microscopy study of $C_{84}$ on the Si(100) surface, *Jpn. J. Appl. Phys.* **32**, L132–L134, (1993).

55. T. Odom, J. Huang, P. Kim, and C. Lieber, Atomic structure and electronic properties of single-walled carbon nanotubes, *Nature.* **391**, 62–64, (1998).

56. J. Wildoer, L. Venema, A. Rinzler, R. Smalley, and C. Dekker, Electronic structure of atomically resolved carbon nanotubes, *Nature.* **391**, 59–62, (1998).

57. N. Naguib, H. Ye, Y. Gogotsi, A. G. Yazicioglu, C. M. Megaridis, and M. Yoshimura, Observation of water confined in nanometer channels of closed carbon nanotubes, *Nano Lett.* **4**, 2237–2243, (2004).

58. S. Stevenson, G. Rice, T. Glass, K. Harich, F. Cromer, M. R. Jordan, J. Craft, E. Hadju, and R. Bible, Small-bandgap endohedral metallofullerenes in high yield and purity, *Nature.* **401**, 55–57, (1999).

59. C. R. Wang, T. Kai, T. Tomiyama, T. Yoshida, Y. Kobayashi, E. Nishibori, M. Takata, M. Sakata, and H. Shinohara, Scandium-carbide endohedral metallofullerenes $(Sc_2C_2)@C_{84}$, *Angew. Chem. Int. Ed.* **40**, 397 – 399, (2001).

60. G. von Helden, M. T. Hsu, P. R. Kemper, and M. T. Bowers, Structures of carbon cluster ions from 3 to 60 atoms: Linears to rings to fullerenes, *J. Chem. Phys.* **95**, 3835–3837, (1991).

61. R. R. Hudgins, P. Dugourd, J. M. Tenenbaum, and M. F. Jarrold, Structural transitions in sodium chloride nanocrystals, *Phys. Rev. Lett.* **78**, 4213–4216, (1997).

62. A. A. Shvartsburg, R. R. Hudgins, P. Dugourd, R. Gutierrez, T. Frauen-heim, and M. F. Jarrold, Observation of "stick" and "handle" intermediates along the fullerene road, *Phys. Rev. Lett.* **84**, 2421–2424, (2000).

63. A. A. Shvartsburg, R. R. Hudgins, R. Gutierrez, G. Jungnickel, T. Frauen-heim, K. A. Jackson, and M. F. Jarrold, Ball-and-chain dimers from a hot fullerene plasma, *J. Phys. Chem. A.* **103**, 5275–5284, (1999).

64. A. G. Nasibulin, P. V. Pikhitsa, H. Jiang, and E. I. Kauppinen, Correla-tion between catalyst particle and single-walled carbon nanotube diameters, *Carbon.* **43**, 2251–2257, (2005).

65. D. Kondo, S. Sato, and Y. Awano, Low-temperature synthesis of single-walled carbon nanotubes with a narrow diameter distribution using size-classified catalyst nanoparticles, *Chem. Phys. Lett.* **422**, 481–487, (2006).

66. C. D. Scott, M. Ugarov, R. H. Hauge, E. D. Sosa, S. Arepalli, J. A. Schultz, and L. Yowell, Characterization of large fullerenes in single-wall carbon nanotube production by ion mobility mass spectrometry, *J. Phys. Chem. C.* **111**, 36–44, (2007).

67. K. Tanaka, H. Waki, Y. Ido, S. Akita, Y. Yoshida, and T. Yoshida, Protein and polymer analyses up to m/z 100000 by laser ionization time-of flight mass spectrometry, *Rapid Commun. Mass Spectrom.* **2**, 151–153, (1988).

68. D. E. Powers, S. G. Hansen, M. E. Geusic, A. C. Puiu, J. B. Hopkins, T. G. Dietz, M. A. Duncan, P. R. R. Langridge-Smith, and R. E. Smalley, Supersonic metal cluster beams: Laser photoionization studies of $Cu_2$, *J. Phys. Chem.* **86**, 2556–2560, (1982).

69. A. Koshio, M. Yudasaka, M. Ozawa, and S. Iijima, Fullerene formation via pyrolysis of ragged single-wall carbon nanotubes, *Nano Lett.* **2**, 995–997, (2002).

70. T. Sugai and H. Shinohara, Production and characterization of thermally-annealed large cluster by high-pressure laser-vaporization technique, *Z. Phys. D.* **40**, 131–135, (1997).

71. N. R. Greiner, D. S. Phillips, J. D. Johnson, and F. Volk, Diamonds in detonation soot, *Nature.* **333**, 440–442, (1988).

72. D. Ugarte, Curling and closure of graphitic networks under electron-beam irradiation, *Nature.* **359**, 707–709, (1992).

73. A. Nakayama, S. Iijima, Y. Koga, , K. Shimizu, K. Hirahara, and F. Kokai, Compression of polyhedral graphite up to 43 GPa and x-ray diffraction study on elasticity and stability of the graphite phase, *Appl. Phys. Lett.* **84**,

5112–5114, (2004).

74. Y. Achiba, T. Wakabayashi, T. Moriwaki, S. Suzuki, and H. Shiromaru, A hypothetical growth mechanism of carbon five- and six-membered ring networks, *Mater. Sci. & Eng. B.* **19**, 14–17, (1993).

75. L. A. Wade, I. R. Shapiro, Z. Ma, S. R. Quake, and C. P. Collier, Correlating afm probe morphology to imag resolution for single-wall caron nanotube tips, *Nano Lett.* **4**, 725–731, (2004).

76. H. Nishijima, S. Kamo, S. Akita, Y. Nakayama, K. Hohmura, S. Yoshimura, and K. Takeyasu, Carbon-nanotube tips for scanning probe microscopy: Preparation by a controlled process and observation of deoxyribonucleic acid, *Appl. Phys. Lett.* **74**, 4061–4063, (1999).

77. K. Yamamoto, S. Akita, and Y. Nakayama, Orientation and purification of carbon nanotubes using ac electrophoresis, *J. Phys. D.* **31**, L34–L36, (1998).

78. H. Shinohara, H. Sato, M. Ohkohchi, Y. Ando, T. Kodama, T. Shida, T. Kato, and Y. Saito, Encapsulation of a scandium trimer in $C_{82}$, *Nature.* **357**, 52–54, (1992).

79. P. Dugourd, R. R. Hudgins, D. E. Clemmer, and M. F. Jarrold, High-resolution ion mobility measurements, *Rev. Sci. Instrum.* **68**, 1122–1129, (1997).

80. T. Kimura, T. Sugai, H. Shinohara, K. Tohji, and I. Matsuoka, Preferential arc-discharge production of higher fullerenes, *Chem. Phys. Lett.* **246**, 571, (1995).

81. T. Kimura, T. Sugai, and H. Shinohara, Production and characterization of boron- and silicon- doped carbon clusters, *Chem. Phys. Lett.* **256**, 269, (1996).

82. T. Sugai and H. Shinohara, Production and mass spectroscopic characterization of ammonium halide clusters, *Chem. Phys. Lett.* **264**, 327–332, (1997).

83. R. Uyeda, Studies of ultrafine particles in japan: Crystallography. methods of preparation and technological applications, *Progress in Materials Science.* **35**, 1–96, (1991).

84. A. Jorio, R. Saito, J. H. Hafner, C. M. Lieber, M. Hunter, T. McClure, G. Dresselhaus, and M. S. Dresselhaus, Structural (n, m) determination of isolated single-wall carbon nanotubes by resonant Raman scattering, *Phys. Rev. Lett.* **86**, 1118, (2001).

85. J. L. Hutchison, N. A. Kiselev, E. Krinichnaya, A. Krestinin, R. Loutfy, A. Morawsky, V. Muradyan, E. Obraztsova, J. Sloan, S. Terekhov, and D. Zakharov, Double-walled carbon nanotubes fabricated by a hydrogen arc discharge method, *Carbon.* **39**, 761–770, (2001).

86. Y. Saito, T. Nakahira, and S. Uemura, Growth conditions of double-walled carbon nanotubes in arc discharge, *J. Phys. Chem. B.* **107**, 931–934, (2003).

87. H. Kataura, Y. Kumazawa, Y. Maniwa, Y. Ohtsuka, R. Sen, S. Suzuki, and

Y. Achiba, Diameter control of single-walled carbon nanotubes, *Carbon.* **38**, 1691–1697, (2000).

88. C. Journet, W. K. Maser, P. Bernier, A. Loiseau, M. L. de la Chapelle, S. Lefrant, P. Deniard, R. Lee, and J. E. Fischer, Large-scale production of single-walled carbon nanotubes by the electric-arc technique, *Nature.* **388**, 756–758, (1997).

89. A. Ayuela, G. Seifert, and R. Schmidt, Electronic structure of lanthanum-carbon clusters, *Z. Phys. D.* **41**, 69–72, (1997).

90. D. L. Strout and M. B. Hall, Structure and stability of lanthanum-carbon cations, *J. Phys. Chem. A.* **102**, 641–645, (1998).

91. M. B. Nardelli, C. Brabec, A. Maiti, C. Roland, and J. Bernholc, Lip-lip interactions and the growth of multiwalled carbon nanotubes, *Phys. Rev. Lett.* **80**, 313–316, (1998).

92. R. Saito, R. Matsuo, T. Kimura, G. Dresselhaus, and M. S. Dresselhaus, Anomalous potential barrier of double-wall nanotube, *Chem. Phys. Lett.* **348**, 187–193, (2001).

93. K. Uchida, S. Okada, K. Shiraishi, and A. Oshiyama, Quantum effects in a double-walled carbon nanotube capacitor, *Phys. Rev. B.* **76**, 155436, (2007).

94. S. Uryu and T. Ando, Electronic intertube transfer in double-wall carbon nanotubes with impurities:tight-binding calculations, *Phys. Rev. B.* **76**, 155434, (2007).

95. Y. Miyata, T. Kawai, Y. Miyamoto, K. Yanagi, Y. Maniwa, and H. Kataura, Chirality-dependent combustion of single-walled carbon nanotubes, *J. Phys. Chem. C.* **111**, 9671–9677, (2007).

96. H. Kurachi, S. Uemura, J. Yotani, T. Nagasako, H. Yamada, T. Ezaki, T. Maesoba, R. Loutfy, A. Moravsky, T. Nakazawa, and Y. Saito, FED with double-walled carbon nanotube emitters, *IDW Proceedings.* pp. 1237–1240, (2001).

97. W. Ren, F. Li, J. Chen, S. Bai, and H.-M. Cheng, Morphology, diameter distribution and Raman scattering measurements of double-walled carbon nanotubessynthesized by catalytic decomposition of methane, *Chem. Phys. Lett.* **359**, 196–202, (2002).

98. L. Ci, Z. Rao, Z. Zhou, D. Tang, X. Yan, Y. Liang, D. Liu, H. Yuan, W. Zhou, G. Wang, W. Liu, and S. Xie, Double wall carbon nanotubes promoted by sulfur in a floating iron catalyst CVD system, *Chem. Phys. Lett.* **359**, 63–67, (2002).

99. S. Bandow, M. Takizawa, K. Hirahara, M. Yudasaka, and S. Iijima, Raman scattering study of double-wall carbon nanotubes derived from the chains of fullerenes in single-wall carbon nanotubes, *Chem. Phys. Lett.* **337**, 48–54, (2001).

100. S. Bandow, G. Chen, G. U. Sumanasekera, R. Gupta, M. Yudasaka, S. Iijima, and P. C. Eklund, Diameter-selective resonant Raman scatter-

ing in double-wall carbon nanotubes, *Phys. Rev. B.* **66**, 075416, (2002).

101. T. Hertel, R. E. Walkup, and P. Avouris, Deformation of carbon nanotubes by surface van der Waals forces, *Phys. Rev. B.* **58**, 13870–13873, (1998).

102. Y. Hirai, S. Nishimaki, H. Mori, Y. Kimoto, S. Akita, Y. Nakayama, and Y. Tanaka, Molecular dynamics studies on mechanical properties of carbon nano tubes with pinhole defects, *Jpn. J. Appl. Phys.* **42**, 4120–4123, (2003).

103. T. Shimada, T. Okazaki, R. Taniguchi, T. Sugai, H. Shinohara, K. Suenaga, Y. Ohno, S. Mizuno, S. Kishimoto, and T. Mizutani, Ambipolar field-effect transistor behavior of $Gd@C_{82}$ metallofullerene peapods, *Appl. Phys. Lett.* **81**, 4067–4069, (2002).

104. T. Okazaki, T. Shimada, K. Suenaga, Y. Ohno, T. Mizutani, J. Lee, Y. Kuk, and H. Shinohara, Electronic properties of $Gd@C_{82}$ metallofullerene peapods: $(Gd@C_{82})n@SWNTs$, *Appl. Phys. A.* **76**, 475–478, (2003).

105. A. Javey, M. Shim, and H. Dai, Electrical properties and devices of large-diameter single-walled carbon nanotubes, *Appl. Phys. Lett.* **80**, 1064–1066, (2002).

106. S. M. Sze, *Physics of Semiconductor Devices.* (Wiley, New York, 1981).

107. Y. Taur and T. Ning, *Fundamentals of Modern VLSI Devices.* (Cambridge University Press, Cambridge, 1998).

108. E. Nishibori, M. Takata, M. Sakata, M. Inakuma, and H. Shinohara, Determination of the cage structure of $Sc@C_{82}$ by synchrotron powder diffraction, *Chem. Phys. Lett.* **298**, 79–84, (1998).

109. K. Kobayashi and S. Nagase, Structures and electronic states of $M@C_{82}$ (M=Sc, Y, La and lanthanides), *Chem. Phys. Lett.* **282**, 325–329, (1998).

110. K. B. Shelimov and M. F. Jarrold, Carbon clusters containing two metal atoms: Structures,growth mechanism, and fullerene formation, *J. Am. Chem. Soc.* **118**, 1139–1147, (1996).

111. J. Lermé, P. Dugourd, R. R. Hudgins, and M. F. Jarrold, High-resolution ion mobility measurements of indium clusters: electron spill-out in metal cluster anions and cations, *Chem. Phys. Lett.* **304**, 19–22, (1999).

112. H. W. Kroto, The stability of the fullerenes $C_n$, with n = 24, 28, 32, 36, 50, 60 and 70, *Nature.* **329**, 529–531, (1987).

113. Y. Iiduka, T. Wakahara, T. Nakahodo, T. Tsuchiya, A. Sakuraba, Y. Maeda, T. Akasaka, K. Yoza, E. Horn, T. Kato, M. T. H. Liu, N. Mizorogi, K. Kobayashi, and S. Nagase, Structural determination of metallofullerene $Sc_3C_{82}$ revisited: A surprising finding, *J. Am. Chem. Soc.* **127**, 12500–12501, (2005).

114. G. A. Eiceman and Z. Karpas, *Ion Mobility Spectroscopy.* (CRC Press, Boca Raton, 2005).

# Theoretical

Chapter 10

# Potential Model for Molecular Dynamics of Carbon

Atsushi M. Ito* and Hiroaki Nakamura

*National Institute for Fusion Science, Toki 509-5292, Japan*
*\*E-Mail address: ito.atsushi@nifs.ac.jp*

A modified reactive force field for classical molecular dynamics simulations of hydrogen interaction with graphite is presented. Based on the Brenner reactive empirical bond order potential (REBO) of 2002, modifications are introduced to eliminate the ambiguity of potential definition, and to suppress numerical errors in computer simulations. Long range interaction terms to describe graphite interlayer interactions are introduced as well. This modified Brenner REBO 2002 is applied to the problem of graphite peeling during hydrogen chemisorption on graphite walls in nuclear plasma fusion reactors.

## 1. History of Potential Model for Carbon

An important issue in classical molecular dynamics (MD) simulation is the choice of the potential model. Simple two-body potential models are used as elemental models of particles by statistical physicists because their goal is to elucidate collective motion which has universality independent of the types of atoms or molecules. On the other hand, a crucial matter of concern to material scientists, chemists and engineers is the reproduction of detailed behavior of atoms and molecules. Consequently, modeling of interactions between atoms has been tackled. In general, potential functions have complicated forms to represent the material structure and chemical interaction. These potential functions are often determined by empirical approaches. Though quantum chemical calculations have developed rapidly with computer technology, classical MD simulations using a potential model will continue to be useful to immediately attack simulations of materials in nano-science, engineering and biology. In this chapter, the potential function models for carbon-based systems are introduced.

The first potential model for a covalent bond was proposed by Morse [1]. He revealed that the Schrödinger equation of the hydrogen molecule, in which the eigenenergy is approximated by the second order of the radial quantum number, has an explicit solution if the potential function between two hydrogen atoms consists of two terms of exponential functions. This potential function is called the Morse potential. These two exponential functions were considered to be repulsive and attractive terms by Abell [2]. The Morse potential was regarded as the basic model of the bond order potential in general. Tersoff adopted multi-body interaction to treat the differences between molecular structures [3, 4]. A multi-body function is given by the function of the bond angle, which is the angle between two covalent bonds, and the cutoff functions, in order to decide the effective range of covalent bonds. He considered that the only attractive term of the Morse potential depends on surrounding atoms because dominant repulsive force is nuclear repulsion from the nearest atoms. Thereby, the coefficient of attractive term was the multi-body function in the Tersoff potential. Consequently, this Tersoff potential enabled simulating carbon, silicon and germanium materials. Currently, many potential models are based on the idea of the Tersoff potential.

Brenner then furthered the concept of the Tersoff potential. He employed a more complicated multi-body function to represent a particular structure of carbon with hydrogen. A graphite structure due to the $sp^2$ state was produced by the use of a dihedral angle, which is the angle between two planes formed by three covalent bonds. Moreover, a large variety of hydrocarbon molecules were realized by the use of cubic spline functions. This potential, which was termed the Brenner reactive empirical bond order (REBO) potential in 1990 [5], became popular coupled with the discoveries of fullerene $C_{60}$ in 1985 [6] and carbon nanotubes (CNTs) in 1991 [7]. The fact that the Brenner REBO potential was used for a number of simulations is a good test of its performance. However, several problems were listed, for example the disagreement of $\pi$ electron effects and amorphous carbon structures. Marks et al. created environment-dependent interaction potential (EDIP) to reproduce $sp^2/sp^3$ fraction in an amorphous carbon. However, the EDIP can apply to carbon atom only. For carbon and hydrogen systems, Brenner et al. proposed a new REBO potential in 2002, hereafter called the Brenner REBO 2002 potential [8].

Graphite has a layered structure, and is a molecular crystal different from diamond, hydrocarbon, silicon and other metals. In the layered structure of graphite, there are no covalent bonds between graphene layers. Since the Tersoff and Brenner potential models adopt cutoff functions to validate a covalent bond at short distances, they cannot treat interlayer intermolecular interaction in graphite, which is a long range interaction. Stuart et al. created an interlayer intermolecular potential suitable for the Brenner REBO potential, called the adaptive intermolecular REBO (AIREBO) potential.

The Brenner REBO 2002 potential was modified to eliminate the ambiguity of potential definition and to suppress numerical errors in actual computer simulations [10]. In the present chapter, the Brenner REBO 2002 potential and the modification points are first introduced. Next, the interlayer intermolecular potential is described along with the cone cutoff method to connect the general bond order potential in short range to the general intermolecular potential in long range [11].

## 2. The Modified Brenner REBO Potential

The Brenner REBO 2002 potential model and the modification points are described here. To eliminate ambiguity, the notations of functions, subscripts and superscripts have been changed from the original Brenner REBO 2002 potential paper. This potential was improved by changes in functions and parameters to relieve numerical errors, and is called the "modified Brenner REBO potential" here.

The modified Brenner REBO potential $U$ is defined as:

$$U \equiv \sum_{i,j>i} \left[ V_{[ij]}^{R}(r_{ij}) - \bar{b}_{ij}(\{r\}, \{\theta^B\}, \{\theta^{DH}\}) V_{[ij]}^{A}(r_{ij}) \right], \tag{1}$$

where $r_{ij}$ is the distance between the $i$th and $j$th atoms. The bond angle $\theta_{jik}^{B}$ is the angle between the vector from the $i$th atom to the $j$th atom and the vector from the $i$th atom to the $k$th atom, derived as follows:

$$\cos\theta_{jik}^{B} = \frac{\vec{r}_{ji} \cdot \vec{r}_{ki}}{r_{ji}r_{ki}}, \tag{2}$$

where $\vec{r}_{ij} \equiv \vec{r}_i - \vec{r}_j$ is the relative position of the $i$th atom with respect to the $j$th atom. The dihedral angle $\theta_{kijl}^{DH}$ is the angle between the plane passing through the $j$th, $i$th, and $k$th atoms and the plane passing through

Table 1. Parameters for the repulsive function $V_{[ij]}^{\mathrm{R}}$ and the attractive function $V_{[ij]}^{\mathrm{A}}$. The parameters depend on the species of the $i$ th and $j$ th atoms.

| Parameter | [ij] | | |
|---|---|---|---|
| | CC | HH | CH or HC |
| $Q_{[ij]}$ | 0.3134602960833 Å | 0.370471487045 Å | 0.340775728 Å |
| $A_{[ij]}$ | 10953.544162170 eV | 32.817355747 eV | 149.94098723 eV |
| $\alpha_{[ij]}$ | 4.7465390606595 Å$^{-1}$ | 3.536298648 Å$^{-1}$ | 4.10254983 Å$^{-1}$ |
| $B_{1[ij]}$ | 12388.79197798 eV | 29.632593 eV | 32.3551866587 eV |
| $B_{2[ij]}$ | 17.56740646509 eV | 0 eV | 0 eV |
| $B_{3[ij]}$ | 30.71493208065 eV | 0 eV | 0 eV |
| $\beta_{1[ij]}$ | 4.7204523127 Å$^{-1}$ | 1.71589217 Å$^{-1}$ | 1.43445805925 Å$^{-1}$ |
| $\beta_{2[ij]}$ | 1.4332132499 Å$^{-1}$ | 0 Å$^{-1}$ | 0 Å$^{-1}$ |
| $\beta_{3[ij]}$ | 1.3826912506 Å$^{-1}$ | 0 Å$^{-1}$ | 0 Å$^{-1}$ |

the $i$ th, $j$ th, and $l$ th atoms. The cosine function of $\theta_{kijl}^{\mathrm{DH}}$ is given by:

$$\cos\theta_{kijl}^{\mathrm{DH}} = \frac{\vec{r}_{ik} \times \vec{r}_{ji}}{r_{ik}r_{ji}} \cdot \frac{\vec{r}_{ji} \times \vec{r}_{lj}}{r_{ji}r_{lj}}. \tag{3}$$

The repulsive function $V_{[ij]}^{\mathrm{R}}(r_{ij})$ and the attractive function $V_{[ij]}^{\mathrm{A}}(r_{ij})$ are defined as:

$$V_{[ij]}^{\mathrm{R}}(r_{ij}) \equiv f_{[ij]}^{\mathrm{c}}(r_{ij}) \left(1 + \frac{Q_{[ij]}}{r_{ij}}\right) A_{[ij]} \exp\left(-\alpha_{[ij]}r_{ij}\right), \tag{4}$$

$$V_{[ij]}^{\mathrm{A}}(r_{ij}) \equiv f_{[ij]}^{\mathrm{c}}(r_{ij}) \sum_{n=1}^{3} B_{n[ij]} \exp\left(-\beta_{n[ij]}r_{ij}\right). \tag{5}$$

The square brackets $[ij]$ indicate that each function and each parameter depend on only the $i$ th and $j$ th atom species, for example, $V_{\mathrm{CC}}^{\mathrm{R}}$, $V_{\mathrm{HH}}^{\mathrm{R}}$, and $V_{\mathrm{CH}}^{\mathrm{R}}$ ($= V_{\mathrm{HC}}^{\mathrm{R}}$). The parameters are listed in Table 1.

The presence of a covalent bond is judged by a distance between two atoms $r_{ij}$ using the cutoff function $f_{[ij]}^{\mathrm{c}}(r_{ij})$. The two atoms are bound by a covalent bond if the distance $r_{ij}$ is shorter than $D_{[ij]}^{\mathrm{min}}$. The two atoms are not bound by a covalent bond if the distance $r_{ij}$ is longer than $D_{[ij]}^{\mathrm{max}}$. The cutoff function $f_{[ij]}^{\mathrm{c}}(r_{ij})$ connects the above two states smoothly as

$$f_{[ij]}^{\mathrm{c}}(x) \equiv \begin{cases} 1 & (x \le D_{[ij]}^{\mathrm{min}}), \\ \frac{1}{2}\left[1 + \cos(\pi \frac{x - D_{[ij]}^{\mathrm{min}}}{D_{[ij]}^{\mathrm{max}} - D_{[ij]}^{\mathrm{min}}})\right] & (D_{[ij]}^{\mathrm{min}} < x \le D_{[ij]}^{\mathrm{max}}), \\ 0 & (x > D_{[ij]}^{\mathrm{max}}). \end{cases} \tag{6}$$

Table 2. Parameters for the cutoff function $f^c_{[ij]}(r_{ij})$. The parameters depend on the species of the $i$th and $j$th atoms.

| [ij] | $D^{\min}_{[ij]}$ (Å) | $D^{\max}_{[ij]}$ (Å) |
|------|-----------------------|-----------------------|
| CC   | 1.7 | 2.0 |
| CH   | 1.3 | 1.8 |
| HH   | 1.1 | 1.7 |

The constants $D^{\min}_{[ij]}$ and $D^{\max}_{[ij]}$ depend on the species of the two atoms (Table 2).

The functions $V^R_{[ij]}$ and $V^A_{[ij]}$ in Eq. (1) generate a two-body force because both are functions of only the distance $r_{ij}$. The multi-body force is used instead of considering the effect of an electron orbital. In this model, $\bar{b}_{ij}(\{r\}, \{\theta^B\}, \{\theta^{DH}\})$ in Eq. (1) gives the multi-body force and is defined by

$$\bar{b}_{ij}(\{r\}, \{\theta^B\}, \{\theta^{DH}\}) \equiv \frac{1}{2}\left[b^{\sigma-\pi}_{ij}(\{r\}, \{\theta^B\}) + b^{\sigma-\pi}_{ji}(\{r\}, \{\theta^B\})\right]$$
$$+ \Pi^{RC}_{ij}(\{r\}) + b^{DH}_{ij}(\{r\}, \{\theta^{DH}\}). \tag{7}$$

The first term $\frac{1}{2}[\cdots]$ mainly generates the three-body force. The second term $\Pi^{RC}_{ij}$ in Eq. (7) is a correction factor to treat molecules with $\pi$ conjugation and radicals. The third term, $b^{DH}_{ij}(\{r\}, \{\theta^{DH}\})$, derives the four-body force relative to the dihedral angle. Since these functions are obtained from cutoff functions $f^c_{[ij]}(r_{ij})$, five- or more-body forces are generated during the chemical reaction.

The function $b^{\sigma-\pi}_{ij}(\{r\}, \{\theta^B\})$ in Eq. (7) is defined by

$$b^{\sigma-\pi}_{ij}(\{r\}, \{\theta^B\}) \equiv \left[1 + \sum_{k\neq i,j} f^c_{[ik]}(r_{ik})\tilde{G}_i(\cos\theta^B_{jik})e^{\lambda_{[ijk]}} + P_{[ij]}(N^H_{ij}, N^C_{ij})\right]^{-\frac{1}{2}}. \tag{8}$$

The function $\tilde{G}_i$ in Eq. (8) depends on the species of the $i$th atom. If $\cos\theta^B_{jik} > \cos(109.47°)$ and the $i$th atom is carbon, $\tilde{G}_i$ is defined by

$$\tilde{G}_i(\cos\theta^B_{jik}) \equiv \left[1 - Q_c(M^t_i)\right]G_C(\cos\theta^B_{jik}) + Q_c(M^t_i)\gamma_C(\cos\theta^B_{jik}). \tag{9}$$

If $\cos\theta^B_{jik} \leq \cos(109.47°)$ and the $i$th atom is carbon, $\tilde{G}_i$ is defined by

$$\tilde{G}_i(\cos\theta^B_{jik}) \equiv G_C(\cos\theta^B_{jik}). \tag{10}$$

Table 3. Parameters for the sixth-order spline function $G_C(\cos\theta^B_{jik})$.

| $\cos\theta^B_{jik}$ | $G_C$ | $G'_C$ | $G''_C$ | $G^{(3)}_C$ |
|---|---|---|---|---|
| $-1$ | $-0.001$ | $0.10400$ | $0$ | $0$ |
| $-1/2$ | $0.05280$ | $0.170$ | $0.370$ | $-5.232$ |
| $\cos(109.47°)$ | $0.09733$ | $0.400$ | $1.980$ | $41.6140$ |
| $1$ | $8.0$ | $0.23622$ | $-166.1360$ | — |

Table 4. Parameters for the sixth-order spline function $\gamma_C(\cos\theta^B_{jik})$.

| $\cos\theta^B_{jik}$ | $\gamma_C$ | $\gamma'_C$ | $\gamma''_C$ | $\gamma^{(3)}_C$ |
|---|---|---|---|---|
| $\cos(109.47°)$ | $0.09733$ | $0.400$ | $1.980$ | $-9.9563027$ |
| $1$ | $1.0$ | $0.78$ | $-11.3022275$ | — |

Table 5. Parameters for the sixth-order spline function $G_H(\cos\theta^B_{jik})$. The parameters are determined when $\cos\theta^B_{jik} = 0$.

| Parameter | Value |
|---|---|
| $G_H(0)$ | $19.06510$ |
| $G'_H(0)$ | $1.08822$ |
| $G''_H(0)$ | $-1.98677$ |
| $G^{(3)}_H(0)$ | $8.52604$ |
| $G^{(4)}_H(0)$ | $-6.13815$ |
| $G^{(5)}_H(0)$ | $-5.23587$ |
| $G^{(6)}_H(0)$ | $4.67318$ |

Also, if the $i$ th atom is hydrogen, $\tilde{G}_i$ is defined by

$$\tilde{G}_i(\cos\theta^B_{jik}) \equiv G_H(\cos\theta^B_{jik}). \tag{11}$$

Here, $G_C$, $\gamma_C$, and $G_H$ are sixth-order polynomial spline functions with seven coefficients in Tables 3–5, respectively. The function $Q_c$ and the coordination number $M^t_i$ in Eq. (9) are defined as:

$$Q_c(x) \equiv \begin{cases} 1 & (x \leq 3.2), \\ \frac{1}{2}\left[1 + \cos\left(2\pi\left(x - 3.2\right)\right)\right] & (3.2 < x \leq 3.7), \\ 0 & (x > 3.7), \end{cases} \tag{12}$$

$$M_i^{\mathrm{t}} \equiv \sum_{k \neq i} f_{[ik]}^{\mathrm{c}}(r_{ik}). \tag{13}$$

The constant $\lambda_{[ijk]}$ in Eq. (8) is defined by:

$$\lambda_{\mathrm{HHH}} = 4.0, \tag{14}$$

$$\lambda_{\mathrm{CCC}} = \lambda_{\mathrm{CCH}} = \lambda_{\mathrm{CHC}} = \lambda_{\mathrm{HCC}}$$
$$= \lambda_{\mathrm{HHC}} = \lambda_{\mathrm{HCH}} = \lambda_{\mathrm{CHH}} = 0. \tag{15}$$

The function $P_{[ij]}$ in Eq. (8), which is a bicubic spline function that depends on the parameters in Table 6, is necessary for describing solid structures. The variables $N_{ij}^{\mathrm{H}}$ and $N_{ij}^{\mathrm{C}}$ are, respectively, the numbers of hydrogen and carbon atoms bound with the $i$th atom as follows:

$$N_{ij}^{\mathrm{H}} \equiv \sum_{k \neq i,j}^{\mathrm{hydrogen}} f_{[ik]}^{\mathrm{c}}(r_{ik}), \tag{16}$$

$$N_{ij}^{\mathrm{C}} \equiv \sum_{k \neq i,j}^{\mathrm{carbon}} f_{[ik]}^{\mathrm{c}}(r_{ik}). \tag{17}$$

Table 6. Parameters for the bicubic spline function $P_{[ij]}(N_{ij}^{\mathrm{H}}, N_{ij}^{\mathrm{C}})$. The parameters that are not stated are zero.

| $P_{[ij]}(N_{ij}^{\mathrm{H}}, N_{ij}^{\mathrm{C}})$ | Value |
| --- | --- |
| $P_{\mathrm{CC}}(1,1)$ | 0.003026697473481 |
| $P_{\mathrm{CC}}(2,0)$ | 0.007860700254745 |
| $P_{\mathrm{CC}}(3,0)$ | 0.016125364564267 |
| $P_{\mathrm{CC}}(1,2)$ | 0.003179530830731 |
| $P_{\mathrm{CC}}(2,1)$ | 0.006326248241119 |
| $P_{\mathrm{CH}}(1,0)$ | 0.2093367328250380 |
| $P_{\mathrm{CH}}(2,0)$ | −0.064449615432525 |
| $P_{\mathrm{CH}}(3,0)$ | −0.303927546346162 |
| $P_{\mathrm{CH}}(0,1)$ | 0.01 |
| $P_{\mathrm{CH}}(0,2)$ | −0.1220421462782555 |
| $P_{\mathrm{CH}}(1,1)$ | −0.1251234006287090 |
| $P_{\mathrm{CH}}(2,1)$ | −0.298905245783 |
| $P_{\mathrm{CH}}(0,3)$ | −0.307584705066 |
| $P_{\mathrm{CH}}(1,2)$ | −0.3005291724067579 |

The second term, $\Pi_{ij}^{\mathrm{RC}}$, in Eq. (7) is defined by a tricubic spline function $F_{[ij]}$ as:

$$\Pi_{ij}^{\mathrm{RC}}(\{r\}) \equiv F_{[ij]}(N_{ij}^{\mathrm{t}}, N_{ji}^{\mathrm{t}}, N_{ij}^{\mathrm{conj}}), \qquad (18)$$

whose parameters are given in Table 7. The variables are defined as:

$$N_{ij}^{\mathrm{t}} \equiv \sum_{k \neq i,j} f_{[ik]}^{\mathrm{c}}(r_{ik}), \qquad (19)$$

$$N_{ij}^{\mathrm{conj}} \equiv 1 + \sum_{\substack{k(\neq i,j)}}^{\mathrm{carbon}} f_{[ik]}^{\mathrm{c}}(r_{ik}) C_{\mathrm{N}}(N_{ki}^{\mathrm{t}}) + \sum_{\substack{l(\neq j,i)}}^{\mathrm{carbon}} f_{[jl]}^{\mathrm{c}}(r_{jl}) C_{\mathrm{N}}(N_{lj}^{\mathrm{t}}), \qquad (20)$$

with

$$C_{\mathrm{N}}(x) \equiv \begin{cases} 1 & (x \leq 2), \\ \frac{1}{2}\left[1 + \cos(\pi(x-2))\right] & (2 < x \leq 3), \\ 0 & (x > 3). \end{cases} \qquad (21)$$

The third term, $b_{ij}^{\mathrm{DH}}(\{r\}, \{\theta^{\mathrm{DH}}\})$, in Eq. (7) is defined as:

$$b_{ij}^{\mathrm{DH}}(\{r\}, \{\theta^{\mathrm{DH}}\}) \equiv T_{[ij]}(N_{ij}^{\mathrm{t}}, N_{ji}^{\mathrm{t}}, N_{ij}^{\mathrm{conj}})$$
$$\times \left[ \sum_{k \neq i,j} \sum_{l \neq j,i} \left(1 - \cos^2 \theta_{kijl}^{\mathrm{DH}}\right) f_{[ik]}^{\mathrm{c}}(r_{ik}) f_{[jl]}^{\mathrm{c}}(r_{jl}) \right], \qquad (22)$$

where $T_{[ij]}$ is a tricubic spline function and has the same variables as $F_{[ij]}$ in Eq. (18). The coefficients for $T_{[ij]}$ are also revised because of the modified $N_{ij}^{\mathrm{conj}}$ (Table 8).

The following points regarding modification from the Brenner REBO 2002 potential are noted. In the original paper describing the Brenner REBO 2002 potential, only six coefficients of the sixth-order polynomial spline functions $G_{\mathrm{C}}$, $\gamma_{\mathrm{C}}$, and $G_{\mathrm{H}}$ of Eqs. (9)–(11) were given. However, the sixth-order polynomial spline functions require seven coefficients. Therefore, the modified Brenner REBO potential includes seven coefficients of the spline functions $G_{\mathrm{C}}$, $\gamma_{\mathrm{C}}$, and $G_{\mathrm{H}}$ in Tables 3–5. The second and third terms on the right-hand side of Eq. (20) are squared in the Brenner REBO 2002 potential, while they are not squared in the modified Brenner REBO potential. The coefficients of $F_{[ij]}$ in Table 7 were revised simultaneously. With this modification, the numerical error becomes smaller than that using the Brenner REBO 2002 potential. We set the time step to $5 \times 10^{-18}$ s in the present simulation. This time step is smaller than that in the general

Table 7. Parameters for the tricubic spline function $F_{[ij]}$. The parameters that are not stated are zero. The function $F_{[ij]}$ satisfies the following rules: $F_{[ij]}(N_1, N_2, N_3) = F_{[ij]}(N_2, N_1, N_3)$, $\partial_{N_1} F_{[ij]}(N_1, N_2, N_3) = \partial_{N_1} F_{[ij]}(N_2, N_1, N_3)$, and $F_{[ij]}(N_1, N_2, N_3) = F_{[ij]}(3, N_2, N_3)$ if $N_1 > 3$, and $F_{[ij]}(N_1, N_2, N_3) = F_{[ij]}(N_1, N_2, 5)$ if $N_3 > 5$, where $\partial_{N_i} \equiv \partial/\partial N_i$.

| Function | Variables | | | Value |
|---|---|---|---|---|
| | $N_1$ | $N_2$ | $N_3$ | |
| $F_{CC}(N_1, N_2, N_3)$ | 1 | 1 | 1 | 0.105000 |
| | 1 | 1 | 2 | −0.0041775 |
| | 1 | 1 | 3 to 5 | −0.0160856 |
| | 2 | 2 | 1 | 0.09444957 |
| | 2 | 2 | 2 | 0.04632351 |
| | 2 | 2 | 3 | 0.03088234 |
| | 2 | 2 | 4 | 0.01544117 |
| | 2 | 2 | 5 | 0.0 |
| | 0 | 1 | 1 | 0.04338699 |
| | 0 | 1 | 2 | 0.0099172158 |
| | 0 | 2 | 1 | 0.0493976637 |
| | 0 | 2 | 2 | −0.011942669 |
| | 0 | 3 | 1 to 5 | −0.119798935 |
| | 1 | 2 | 1 | 0.0096495698 |
| | 1 | 2 | 2 | 0.030 |
| | 1 | 2 | 3 | −0.0200 |
| | 1 | 2 | 4 to 5 | −0.030133632 |
| | 1 | 3 | 2 to 5 | −0.124836752 |
| | 2 | 3 | 1 to 5 | −0.044709383 |
| $\partial_{N_1} F_{CC}(N_1, N_2, N_3)$ | 2 | 1 | 1 | −0.052500 |
| | 2 | 1 | 3 to 5 | −0.054376 |
| | 2 | 3 | 1 | 0.0 |
| | 2 | 3 | 2 to 5 | 0.062418 |
| $\partial_{N_3} F_{CC}(N_1, N_2, N_3)$ | 2 | 2 | 4 | −0.006618 |
| | 1 | 1 | 2 | −0.060543 |
| | 1 | 2 | 3 | −0.020044 |
| $F_{HH}(N_1, N_2, N_3)$ | 1 | 1 | 1 | 0.249831916 |
| $F_{CH}(N_1, N_2, N_3)$ | 0 | 2 | 3 to 5 | −0.009047787516128811 |
| | 1 | 3 | 1 to 5 | −0.213 |
| | 1 | 2 | 1 to 5 | −0.25 |
| | 1 | 1 | 1 to 5 | −0.5 |

MD simulation because the potential model has a complicated form with cutoff functions and the spline function. However, if using the Brenner REBO 2002 potential, a smaller time step is needed for numerical accuracy. In particular, the square formulas used in Eq. (20) of the Brenner REBO 2002 potential often generate numerical errors because the cutoff functions $f^c_{[ij]}(r_{ij})$, which result in steep slopes of the potential energy curves for $D^{min}_{[ij]} < r < D^{max}_{[ij]}$, are squared.

Table 8. Parameters for the tricubic spline function $T_{CC}$. The parameters that are not stated are zero. The function $T_{CC}$ satisfies the following rule: $T_{CC}(N_1, N_2, N_3) = T_{CC}(N_1, N_2, 5)$ if $N_3 > 5$.

| Function | Variables | | | Value |
|---|---|---|---|---|
| | $N_1$ | $N_2$ | $N_3$ | |
| $T_{CC}(N_1, N_2, N_3)$ | 2 | 2 | 1 | $-0.070280085$ |
| | 2 | 2 | 5 | $-0.00809675$ |

Actually, the binding energy of hydrogen atoms on a graphene surface, as calculated by the modified Brenner REBO potential (as well as the Brenner REBO 2002 potential), is larger than the actual binding energy. The parameter $T_{CC}(2, 2, 5)$ affects the strength of the flat structure of graphene, and the parameters $T_{CC}(2, 2, 3)$ and $T_{CC}(2, 2, 4)$ determine the binding energy of hydrogen atoms on a graphene surface. As the parameters $T_{CC}(2, 2, 3)$ and $T_{CC}(2, 2, 4)$ decrease, the binding energy of a hydrogen atom on a graphene surface becomes closer to the actual binding energy. However, there is a potential barrier in the potential energy curve as a function of the distance between the hydrogen atom and the graphene surface. The height of the potential energy barrier of the modified Brenner REBO potential, which agrees with the actual potential energy barrier, increases as the parameters $T_{CC}(2, 2, 3)$ and $T_{CC}(2, 2, 4)$ decrease. Therefore, in the modified Brenner REBO potential, the parameters $T_{CC}(2, 2, 3)$ and $T_{CC}(2, 2, 4)$ are set to zero, equal to that in the Brenner REBO 2002 potential.

As a benchmark of the modified REBO potential model, the potential energies of structures of $C_{28}$ were compared with Density Functional Tight Binding (DFTB), B3LYP/6-31G(d), B3LYP/6-31G, AM1 and PM3 calculations [12, 13]. The structures of $C_{28}$ shown in Fig. 1 were considered by Portmann et al. [14]. The potential energies were obtained using individually optimized molecular structures. The results of potential energies calculations are shown in Table 9, wherein the reference value is the potential energy of buckyD$_2$. This comparison implies that calculation of the modified REBO potential is close to that of DFTB. More precisely, relative isomer energies as calculated by the modified REBO potential for structures containing mainly six- and five-membered rings agree well with those calculated using the first principles B3LYP method.

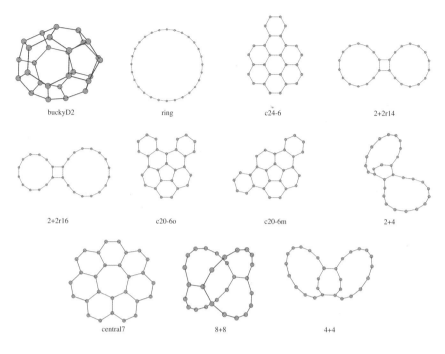

buckyD2    ring    c24-6    2+2r14

2+2r16    c20-6o    c20-6m    2+4

central7    8+8    4+4

Fig. 1. The structures of $C_{28}$.

## 3. Interlayer Intermolecular Potential

The interlayer intermolecular potential in graphite is described in this section. As described in the previous section, the REBO potentials adopt the cutoff function of a short distance to valid covalent bonds, which is 2 Å or less. Therefore, in order to treat long range interactions, such as the interlayer intermolecular potential, the intermolecular potential effective over a long range should be used in addition to the REBO potentials. The simple two-body function is sufficient for the intermolecular potential. In general, the attractive term of the intermolecular potential is proportional to $r^{-6}$, where $r$ is the distance between atoms. The repulsive term is selected from convenient functions such as the Lennard-Jones potential and exponent-6 potential.

A simple method to connect the intermolecular and REBO potentials is to consider that these potentials are switched by only the distance between atoms. However, this simple method causes the following problem when versatility is considered. Here, the distance to switch the intermolecular

Table 9. Relative energies of $C_{28}$ in eV for comparison with the modified REBO potential, DFTB, B3LYP/6-31G(d), B3LYP/6-31G, AM1, and PM3 calculations.

| Structures of $C_{28}$ | the modified REBO | DFTB | B3LYP/ 6-31G(d) | B3LYP/ 6-31G | AM1 | PM3 |
|---|---|---|---|---|---|---|
| buckyD$_2$ | 0.00 | 0.00 | 0.00 | 0.00 | 0.00 | 0.00 |
| ring | 8.01 | 8.10 | 3.32 | 0.78 | −7.69 | −2.15 |
| c24-6 | 2.47 | 3.56 | 3.17 | 1.99 | 0.43 | 1.77 |
| 2+2r14 | 9.64 | 9.66 | 5.08 | 2.90 | −3.34 | 0.91 |
| 2+2r16 | 9.67 | 10.25 | 6.01 | 3.87 | −3.37 | 0.90 |
| c20-6o | 4.41 | 5.52 | 5.41 | 4.34 | 3.42 | 4.23 |
| c20-6m | 4.43 | 5.62 | 5.57 | 4.48 | 3.43 | 4.24 |
| 2+4 | 8.81 | 10.28 | 7.97 | 6.00 | 0.10 | 3.60 |
| central7 | 5.53 | 6.07 | 5.86 | 4.84 | | |
| 8+8 | 8.98 | 9.43 | 7.43 | 5.31 | −3.24 | 0.79 |
| 4+4 | 9.05 | 14.27 | 9.91 | 8.52 | 1.52 | 4.97 |

and REBO potentials is defined by $c$. The REBO potential becomes valid when $r < c$, while the intermolecular potential becomes valid when $r > c$. The intermolecular potential acts on atoms that are members of the same molecule only if $r > c$. Therefore, if the REBO potential can reproduce the ideal molecular structures completely, the intermolecular potential disturbs the ideal molecular structures. The binding energy of the intermolecular potential is weaker than that of the REBO potential in general. However, this weak binding energy of the interlayer potential is sufficient to change the behavior of atoms at low temperatures.

The intermolecular potential is often used along with the grouping of atoms in molecules. For instance, one carbon atom in graphite interacts with atoms located in the same graphene layer due to the REBO potential, while it interacts with atoms located in other graphene layers due to the intermolecular potential. Thus, classification in molecules specifies which of the REBO or the intermolecular potential should be used. However, this method necessitates the grouping of atoms in molecules as often as molecules change species due to chemical reactions. The calculation for the grouping of atoms requires a high cost. Moreover, in the cases of polymer molecules, the intermolecular interaction should occur between atoms located in the same molecule. Therefore, this method by grouping of atoms in molecules is also not versatile.

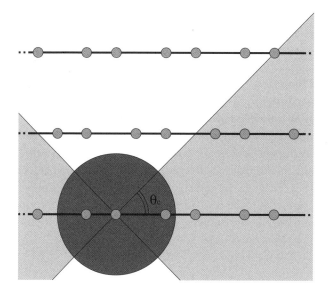

Fig. 2. The effective region of the cone cutoff method. Grey spheres are carbon atoms in the layer structure of graphite. The angle $\theta_c$ is half the vertical angle of the circular cone.

The cone cutoff method solves these problems. The cone cutoff method employs the cutoff functions of not only distances but also angles. The interactions with one atom, called the center atom here, are classified in Fig. 2 showing a graphite structure. First, atoms located in the region within the cutoff distance $c$ from the center atom, which is the dark gray region in Fig. 2, are not subjected to the intermolecular potential with the center atom. Next, the circular cone centered on the axis of the covalent bond from the center atom is considered. If an atom is located in the region within the circular cone painted light gray, the atom also does not experience the intermolecular potential. If an atom is not located in both regions within the cutoff distance $c$ and the circular cone, the atom is then subjected to the intermolecular force with the center atom. In this method, it is considered that there are atoms of same molecule in the direction of the covalent bond from the center atom. Then, the circular cone encloses these atoms of the same molecule and the intermolecular potential is disabled between the center atom and enclosed atoms.

The cone cutoff method defines the effective region of the intermolecular potential according to the vertical angle of the circular cone. Since

graphene has a two-dimensional structure and each carbon atom has three
covalent bonds, three circular cones with the vertical angle greater than
$2\pi/3$ fully cover whole atoms in the graphene. For general use of classical
MD simulation in three-dimensional space, the vertical angle of the circular
cone should be selected out of the range from $2\pi/3$ to $\pi$.

The above description is the basic idea of the cone cutoff method. Here,
one function of the cone cutoff method is introduced. If a better function
for the circular cone can be discovered, then it should be used.

The following conditions are imposed on a function to achieve the cone
cutoff method. If the distance between atoms is less than the cutoff length
of the REBO potential, the intermolecular potential becomes zero. If one
atom is located within the cone of the other atom, then the intermolecular
potential between these atoms becomes zero. The cone cutoff function
which satisfies these conditions is defined as:

$$C_{ij}(\{r\},\{\theta^{\mathrm{B}}\}) \equiv \frac{1}{2}\left\{\prod_{k\neq i}\left[1 - f^{\mathrm{c}}_{[ik]}(r_{ik})f^{\mathrm{a}}(\cos\theta^{\mathrm{B}}_{jik})\right]\right. \tag{23}$$

$$\left. + \prod_{l\neq j}\left[1 - f^{\mathrm{c}}_{[jl]}(r_{jl})f^{\mathrm{a}}(\cos\theta^{\mathrm{B}}_{jil})\right]\right\}, \tag{24}$$

where $\vec{x}_{ij} \equiv \vec{r}_i - \vec{r}_j$, $r_{ij} = |\vec{r}_{ij}|$ and $\cos\theta^{\mathrm{B}}_{jik} = (\vec{r}_{ji}\cdot\vec{x}_{ki})/(r_{ji}r_{ki})$. The
functions $f^{\mathrm{a}}(\cos\theta)$ are given by

$$f^{\mathrm{a}}(\cos\theta) \equiv \begin{cases} 0 & (\cos\theta \leq c_{\mathrm{on}}), \\ \frac{(2\cos\theta - 3c_{\mathrm{off}} + c_{\mathrm{on}})(\cos\theta - c_{\mathrm{on}})^2}{(c_{\mathrm{on}} - c_{\mathrm{off}})^3} & (c_{\mathrm{on}} < \cos\theta \leq c_{\mathrm{off}}), \\ 1 & (c_{\mathrm{off}} < \cos\theta), \end{cases} \tag{25}$$

$$\tag{26}$$

where $c_{\mathrm{on}} = 0.25$ and $c_{\mathrm{off}} = 0.35$. The function $f^{\mathrm{c}}_{[ij]}(r)$ is equal to the cutoff
function of the modified Brenner REBO potential:

$$f^{\mathrm{c}}_{[ij]}(r) \equiv \begin{cases} 1 & (r \leq D^{\mathrm{min}}_{[ij]}), \\ \frac{1}{2}\left[1 + \cos(\pi\frac{r - D^{\mathrm{min}}_{[ij]}}{D^{\mathrm{mar}}_{[ij]} - D^{\mathrm{min}}_{[ij]}})\right] & (D^{\mathrm{min}}_{[ij]} < r \leq D^{\mathrm{max}}_{[ij]}), \\ 0 & (r > D^{\mathrm{max}}_{[ij]}), \end{cases} \tag{27}$$

where the parameters are denoted in Table 2.

Using the cone cutoff function $C_{ij}(\{r\}, \{\theta^{\mathrm{B}}\})$, the interlayer intermolecular potential for graphite is defined as:

$$U_{\mathrm{IL}} = \sum_{i,j>i} C_{ij}(\{r\}, \{\theta^{\mathrm{B}}\}) V_{\mathrm{IL}}(r_{ij}). \tag{28}$$

The function $V_{\mathrm{IL}}(r)$ is the main term of the intermolecular potential as a two-body function of the distance between atoms defined by:

$$V_{\mathrm{IL}}(r) = A \left\{ \frac{n}{\alpha} e^{-\alpha\left(\frac{r}{c}-1\right)} - \left(\frac{c}{r}\right)^{n} \right\}, \tag{29}$$

where $r$ is the distance between two carbon atoms, $n$ is the exponent of attraction, and $A, \alpha, c$ are the parameters to determine binding energy. If $n > \alpha$, the potential function has a positive local maximum at $r = c$. The parameter $c$ is set to the minimum cutoff length of a C–C bond $D_{\mathrm{CC}}^{\min}$ in the modified Brenner REBO potential. The parameter $n$ is generally 6, as the exponent of the Van der Waals potential.

In research efforts on the interlayer intermolecular interaction in graphite, the binding energy has been investigated [15]. The binding energy was calculated to be 53.9 meV/atom and the interlayer distance of graphite of 3.35 Å determines the local minimum of the interlayer intermolecular potential. The activation energy of the transition from graphite to diamond, which is 0.347–0.414 eV/atom [16], serves as a reference to the local maximum energy of the interlayer intermolecular potential at $r = c$. As a result, the parameters of the interlayer intermolecular potential for graphite are determined as follows. The interlayer distance depends on the parameters $c$ and $\alpha$. The binding energy depends on the parameter $A$. The parameters are determined by $c = 1.8$ Å, $\alpha = 4.84$, and $A = 4.7247772$ eV (see Table 10). Consequently, the binding energy and the local maximum energy become 53.9 meV/atom and 0.340–0.435 eV/atom in a graphite structure composed of 5 layers, respectively (see Fig. 3).

Table 10. The parameters of the potential model of the interlayer intermolecular interaction. The parameters $A_{20}, A_{60}$ and $A_{100}$ correspond to the coefficient $A$ in $V_{\mathrm{IL}}(r)$ which determines the interlayer binding energy per atom to 20 meV, 40 meV and 60 meV, respectively.

| $C_{ij}(\{r\}, \{\theta^{\mathrm{B}}\})$ | $V_{\mathrm{IL}}(r_{ij})$ | |
|---|---|---|
| $c_{\mathrm{on}} = 0.25$ | $\alpha = 4.84$ | $A = 4.7247772$ eV |
| $c_{\mathrm{off}} = 0.35$ | $n = 6$ | $c = 1.8$ Å |

The most stable structure of graphite is the "ABAB" stacking structure shown in Fig. 4. As a benchmark of the interlayer intermolecular potential, the potential energy curves of the "ABAB" stacking structure are compared with other stacking structures in Fig. 4. From Fig. 3, structure (a) becomes the most stable structure of graphite. The stacking structure (b) is more unstable than the stacking structure (c).

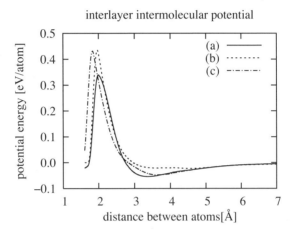

Fig. 3. The interlayer intermolecular potentials as functions of distance between atoms. The symbols (a), (b) and (c) correspond to the three kinds of stacking of graphite in Fig. 4.

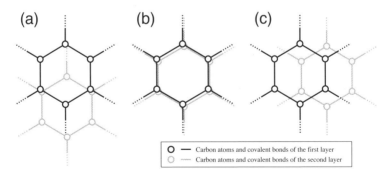

Fig. 4. Three samples of the stacking of graphite. In general, the most stable structure for graphite is the "ABAB" stacking structure (a).

## 4. Application to Plasma-Wall Interaction

The bond order potential models which are used to treat the chemical reactions of carbon are often used for simulations of the formation of nano-carbon such as fullerenes and carbon nanotubes. Since the intermolecular force is not effective in nano-carbon, simulations were carried out without using the intermolecular potential models. The intermolecular potential models are mainly used for simulations of graphite. As a sample case wherein the bond order and intermolecular potential models are concurrently required, the chemical erosion of graphite surfaces by hydrogen atom bombardment is described in the present section.

In nuclear fusion research, investigation of plasma wall/surface interactions (PWI/PSI) on the inside wall of a nuclear fusion device is important. In plasma confinement in nuclear fusion devices, a part of inside walls called "divertor" are bombarded by hydrogen ions and atoms guided by magnetic line. The divertor consists of graphitic material that has a high heat resistance. However, the hydrogen ions and atoms erode these graphite tiles and can form hydrocarbon molecules, such as $CH_x$ and $C_2H_x$. These processes are termed chemical erosion and chemical sputtering, respectively. The hydrocarbon molecules affect plasma confinement. Therefore, clarification of chemical erosion and sputtering is hoped. Moreover, hydrogen (isotope) retention in graphite is also a problem. Tritium and deuterium will be used as fuel in future nuclear fusion devices such as the ITER. Since the cost of tritium production is very high, it is anticipated that tritium will be recycled in nuclear fusion devices. However, tritium retention in graphite reduces the efficiency of tritium recycling. MD simulation is thus very convenient for investigating these PWI/PSI processes of graphite and hydrogen (isotopes) [11, 17–28]. The MD simulations can display the motions of atoms and consequently clarify the mechanisms of chemical erosion and sputtering. In addition, experimental treatment and handling of tritium is dangerous, and MD simulations can investigate tritium effects safely.

To deal with chemical erosion and sputtering in nano-scale, we should be aware that layered graphite has three different surface types, flat (0 0 0 1), armchair (1 1 $\bar{2}$ 0) and zigzag (1 0 $\bar{1}$ 0). These three types of surfaces result in differences in chemical erosion and sputtering of the graphite. In the present work, hydrogen atom bombardment effects on the three kinds of graphite surface have been investigated using MD simulations.

The three graphite surface types were prepared as follows. Hydrogen atoms were injected perpendicular to the graphite surface measuring about

$2.0 \times 2.0$ nm$^2$, where periodic boundary conditions were imposed on directions parallel to the graphite surface. In the case of the flat $(0\ 0\ 0\ 1)$ surface, eight graphene layers were located parallel to the graphite surface, while in the case of the armchair $(1\ 1\ \bar{2}\ 0)$ and zigzag $(1\ 0\ \bar{1}\ 0)$ surfaces, six graphene layers were located perpendicular to the graphite surfaces. For all of these, the interlayer distance of graphite was set to 3.35 Å and the graphene layers were stacked in an "ABAB" pattern, initially. Two carbon atoms in bottom layer of the flat $(0\ 0\ 0\ 1)$ surface and six carbon atoms in bottom edges of the armchair $(1\ 1\ \bar{2}\ 0)$ and zigzag $(1\ 0\ \bar{1}\ 0)$ surfaces were fixed during time development to keep graphite position in simulation box.

Hydrogen atoms were injected into the graphite surface vertically. The incident positions of the hydrogen atoms in the direction parallel to the graphite surface were determined at random with uniform distribution. The hydrogen atoms were injected at intervals of 0.1 ps and the incident flux was about $2.5 \times 10^{30}$ m$^{-2}$s$^{-1}$. The incident energy was varied from 1 to 30 eV, which is considered to be the region wherein chemical sputtering occurs.

In the present simulation, two types of potential models were used to deal with chemical interactions at short distances and interlayer intermolecular interactions at long distances. Chemical interaction is represented by the modified Brenner REBO potential. The layered structure of graphite is supported by the interlayer intermolecular potential with the cone cutoff method. The interlayer intermolecular potential can produce an interlayer distance of 3.35 Å and "ABAB" stacking.

The time evolution of the equation of motion was solved by second-order symplectic integration with a time step of $5 \times 10^{-18}$ s. The MD simulations were performed under the NVE condition, in which the number of particles (N), volume (V), and total energy (E) are conserved.

The simulation results of hydrogen atom bombardment onto the flat $(0\ 0\ 0\ 1)$ surface of graphite is described first. The graphite surface adsorbed many hydrogen atoms. The graphene layer on the surface side was eroded. However, the layered structure of graphite was maintained. As the injection of hydrogen atoms onto the flat surface was continued, the graphene layers were peeled off from the surface sheet by sheet (see Fig. 5). This phenomenon is called "graphite peeling". During the graphite peeling, the layered structure of graphite under the surface was maintained. The peeled graphene was broken up and small hydrocarbon molecules were produced. These small hydrocarbon molecules mainly had chain structures in which hydrogen atoms terminate at the edges of the chains.

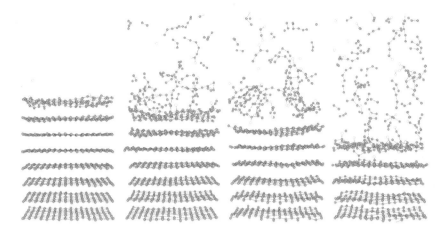

Fig. 5. Snapshots of graphene peeling from the graphite (0 0 0 1) surface by hydrogen atom bombardment. The white spheres represent hydrogen atoms and the grey spheres represent carbon atoms. From left to right, it is demonstrated that the graphene layers were peeled off one sheet by one sheet. Such graphene peeling is particular to the layered structure of graphite maintained by interlayer intermolecular interaction.

Graphite peeling due to the hydrogen atom bombardment is explained as follows. Almost all of incident hydrogen atoms for incident energy of less than 30 eV are adsorbed or reflected by a surface graphene layer. The incident energy and the binding energy of hydrogen adsorption change into the kinetic energy of the surface graphene layer. Since the distance between graphene layer is kept by the interlayer intermolecular forces, a new C-C bond is hardly created between the graphene layers. The raised kinetic energy stays in the surface graphene layer because there is not a covalent bond between the graphene layers to transport kinetic energy into underlying graphene layers. Since the binding energy of the interlayer intermolecular potential is weaker than that of the C-C bond, the surface graphene layer that has the raised kinetic energy escapes from the underlying graphene layers easily. Consequently, the surface graphene layer is peeled off while maintaining the layered structure of graphite under the surface.

Actually, graphite peeling does not occur in MD simulations of hydrogen atom bombardment onto the graphite surface in the absence of the interlayer intermolecular potential. When the interlayer intermolecular potential is not used, the graphite changes into amorphous carbon due to the hydrogen atom bombardment. Moreover, in the case where vacancies are present in the graphite with the interlayer intermolecular potential, graphite peeling occurs, however, the layered structure of graphite under the surface is not maintained. Rather, the graphite under the surface is amorphized by the existence of vacancies [27].

Next, hydrogen atom bombardment onto armchair $(1\,1\,\bar{2}\,0)$ and zigzag $(1\,0\,\bar{1}\,0)$ surfaces of graphite were simulated. The armchair and zigzag edges of graphene layers were easily terminated by the incident hydrogen atoms. The layered structures of the armchair and zigzag surfaces were also maintained by the interlayer intermolecular potential. Graphite peeling did not occur in these cases because the graphene layers of the armchair and zigzag surfaces are perpendicular to the direction of the graphite surface.

As hydrogen atoms were attached to the armchair and zigzag edges of the graphene layers, hydrocarbon molecules were produced. These were mainly $C_2H_2$ and $H_2$. In particular, the erosion yield of $C_2H_2$ from the armchair surface, which is the amount of $C_2H_2$ produced, was larger than that from the zigzag surface, while the erosion yield of $H_2$ from the armchair surface was smaller than that from the zigzag surface. This difference was understood as follows.

In the armchair $(1\,1\,\bar{2}\,0)$ surface, the production mechanism of $C_2H_2$ can be explained using Fig. 6 (a) When the hydrogen atoms (I) are adsorbed by the carbon atoms (III), the incident energy and the binding energy of a C-H bond (i), created between the hydrogen atom (I) and the carbon atom (III), are diffused into the carbon atoms (IV) and (V). The increase in the kinetic energy of the carbon atom (V) is greater than that of the carbon atom (IV) because the carbon atom (V) located in the incident direction of the hydrogen atom (I) is pushed by the carbon atom (III) more strongly than the carbon atom (IV). Therefore, the C-C bond (iii) is broken easily, more than the C-C bond (ii). In the same way, when the hydrogen atom (II) is adsorbed by the carbon atom (IV), the C-C bond (iv) is broken. As a result, $C_2H_2$, which consists of atoms (I) to (IV), is detached from the armchair surface.

On the other hand, the production mechanism of $H_2$ on the zigzag $(1\,0\,\bar{1}\,0)$ surface can be explained using Fig. 6 (b). When the hydrogen atom (VII) is adsorbed by the carbon atom (VIII), the incident energy and the binding energy of the C-H bond (v) created are diffused into the carbon atoms (IX) and (X). Since the positions of carbon atoms (IX) and (X) are symmetric with respect to the incident axis of the hydrogen atom (VII), increases in the kinetic energies of (IX) and (X) are equal. Therefore, breaking of C-C bonds (vi) and (vii) is difficult on the zigzag surface, as compared to breaking of C-C bonds (iii) and (iv) on the armchair surface.

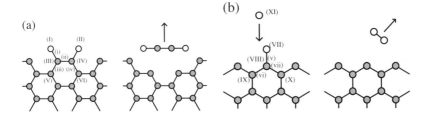

Fig. 6. Production mechanism of $C_2H_2$ from the armchair (1 1 $\bar{2}$ 0) surface (a) and the production mechanism of $H_2$ from the zigzag (1 0 $\bar{1}$ 0) surface (b).

When hydrogen atoms (XI) approach the zigzag surface subsequently, they create H-H bonds with the hydrogen atoms (VII). The C-H bond (v) is broken using the incident energy of the hydrogen atom (XI) and the binding energy of the H-H bond created. Consequently, $H_2$ is desorbed from the zigzag surface.

From the above scenario, the dominant hydrocarbon molecules produced from the armchair and zigzag surfaces are $C_2H_2$ and $H_2$, respectively.

Thus, the combination of the modified REBO potential and the interlayer intermolecular potential with the cone cutoff method enable us to treat chemical erosion of graphite which has a layered structure. As a result, the mechanisms of chemical erosion and sputtering were clarified. The MD simulations using these potential models thus achieved successful outcomes in the investigation of PWI/PSI for nuclear fusion.

## 5. Concluding Remarks

In this chapter, the modified Brenner REBO potential and the interlayer intermolecular potential with the cone cutoff method were introduced. Classical MD simulations using these potential models were demonstrated in the PWI investigation of hydrogen atom bombardment onto a graphite surface.

The bond order potential models for atoms other than carbon atom were introduced here. The Brenner REBO 2002 potential was expanded to a model in which oxygen can be treated in addition to carbon and hydrogen [29]. In the PWI scene, the potential models for beryllium, helium, tungsten and iron on the basis of the Tersoff potential were proposed to investigate plasma facing material substitutes for graphite [30–33]. Thus, the bond order potential was actively investigated.

However, the bond order potentials leave some problems.

Though the binding energy and stable bond length could be well fitted to the energies obtained by quantum chemical calculations and experiments, the potential energy curves (PECs) are not sufficient. One cause of this problem is that the cutoff length of the bond order potential is short. The PECs drawn by the potential function are warped by the cutoff function of the short range. There have been attempts to solve this problem, in which the bond order potential is constructed from the functions extended to a long range [34].

The bond order potential determines the potential energy of the molecular structures by taking into account the bond number of the surrounding atoms. Thus, in the structure which has an irregular network of covalent bonds, such as an amorphous structure, a difference between the potential energies calculated by the bond order potential and quantum chemical calculation easily occurs.

In addition, the current potential model in classical MD simulation can only deal with neutral atoms. Treatment of ionization from neutral atoms or neutralization from ions are hoped for in the arena of surface science using ion beams.

Thus, progress in nano-science, engineering and biology expands applications of classical MD simulation. This demands new ability of the potential model. The development of the potential model in classical MD simulations needs to be continued.

Although the functions of bond order potentials were empirically determined in general, similar formula to the bond order potentials was led by Pettifor et al. from quantum mechanics [35, 36]. Quantum chemical calculations play an important role to derive the parameters of potential functions. For instance, the "thermodynamic downfolding" method removes the burden of parameter fitting [37]. In this method, the parameters of potential functions are automatically set by comparison with quantum chemical calculations. Consequently, we can concentrate on efforts to find the best form of the potential functions. Furthermore, next-generation potential models might go beyond the concept of "bond order" which will be very interesting.

## References

[1] Morse, P. M. (1929). Diatomic molecules according to the wave mechanics. II. vibrational levels, *Phys. Rev.* **34**, 1, pp. 57–64.

[2] Abell, G. C. (1985). Empirical chemical pseudopotential theory of molecular and metallic bonding, *Phys. Rev. B* **31**, 10, pp. 6184–6196.

[3] Tersoff, J. (1988). New empirical approach for the structure and energy of covalent systems, *Phys. Rev. B* **37**, 12, pp. 6991–7000.

[4] Tersoff, J. (1989). Modeling solid-state chemistry: Interatomic potentials for multicomponent systems, *Phys. Rev. B* **39**, 8, pp. 5566–5568 [Errata; **41**, 5, p. 3248].

[5] Brenner, D. W. (1990). Empirical potential for hydrocarbons for use in simulating the chemical vapor deposition of diamond films, *Phys. Rev. B* **42**, 15, pp. 9458–9471 [Errata; **46**, 3, p. 1948].

[6] Kroto, H. W., Heath, J. R., O'Brien, S. C., Curl, R. F. and Smalley, R. E. (1985). $C_{60}$: Buckminsterfullerene, *Nature* **318**, 14, pp. 162–163.

[7] Iijima, S. (1991). Helical microtubules of graphitic carbon, *Nature* **354**, 7, pp. 56–58.

[8] Brenner, D. W., Shenderova, O. A., Harrison, J. A., Stuart, S. J., Ni, B. and Sinnott, S. B. (2002). A second-generation reactive empirical bond order (REBO) potential energy expression for hydrocarbons, *J. Phys.: Condens. Matter* **14**, pp. 783–802.

[9] Marks, N. A. (2001). Generalizing the environment-dependent interaction potential for carbon, *Phys. Rev. B* **63**, 035401.

[10] Ito, A., Nakamura, H. and Takayama, A. (2008). Molecular dynamics simulation of the chemical interaction between hydrogen atom and graphene, *J. Phys. soc. Jpn.* **77**, 11, 114602.

[11] Ito, A. and Nakamura, H. (2008). Molecular dynamics simulation of bombardment of hydrogen atoms on graphite surface, *Commun. Comput. Phys.* **4**, 3, pp. 592–610.

[12] Zheng, G., Irle, S., Elstner, M. and Morokuma, K. (2004). Quantum chemical molecular dynamics model study of fullerene formation from open-ended carbon nanotubes, *J. Phys. Chem. A* **108**, pp. 3182–3194.

[13] Irle, S., Zheng, G., Wang, Z. and Morokuma, K. (2006). The $C_{60}$ formation puzzle "solved": QM/MD simulations reveal the shrinking hot giant road of the dynamic fullerene self-assembly mechanism, *J. Phys. Chem. B* **110**, pp. 14531–14545.

[14] Portmann, S., Galbraith, J. M., Schaefer, H. F., Scuseria, G. E. and Lüthi, H. P. (1999). Some new structure of $C_{28}$, *Chem. Phys. Lett.* **301**, pp. 98–104.

[15] Hasegawa, M. and Nishidate, K. (2004). Semiempirical approach to the energetics of interlayer binding in graphite, *Phys. Rev. B* **70**, 205431.

[16] Tateyama, Y., Ogitsu, T., Kusakabe, K. and Tsuneyuki, S. (1996). Constant-pressure first-principles studies on the transition states of the graphite-diamond transformation, *Phys. Rev. B* **54**, 21, pp. 14994–15001.

[17] Salonen, E., Nordlund, K., Tarus, J., Ahlgren, T., Keinonen, J. and Wu, C. H. (1999). Suppression of carbon erosion by hydrogen shielding during high-flux hydrogen bombardment, *Phys. Rev. B* **60**, 20, R14005.

[18] Salonen, E., Nordlund, K., Keinonen, J. and Wu, C. H. (2001). Swift chemical sputtering of amorphous hydrogenated carbon, *Phys. Rev. B* **63**, 195415.

[19] Alman, D. A. and Ruzic, D. N. (2003). Molecular dynamics calculation of carbon/hydrocarbon reflection coefficients on a hydrogenated graphite surface, *J. Nucl. Mater.* **313–316**, pp. 182–186.

[20] Alman, D. A. and Ruzic, D. N. (2004). Molecular dynamics simulation of hydrocarbon reflection and dissociation coefficients from fusion-relevant carbon surfaces, *Phys. Scr.* **T111** pp. 145–151.

[21] Marian, J., Zepeda-Ruiz, L. A., Gilmer, G. H., Bringa, E. M. and Rognlien, T. (2006). Simulations of carbon sputtering in amorphous hydrogenated samples, *Phys. Scr.* **T124**, pp. 65–69.

[22] Yamashiro, M., Yamada, H. and Yamaguchi, S. (2007). Molecular dynamics simulation study on substrate tenperature dependence of sputtering yields for an organic polymer under ion bombardment, *J. Appl. Phys.* **101**, 046108.

[23] Ohya, K., Kikuhara, Y., Inai, K., Kirchner, A., Borodin, D., Ito, A., Nakamura, H. and Tanabe, T. (2009). Simulation of hydrocarbon reflection from carbon and tungsten surfaces and its impact on codeposition patterns on plasma facing components, *J. Nucl. Mater.* **390-391**, 15, pp. 72-75.

[24] Kikuhara, Y., Inai, K., Ito, A., Nakamura, H. and Ohya, K. (2008). Hydrocarbon Reflection and Redeposition on Plasma-Facing Surfaces, *2nd Japan-China Workshop on blanket and tritium technology*, proceedings.

[25] Nakamura, H. and Ito, A. (2007). Molecular dynamics simulation of sputtering process of hydrogen and graphene sheets, *Mil. Sim.* **33**, 1-2, pp. 121-126.

[26] Ito, A. and Nakamura, H. (2008). Hydrogen isotope sputtering of graphite by molecular dynamics simulation, *Thin Solid Films* **516**, 3, pp. 6553-6559.

[27] Ito, A., Wang, Y., Irle, S., Morokuma, K. and Nakamura, H. (2009). Molecular dynamics simulation of hydrogen atom sputtering on the surface of graphite with defect and edge, *J. Nucl. Mater.* **390-391**, 15, pp. 183-187.

[28] Ito, A., Wang, Y., Irle, S., Morokuma, K. and Nakamura, H. (2008). Molecular dynamics simulation of chemical sputtering of hydrogen atom on layer structured graphite, *The 22nd IAEA Fusion Energy Conference*, proceedings, TH/7-1.

[29] Ni, B., Lee, K.-H. and Sinnott, S. B. (2004). A reactive empirical bond order (REBO) potential for hydrocarbon-oxygen interactions. *J. Phys.: Condens. Matter* **16**, pp. 7261–7275.

[30] Ueda, S., Ohsaka, T., Kuwajima, S. (1998). Molecular dynamics evaluation of self-sputtering of beryllium, *J. Nucl. Mater.* **258–263**, pp. 713–718.

[31] Ueda, S., Ohsaka, T., Kuwajima, S. (2000). Sputtering studies of beryllium with helium and deuterium using molecular dynamics approach, *J. Nucl. Mater.* **283–287**, pp. 1100–1104.

[32] Juslin, N., Erhart, P., Träskelin, P., Nord, J., Henriksson, K. O. E., Nordlund, K., Salonen, E. and Albe, K. (2005). Analytical interatomic potential

for modeling nonequilibrium processes in the W-C-H system, *J. Appl. Phys.* **98**, 123520.

[33] Juslin, N. and Nordlund, K. (2008). Pair potential for Fe-He, *J. Nucl. Mater.* **382**, pp. 143–146.

[34] Los, J. H., Ghiringhelli, L. M., Meijer, E. J. and Fasolino, A. (2005). Improved long-range reactive bond-order potential for carbon. I. Construction, *Phys. Rev. B* **72**, 214102. [errata : *Phys. Rev. B* **73**, 229901].

[35] Pettifor, D. G. and Oleinik, I. I. (1999). Analytic bond-order potentials beyond Tersoff-Brenner. I. Theory, *Phys. Rev. B* **59**, pp. 8487–8499.

[36] Oleinik, I. I. and Pettifor, D. G. (1999). Analytic bond-order potentials beyond Tersoff-Brenner. II. Application to the hydrocarbons, *Phys. Rev. B* **59**, pp. 8500–8507.

[37] Yoshimoto, Y. (2006). Extended multicanonical method combined with thermodynamically optimized potential: Application to the liquid-crystal transition of silicon, *J. Chem. Phys.* **125**, 184103.

Chapter 11

# Electronic and Molecular Structures of Small- and Medium-Sized Carbon Clusters

Vudhichai Parasuk

*Department of Chemistry, Faculty of Science, Chulalongkorn University, Phyathai Rd., Patumwan Bangkok, 10330, Thailand*

Electronic and molecular structures of small- and medium-sized carbon clusters, $C_n$ ($n = 4$–18), have been surveyed. Theoretical calculations have played important roles in the elucidation of their molecular structures. Geometries of these carbon clusters depend on their electronic structures which related directly to relative stabilities. Even-numbered $C_n$ are more stable than their odd-numbered counterparts. Two types of geometries, linear and monocyclic ring, have been proposed for these carbon clusters. Originally, linear structures were assigned for small even-numbered clusters and most odd-numbered clusters. There are two possibilities for the linear form, i.e. cumulene and polyacetylene in which the cumulene is the most stable. With suggestions from theoretical calculations and later more support from experiments, monocylic ring was found to be the ground state structure for most even-numbered clusters while it is the minimum structure for larger odd $C_n$ ($n \geq 11$). On the contrary, the linear structure was found as the minimum for most negative ion carbon clusters. This might be the reason while originally the linear structure was proposed for these compounds. However, the stability of positive ion clusters is similar to the neutral compounds in which the monocyclic structure was found as the minimum.

## 1. Introduction

Carbon clusters $C_n$, compounds or molecules whose components are only carbon atoms, have been identified in atmospheres of carbon stars, intergalactic dust, and comet tails.[1,2] They are also found during the combustion of hydrocarbons.[3] In addition, they are precursors to "fullerene", a novel carbon compounds. Understanding electronic and molecular structures as well as properties of this class of compounds would benefit vast areas of research in chemistry and physics. Numerous researches both experimentally and theoretically have been carried out in this light. Carbon clusters can be prepared by evaporating graphite.[4,5] A wide range of carbon clusters from the size of 2 to more than 600 carbon atoms can be produced.[5,6] Using mass spectroscopy, the abundant spectra which displayed the bimodal distribution in the region of $1 \leq n \leq 30$ and $20 \leq n \leq 100$ were observed.[6-9] For small and medium-sized clusters ($1 \leq n \leq 30$), the abundance spectra reveal an alteration between odd- and even-numbered clusters. For the large clusters ($20 \leq n \leq 100$), only the even cluster were observed. The abundance spectra suggest that the small and large carbon clusters are distinct in their structures and properties. This special pattern of the abundance spectra was called "magic number" and represents the stability of particular $C_n$ species. In this article, we will focus only on electronic and molecular structures of small and medium-sized carbon clusters. Electronic and molecular structures of large carbon clusters have already been reviewed in "DFT Calculations on Fullerene and Carbon Nanotubes".[10]

## 2. Small and Medium-Sized Carbon Clusters

Structures of small and medium-sized carbon clusters have been the subject of interest for many years. Owing to their very short lifetime, theoretical calculations have played quite an important role in the elucidation of electronic and molecular structures of small- and medium-sized carbon clusters. Early theoretical calculations predicted small carbon clusters, $C_n$ $n < 10$, to have the linear structure whereas, for those with $n > 10$, the monocyclic ring structure is preferred.[11-13] The proposed ring structure would explain the disappearance of the odd-even alteration in the abundance spectra of negative ions for clusters larger than 8 atoms. The ring structure also suggests that the clusters $C_{10}$, $C_{14}$, $C_{18}$, $C_{22}$, ..., $C_{4n+2}$ are particularly stable (Hückel's $4n + 2$ $\pi$ electrons rule). However, the abundance spectra of positive ions do not support this assumption, and magic numbers at $n = 11$, 15, 19, 23 are instead observed.[6,7] It has later been discovered that carbon clusters are photo-fragmented during the photo-ionization process, and the observed magic numbers correspond to the fragment $C_{n-3}^+$ rather than the initial clusters $C_n^+$.[11,14] The observed magic numbers are, therefore, related to clusters $C_n$ $n=14$, 18, 22, 26. The even members of $C_n$ with $n < 10$ are of particular interest both experimentally and theoretically. This is because the even clusters with the monocyclic ring geometry would have closed-shell electronic structures, and the particular clusters with $4n + 2$ $\pi$ electrons could become especially stable according to Hückel's rule. Like small even-numbered $C_n$, originally odd-numbered $C_n$ were proposed to have only the linear structure. However, many modern theoretical calculations have shown cyclic odd-numbered carbon clusters to possess a particular stability.[15-18] Except, $C_2$ and $C_3$ which have linear structures and $^1\Sigma_g^+$ ground states, electronic and molecular structures of other carbon clusters are not well characterized.

## 2.1. Linear $C_n$

Two possible structures are proposed for linear $C_n$, "cumulene" and "polyacetylene". Cumulene is referred to the structure with all equivalent carbon-carbon bonds and polyacetylenes is referred to the structure with alternative carbon-carbon bonds. The cumulene has the electronic structure similar to that of $O_2$ but with extended $\pi$ system. The polyacetylene has two localized open-shell electrons at terminal carbons. Electronic structures of cumulenes and polyacetylenes are displayed in Fig. 1.

## 2.2. Cyclic $C_n$

The monocyclic ring structure was suggested not only for clusters with $n > 10$ but also for small $C_{2n}$ (even-numbered) such as $C_4$, $C_6$, and $C_8$.[14,19-29] Electronic structures of monocyclic $C_{2n}$ are similar to those of annulenes $(C_nH_n)$. However, they have, in addition to $2n$ $\pi$ electrons (out-of-plane $\pi$), $2n$ pseudo $\pi$ electrons (in-plane $\pi$). Due to similarity to the annulenic system, one would expect the cyclic $C_{2n}$ to possess $D_{(2n)h}$-symmetric structure. From their electronic structures, the cyclic $C_{2n}$ can be classified into aromatic $(4p + 2$ $\pi$ electrons) and anti-aromatic $(4p$ $\pi$ electrons). The aromatic $C_{2n}$ would have its highest occupied molecular orbitals (HOMOs), $e_u$ and $e_g$ orbitals, completely filled, and a closed-shell electronic structure. The anti-aromatic $C_{2n}$ would have half-filled HOMOs and open-shell electronic state with quintet spin $(S = 2)$. By breaking the degeneracy of orbitals, open-shell electrons in anti-aromatic $C_{2n}$ can be paired and lead to lowering of the molecular symmetry, *i.e.* from $D_{(2n)h}$ to $C_{nh}$. A schematic illustration of electronic configurations of the aromatic and anti-aromatic $C_{2n}$ is shown in Fig. 2.

For odd clusters, the cyclic $C_{2n+1}$ should have singlet state with $D_{nh}$ symmetry ($^1A_1'$). However, it is proposed that the $D_{nh}$ ring will undergo Jahn-Teller distortion which will result in the lowering $C_{2v}$ symmetry.

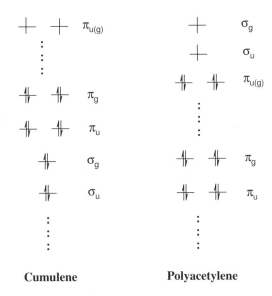

Fig. 1. Comparison of electronic structures of linear cumulene and polyacetylene.

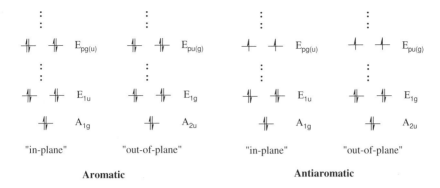

"in-plane"     "out-of-plane"     "in-plane"     "out-of-plane"

**Aromatic**                    **Antiaromatic**

Fig. 2. Electronic structures of aromatic and anti-aromatic $C_{2n}$.

## 3. Even $C_{2n}$

In the beginning, only the linear structure has been observed experimentally for carbon clusters. This implies that carbon clusters exists only in the linear form. However, theoretical calculations found the monocyclic ring structure as a possible candidate. The first experiment which supports the existence of the cyclic structure is the Coloumb Explosion (CE) experiment of $C_4$, where the existence of the rhombic structure was reported.[30-31] Later on, the CE experiment again suggested the cyclic structure for $C_6$.[32] Recently, in conjunction with theoretical calculations IR frequencies were reassigned and many previously assigned peaks for linear $C_n$ have been reallocated for cyclic structures.[24,27]

### 3.1. $C_4$

The earliest theoretical calculations were performed by Pitzer and Clementi[11] who carried out semi-empirical calculations and they confirmed that $C_4$ and even-numbered $C_{2n}$ have triplet ground states and the linear geometries with all equal bonds, called "cumulene". The ${}^3\Sigma_g^-$ states of cumulenes have large and positive electron affinities which suggested favoritism for negative ions. This finding supports the abundance spectra of negative ions where even-numbered clusters are more abundant. Later on, Pitzer and Strickler refined the earlier calculation and showed that $C_4$ would rather have a conjugated triple bond structure called "polyacetylene" as the minimum structure instead of the earlier proposed cumulene structure.[12] This prediction was confirmed by Hoffmann who carried out extended Hückel calculations for the same system.[13] The structures proposed by Pitzer *et al.* which successfully described the characteristic of abundance spectra of carbon clusters have been well accepted for quite some times. In a hope to clarify molecular structures of $C_n$ $n \leq 10$, spectroscopic measurements such as infrared spectroscopy (IR)[34,35] and electron spin resonance spectroscopy (ESR)[36] of carbon clusters have been carried out in rare-gas matrices. However, while the IR spectrum yielded an asymmetric stretching frequency at 2164 cm$^{-1}$ suggesting the polyacetylene structure, the ESR spectrum indicated very strong $\pi\pi$ character of the unpaired electrons suggesting the cumulene structure. The first hint for the monocyclic structure came from the semi-empirical calculation of $C_4$ by Slanina and Zahradnik.[33] Nonetheless, no experimental evidence has been observed for such a monocyclic structure, not until Algranati *et al.* performed Coulomb Explosion (CE) experiment and the existence of rhombic $C_4$ was confirmed.[31] However, it is known that CE cannot easily distinguish between truly nonlinear and linear, but "floppy" molecules.[37,38] Due to its small size, a number of high levels theoretical calculations have been carried out for $C_4$. The summary is given in Table 1.

Three low-lying states of cumulene were compared. The $^3\Sigma_g^-$ state was found to be the lowest state. It lies below the $^1\Delta_g$ and $^1\Sigma_g^+$ states around 7 and 7 – 10 kcal/mol, respectively. This energy difference was already accurately determined at CASSCF level. The dynamic correlation has very little effect on the energy difference. The density functional theory seems to overestimate the stability of the triplet state and would not be suitable to use for the characterization of the low-lying states of cumulene $C_4$.

The polyacetylene structure was predicted to be higher in energy than the cumulene one. Without dynamic correlation, the $^3\Sigma_u^+$ of polyacetylene was found to be 7 kcal/mol above the $^3\Sigma_g^-$ state of cumulene. The energy difference is enhanced to 25.7 kcal/mol

Table 1. Energies (kcal/mol) relative to $^3\Sigma_g^-$ cumulene of different states and geometries of $C_4$ computed at various levels of theory.

| methods | basis set | Cumulene | | Polyacetylene | rhombic | Ref. |
|---|---|---|---|---|---|---|
| | | $^1\Delta_g$ | $^1\Sigma_g^+$ | $^3\Sigma_u^+$ | $^1A_g$ | |
| CAS(10,10) | [13 8 4]/(431) | 6.9 | 9.5 | 7.3 | 25.7 | 20 |
| 10-MRCI | | 6.4 | | 23.5 | 35.0 | |
| 16-SDCI | | | | | 6.2 | |
| 16-MCPF | | | | | 4.7 | |
| 16-MRCI | | | | | 4.1 | |
| HF | | | | | 9.8 | |
| MP2 | | | | | −6.7 | |
| CAS(10,10) | [13 8 6 4]/(4421) | 6.9 | 9.6 | 6.7 | 23.2 | |
| 10-MRCI | | 8.0 | | 25.7 | 34.5 | |
| 16-SDCI | | | | | 3.5 | |
| HF | | | | | 7.6 | |
| MP2 | | | | | −9.8 | |
| 16-MCPF | [13 8 4]/(542) | | | | 2.3 | |
| 16-MRCI | | | | | 1.6 | |
| HF | 6-31g(d) | | | | −9.3 | 52 |
| MP2 | | | | | −13.1 | |
| MP4(SDTQ) | | | | | −13.8 | |
| QCISD(T) | | | | | −10.4 | |
| CCSD(T) | cc-pVDZ | | | | −0.69 | |
| B3LYP | 6-31g(d) | | | | 37.7 | 43 |
| CCSD(T) | aug-cc-pVDZ | | 7.2 | | −2.8 | 21 |
| BP88 | [10 6 1](431) | | | | 13.8 | 15 |
| 16-SDCI | [951]/(421) | 9.8 | | 23.5 | | 44 |
| CCSDT-1 | | | | | −5.0 | 58 |
| 2R-CI | | | | | −2.7 | 54 |
| CCSDT-1 | [951]/(431) | | | | 0.49 | 40 |
| MP4 | 6-31g(d) | | | | −0.7 | 55 |
| B3LYP | aug-cc-pVTZ | | 17.9 | | 16.7 | 53 |
| CCSD(T) | | | 8.8 | | −1.4 | |
| B3LYP | cc-pVDZ | | | | 13.6 | 44 |
| CCSD(T) | | | | | −0.69 | |
| CCSD(T) | cc-pVTZ | | | | −0.09 | |
| CCSD(T) | cc-pVQZ | | | | −0.93 | |

(16e$^-$ correlated MRCI) when dynamic correlation was included. From theoretical calculations, it is very conclusive that linear $C_4$ is cumulene, the finding which is contradicted to the IR study. However, IR frequency calculations of $C_4$ suggested that the experimental value which originally assigned for $C_4$, 2164 cm$^{-1}$, should actually belong to $C_5$ and the observed frequency of 1544 cm$^{-1}$ of $C_5$ should be reassigned for $C_4$.[39,40] This reassignment confirms the cumulene structure.

Large numbers of calculations have been carried out to verify whether $C_4$ has linear or cyclic (rhombic, $D_{2h}$) ground state. Unfortunately, it is thus far from conclusive. The MP2 calculations find the cyclic to be 7 – 10 kcal/mol below the cumulene. At MP4 level of theory, the stability over the cumulene of the rhombic $C_4$ is reduced to only 0.7 kcal/mol. Predictions based on CCSD(T)/CCSDT-1 calculations were ranging from 5 kcal/mol below to 0.49 kcal/mol above for the cyclic structure. The single-reference CI based methods (SDCI, MCPF) predicted the cumulene to be more stable than the monocyclic by 3.5 – 6.2 kcal/mol. With multi-reference methods, the cumulene remains more stable but the barrier reduces to 1.6 – 4.1 kcal/mol. The density functional theoretical based methods, however, give the energy of the rhombic form to be more than 10 kcal/mol above that of the cumulene. For CI based methods, the number of correlated electrons is very important. The 10e$^-$ correlated MRCI suggested the cyclic form to be 35 kcal/mol above the cumulene while this gap is diminished to 4.1 kcal/mol at 16e$^-$ correlated MRCI. The energy gap between the rhombic and the cumulene is also sensitive to basis set used in favor of the cumulene. The gap decreases from 4.1 to 1.6 kcal/mol at MRCI level, when changing from 4s3p1d to 5s4p2d basis. Similarly, at CCSD(T) level it is reduced from −2.8 to −1.4 kcal/mol when changing from aug-cc-pVDZ to aug-cc-pVTZ basis. Up to today, it is still not clear whether $C_4$ has the linear or the cyclic ground state. We can only conclude that the 2 forms are possibly nearly isoenergetic. This conclusion leads to the question why cyclic $C_4$ could not be deduced from both cationic and anionic abundance spectra of $C_4$. To answer this, calculations on positive and negative ions of $C_4$ were carried out.

For negative ion, all calculations predicted the linear cumulene to be more than 27 kcal/mol lower in energy, a much higher energy difference as compared to the neutral cluster. This could explain why only linear form was found in the abundance spectra of negative ion clusters. Regardless of this difficulty, Blanksby et al.[21] has shown how to prepare the rhombic $C_4$, experimentally. The situation for the positive ion is similar to the neutral cluster. It is not conclusive whether the linear or the cyclic form is more stable. The MRCI calculation predicted the cyclic form to be more stable by 6.1 kcal/mol while the CCSD(T) gave the cumulene to be 3.7 kcal/mol more stable. The calculated results are listed in Table 2. Geometries of cumulene, polyacetylene, and rhombic $C_4$ are illustrated in Fig. 3.

Table 2. Relative energies (kcal/mol) of different geometries of $C_4^+$ and $C_4^-$ computed at various levels of theory.

| $C_4^+$ | basis set | cumulene $^2\Pi_g$ | rhombic $^2B_{3u}/^2A$ | Ref. |
|---|---|---|---|---|
| 15-MRCI | [13 8 4]/(431) | 0.0 | 6.6 | 20 |
| B3LYP | aug-cc-pVTZ | 0.0 | 5.7 | 53 |
| CCSD(T) | | 0.0 | −2.4 | |
| CCSD(T) | aug-cc-pVDZ | 0.0 | −3.7 | 21 |
| B3LYP | cc-pVDZ | 0.0 | 1.2 | 18 |
| B3PW91 | | 0.0 | −3.8 | |
| CCSD(T) | | 0.0 | −2.9 | |
| $C_4^-$ | | cumulene $^2\Pi_g$ | rhombic $^2B_{2g}$ | Ref. |
| CCSD(T) | aug-cc-pVDZ | 0.0 | 27.2 | 21 |
| ROHF | | 0.0 | 37.8 | 17 |
| B3LYP | | 0.0 | 38 | |
| CCSD(T) | | 0.0 | 29.6 | |
| B3LYP | aug-cc-pVTZ | 0.0 | 39.9 | 53 |
| CCSD(T) | | 0.0 | 29.9 | |

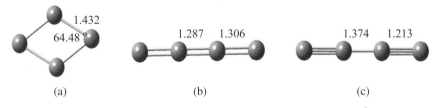

(a)  (b)  (c)

Fig. 3. Geometries with geometrical parameters (in angstrom and degree) for (a) rhombic, (b) cumulene and (c) polyacetylene structures of $C_4$.

## 3.2.  $C_6$

Earlier IR experiments suggested the cumulene-like structure for $C_6$.[41,42] Later on, Szczepanski and Vala made conclusion that the IR bands at 1695 cm[-1] originally assigned for linear $C_8$ should belong to cyclic $C_6$.[43] This was subsequently verified by Presilla-Marquez *et al.*[24] Thus, the two forms can actually co-exist. With double 6 π electrons, $C_6$ is aromatic ($4n + 2$ π electron) with perfect "$D_{6h}$" symmetry. This geometry contains all equivalent carbon-carbon bonds, referred to "cyclo-cumulene" in analogy to the cumulene structure. The $D_{6h}$ structure is supposed to be the lowest cyclic form, since it allows maximum electronic delocalization. However, theoretical calculations also suggested another cyclic geometry with "$D_{3h}$" symmetry. This structure has all equivalent bonds but with bond angle deviates from 120°. Thus, it is named "distorted cyclo-cumulene". Several theoretical calculations were carried out to compare energies of the cyclic and the linear forms of $C_6$. The results are summarized in Table 3.

At CASSCF level, the $^3\Sigma_g^-$ of cumulene is 9.5 kcal/mol more stable than the $^3\Sigma_u^+$ of polyacetylene. This difference increases to 20.9 kcal/mol at 10e[-] correlated MRCI, the number which is consistent to 19.8 kcal/mol of 24e[-] correlate SDCI method. Thus, the cumulene is much stable than the polyacetylene. The dynamic correlation seems to be more important than the near-degeneracy one for this case, while the number of correlated electrons do not seem to have any effect. When compared to $C_4$, it was found that the energy gap between the cumulene and the polyacetylene is reduced as the linear chain elongates (23.5 and 20.9 kcal/mol for $C_4$ and $C_6$, respectively). This is in agreement with Liang and Schaeffer's suggestion.[44] For cumulene structure, $^3\Sigma_g^+$ is the lowest state. The $^1\Delta_g$ and the $^1\Sigma_g^+$ are found to be 3.4 and 6.4 kcal/mol higher in energy at MRCI/4s3p1d level. They are 7 and 12.4 kcal/mol, respectively, when computed using CASSCF method. Thus, the conclusion made earlier for the cumulene-polyacetylene system is still applied.

As mentioned earlier, the cyclo-cumulene structure, which allows maximum electronic delocalization, is supposed to be the lowest cyclic form. However, all theoretical calculations predicted the distorted

Table 3. Energies (kcal/mol) relative to $^3\Sigma_g^-$ cumulene of different $C_6$ states and geometries computed at various levels of theory.

| methods | basis set | cumulene | | polyacetylene | $D_{6h}$[a] c-c | $D_{3h}$[b] d-c-c | Ref. |
|---|---|---|---|---|---|---|---|
| | | $^1\Delta_g$ | $^1\Sigma_g^+$ | $^3\Sigma_u^+$ | $^1A_{1g}$ | $^1A_1$ | |
| CAS(10,10) | [13 8 4]/(431) | 7.0 | 12.4 | 9.5 | 14.9 | | 22 |
| 10-MRCI | | 3.4 | 6.4 | 20.9 | 37.2 | | |
| 24-SDCI | | | | | 30.6 | 2.1 | |
| 24-MCPF | | | | | 7.7 | −3.7 | |
| 24-MRCI | | | | | 14.7 | −0.7 | |
| HF | | | | | 58.1 | | |
| MP2 | | | | | −21.7 | | |
| MCPF | [13 8 4]/(542) | | | | 1.3 | −9.5 | |
| HF | | | | | 55.5 | | |
| MP2 | | | | | −29.4 | | |
| MCPF | 6-31g(d) | | | | 1.7 | −6.7 | |
| HF | | | | | 58.2 | | |
| MP2 | | | | | −24.7 | | |
| HF | [13 8 4]/(654) | | | | 55.4 | | |
| MP2 | | | | | −30.7 | | |
| HF | 6-31g(d) | | | | 50.9 | | 52 |
| MP2 | | | | | −49.7 | | |
| MP4 | | | | | 28.9 | | |
| QCISD(T) | | | | | −10.2 | | |
| CCSD(T) | cc-pVDZ | | | | 7.1 | | |
| CCSD(T) | cc-pVTZ | | | | 11.0 | | |
| B3LYP | cc-pVDZ | | | | 59.5 | 35.8 | 43 |
| 24-SDCI+DVD | [951]/(421) | 4.5 | | 19.8 | | | 44 |
| MP4 | 6-31G(d) | | | | −16.3 | −20.4 | 14 |
| MP2 | | | | | 80.2 | −31.1 | |
| MP3 | | | | | 9.7 | | |
| B3LYP | cc-pVDZ | | | | | 5.92 | 29 |
| CCSD(T) | | | | | | −7.07 | |
| CCSD(T) | cc-pVTZ | | | | | −11.0 | |
| CCSD(T) | aug-cc-pVDZ | | | | | | 28 |
| CCSD(T) | 432 | | | | | | |
| CCSDT-1 | | | | | | | |
| CCSDT | cc-pVDZ | | | | | | |
| B3LYP | aug-cc-pVTZ | | | | | | |
| CAS(12,12) | cc-pVDZ | | | | | | |
| CASPT2(12,12) | cc-pVTZ | | | | | | |
| CASSCF | (321) | | | | | −5.3 | 15 |
| CASPT2 | | | | | | 14.8 | |
| CCSD(T) | | | | | | 3.8 | |
| CCSD | (4321) | | | | | 8.6 | |
| CCSD(T) | | | | | | 10.7 | |

[a]cyclo-cumulene $D_{6h}$, [b]distorted cyclo-cumulene $D_{3h}$

cyclo-cumulene to be more stable. Using 6-31g(d) basis, the distorted cyclo-cumulene is 8.4 and 7.2 kcal/mol lower in energy than the cyclo-cumulene at MCPF and CCSD(T) levels of theory while at MP2 and MP4 levels the predicted energy gap are 111.3 and 4.1 kcal/mol, respectively. The vast difference between MP2 and MP4 results suggests the convergence problem of the perturbation theory. At 4s3p1d basis, energy gaps between $D_{6h}$ and $D_{3h}$ cyclic forms are 28.5, 11.4, and 15.4 kcal/mol, respectively, for SDCI, MCPF, and MRCI calculations. The lack of size-extensivity causes an error of around 17 kcal/mol while the lack of the static correlation gains an error of 13 kcal/mol. Changing basis set from 4s3p1d to 5s4p2d, the energy difference is reduced by 0.6 kcal/mol. Considering all effects, the barrier between $D_{3h}$ and $D_{6h}$ monocyclic should be less than 10 kcal/mol.

Recently, CASPT2/cc-pVTZ calculations have been performed and the energy gap of 0.2 kcal/mol was reported. In addition, frequency calculations of cyclo-cumulene ($D_{6h}$ symmetry) suggested that the structure is not the minimum but rather the saddle point.

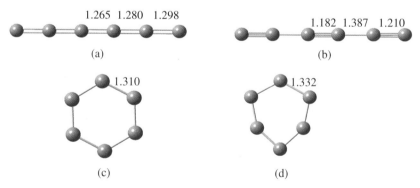

Fig. 4. Geometries with geometrical parameters (in angstrom and degree) for (a) cumulene, (b) polyacetylene, (c) cyclo-cumulene $D_{6h}$, and (d) distorted cyclo-cumulene $D_{3h}$ structures of $C_6$.

Comparing to cumulene, the distorted cyclo-cumulene is more stable. Only B3LYP calculations predicted the cumulene to be up to 36 kcal/mol lower in energy. With 4s3p1d basis, the energy differences between the cyclic $D_{3h}$ and the linear structure are 2.1, −3.7, and −0.7 kcal/mol for

SDCI, MCPF, and MRCI, respectively. (The negative sign suggests the cyclic form to be favor.) Errors causing by the lack of size-extensivity and near-degeneracy correlation are noticed. Both errors are in favor of the cyclic structure. The effect of basis set is also in the same direction (in favor of the cyclic form). Changing from 4s3p1d to 5s4p2d, the linear-cyclic energy gap increases from −3.7 to −9.5 kcal/mol at MCPF level. Thus, when considering all effects it is likely that the distorted-cyclo cumulene is the minimum structure of $C_6$. Interestingly, the energy difference between the cumulene and the cyclo-cumulene is 59.5 kcal/mol at B3LYP and 50.89 kcal/mol at HF. It is apparent that the correlation at DFT level could not describe the electronic delocalization for this system. The convergence problem of the MP series is also observed for the linear-cyclic system. Using 6-31g(d) basis, relative energies between the $^3\Sigma_g^-$ state of cumulene and the $^1A_{1g}$ state of cyclo-cumulene ($E(^1A_{1g}) - E(^3\Sigma_g^-)$) for MP2, MP3, and MP4 calculations are 80.2, 9.7, and −16.3 kcal/mol, respectively. Geometries of cumulene, polyacetylene, cyclo-cumulene ($D_{6h}$), and distorted cyclo-cumulene ($D_{3h}$) are given in Fig. 4. Relative energies between the linear and the cyclic structures of positive and negative $C_6$ clusters are listed in Table 4.

Table 4. Relative energies (kcal/mol) of different geometries of $C_6^+$ and $C_6^-$ computed at various levels of theory.

| $C_6^+$ | basis set | cumulene $^2\Pi$ | cyc. $^2A$ | Ref. |
|---|---|---|---|---|
| B3LYP | cc-pVDZ | 0.0 | 11.2 | 18 |
| B3PW91 | | 0.0 | 0.8 | |
| CCSD(T) | | 0.0 | 2.9 | |

| $C_6^-$ | basis set | cumulene $^2\Pi_u$ | cyc. $D_{3h}$ $^2A_1$ | Ref. |
|---|---|---|---|---|
| ROHF | aug-cc-pVDZ | 0.0 | 50.1 | 17 |
| B3LYP | | 0.0 | 35.4 | |
| CCSD(T) | | 0.0 | 25.5 | |
| B3LYP | 6-31g(d) | 0.0 | 35.8 | 57 |

For negative $C_6$ ion, it is conclusive that the linear form is the most stable. All methods (ROHF/aug-cc-pVDZ, B3LYP/aug-cc-pVDZ, CCSD(T)/aug-cc-pVDZ) predicted the cyclic form to be at least 20 kcal/mol above the linear structure. The situation is not so obvious for the cationic $C_6$. Using cc-pVDZ basis, B3PW91 and CCSD(T) suggested the two forms (cyclic and linear) to be almost isoenergetic (0.8 and 2.9 kcal/mol). Only B3LYP calculations gave the linear to be more stable by 11.2 kcal/mol. Probably, B3LYP calculations overshoot the relative energy.

### 3.3. $C_8$

ESR experiments by Van Zee *et al.* identified $C_8$ as linear compound. IR studies have also assigned frequencies at 2071.5 and 1710.5 $cm^{-1}$ for the linear $C_8$.[45] However, the frequency at 1710.5 $cm^{-1}$ was later reassigned to the cyclic $C_6$,[24] while the absorption at 1818 $cm^{-1}$, which was originally unassigned, was then given to the cyclic $C_8$.[46] According to Liang and Schaefer, the lowest form of linear $C_8$ is cumulene.[47] Also from the same work, the $^3\Sigma_g^-$ state was found to be the lowest state of cumulene. The lowest state and structure of the linear $C_8$ appears to be conclusive. The situation, however, seems to be less evident for the cyclic $C_8$. The cyclic $C_8$ is anti-aromatic and have singlet spin with $C_{4h}$ symmetry ($^1A_g$ state). This structure has alternated bonds and can be referred to as "cyclo-polyacetylene". In comparison to the linear structure, various theoretical calculations predicted the cyclic $C_{4h}$ to be between $-12.53$ to 16.60 kcal/mol of the linear $C_8$. (The negative number reflects the stability of the cyclic over the linear.) The summary of theoretical calculations and geometries of cyclic $C_{4h}$ and linear cumulene of $C_8$ are given in Table 5 and Fig. 5, respectively.

Using 4s3p1d basis, HF and MP2 methods found the linear $C_8$ to be more stable than the cyclic form by 5.38 and 4.69 kcal/mol, respectively. However, at SDCI and MCPF levels the $C_{4h}$ cyclic is favor by 1.05 and 3.41 kcal/mol. The dynamic correlation improves the stability of the cyclic form by 6 – 8 kcal/mol. The SDCI and MCPF calculations were in agreement with CCSD(T)/cc-pVDZ calculations which gave the cyclic

Table 5. Relative energies (kcal/mol) of different geometries of $C_8$ computed at various levels of theory.

| | basis set | cumulene $^3\Sigma_g^-$ | polyacetylne $^3\Sigma_u^+$ | cyclic $^1A_g$ | Ref. |
|---|---|---|---|---|---|
| SDCI | [13 8 4]/(431) | 0 | | −1.05 | 25 |
| MCPF | | 0 | | −3.41 | |
| HF | | 0 | | 5.38 | |
| MP2 | | 0 | | 4.69 | |
| 32-MRCI | | 0 | | 7.22 | |
| SDCI | [13 8 4]/(542) | 0 | | −12.53 | |
| MCPF | | 0 | | −10.04 | |
| HF | | 0 | | 1.74 | |
| MP2 | | 0 | | −3.39 | |
| HF | [13 8 4]/(654) | 0 | | 1.74 | |
| MP2 | | 0 | | −4.86 | |
| B3LYP | cc-pVDZ | 0 | | 16.6 | 29 |
| CCSD(T) | | 0 | | −4.09 | |
| CCSD(T) | cc-pVTZ | 0 | | −6.07 | |

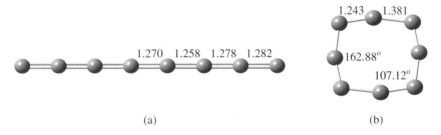

(a)        (b)

Fig. 5. Geometries with geometrical parameters (in angstrom and degree) for (a) cumulene, (b) cyclic $C_{4h}$ structures of $C_8$.

form 4.09 kcal/mol more stable. When including the near-degeneracy correlation such as in MRCI calculations, this cyclic-linear energy difference is 7.22 kcal/mol (in favor of the linear form). Changing basis set from 4s3p1d to 5s4p2d, the cyclic form becomes even more stable. The SDCI and MCPF calculations predicted this linear-cyclic stability to be −12.53 and −10.04 kcal/mol, while it is 1.74 and −3.39 kcal/mol for HF and MP2 calculations. (Again, the negative values implies that the cyclic form is more stable.) This is 6.5 kcal/mol increases in the stability for MCPF and 8 kcal/mol for MP2. When further increasing the size of basis set to 6s5p4d, the cyclic-linear stability becomes 1.74 (HF) and −4.86 (MP2) kcal/mol. Apparently, at 6s5p4d basis set HF calculations reaches the basis-set limit while at MP2 level the basis set effect is reduced to a mere 1.5 kcal/mol. Thus, we expect that the effect of the size of basis set will not exceed 10 kcal/mol favoring the cyclic structure. Accounting for all deficiencies *i.e.* size of basis set and lacks of dynamic and near-degeneracy correlations, the cyclic $C_{4h}$ structure of $C_8$ should be more stable than the linear form by 5 − 6 kcal/mol. There are other suggested cyclic structures such as $D_{4h}$ (distorted cyclo-cumulene) and $D_{8h}$ (cyclo-cumulene) which were proposed. Among these structures, the cyclic $D_{4h}$ is the most stable but it is 20.1 kcal/mol above the cyclic $C_{4h}$ structure at CCSD(T)/cc-pVDZ. It is, thus, concluded that cyclic $C_{4h}$ is the lowest cyclic form and also the minimum structure for $C_8$.

For the positive $C_8$ ion, the situation is similar to the neutral cluster. Theoretical calculations found the cyclic form to be more stable than the linear one. The energy difference between −2.2 to −13.2 kcal/mol was reported. The cyclic $C_8{}^+$ has doublet spin state and $C_s$ symmetry. For the negative ion, the cyclic structure has doublet spin and $C_{2v}$ symmetry and is 50 (B3LYP/aug-cc-pVDZ) and 49.7 kcal/mol (CCSD(T)/aug-cc-pVDZ) less stable than the corresponding linear form.

### 3.4. $C_{10}$

Electronic and molecular structures of $C_{10}$ have not been widely investigated experimentally as they have been for other small- and medium-sized carbon clusters. The only experimental evidence is the IR studies which proposed a linear or near linear structure for $C_{10}$.[41,42] Although there is no experimental support for the existence of the cyclic form, it is quite rather well accepted that $C_{10}$ is monocyclic ring. This is owing to the very large relative energy between cyclic and linear structures of $C_{10}$. The cyclic $C_{10}$ has 10 π electrons and belongs to the aromatic category. Thus, it has the closed-shell electronic structure and possesses $D_{10h}$ symmetry with all equivalent carbon-carbon bonds called "cyclo-cumulene". Like $C_6$, it was found that 2 lower-symmetry cyclic structures with $D_{5h}$ symmetry could also be possible candidates for the cyclic $C_{10}$. The two possible $D_{5h}$ structures are "distorted cyclo-cumulene" or "$D_{5h}$ cyclo-cumulene" (with all equivalent bonds and deviated bond angles) and "cyclo-polyacetylene" (with alternated bonds). The summary of theoretical calculations and illustrations of geometries of the linear and the cyclic $C_{10}$ are given in Table 6 and Fig. 6, respectively.

Table 6. Relative energies (kcal/mol) of different geometries of $C_{10}$ computed at various levels of theory.

| | basis set | cumulene | | | poly-acetylene $^3\Sigma_u^+$ | $D_{10h}$ c-c $^1A_{1g}$ | $D_{5h}$ d-c-c $^1A_1{}'$ | $D_{5h}$ c-a $^1A_1{}'$ | Ref. |
| | | $^1\Delta_g$ | $^1\Sigma_g^+$ | $^3\Sigma_g^-$ | | | | | |
|---|---|---|---|---|---|---|---|---|---|
| SDCI | [1384]/(431) | | | 66.57 | | 15.75 | 0 | 18.21 | 25 |
| MCPF | | | | 61.1 | | 5.19 | 0 | 3.46 | |
| CCSD | | | | | | 6.97 | 0 | 4.28 | |
| CCSD(T) | | | | | | 1.24 | 0 | 0.76 | |
| HF | | | | | | 31 | 0 | 29.78 | |
| MP2 | | | | | | −35.79 | 0 | −24.75 | |
| HF | [1384]/(542) | | | | | 30.78 | 0 | 29.59 | |
| MP2 | | | | | | −36.77 | 0 | −25.14 | |
| HF | [73]/(32) | | | | | 2.3 | 0 | 1.77 | |
| MP2 | | | | | | −52.07 | 0 | −49.99 | |
| MP3 | | | | | | −16.41 | 0 | −25.58 | |
| MP4(SDQ) | | | | | | −22.57 | 0 | −28.21 | |
| MP4(SDTQ) | | | | | | −40.33 | 0 | −36.79 | |
| HF | [731]/(321) | | | | | 30.31 | 0 | | |
| MP2 | | | | | | −31.46 | 0 | | |
| MP3 | | | | | | 13.11 | 0 | | |
| MP4(SDQ) | | | | | | 6.54 | 0 | | |
| MP4(SDTQ) | | | | | | −21.37 | 0 | | |
| CCSD | | | | | | 8.13 | 0 | 5.1 | |
| CCSD(T) | | | | | | 3.27 | 0 | 2.51 | |
| MCPF | | | | | | 8.55 | 0 | 2.51 | |
| CCSD(T) | aug-cc-pVDZ | | | | | 0.2 | 0 | | 28 |
| B3LYP | aug-cc-pVTZ | | | | | 3.6 | 0 | | |

c-c = cyclo-cumulene
d-c-c = distorted cyclo-cumulene
c-a = cyclo-polyacetylene

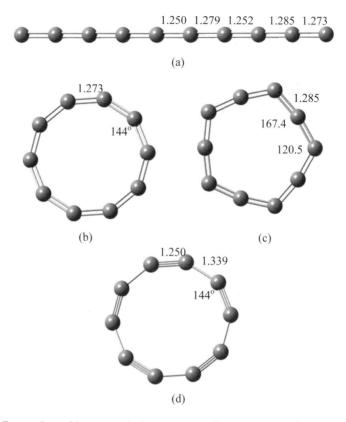

Fig. 6. Geometries with geometrical parameters (in angstrom and degree) for (a) cumulene, (b) cyclo-cumulene $D_{10h}$, (c) distorted cyclo-cumulene $D_{5h}$, and (d) cyclo-polyacetylene $D_{5h}$ structures of $C_6$.

From Table 7, the most stable $C_{10}$ is the $D_{5h}$ cyclo-cumulene. The structure was found to be 66.57 and 61.1 kcal/mol below the $^3\Sigma_g^-$ cumulene by SDCI and MCPF calculations. Compared to $D_{10h}$ cyclo-cumulene, using 4s3p1d basis the SDCI method predicted the $D_{5h}$ cyclo-cumulene structure to be 15.75 kcal/mol more stable, whereas MCPF and CCSD methods found this $D_{5h}$ structure 5.19 and 6.97 kcal/mol lower in energy. The correction to the size-extensivity reduces the energy difference between $D_{5h}$ and $D_{10h}$ cyclo-cumuelenes. With inclusion of triples, the $D_{10h}$-$D_{5h}$ relative energy shrinks to 1.24 kcal/mol at CCSD(T) level. It is also found that the $D_{10h}$ cyclic structure has 1 imaginary frequency which suggests the structure as the saddle point not the minimum. The convergence problem of the MP series was also observed. Using, 3s2p1d basis, relative energies of 30.31, −31.46, 13.11, and −21.37 kcal/mol were observed for HF, MP2, MP3, and MP4(SDTQ) calculations. It seems that the perturbation theory has again problem in describing the extended electronic delocalization like that found for $C_6$. Compared to cyclo-polyacetylene, using 4s3p1d basis the energy of the $D_{5h}$ cyclo-cumulene is lower than that of the cyclo-polyacetylene by 3.46, 4.28, and 0.76 kcal/mol, at MCPF, CCSD, and CCSD(T) levels, respectively. Thus, the two forms are nearly isoenergetic.

For the positive ion, the minimum structure of $C_{10}^+$ is cyclic and it has doublet spin and $C_{2v}$ symmetry. At CCSD(T)/cc-pVDZ, its energy is 46.5 kcal/mol below the linear form. For the negative ion, the linear and the cyclic structures are almost isoenergetic, 0.06 kcal/mol at CCSD(T)/aug-cc-pVDZ. The cyclic $C_{10}^-$ has doublet spin and $C_2$ symmetry.

### 3.5. $C_{2n}$, $n = 6 - 9$ ($C_{12}$, $C_{14}$, $C_{16}$, and $C_{18}$)

Experimental evidences for $C_{12}$, $C_{14}$, $C_{16}$, and $C_{18}$ are very scarce and not conclusive. Moreover, owing to their sizes almost no high level theoretical calculations have been carried out. The minimum structures for these carbon clusters are believed to possess the monocyclic shape, since the linear structures, as in the case of $C_{10}$, would be highly unstable comparing to corresponding cyclic structures.[15] The summary of theoretical calculations for $C_{2n}$, $n = 6 - 9$ is listed in Table 7. Geometries of the most stable $C_{2n}$ are given in Fig. 7.

$C_{12}$ and $C_{16}$ have $4n$ $\pi$ electrons, the electronic structure which is classified as "anti-aromatic" whereas $C_{14}$ and $C_{18}$ have $4n + 2$ $\pi$ electrons, the electronic structure which is classified as "aromatic". The anti-aromatic $C_{2n}$ has $C_{nh}$ symmetry as in the case of $C_8$ while the aromatic $C_{2n}$ has $D_{2nh}$ or $D_{nh}$ symmetry as in the case of $C_6$ and $C_{10}$. Yousaf and Taylor carried out CCSD(T)/4s3p2d and B3LYP/cc-pVTZ calculations on 2 cyclic forms, $D_{6h}$ and $C_{6h}$, of $C_{12}$.[28] The $C_{6h}$ cyclic was found to be at the minimum with energy of either 39.7 (CCSD(T)) or 15.2 (B3LYP) kcal/mol lower than the $D_{6h}$ structure depending on the methods employed. The article also suggested that the $D_{6h}$ structure should not be the minimum but rather the first order saddle point. For $C_{16}$, Martin *et al.* found $C_{8h}$ as the minimum structure at the SCF level.[48] However, this $C_{8h}$ structure slightly differs from HF's $D_{8h}$ structure but largely differs from the structure obtained by B3LYP method. No energy comparison was given in that article. For $C_{14}$, the high symmetric $D_{14h}$ structure (cyclo-cumulene) is compared to $D_{7h}$ structure (distorted cyclo-cumulene). At BLYP/6-31G* and B3LYP/aug-cc-pVDZ, the $D_{7h}$ structure is, in respective order, 2.8 and 0.1 kcal/mol more stable than the $D_{14h}$ structure. Similar to all aromatic $C_{2n}$, high symmetric structures were found to possess 1 imaginary frequency and, thus, cannot be the minimum structure. Interestingly, at HF level the cyclo-polyacetylene $C_{7h}$ was found as the minimum structure and the energy difference between $D_{14h}$ and $C_{7h}$ structure is 39.1 kcal/mol. (With the distorted

Table 7. Relative energies (kcal/mol) of different geometries of $C_{12}$, $C_{14}$ and $C_{18}$ computed at various levels of theory.

| $C_{12}$ | | basis set | $C_{6h}$ | $D_{6h}$ | | | Ref. |
|---|---|---|---|---|---|---|---|
| | CCSD(T) | 4s3p2d | 0 | 15.2 | | | 28 |
| | B3LYP | aug-cc-pVTZ | 0 | 39.7 | | | |
| | | | $D_{7h}$ | $D_{14h}$ | | | |
| $C_{14}$ | | | d-c-c | c-c | | | |
| | B3LYP | aug-cc-pVTZ | 0 | 0.1 | | | 28 |
| | HF | aug-cc-pVDZ | 0 | 30.6 | | | |
| | BLYP | 6-31g(d) | 0 | 2.8 | | | 50 |
| | | | $D_{9h}$ | $D_{9h}$ | $D_{18h}$ | $C_{9h}$ | |
| $C_{18}$ | | | d-c-c | c-a | c-c | d-c-a | |
| | HF | 3-21g | 37.6 | 0 | 53.8 | | 56 |
| | HF | 431 | 35.2 | 0 | 68.7 | | 49 |
| | MP2 | | 53.7 | 0 | −54.6 | | |
| | HF | 5432 | | 0 | 71.5 | | |
| | MP2 | | | 0 | −85.7 | | |
| | HF | 6-31g(d) | 32.1 | 0 | 68.4 | −0.3 | 50 |
| | BLYP | | −17.7 | 0 | −20.1 | −1.7 | |

c-c = cyclo-cumulene
d-c-c = distorted cyclo-cumulene
c-a = cyclo-polyacetylene
d-c-a = distorted cyclo-polyacetylene

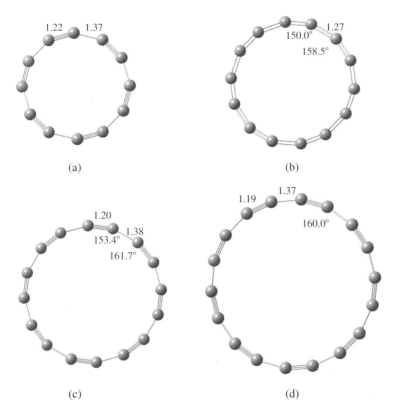

Fig. 7. Geometries with geometrical parameters (in angstrom and degree) for (a) cyclic $C_{12}$, (b) cyclic $C_{14}$, (c) cyclic $C_{16}$, and (d) cyclic $C_{18}$.

cyclo-cumulene, it is 30.6 kcal/mol.) As discussed earlier, both HF and DFT methods fail to describe the system with extensive delocalization. Thus, the relative energy between the two structures is not yet beyond certainty. The incapability of theoretical methods in describing double resonance system proves to be even more serious for $C_{18}$. While HF predicted the cyclo-polyacetylene $D_{9h}$ to be the minimum with cyclo-cumulene $D_{18h}$ and distorted cyclo-cumulene $D_{9h}$ being 35.2 and 68.7 kcal/mol less stable, B3LYP and MP2 suggested the $D_{18h}$ structure with energy of 54.6 and 108.8 kcal/mol below the cyclo-polyacetylene and the distorted cyclo-cumulene structures. The potential energy surfaces obtained by HF and MP2 methods are totally contrary as pointed out by Parasuk and Almlöf.[49] We have shown earlier for $C_6$ and $C_{10}$ that there is the convergence problem for the MP series in describing electronic delocalization of aromatic carbon clusters. No conclusion should be drawn regarding the minimum structure of $C_{18}$ from HF and MP2 results. Platner and Houk[50] proposed the correction scheme for the relative energy computed by BLYP method. With this correction, the minimum is the cyclo-polyacetylene $C_{9h}$. However, there is a tendency for HF as well as BLYP to artificially yielding lower symmetry structures. Thus, the most stable structure of the cyclic $C_{18}$ is still debatable. However, Hutter, Luthi, and Diederich commented that there is a possibility that the $D_{18h}$ structure is the saddle point and the distorted cyclo-cumulene $D_{9h}$ should be the minimum structure for $C_{18}$.[15]

To our best knowledge, no theoretical calculation has been reported regarding the minimum structure of $C_{2n}^-$ when n is between 6 and 9. It is generally accepted that negative ion clusters of these sizes possess linear forms when their corresponding cyclic structures are anti-aromatic and they have monocyclic ring structures when their corresponding cyclic structures are aromatic. For positive ion clusters, cyclic structures are much lower in energy than the corresponding linear forms. For $C_{12}^+$, the cyclic form is 37.5 and 40.6 kcal/mol lower in energy at B3LYP/cc-pVDZ and CCSD(T)/cc-pVDZ level of calculations, respectively. For $C_{14}^+$, using B3LYP/cc-pVDZ the cyclic form was found to be 64.0 kcal/mol below the linear structure.

## 4. Odd $C_{2n+1}$

Odd carbon clusters are less stable than the even-numbered counterparts, the suggestion which corresponds to the abundance spectra of negative ions. Formerly, it is believed that small and medium-sized odd-numbered carbon clusters can have only linear structures. With high level theoretical calculations, monocyclic structures were also found to possess high stability as compared to linear forms.

## 4.1. $C_5$

Undoubtedly, the minimum structure of $C_5$ is linear. All calculations, as seen in Table 8, gave the linear form to be between 58.1 and 73.4 kcal/mol more stable than the corresponding cyclic structure. All methods seem to provide similar values for the relative energy. The relative energy between the linear and the cyclic forms could be well described even at the low level theory such as HF. The cyclic $C_5$ is proposed to be non-planar and have $C_2$ symmetry. Geometries of the linear and the cyclic forms of $C_5$ are illustrated in Fig. 8.

Table 8. Relative energies (kcal/mol) of different geometries of $C_5$ computed at various levels of theory.

| $C_5$ | basis set | cyc $C_2$ non-plan | lin | Ref. |
|---|---|---|---|---|
| B3PW91 | cc-pVDZ | 60.4 | 0 | 16 |
| B3LYP | | 70.4 | 0 | |
| HF | | 74.3 | 0 | |
| MP2 | | 69.5 | 0 | |
| CCSD | | 58.1 | 0 | |
| CCSD(T) | | 58.8 | 0 | |

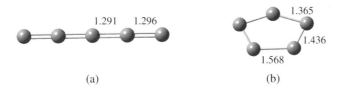

(a)          (b)

Fig. 8. Geometries with geometrical parameters (in angstrom and degree) for (a) linear and (b) cyclic $C_5$.

For the negative $C_5$ ion, the linear form is 67.4 kcal/mol below the cyclic structure at CCSD(T)/aug-cc-pVDZ. Using the same basis set, HF and B3LYP methods gave the relative energies of 84.3 and 78.8 kcal/mol, respectively. The cyclic $C_5^-$ has doublet spin state and $C_2$ symmetry. The situation is quite different for the positive $C_5$ ion. CCSD(T)/cc-pVDZ calculations predicted the cyclic form to be slightly more stable than the linear form (−2.9 kcal/mol) while the energy differences between the 2 forms are −3.8 and 1.2 kcal/mol at B3PW91 and B3LYP, respectively. It can be concluded that for the positive ion cluster, linear and cyclic $C_5$ is isoenergetics. The cyclic $C_5^+$ is $C_s$ symmetric.

## 4.2. $C_7$

Two forms of cyclic structures were proposed for $C_7$, see Fig. 9, but they all are higher in energy than the corresponding linear structure. At CCSD(T)/cc-pVDZ level of theory, both cyclic forms are 14.0 and 17.5 kcal/mol less stable than the linear form. Using the same basis set, HF and density functional theory overestimate the stability of the linear form. The relative energies between two cyclic forms and the linear $C_7$ were given in Table 9. Similar to the neutral $C_7$, the structure of the corresponding negative ion is also linear but with energy of 37.6 kcal/mol below the cyclic $C_7^-$ at CCSD(T)/aug-cc-pVDZ. This energy difference is, however, 49.7 kcal/mol at B3LYP/aug-cc-pVDZ and 66.3 kcal/mol at HF/aug-cc-pVDZ. Interestingly, for $C_7^+$ all methods at cc-pVDZ basis set predicted the cyclic structure to be more stable (−20.9 kcal/mol at CCSD(T)), in exception of HF which reports the linear form to be slightly more stable (4.9 kcal/mol). For neutral, negative, and positive $C_7$, the discrepancy between HF and CCSD(T) might be due to the complication in describing electron delocalization.

Table 9. Relative energies (kcal/mol) of different geometries of $C_7$ computed at various levels of theory.

| $C_7$ | basis set | cyc. | | lin | Ref. |
|---|---|---|---|---|---|
| | | $C_{2v}$ | elong $C_{2v}$ | | |
| HF | cc-pVDZ | 24.5 | 52.7 | 0 | 16 |
| MP2 | | 18 | 4.5 | 0 | |
| CCSD | | 10 | 17.6 | 0 | |
| CCSD(T) | | 14 | 17.5 | 0 | |
| B3LYP | | 24.5 | 31.5 | 0 | |
| B3PW91 | | 15.4 | 21.7 | 0 | |

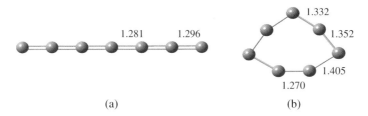

1.332

1.281 1.296

1.352

1.405

1.270

(a)                              (b)

Fig. 9. Geometries with geometrical parameters (in angstrom and degree) for (a) linear and (b) cyclic $C_7$.

## 4.3. $C_9$

Four cyclic structures, singlet $C_{2v}$, singlet elongated $C_{2v}$, singlet non-planar, and triplet $D_{9h}$, have been proposed for $C_9$. According to theoretical calculations, these cyclic structures were higher in energy than the corresponding linear form. Among cyclic structures the singlet $C_{2v}$ was proposed to be the lowest cyclic structure by HF, CCSD, CCSD(T), B3LYP, and B3PW91 methods. However, these methods did not agree on relative energies and for some methods also on the relative ordering. While CCSD(T) predicted the singlet $C_{2v}$ to be 11.3 kcal/mol higher than the linear form which followed by, in the relative ordering, the singlet non-planar, the singlet elongated $C_{2v}$, and then the triplet $D_{9h}$, HF gave the singlet $C_{2v}$ to be 30.4 kcal/mol less stable which followed by the triplet $D_{9h}$, the singlet elongated $C_{2v}$, and then the singlet non-planar. The B3PW91 prediction is in very good agreement with CCSD(T) both in ordering and the relative energies. Energies of these cyclic structures relative to the linear $C_9$ computed at various methods and 6-31G(d) basis are given in Table 10. The geometries of the cyclic and the linear $C_9$ were shown in Fig. 10.

For negative $C_9$ ion, the linear form has the lowest energy. It is 24.6 kcal/mol at CCSD(T) and 32.1 kcal/mol at B3LYP below the $C_2$-symmetric cyclic structure when computed using aug-cc-pVDZ basis. For the positive ion, it is the $C_{2v}$-cyclic which is the lowest, 13.1 kcal/mol (CCSD(T)/cc-pVDZ) below the linear.

Table 10.  Relative energies (kcal/mol) of different geometries of $C_9$ computed at various levels of theory.

| $C_9$ | basis set | | | cyc | | lin | Ref. |
|---|---|---|---|---|---|---|---|
| | | $C_{2v}$ | elong $C_{2v}$ | non-planar | $D_{9h}$, triplet | | |
| HF | 6-31g(d) | 30.4 | 59 | 62.9 | 39.4 | 0 | 16 |
| MP2(ALL) | | 12.6 | 7.3 | 6.3 | 4.1 | 0 | |
| MP2(FC) | | 12.8 | 7.3 | 6.1 | 4 | 0 | |
| CCSD | | 6.6 | 22.8 | 23 | 25.4 | 0 | |
| CCSD(T) | | 11.3 | 23 | 21.7 | 25.4 | 0 | |
| B3LYP | | 22.2 | 35.2 | 36.1 | 35.3 | 0 | |
| B3PW91 | | 14.8 | 25.4 | 26.4 | 27.4 | 0 | |

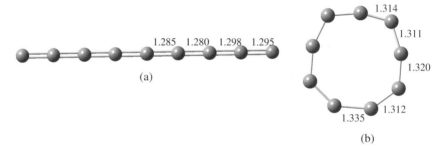

Fig. 10.  Geometries with geometrical parameters (in angstrom and degree) for (a) linear and (b) cyclic $C_9$.

## 4.4. $C_{2n+1}$, $n = 5 - 7$ ($C_{11}$, $C_{13}$, and $C_{15}$)

Unlike the smaller odd-numbered $C_n$ (neutral) which has the linear form as the most stable structure, the odd-numbered $C_n$ for $n > 10$ has the cyclic form as the minimum. The cyclic form remains the most stable structure of the positive $C_n$ for $n > 10$, however with larger stabilization energy. To our best knowledge, no information regarding relative stabilites and geometries is available for negative $C_n$ with $n > 10$. In addition, we also could not find any information on electronic and molecular structures of neutral, positive ion, and negative ion of $C_{17}$. The relative stabilities of neutral and positive ion $C_{11}$, $C_{13}$, and $C_{15}$ computed at different levels of theory are displayed in Table 11. Geometries of linear and cyclic $C_n$ and $C_n^+$ are shown in Fig. 11.

For $C_{11}$, using cc-pVDZ basis set all methods predicted the cyclic form to be more stable than its corresponding linear structure. At CCSD(T), this relative stability is 26.9 kcal/mol. The density functional theoretical based and HF methods underestimated this relative energy by 7 – 10 kcal/mol while the underestimation was 17 kcal/mol at the MP2

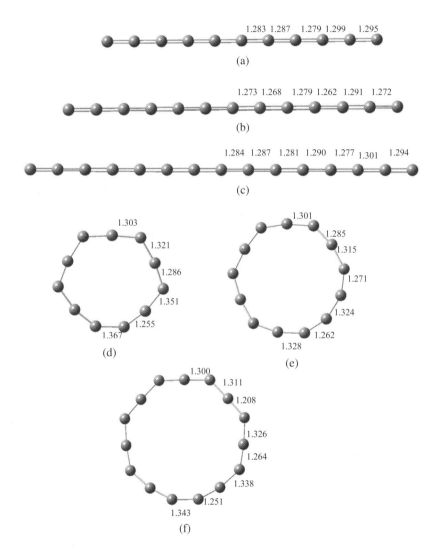

Fig. 11. Geometries with geometrical parameters (in angstrom and degree) for (a) linear $C_{11}$, (b) linear $C_{13}$, (c) linear $C_{15}$, (d) cyclic $C_{11}$, (e) cyclic $C_{13}$, and (f) cyclic $C_{15}$.

Table 11. Relative energies (kcal/mol) of different geometries of $C_{11}$, $C_{13}$ and $C_{15}$ computed at various levels of theory.

| $C_{11}$ | | basis set | cyc $C_{2v}$ | lin | Ref. |
|---|---|---|---|---|---|
| | B3PW91 | cc-pVDZ | −21.8 | 0 | 16 |
| | B3LYP | | −16.4 | 0 | |
| | HF | | −23 | 0 | |
| | MP2 | | −10 | 0 | |
| | CCSD | | −35.8 | 0 | |
| | CCSD(T) | | −26.9 | 0 | |
| $C_{13}$ | | | cyc $C_{2v}$ | lin | Ref. |
| | B3PW91 | cc-pVDZ | −28 | 0 | 16 |
| | B3LYP | | −19.6 | 0 | |
| $C_{15}$ | | | cyc $C_{2v}$ | lin | Ref. |
| | B3LYP | cc-pVDZ | −40.1 | 0 | 16 |

level. This might suggest again the poor convergence of the MP series for predicting relative stabilities between cyclic and linear Cn. For $C_{11}^{+}$, while using cc-pVDZ basis CCSD(T) predicted the cyclic form to be 56.2 kcal/mol more stable, other methods underestimated the relative stability by 6 – 20 kcal/mol.

Only B3PW91 and B3LYP calculations have been carried out for the relative stability of linear and cyclic $C_{13}$ and it was found that the cyclic $C_{13}$ is 19.6 (B3LYP/cc-pVDZ) or 28.0 (B3PW91/cc-pVDZ) kcal/mol more stable than the linear form. For the positive ion cluster, at CCSD(T)/cc-pVDZ the cyclic $C_{13}^{+}$ is 47.5 kcal/mol more stable. One can observe that the relative energies computed using B3PW91 and CCSD(T) are very close. A similar observation can also be found for other carbon clusters.

The cyclic forms, at B3LYP/cc-pVDZ, were found to be 40.1 and 64.9 kcal/mol more stable than the linear structures for the neutral and cation cluster of $C_{15}$. However, the relative energy of the positive ion at B3PW91/cc-pVDZ is 71.5 kcal/mol. Again, B3LYP method gave too small relative energy.

## 5. Conclusion

Small- and medium-sized carbon clusters are originally believed to possess only linear structure. With guidance from theoretical calculations, more and more experiments have confirmed the existence of the monocyclic ring structure for these clusters. For even-numbered carbon clusters $C_{2n}$, cyclic forms are found to be energetically favorable. The cyclic $C_{2n}$ can be classified as aromatic and anti-aromatic. The aromatic $C_{2n}$ such as $C_6$, $C_{10}$, $C_{14}$, and $C_{18}$ has $D_{nh}$ distorted cyclo-cumulene structure rather than $D_{2nh}$ cyclo-cumulene structure. The anti-aromatic $C_{2n}$ such as $C_4$, $C_8$, $C_{12}$, and $C_{16}$ has $C_{nh}$ structure. The linear and cylic $C_4$ are almost isoenergetic. The relative stability of cyclic $C_{2n}$ to the linear form increases as "$n$" increases. The relative stability to the linear form as large as 50 kcal/mol was observed. Molecular structures of position ion $C_{2n}$ have similar tendency to the neutral clusters. The cyclic and the linear forms of $C_4^+$, $C_6^+$, and $C_8^+$ are nearly isoenergetic whereas for larger $C_{2n}^+$ ($C_{10}^+$, $C_{12}^+$, $C_{14}^+$, $C_{16}^+$, and $C_{18}^+$) the cyclic form becomes energetically more stable than the linear. The situation is different for $C_{2n}^-$ clusters. The linear form is more stable than the cyclic structure for $C_4^-$, $C_6^-$, $C_8^-$, $C_{12}^-$, and $C_{16}^-$. For $C_{10}^-$, the linear and the cyclic forms are isoenergetic. The cyclic form becomes more stable than the linear form for $C_{14}^-$ and $C_{18}^-$. The stability of the linear form over the cyclic form for negative ion $C_{2n}$ is one of the reason for the earlier conclusion which stated that small- and medium-sized carbon clusters should have the linear structure.

The monocyclic ring structure was found as the minimum structure for odd-numbered clusters as well, however only for large $C_{2n+1}$ when $n \geq$ 5. Relative stabilities of cyclic $C_{2n+1}$ as compared to their corresponding linear one could be as large as 40 kcal/mol ($C_{15}$). These stabilities increases as "$n$" increases. The trend of the cyclic-linear relative stability of negative ion $C_{2n+1}$ is similar to those of neutral clusters, *i.e.* linear is more stable for smaller $n$ ($n < 5$) and cyclic is more stable for larger $n$ ($n \geq 5$). However, for odd-numbered positive ion carbon clusters the cyclic form is more stable than the linear structure even for $C_5^+$. The trend for molecular structures of neutral, positive ion, and negative ion of odd-numbered carbon clusters is not the same as even-numbered clusters.

## Acknowledgements

Thanks to Dr. Stephan Irle for inviting the author to write this review article. Thanks to the department of chemistry and the supercomputer institute at the University of Minnesota for which most parts of author's research regarding to this topic were carried out there. Special thanks to the late Dr. Jan Almlöf who was the graduate advisor of the author during the years in Minnesota and the author would like to contribute this review article in memory of him.

## References

1. E. Herbst and W. Klemperer, *Astrophys. J.*, **185**, 505 (1973).
2. A. Dalgarno and J.H. Black, *Rep. Prog. Phys.*, **39**, 573 (1976).
3. R.G.W. Norrish, G. Porter and B.A. Thrush, *Nature*, **169**, 582 (1952).
4. H. Feld, R. Zurmühlen, A. Leute and A. Bennighaven, *J. Phys. Chem.*, **94**, 4595 (1990).
5. H.Y. So and C.L. Wilkins, *J. Phys. Chem.*, **93**, 1184 (1990).
6. E.A. Rohlfing, D.M. Cox and A. Kaldor, *J. Chem. Phys.*, **81**, 3322 (1984).
7. N. Fuerstenau and F.Hillenkamp, *Int. J. Mass Spectrom. Ion Phys.*, **35**, 201 (1981).
8. R.E. Honig, *J. Chem. Phys.*, **22**, 126 (1954).
9. B.K. Rao, S.N. Khanna and P. Jena, *Solid State Comm.*, **58**, 53 (1986).
10. V. Parasuk, in *DFT Calculations on Fullerenes and Carbon Nanotubes*, Ed. V.A. Basiuk, and S. Irle (Research Signpost, Singapore, 2008), p 31.
11. K.S. Pitzer, E. Clementi, *J. Am. Chem. Soc.*, **81**, 4477 (1959).
12. S.J. Strickler and K.S. Pitzer, in *Molecular Orbital Chemistry, Physics, and Biology* Ed. P.O. Löwdin and B. Pullman (Academic Press, New York, 1964).
13. R. Hoffmann, *Tetrahedron*, **22**, 521 (1966).
14. K. Raghavachari, R.A. Whiteside and J.A. Pople, *J. Chem. Phys.*, **85**, 6623 (1986).
15. J. Hutter, H.P. Lüthi and F. Diederich, *J. Am. Chem. Soc.*, **116**, 750-756 (1994).
16. J.M.L. Martin, J. El-Yazal and J.-P. François, *Chem. Phys. Lett.*, **252**, 9 (1996).
17. M.G. Giuffreda, M.S. Deleuze and J.-P. François, *J. Phys. Chem. A*, **103**, 5137 (1999).
18. M.G.Giuffreda, M.S. Deleuze and J.-P. François, *J. Phys. Chem. A*, **106**, 8659 (2002).
19. K. Raghavajari and J.S. Binkley, *J. Chem. Phys.*, **87**, 2191 (1987).
20. V. Parasuk and J. Almlöf, *J. Chem. Phys.*, **94**, 8172 (1991).
21. S.J. Blanksby, D. Schröder, S. Dua, J.H. Bowie and H. Schwarz, *J. Am. Chem. Soc.*, **122**, 7105-7113 (2000).
22. V Parasuk and J. Almlöf, *J. Chem. Phys.*, **91**, 1137 (**1989**).

23. X.-L. Zhou, Y.-L. Bai, X.-R. Chen and X.-D. Yang, *Chin. Phys. Lett.*, **21**, 283 (2004).

24. J.D. Presilla-Márquez, J.A. Sheehy, J.D. Mills, P.G. Carrick and C.W. Larson, *Chem. Phys. Lett.*, **274**, 439 (1997).

25. V. Parasuk and J. Almlöf, *Theor. Chim. Acta*, **83**, 227 (1992).

26. X.-R. Chen, L.-Y. Bai, X.-L. Zhou and X.-D. Yang, *Chem. Phys. Lett.*, **380**, 330 (2003).

27. J.D. Presilla-Márquez, J. Harper, J.D. Sheehy, P.G. Carrick and C.W. Larson, *Chem. Phys. Lett.*, **300**, 719 (1999).

28. K.E. Yousaf and P.R. Taylor, *Chemical Physics*, **349**, 58 (2008), in press.

29. Martin, J.M.L., Taylor, P. *J. Phys. Chem.* 1996, **100**, 6047-6056.

30. D. Zajfman, H. Feldman, O. Heber, D. Kella, D. Majer, Z. Vager and R. Naaman, *Science*, **258**, 1129 (1992).

31. M. Algranati, H. Feldman, D. Kella, E. Malkin, E. Miklazky, R. Naaman, Z. Vager and J. Zajfman, *J. Chem. Phys.*, **90**, 4617 (1989).

32. H. Feldman, D. Kella, E. Malkin, E. Miklazky, Z. Vager, J. Zajfman and R. Naaman, *J. Chem. Soc. Faraday Trans.*, **86**, 2469 (1990).

33. Z. Slanina and R. Zahradnik, *J. Phys. Chem.*, **81**, 2252 (1977).

34. W. Weltner Jr. and D. McLeod Jr., *J. Chem. Phys.*, **45**, 3096 (1966).

35. K.R. Thomson, R.L. Dekock and W.Weltner Jr., *J. Am. Chem. Soc.*, **93**, 4688 (1971).

36. H.M. Cheung and W.R.M Graham,. *J. Chem. Phys.*, **91**, 6664 (1989).

37. A. Faibis, E.P. Kanter, L.M. Tack, E. Bakke and B.J. Zabransky, *J. Phys. Chem.*, **91**, 6645 (1987).

38. Z. Vager and E.P Kanter, *J. Phys. Chem.*, **93**, 7745 (1989).

39. J.M.L. Martin, J.-P. François and R. *J.* Gijbels, *Chem. Phys.*, **90**, 3403 (1989).

40. D.E. Bernholdt, D.H Magers and R.J. Bartlett, *J.Chem. Phys.*, **89**, 3612 (1988).

41. K.R. Thompson, R.L. Dekock and W. Weltner Jr., *J. Am. Chem. Soc.*, **93**, 4688 (1971).

42. M. Vala, T.M. Chadrasekhar, J. Szczepanski and R. Pellow, *High Temp. Sci.*, **27**, 19 (1990).

43. J. Szczepanski and M.Vala, *J. Phys. Chem.*, **95**, 2792 (1991).

44. C. Liang and H.F. Schaefer III; *J. Chem. Phys.*, **93**, 8844 (1990).

45. R.J. Van Zee, R.F. Ferrante, K.J. Zeringue, W. Weltner Jr.; Ewing, D.W. *J. Chem. Phys.*, **88**, 3465 (1988).

46. J.D. Presilla-Márquez, J.A. Sheehy, J.D. Mills, P.G. Carrick and C.W. Larson, *Chem. Phys. Lett.*, **274**, 439 (1997).

47. C. Liang and H.F., Schaefer III; J. *Chem. Phys. Lett.*, **169**, 150 (1990).

48. J.M.L. Martin, J. El-Yazal and J.-P. François, *Chem. Phys. Lett.*, **242**, 570 (1995).

49. V. Parasuk and J. Almlöf, *J. Am. Chem. Soc.*, **113**, 1049-1050 (1991).

50. D.A. Plattner and K.N. Houk, *J. Am. Chem. Soc.*, **117**, 4405 (1995).

51. A.F. Jalbout, S.; Fernanez, and H. Chen, *J. Mol. Struct. (Theochem)*, **584**, 143 (2002).
52. A.F. Jalbout and S. Fernandez, *J. Mol. Struct. (Theochem)*, **584**, 169-182 (2002).
53. A. Fura, F. Tureček, and F.W. McLafferty, *Int. J. Mass. Spect.*, **217**, 81-96 (2002).
54. J.P. Ritchie, H.F. King and W.S. Young, *J. Chem. Phys.*, **85**, 5157 (1986).
55. R.A. Whiteside, R. Krishnan, D.J. Defrees, J.A. Pople and P.v.R. Schleyer, *Chem. Phys. Lett.*, **80**, 547 (1981).
56. F. Diederich, Y. Rubin, C.B. Knobler, R.L. Whetten, K.E. Schriver, K.N. Houk and Y. Li, *Science*, **245**, 1088 (1989).
57. J. Szczepanski, S. Ekern and M. Vala, *J. Phys. Chem. A*, **101**, 1841 (1997).
58. D.H. Magers, R.J. Harrison and R.J. Bartlett *J. Chem. Phys.*, **89**, 3284 (1986).

Chapter 12

# Vibrational Spectroscopy of Linear Carbon Chains

Chien-Pin Chou, Wun-Fan Li, and Henryk A. Witek*

*Department of Applied Chemistry and Institute of Molecular Science,*
*National Chiao Tung University, Hsinchu, Taiwan*
*\* E-mail address: hwitek@mail.nctu.edu.tw*


Marcin Andrzejak

*Department of Theoretical Chemistry, Faculty of Chemistry,*
*Jagiellonian University, Krakow, Poland*
*E-mail address: andrzeja@chemia.uj.edu.pl*

A detailed theoretical study of geometric, electronic, vibrational, and spectroscopic properties of linear carbon chains is presented. The study is supplemented with an extensive survey of available experimental and theoretical results. Our calculations constitute a bridge between the quantum-chemical and solid-state simulations, using the SCC-DFTB (self-consistent-charge density-functional tight-binding) methodology. The computed equilibrium geometry, electronic band structure, and phonon dispersion curves of infinite carbon chains are compared with analogous results obtained for finite oligomers. A surprisingly fast convergence of all the studied properties of the finite systems to the infinite limit and a rather short-ranged influence of the terminal sections of the chain are observed. The molecular calculations display analogues of well known solid state physics phenomena, such as Peierls distortion or Kohn anomaly. The presented IR and Raman spectra of finite chains show that the infinite limit is approached rapidly both in the frequency and intensity domain. For a constant mass sample, the intensity of the IR signal is inversely proportional to the number of carbons atoms in the chain, while the intensity of the main Raman band becomes independent on the chain length. The satellite bands and terminal group vibrations disappear from the Raman spectra already for relatively short chains.

## 1. Introduction

The simplest molecular realization of a periodic carbon chain is a linear system of carbon atoms connected by adjacent double bonds. The common name for this system is cumulene. The unit cell of cumulene — shown in Fig. 1 — consists of a single carbon atom located in its center. Because the cumulene is a linear one-dimensional (1D) structure, the unit cell is fully characterized by a single lattice constant $a$, which is comparable to the length of a typical double bond found in organic molecules (1.34 Å). This name cumulene was derived from the trivial (i.e., non-systematic) name of one of the simplest hydrocarbons displaying such a structure, butatriene. The infinite cumulene chain can be thought of as the limit of a sequence of hydrocarbons with cumulative double bonds. The first three members of this sequence are propadiene ($C_3H_4$ aka allene), butatriene ($C_4H_4$ aka cumulene), and pentatetraene ($C_5H_4$). More examples are given in Fig. 2. Cumulene is a typical example of a 1D metallic system, for which the Peierls distortion[1] should be observed. The cumulene lattice would be at equilibrium were it not for the presence of the $\pi$ electrons. Being strongly delocalized, they can be approximately treated as a gas of nearly-free, non-interacting particles, with energies equal to $h^2 k^2 / 2m_e$, $k$ representing the quasimomentum of an electron. Peierls distortion arises as a result of the interaction of the $\pi$ electrons with periodic lattice deformations along the atomic chain (longitudinal phonons). Since a deformation with a wave number $q = 2\pi/l$ (corresponding to the periodicity $l$ in a physical space) would produce a gap around $k = \pi/l$ in the electronic energy band, the effective lowering of the electronic energy can be caused only by phonons of quasimomentum close to $2k_f$. If the electron-phonon coupling is strong enough to overcome the increase of the repulsion energy between the atomic cores, the lattice adopts new periodicity equal to $\pi/k_f$ and the Peierls distortion is observed. It should be noted in passing that in linear carbon chains two degenerate, symmetry-independent $\pi$ electron systems take part in the electron-phonon coupling, both contributing to a considerable lattice distortion. Since in the case of the cumulene chain only half of each $\pi$ band is occupied ($k_f = \pi/2a$), the new lattice period is $2a$, and the system of cumulative double bonds is replaced by an alternating pattern of shorter

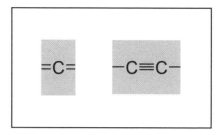

Fig. 1. One-dimensional unit cells used in the solid-state SCC-DFTB calculations of the infinite cumulene (left) and polyyne chains (right).

(approximately triple) and longer (approximately single) bonds. The common name for the resulting carbon chain is polyyne. Again, the infinite polyyne can be thought of as the limit of a sequence of hydrocarbons with alternated single-triple bonds. The first three members of this sequence are ethyne ($C_2H_2$ aka acetylene), butadiyne ($C_4H_2$ aka diacetylene), and hexatriyne ($C_6H_2$ aka triacetylene). More examples are given in Fig. 2. A schematic comparison between the unit cells of cumulene and polyyne is given in Fig. 1. The fact that the unit cell of polyyne contains twice as many atoms than that of cumulene brings about important physical consequences, some of which we discuss in detail in the next section.

Linear carbon chains have been a subject of interest for scientists since the first synthesis of their finite analogues in the second half of the 19th century.[2] They have been found in interstellar clouds[3] and cool carbon stars,[4] as well as in various biological objects, like fungi[5] or certain tropical weeds.[6] Since 1953, a number of new synthetic methods were successfully applied to produce carbon chains of various lengths.[7–10] Recently, the development of organic electronics, rapid advances in computational chemistry and physics, and a discovery of new allotropic forms of carbon rekindled interest in the linear carbon chains.[11] Their relative structural simplicity designates them as an ideal playground for testing various models of the solid-state physics. On the other hand, their semiconducting properties suggest potential applicability of linear carbon chains as molecular nanowires. New ways of producing polyynes, such as arc discharge and laser ablation techniques have been developed and used to generate linear carbon chains containing from 8 to 20 acetylenic units and terminated with a variety of ligands.[12–17] Polyynes with bulky terminal groups were used to obtain single crystals containing isolated carbon chains, which allowed for experimental determination of their geometries and physical proper-

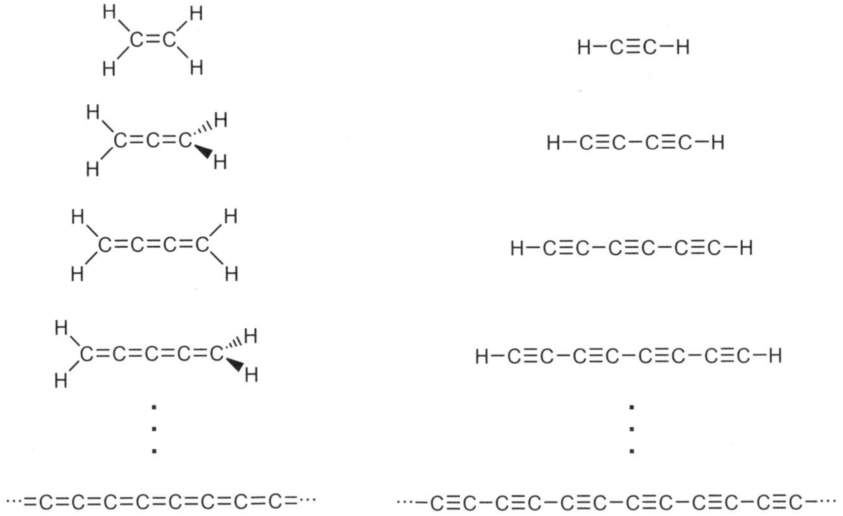

Fig. 2.   Sequences of finite-length hydrocarbons, $C_nH_2$ and $C_nH_4$, converging in the limit $n \to \infty$ to an infinite cumulene (left) and polyyne chain (right plot).

ties.[14,16] H-terminated polyynes containing up to 13 acetylenic units were studied in the gas phase,[11] while slightly shorter chains were placed in neon matrices[18] or dissolved in hexane.[16,19] Various experimental techniques, like UV-Vis spectroscopy, Raman spectroscopy, surface enhanced Raman spectroscopy (SERS), and resonance Raman spectroscopy were applied to study the nature of electronic and vibrational states of polyynes. The experimental investigations of the C–C bond lengths and the frequencies of the longitudinal vibrations of the carbon chain are of special interest for the theoretical physicists and computational chemists. The experimental values of these quantities are not only directly comparable with the outcome of the quantum mechanical calculations but they are also used to estimate input parameters for models of infinite polyyne chains.

The optical energy gap for finite polyyne chains, associated in experiment with the energy of the 0–0 line of the optically allowed transition, was measured in the hexane solution,[16] in the neon matrix,[18] and in the gas phase.[11] The results of the matrix and gas phase measurements are rather similar, with the former set of energies nearly uniformly shifted to lower values by 0.15–0.17 eV as a result of guest-host interactions within the matrix. Also, both sets of energies display a linear dependence on the reciprocal of the number of carbon atoms $(1/N)$, which, extrapolated to infinite chains, produce nearly the same energy gap of about 2.4 eV. The results from the hexane solution, however, are somewhat different; the gas to solution shift markedly changes with the increasing length of the carbon chain and the measured energies do not follow a clear, linear trend with respect to $1/N$. According to Tykwinski *et al.* the energies are proportional to $n^{-0.379\pm0.002}$, $n$ being the number of acetylenic units constituting the carbon chain. This relation yielded 2.2 eV as the value of the energy gap for the infinite polyyne chain. Zhao *et al.*[20] performed resonance Raman measurements for very long carbon chains (containing over a hundred atoms) confined within multi-wall carbon nanotubes (MWCNT). The maximum resonance enhancement of the Raman scattering occurred at 2.5 eV. Other measurements for shorter oligomers ($C_{10}$ and $C_{12}$) in single-wall carbon nanotubes (SWCNT) yielded a significantly lower value of the resonance energy between 2.0–2.3 eV.[21] The surprisingly low value of the resonance energy indicates that the interactions between the nanotube and the confined carbon chain may be strong enough to break the symmetry-based selection rules and enable the low-energy electronic excitations of the polyynic chain. This conjecture has been corroborated by the theoretical work of Rusznyak *et al.*,[22] which showed significant changes in the bond length difference, owing to charge transfer from the tube to the chain. Such influence of the nanotube on the confined carbon chains may also affect other properties of the guest molecule, particularly the frequencies of the longitudinal, optical vibrations. The resonance Raman measurements of Malard and co-workers[21] reported Raman bands at 2066 cm$^{-1}$ for $C_{10}H_2$, and at 2020 and 2050 cm$^{-1}$ for $C_{12}H_2$. The frequencies of the main Raman band obtained for the same compounds in the off-resonance Raman study of polyynes in hexane solution were indeed shifted towards higher frequencies by some 50–60 cm$^{-1}$ (2123 and 2096 cm$^{-1}$) and the frequency of the satellite band (for $C_{12}H_2$) was up-shifted by 31 cm$^{-1}$. Therefore, the frequencies of the longitudinal normal modes for long polyynic chains in MWCNT located by the resonance Raman measurements between 1780 and 1860 cm$^{-1}$ should probably be treated as the lowest limit for the corresponding frequencies of the free molecules. Regardless of the absolute location of the longitudinal vibrations of the polyynic chains in various environments, the LO frequencies show interesting dependence on the length of the chain. The strongest band in the Raman spectra of polyynes (the so called $\alpha$ band) corresponds to the longitudinal out-of-phase motion of the triple and single bonds throughout the whole molecule. Its relative intensity with respect to other bands grows with increasing length of the carbon chain. On the other hand, for short chains, the frequency of the α band has been found inversely proportional to the number of acetylenic units.[23,24] The satellite bands show neither of these trends; their intensities and frequencies undergo much less pronounced and more irregular changes. The vibrations of polyynes were also investigated by means of the surface enhanced Raman scattering (SERS). Lucotti *et al.*[25] reported a study of a mixture of polyynes containing from 8 to 16 carbon atoms in the presence of the colloidal silver particles. They noted an 60 cm$^{-1}$ down-shift of the Raman α-band with the enhancement factor of about $10^6$. Also, after adding the colloidal silver a new strong band appeared at 1975 cm$^{-1}$ and a weaker one at 2040 cm$^{-1}$. The authors attributed these bands either to cumulenic structures stabilized by binding to the Ag particles or to an Ag-catalyzed conversion of the $C_8$ chains into longer ones ($C_{16}$). A more elaborate study of the Raman spectra of polyynes was presented by Tabata *et al.*[23] The polyynes of different sizes were isolated and purified prior to Raman measurements. The authors recorded both normal and surface-enhanced spectra. They observed a uniform down-shift of the α-band frequencies by some 80 cm$^{-1}$ after immersion of the Ag island film in the polyyne solution. New bands of comparable intensity appear at lower energies and are rationalized as the SERS counterparts of the satellite bands present in the normal Raman spectra. Their frequency shift is not as regular as for the α-band and further investigation (experimental or theoretical) is necessary to verify the origin of those bands. The overall enhancement is, however, so strong ($> 10^6$) that SERS is used as a sensitive tool for detecting the presence of the polyynes in organic synthesis.[26]

The first thorough *ab initio* studies of the linear polyyne chain, including numerical geometry optimization were performed by Kertesz *et al.*[27,28] and Karpfen *et al.*[29,30] Their calculations showed the structure with an alternating pattern of bond lengths to be more stable than the equidistant one. Karpfen also calculated dispersion curves for both linear and transversal phonon branches.[30] The results resemble closely the standard outcome of the phonon calculations based on the nearest neighbor approximation for the force constants.[31] The frequency of the longitudinal optical phonon in the center of the Brillouin zone (Γ point) was established at about 2400 cm$^{-1}$, which indicated some inadequacy of the applied methodology when compared to experimentally based estimates[32,33] of about 1850–1900 cm$^{-1}$. However, the most interesting behavior was predicted for the transversal phonon dispersion curves, which showed nearly quadratic dependence of the frequency on the wave vector for small values of $k$. (Note that usually a standard linear dependence is expected for acoustic phonons around the $\Gamma$ point.) Similar observations were reported much later by Seitz[34] for finite length polyyne chains (HF, MP2, and DFT/B3LYP calculations). The authors rationalized these results by adopting the classical string theory to the case in which the initial tension of the string is zero. The only tension appears as a result of small elongation of the molecular string upon bending (usually regarded as negligible in comparison with the initial tension of a string). Such assumption leads to time dependence of the wave velocity and, consequently, to the square dependence of the frequency on the number of a harmonic. Unfortunately, further verification of this conjecture was difficult as little attention has been paid so far to transversal vibrations of the carbon chains.

The early calculations, although qualitatively correct, failed to reproduce quantitatively the electronic energies. In consequence, the electronic energy gap was greatly overestimated (11.1 eV[28] or exceeding 7.5 eV,[30] as compared to 2.4 eV, given by extrapolation of the experimental results from the gas phase measurements[11] to infinite number of carbon atoms or to 2.15 eV, given by similar extrapolation of the measurements in the hexane solution[16]). Improved calculations, performed by Springborg[35] used a combined approach dividing the space into non-overlapping spheres around each carbon atom and the interstitial region outside the spheres. Inside the atomic spheres the potential, basis set and charge density were described numerically whereas in the outside region analytical solution was obtained with the use of Hankel functions of the first kind. The main advantage of this approach was relatively accurate description of the wave function close to the nuclei with very few basis functions. The results were considerably better than those of Kertesz or Karpfen, yielding 2.8 eV, a value much closer to the experimental estimate. In fact a recent work of Tommasini *et al.*[36] reports a value of 2.7 eV, the result of model calculations based on the Hückel formalism with parameters evaluated from experimental band gaps for finite polyynes.[18] The authors were able to find a consistent parametrization, which reproduced the experimental band gaps for the polyynes containing from 12 to 24 carbon atoms. The same parameters were used to calculate the band gap for an infinite polyynic chain. The value of 2.7 eV is larger than 2.4 eV obtained from a simple extrapolation of the experimental values for finite polyynes to the infinite-length chain. However, the authors point out that for very long polyynes (over 40 carbon atoms) the dependence of the energy gap on $1/N$ ($N$ being the number of carbon atoms in the chain) becomes non-linear as a result of the flattening of the $\pi$ electron bands near the Brillouin zone boundary. Consequently, linear extrapolation is bound to produce a too low value of the electronic energy gap for the infinite polyynic chain. The resonance Raman measurements performed by Zhao *et al.*[20] for long carbon chains confined in the multi-wall carbon nanotubes showed that the maximum resonance enhancement occurs at 2.5 eV. This value is slightly lower than 2.7 eV proposed by Tommasini *et al.* The difference may be tentatively attributed to the interactions of the carbon chain with the surrounding nanotube. In this context the value of resonance energies between 2.0–2.3 eV reported by Malard *et al.* for relatively short ($C_{10}$ and $C_{12}$) polyynic chains confined in SWCNT cannot be associated with the optical band gap but rather, as the authors suggested, with the forbidden excitations of the polyynic chain partly allowed by interactions with the surrounding nanotube.

As far as the molecular geometry is concerned, experimental verification of the theoretical results is more difficult since the available experimental data concern only relatively short polyynes (from 4 to 8 polyynic units) and the bond lengths do not follow a clear trend as the number of polyynic units increases.[14] However, one can expect the bond lengths in long polyyne chains to approach 1.34–1.35 Å for the single bond and 1.21–1.22 Å for the triple bond. High-level correlated calculations (MP2, CCSD, CCSD(T)) for finite carbon chains[37] yielded 1.34–1.36 Å for the single bond and 1.20–1.22 Å for the triple bond. Analogous DFT/LDA calculations gave 1.29 Å for the single bond and 1.25 Å for the triple bond.[37] Such a surprisingly small bond length difference was obtained also by Bylaska *et al.*[38] and Yang and Kertesz.[39] The latter authors investigated in detail the CC bond lengths and the bond length difference (BLD = $\Delta r = r_{C-C} - r_{C\equiv C}$) in polyynic chains obtained from DFT calculations with a number of popular exchange-correlation functionals and a moderate-size 6–31G basis set. The bond length difference for the LDA calculations for an oligomer containing 36 acetylenic units slightly exceeded 0.02 Å, contrary to the results of correlated *ab initio* calculations and the experimental estimates. The calculated LDA geometry of the finite polyyne is far too close to the geometry of cumulene. Moving to an infinite carbon chain (i.e., to calculations with periodic boundary conditions) brought only slight improvement to the BLD pushing it up to 0.028 Å. Gradient corrected BLYP[40,41] functional did not change the qualitative picture, although the BLD reached 0.028 and 0.036 Å for the oligomer and the infinite chain, respectively. Introducing the Hartee-Fock (HF) non-local exchange, however, brought about considerable changes: BLD was 0.067, 0.088, 0.093, 0.134 and 0.135 Å for O3LYP[42] (12% of the exact exchange), B3LYP[43] (20%), PBE1PBE[44] (25%), BHandHLYP[45] (50%), and KMLYP[46] (56%), respectively. Pure HF calculations produce a value of 0.183 A, seemingly overestimating the BLD. Interestingly, geometry optimization with the MP2 method yielded a rather low BLD of 0.060 Å. Clearly, for polyynic chains the pure DFT calculations (no HF-type exchange) seriously underestimate the BLD, which increases for hybrid functionals proportionally to the amount of the included HF exchange. It seems that BHandHLYP and KMLYP are the best choice for geometry optimization of linear carbon chains. It also turned out that these functionals provide the best estimation of the energy gap for the infinite chain, if the geometry optimization is followed by single point calculations with the B3LYP functional. Extrapolation of the results for oligomers yielded about 2.2 eV as the energy gap of an infinite polyynic chain. However, since the dependence of the energy gap on the reciprocal of the number of carbon atoms is not linear for long chains,[36] the actual value of energy gap for the infinite chain correctly derived from the calculated energy gaps for oligomers is probably somewhat closer to the experimental estimate (2.5 eV[20]) than the value given by Yang and Kertesz (2.2 eV[39]).

The hybrid functionals are also expected to produce reliable results for vibrational frequencies. For short polyynes (4–8 carbon atoms) performance of the B3LYP functional is indeed satisfactory.[47] However, a decrease of the calculated $\alpha$ mode frequency with the growing number of carbon atoms in the chain is too fast as compared with the experimental findings.[23,24] Interestingly, the frequency of the satellite bands is reproduced with similar accuracy for both short and long polyynes. The $\alpha$ mode of the finite polyynes corresponds to the only Raman active longitudinal phonon of the infinite chain, located in the optical branch at the center of the Brillouin zone ($\Gamma$ point). It is a stationary wave for which the single and triple bonds simultaneously oscillate out of phase throughout the whole chain and can be regarded as a collective oscillation of the bond length difference (BLD). This kind of vibration corresponds to the $\mathcal{R}$ mode of the effective conjugation coordinate theory[48] developed originally for studying vibrational dynamics of polyacetylene and its oligomers. The frequency of that phonon is given by a simple formula

$$\omega^2 = \frac{4F_{\mathcal{R}}}{m}, \tag{1}$$

in which $F_{\mathcal{R}}$ represents the effective force constant defined below. However, owing to the coherence of the motion of nuclei and to the presence of the

$\pi - \pi$ conjugated electron system, the long-range coupling along the carbon chain becomes the key factor for determining the vibrational frequency of the longitudinal phonons. The long-range coupling modifies[33,49] the force constant for the phonon $\mathcal{R}$ according to the formula

$$F_{\mathcal{R}} = \frac{k_1 + k_2}{2} + \sum_{n \geq 1} \left( f_1^n + f_2^n - 2f_{12}^n \right), \tag{2}$$

where, in addition to the diagonal stretching force constants $k_1$ and $k_2$ related to the single and triple bonds, respectively, there appears a sum of terms $f^n$, which represent the interaction stretching force constants at increasing distances $(n)$ along the chain. The lower indices refer to the coupling between the same (1 or 2) or different (12) types of bonds placed at the distance of $n$ unit cells apart. According to Milani *et al.*,[49,50] the $f_1^n$ and $f_2^n$ terms are negative for all $n \geq 1$, whereas the $f_{12}^n$ is always positive, which makes the whole sum always negative and thus reduces the effective force constant for longitudinal phonon $\mathcal{R}$. The authors observed[50] that it is a too slow convergence of this sum that is responsible for overestimated softening of the $\alpha$ band in DFT calculations for finite length polyynes. The rate of decay of the stretching force constants $f^n$ can be associated with the bond length difference (BLD) in the polyynic chain. The long-range contribution to the effective force constant is expressed in terms of the Hückel formalism by so-called bond-bond polarizabilities $\Pi_i^n$ as

$$F_{\mathcal{R}}^{\Pi} = 4 \left( \frac{\partial \beta}{\partial r} \right)^2 \left[ \frac{\Pi_{11}^0 + \Pi_{22}^0}{2} + \sum_{n \geq 1} \left( \Pi_{11}^n + \Pi_{22}^n - 2\Pi_{12}^n \right) \right]. \tag{3}$$

The bond-bond polarizability (BBP) is defined as the second-order partial derivative of the Hückel energy with respect to the hoping integrals $\beta_i$ and $\beta_j$.[51-53] As follows from the definition, BBPs arise solely from the presence of the conjugated system of $\pi$ electrons. The expression $\partial \beta / \partial r$ denotes the electron-phonon coupling parameter, which determines the magnitude of the $\pi$-electron contribution to the effective force constant. BBP is a function of bond length difference $(\Delta r)$ of the polyynic chain through a dimerization parameter $b$ defined as

$$b = \left( \frac{\partial \beta}{\partial r} \right) \frac{\Delta r}{2\beta_0}, \tag{4}$$

where $\beta_0$ is the average of the hopping integrals for the triple and single bonds. The dependence of $\Pi_i^n$ on $b$ is very strong. For the cumulenic case, in which $b = 0$, the extremely slow convergence of the sum in Eq. (3)

(and, consequently, in Eq. (2)) leads to dramatic decrease of the phonon frequency, which manifests itself as the well known Kohn anomaly[54] in the longitudinal phonon branch near the Brillouin zone boundary. Since the Peierls distortion leads to doubling of the cumulenic unit cell and to appearing of the optical phonons in the reduced Brillouin zone, the "memory" of the Peierls distortion remains as the frequency lowering of the longitudinal, optical phonon close to the center of the Brillouin zone. This frequency lowering is not so strong as for the cumulenic chain since the Peierls distortion brings about non-zero BLD and speeds up the convergence of the BBPs sum. Milani *et al.*[49] estimated the frequencies of the phonon $\mathcal{R}$ for different values of BLD; $\Delta r$ equal to 0.038 Å (corresponding to DFT calculations with the PBE functional[44]) led to $\omega_{\mathcal{R}} = 1200$ cm$^{-1}$. Increasing $\Delta r$ to 0.09 Å resulted in $\omega_{\mathcal{R}} = 1800$ cm$^{-1}$, a value much closer to the experimental frequency of about 1900 cm$^{-1}$. However, using the experimental value of $\Delta r = 0.135$ Å brought about an overestimated frequency of about 2040 cm$^{-1}$. Similar predictions were made by Yang *et al.*[47] who extrapolated the vibrational frequencies calculated for long polyynes (containing from 30 to 72 carbon atoms) to infinite chains. The frequency obtained for B3LYP calculations was equal to 1800 cm$^{-1}$. The calculations for the BHandHLYP functional, which reproduces correctly the experimental BLD, yielded a seriously overestimated frequency $\omega_{\mathcal{R}} = 2215$ cm$^{-1}$. By adopting a linear/exponential scaling scheme for the calculated force constants the authors were able to modify the B3LYP results so that they correctly predicted both $\omega_{\mathcal{R}}$ and BLD for an infinite polyynic chain. In this scheme the local force constants are scaled with constant factors, whereas for the long-range ones the scaling parameter decays exponentially with distance along the chain. Tommasini *et al.*[55] tested this scaling procedure in frequency calculations for finite polyynes, using the PBE1PBE hybrid functional. They found that the linear scaling scheme works satisfactorily well for the $\alpha$ modes, but the frequencies of the satellite bands are reproduced less accurately. They proposed an alternative, linear adaptive procedure of scaling the force constants. The local force constants are again scaled by constant parameters, while the long-range ones are multiplied by a factor proportional to the difference in the force constants $f_1^1$ and $f_2^1$. Since this difference depends on the length of the polyynic chain, the long-range scaling factor adapts itself to the size of the molecule. With this scaling procedure the authors were able to reproduce the experimental spectra of polyynes containing from 4 to 10 acetylenic units with excellent accuracy.

In this article we explore the spectroscopic properties of infinite carbon chains by a systematic study of a series of finite hydrocarbons. We introduce two sequences of finite hydrocarbons that converge in the limit of infinite number of carbon atoms to cumulene and polyyne, respectively. The series of such molecules are shown schematically in Fig. 2. The geometry of each of these finite molecules is optimized using the SCC-DFTB method. The equilibrium character of each of the structures is further confirmed by determination of harmonic vibrational frequencies. The distributions of carbon–carbon bond distances for the equilibrium structures display several interesting features, which are analyzed in details in Sec. 4.1. The most important outcome of this analysis is the observed instability of the infinite cumulene chain. A gradual transition from the cumulene to polyyne bonding pattern can be observed for the finite chains when the number of carbon atoms exceeds 20. The transition occurs in the center of finite molecules while the edges retain the cumulenic character. This effect is the most distinct for long chains, in which the whole middle section displays polyynic character. This finding shows that the infinite cumulene chains are indeed unstable and confirms the theoretically predicted Peierls distortion at the molecular level. This is a very important observation, since it has allowed us to use the finite, polyynic model to represent infinite carbon chains in the further analysis. In the remainder of this study, the electronic and vibrational structure of the finite series of acetylene-like hydrocarbons is compared with the analogous information obtained from solid-state calculations. The convergence of the aforementioned properties is inspected, showing that they approach the properties of an infinite polymer for surprisingly short finite chains. Subsequently, in Sec. 5, we use this finding to analyze the infrared (IR) and Raman spectra of polyyne. These spectra are accessed as the limit of the IR and Raman spectra computed for finite acetylene-like hydrocarbons. We show that the limit is actually reached and the shape and intensity pattern of the spectra does not change upon further elongation of the finite acetylene-like chains. The conclusions are given in Sec. 6.

## 2. Computational Details

All theoretical simulations presented in this study are performed using the self-consistent-charge density-functional tight-binding (SCC-DFTB) method.[56,57] The SCC-DFTB method is a semiempirical quantum chemical technique, based on the second-order expansion of the total DFT energy.

SCC-DFTB was successfully applied for reproducing molecular geometries and computing reaction energetics for a wide class of molecules and compounds.[58-61] The SCC-DFTB method considers explicitly only the valence electrons; the chemical interactions of core electrons and nuclei is accounted for in an effective manner via pairwise repulsive potentials. A strong advantage of SCC-DFTB is its transferability. The method contains a number of parameters that are usually determined from accurate DFT calculations for a group of test molecules. A careful choice of the molecules constituting the testing set, i.e., a choice of molecules representing all possible bonding scenarios for a given element, ensures that the the SCC-DFTB method can be applied for modeling a wide group of molecules and materials. It is of course impossible to avoid completely the arbitrariness in the choice of the test set. However, we found no strong dependence of the final SCC-DFTB results on the selection of molecules in the parameterization set. We have recently extended the applicability of the SCC-DFTB method by deriving and coding the analytical Hessian that can be used for fast determination of harmonic vibrational frequencies.[62] This development was further augmented by including in the formalism the derivatives of the SCC-DFTB energy with respect to the external electric field, which enables calculations of the intensities in IR and Raman spectra.[57,63,64] The solid-state SCC-DFTB calculations are performed using the recently released DFTB+ code.[65] The SCC-DFTB method was extensively tested using various carbon molecular materials. The tests showed that the SCC-DFTB method is capable of reproducing the vibrational spectra of carbon fullerenes with good accuracy provided that the HOMO-LUMO gap is not negligible. Since the SCC-DFTB method is not parametrized for a particular molecular system, it usually displays some systematic errors for computing various molecular properties, similarly to standard *ab initio* methods of quantum chemistry. We have tested[66] the systematic errors obtained for the harmonic vibrational frequencies for a large group of chemical compounds (66 molecules with the total number of 1304 distinct vibrational frequencies). This study shows that the mean absolute deviation from experiment for the SCC-DFTB method is approximately 56 cm$^{-1}$. This value is approximately twice larger than analogous results for the scaled DFT calculations. In order to minimize this value, we have performed an optimization of the CC and CH repulsive SCC-DFTB potentials.[67] While using the optimized potentials we were able to reduce the error significantly; mean absolute deviation from experiment computed for a group of 14 hydrocarbons was reduced from 59 to 33 cm$^{-1}$ and the maximal absolute deviation, from 436 to 140 cm$^{-1}$. These numbers show that for the spectra presented in this study we can expect the vibrational levels to be reproduced with an error not exceeding 50 cm$^{-1}$. In practice, such a behavior should not introduce serious problems since the error is likely to be similar for all the members of the studied series of molecules. The second important issue is the relative intensity pattern in the reproduced spectra. The presented IR and Raman spectra are computed for the gas-phase molecules and should be compared to analogous experimental data. We have shown previously that comparing the gas-phase IR spectra with the experimental spectra recorded for the liquid or solid samples can lead to large discrepancies both in the position of bands and their relative intensity.[64] The latter differences are usually more pronounced. The main objective of this study is to investigate the convergence of the IR and Raman spectra for the infinite linear carbon chain from the corresponding spectra computed for the finite-length chains. We believe that despite of the communicated problems with somewhat displaced vibrational levels and a lack of direct correspondence with the experimental liquid and solid vibrational spectra, our study still provides an interesting insight in the vibrational spectra of infinite linear carbon chains.

## 3. Infinite Chains

Two periodic models of infinite linear carbon chains are considered explicitly in our work: cumulene and polyyne. It is certainly interesting to investigate the possibility of the existence of other forms (e.g., bent chains or chains with more complicated bond alternation patters within the unit cell), but such a study would be beyond the scope of the present analysis. In this Section we discuss the geometric, electronic, and vibrational properties of infinite polyyne and cumulene chains obtained from one-dimensional solid-state SCC-DFTB calculations.

### 3.1. *Geometric Structure*

The unit cells for both systems are shown in Fig. 1. The optimized value of the lattice constant for the cumulene unit cell — that can be interpreted also as the length of the double bond in cumulene — is 1.277 Å. This value is shorter than the length of the analogous bond in the related finite hydrocarbons (1.330 Å in ethylene and 1.308 Å in allene).[68] The optimized value of the lattice constant for the polyyne unit cell is 2.584 Å, which is similar to the doubled lattice constant of cumulene (2.554 Å). The lattice constant of polyyne can be decomposed into the single bond and triple bond contributions. The equilibrium length of the single bond in the polyyne unit cell is 1.380 Å; the analogous value for the triple bond is 1.204 Å. The latter value is a very typical length of a triple bond found in organic molecules ($r_{C \equiv C}$ in acetylene is 1.203 Å).[68] However, the length of the single bond in polyyne is very different from a typical length of a single bond in organic chemistry. For example, the experimental length of the $r_{C-C}$ bond is 1.522 Å for ethane ($C_2H_6$), 1.501 Å for cyclopropane ($C_3H_3$), and 1.562 Å for cubane ($C_8H_8$).[68] These values are much larger than 1.380 Å computed for polyyne. Instead, the equilibrium length of the longer bond in polyyne unit cell is similar to a typical length of an aromatic bond (*vide* 1.390 Å in benzene[68]). For the sake of consistence with previous studies and with chemical intuition, we will use the term "single bond" rather than "aromatic bond", remembering that it would probably be more legitimate to use the latter term while referring to the longer CC bond in the polyyne unit cell. The values computed by us, 1.380 and 1.204 Å, give the bond length difference (BLD) of 0.176 Å. As we mentioned earlier (Sec. 1), high-level correlated calculations (MP2, CCSD, CCSD(T)) for finite-length carbon chains yielded 1.34–1.36 Å for the single bond and 1.20–1.22 Å for the triple bond.[37] The earlier numerical, muffin-tin calculations of Springborg[35] yielded 1.44 Å for the single bond and 1.25 Å for the triple bond in polyyne, while the length of the double bond in cumulene was found to be 1.33 Å. Most of the DFT calculations substantially underestimate BLD, yielding a similar length of the single and triple bond in polyyne. It is interesting, that the pure DFT calculations utterly failed to reproduce BLD correctly, while our SCC-DFTB theory — being an approximation to DFT — is able to provide much better estimate of the experimental BLD values for the polyynes.

## 3.2. *Electronic Structure*

Within the framework of the SCC-DFTB method, only the valence electrons and orbitals are explicitly considered. Therefore, for both cumulene and polyyne we have been able to calculate only the electronic bands originating from the valence $2s$ and $2p$ atomic orbitals of carbons. The cumulene unit cell consists of a single carbon atom and the polyyne unit cell contains two carbon atoms. Consequently, the SCC-DFTB band structure of cumulene consists of four electronic bands (two $\sigma$ bands and a single degenerated $\pi$ band) and the SCC-DFTB band structure of polyyne consists of eight

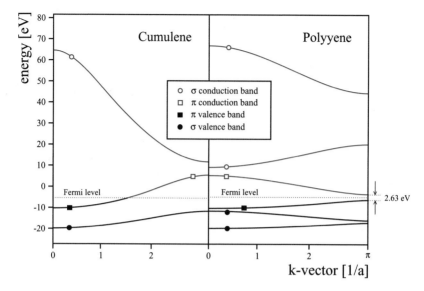

Fig. 3.   Electronic band structure for the infinite cumulene chain (left plot) and the
infinite polyyne chain (right plot) obtained from solid-state SCC-DFTB calculations.

bands (four $\sigma$ bands and two degenerated $\pi$ bands). The electronic band
structures of cumulene and polyyne are shown in Fig. 3. The occupied
electron levels (valence bands) are depicted in black and the virtual levels
(conduction bands), in red. The $\sigma$ bands are marked with a circle, and the
$\pi$ bands with a square. The Fermi energy — defined as the averaged energy
of the highest occupied and the lowest unoccupied levels — is plotted using
a dotted line. It is clear that there exists a close structural similarity between the band structures of cumulene and polyyne. The similarity is most
obvious if one folds the Brillouin zone of cumulene into half and allows the
degenerate bands at the zone boundary to interact. The interaction intro-
duces a coupling of the otherwise degenerate bands that results in a band
gap, which turns the metallic cumulene into a semiconductor (polyyne).
The value of the energy gap obtained in our SCC-DFTB calculations is
2.63 eV, which is in very good agreement with its experimental and theoretical estimates (2.4–2.8 eV) that have been discussed earlier in Sec. 1.

The close structural similarity between the band structures of cumulene and polyyne is very obvious in the computed density of electronic
states (DOS) of both systems. DOSs for cumulene and polyyne are shown
in Fig. 4. Each of the bands of cumulene is split into two parts. It is interesting that the shape of DOS corresponding to the high and low energy limits

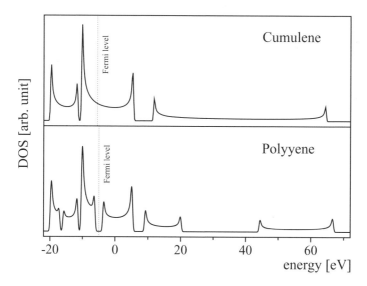

Fig. 4. Density of electronic states (DOS) for the infinite cumulene chain (left plot) and the infinite polyyne chain (right plot) obtained from solid-state SCC-DFTB calculations.

of the two lower bands is almost intact under the Peierls distortion. Only in the middle section of each cumulene band does an energy gap appear, flanked by local maxima of the electronic DOS.

### 3.3. *Phonon Structure*

The phonon dispersion relation for polyyne is shown in Fig. 5. The unit cell of polyyne contains two atoms. Therefore, there exist only six distinct different vibrations of these two atoms for each value of the quasimomentum $k$. Owing to a high symmetry of the system, these six vibrations produce only four phonon branches: transversal optical (TO), transversal acoustic (TA), longitudinal optical (LO), and longitudinal acoustic (LA). The transversal phonon branches are doubly degenerate. The phonon frequencies have been calculated using a standard supercell approach.[69] We have replaced the usual unit cell of polyyne, which contains just two atoms, by a supercell containing 42 atoms. Such a supercell can be viewed as the original two-atom unit cell enveloped by its ten translational copies on each side. We have computed Hessian for the supercell as a matrix of dimension $126 \times 126$ obtained by numerical differentiation of analytical SCC-DFTB forces computed with the DFTB+ code. It has been confirmed

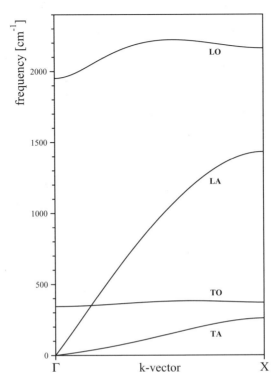

Fig. 5. Phonon dispersion relations for the infinite polyyne chain obtained from the solid-state SCC-DFTB calculations. The following abbreviations are used: TA is the transversal acoustic phonon, TO is the transversal optical phonon, LA is the longitudinal acoustic phonon, and LO is the longitudinal optical phonon.

that using ten enveloping unit cells is sufficient; the Hessian matrix elements for cells located further than nine neighbors apart have been found to be equal to zero. The $k$-dependent $6 \times 6$ dynamical matrix has been constructed by Fourier transformation of a single superrow of the Hessian. Subsequently, the diagonalization of the mass-weighted dynamical matrix yielded the squares of the phonon frequencies for each value of $k$.

The phonon dispersion relation for cumulene, shown in Fig. 6, was obtained in a similar fashion. Since the unit cell of cumulene contains only a single atom, the number of independent vibrations per unit cell is three for each value of $k$. The corresponding number of phonon branches is two: transversal acoustic (TA) and longitudinal acoustic (LA). The transversal branch is doubly degenerate. In contrast to polyyne, we have found that the convergence of the supercell Hessian matrix with respect to the num-

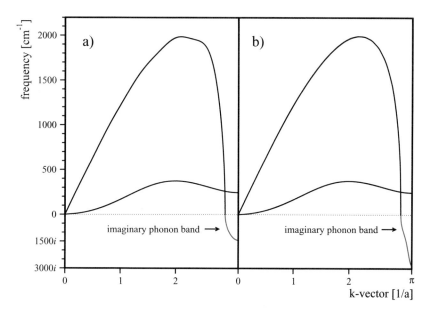

Fig. 6. Phonon dispersion relations for the infinite cumulene chain obtained from the solid-state SCC-DFTB calculations using a supercell containing 21 carbon atoms (a) and a supercell containing 101 carbon atoms (b). The instability of the infinite cumulene chain related to the Peierls distortion can be observed for the longitudinal phonon around $k = \pi/a$.

ber of enveloping unit cells is very slow, which agrees with earlier findings of Tommasini *et al.*[33] and Milani *et al.*[49,50] In Fig. 6 we have shown the phonon dispersion relations obtained with 10 enveloping unit cells on each side (a) and with 50 enveloping unit cells on each side (b). None of these supercells is found to be sufficiently large to ensure vanishing of the Hessian element for distant unit cells, which shows that electron-phonon coupling in cumulene is indeed very pronounced. The magnitude of the largest neglected Hessian element is $10^{-3}$ for (a) and $10^{-5}$ for (b). We have decided to show both of the computed dispersion relation to visualize the magnitude of changes introduced by the neglected terms in the Hessian. As can be seen from Fig. 6, the neglected Hessian elements modify only the shape of the longitudinal phonon branch. The changes are most distinct around $k = \pi/a$. In general, the behavior of the longitudinal phonon dispersion curve is very interesting. It starts as an acoustic phonon (i.e., it has frequency of 0 cm$^{-1}$ at the $\Gamma$ point), it rises rapidly to 2000 cm$^{-1}$ in the middle of the Brillouin zone, and then falls even more rapidly to zero.

The slope of the phonon curve in this region has an almost vertical tangent, which can be interpreted as a very strong electron-phonon coupling known as the Kohn anomaly.[54] At $k = 0.934\pi/a$, the frequency of the longitudinal phonon becomes imaginary, which shows that the cumulene unit cell does not correspond to a minimum but rather to a saddle point on a generalized potential energy surface. Despite of the fact that we are not able to reproduce accurately the exact shape of the longitudinal phonon in the imaginary region due to the lack of convergence in the supercell Hessian approach, it is clear that an analogous effect would be also present in converged calculations. A geometry distortion associated with the normal vector corresponding to the longitudinal phonon at $k = \pi/a$ leads to a doubling of the unit cell and directly designates the polyyne as a lower energy structure. The observed imaginary longitudinal phonon branch is a spectacular manifestation of the Peierls distortion in direct quantum chemical calculations.

It is interesting to compare the phonon dispersion relations of cumulene and polyyne (Figs. 6 and 5). The acoustic phonon branches of polyyne are similar to the corresponding segments of the cumulene's phonon dispersion relations between 0 and $\pi/2a$. Also the shape of the optical transversal phonon is similar to the segment of the cumulene's transversal acoustic phonon between $\pi/2a$ and $\pi/a$. It is only the longitudinal optical phonon branch of polyyne that is distinctly different from the corresponding segment of the phonon dispersion relations of cumulene. Another interesting feature, which was already signalized earlier in Sec. 1, is the LO phonon softening around the $\Gamma$ point. The specific shape of the LO phonon branch in this region can be considered as a "memory effect" of the Kohn anomaly[54] present in the Brillouin zone of cumulene around the $k = \pi/a$ point. In the "folded" Brillouin zone of polyyne, the remaining of this anomaly can be observed as characteristic lowering of the LO phonon frequency.

## 4. Finite Approach — Will Finite Become Infinite?

It is obvious that none of the existing molecular systems is really infinite. Strikingly good performance of periodic models used in solid-state physics shows that the physical properties of systems containing a very large number of atoms or molecules converge toward common limits. It is of ultimate importance to ask how fast this convergence is realized in practice. It is also very important to establish a correspondence between the language of solid-state physics usually employed to describe infinite structures and the terminology of quantum chemistry that is most often used to discuss various molecular properties of finite systems. In the following sections of this study we attempt to establish such a correspondence between the infinite and finite description of the geometric, electronic, and vibrational properties of linear carbon chains. We also try to find the limiting length of the finite carbon chains necessary to observe properties characteristic for the infinite regime. We believe that the presented results are very instructive.

### 4.1. Geometric Structure of Finite Chains

In the infinite polyyne chain there exist only two types of bonds that can be classified in standard organic chemistry terminology as single and triple. According to our solid-state SCC-DFTB calculations, the length of the triple bond is 1.204 Å and the length of the single bond is 1.380 Å. These interatomic separations show that in fact the single CC bond in the infinite polyyne is very short and corresponds rather to an aromatic carbon-carbon bond. In the infinite cumulene chain, there exists only a single type of bond (1.277 Å), which can be classified according to the standard organic chemistry rules as a double bond. It would be very interesting to ask if these bond patterns are changed while the infinite polyyne and cumulene chains are replaced by their finite analogues. The finite analogue of cumulene would be a chain of carbons connected by consecutive double bond terminated on both sides by a single $CH_2$ unit. The shortest possible finite chains obtained in this way would be ethylene, allene, and butatriene. It is easy to generalize this sequence to longer chains. We show this generalization in Fig. 2. In ethylene and butatriene, the terminal $CH_2$ groups are located in the same plane, while in allene, the $CH_2$ planes of the terminal groups are perpendicular. For every finite cumulene-like chain, the relative orientation of these planes is parallel for the molecules with an even number of carbon atoms (point group symmetry $D_{2h}$) and perpendicular for the molecules with an odd number of carbon atoms (point group symmetry $D_{2d}$). A potential problem occurs for very long chains, where the terminal groups are very distant. In principle, the two very long finite cumulene-like chains differing by a single carbon atom should have different point group symmetry. On the other hand, the large spatial separation of the terminal groups should prevent any interactions and enable their almost free mutual rotation. We see later that this issue is resolved in a natural — but rather surprising — way. The finite analogue of polyyne would be a chain of carbon atoms connected by the alternated sequence of single and triple bonds. Again, a schematic representation of this sequence is given in Fig. 2.

To investigate the finite-size effect of the distribution of the carbon-carbon bonds in the polyyne and cumulene chains, we have performed geometry optimization using the SCC-DFTB method for finite polyyne and cumulene chains containing up to 300 carbon atoms. The distribution of equilibrium bond lengths in both series of molecules is shown in Fig. 7. To facilitate comparisons, we have normalized the distributions with respect to the number of carbon atoms present in each analyzed structure. Two interesting features can be observed: (i) long finite cumulenic chains lose their cumulene-like character and show a transition toward the polyyne chain, (ii) the interior of long, finite cumulene and polyyne chains shows a high degree of uniformity characteristic of the infinite polyyne chain. We discuss these two observations in detail below.

The finite cumulene chains with more than 30 carbon atoms are found to be almost entirely polyynic in their character. The cumulene bonding pattern, enforced by the $sp^2$ hybridization of the carbon atom in the terminal $CH_2$ groups, is observed only at the ends of each finite chain. The whole interior of a finite cumulene chain is polyynic. This finding is a very clear illustration of the Peierls distortion observed at the molecular level. In some sense, the Peierls distortion can be seen as a variation of the Jahn-Teller distortion, which lowers the local symmetry and stabilizes the total energy of the molecule. In a solid it is the translational symmetry that is lowered by the Peierls distortion. The translational symmetry is missing in finite chains, but an analogous effect can still be observed. This allows us to state that even relatively short linear chains locally experience translational symmetry and can be treated to a good approximation as infinite systems. It is interesting to discuss in detail the distribution of bond lengths in long, finite cumulene chains. The majority of bonds have identical length to these observed in polyyne. Only the bonds that are close to the chain boundaries have different lengths. Remarkably, the bonds with length of approximately 1.3 Å, which dominate in short finite cumulene chains are completely missing in long chains. A comparison of the bond alternating patterns for the equilibrium SCC-DFTB geometries of finite polyyne and cumulene molecules is shown in Fig. 8. This plot corresponds to finite polyyne and cumulene chains containing 150 carbon atoms. It is quite surprising how fast the finite cumulene chains change their character and become identical to the corresponding polyyne chains.

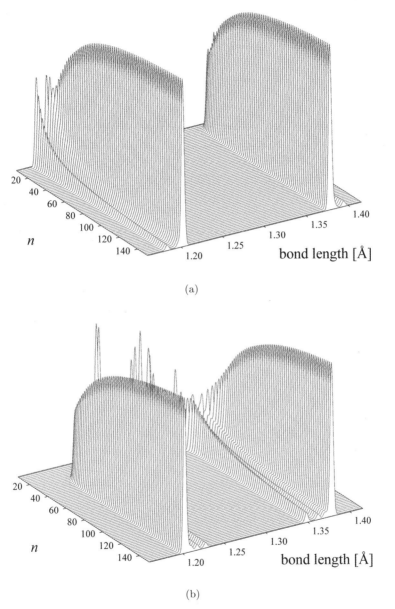

(a)

(b)

Fig. 7. Distribution of the CC bond lengths for the finite-length polyyne chains $C_nH_2$ (a) and for the finite-length cumulene chains $C_nH_4$ (b) as a function of the number of carbon atoms $n$.

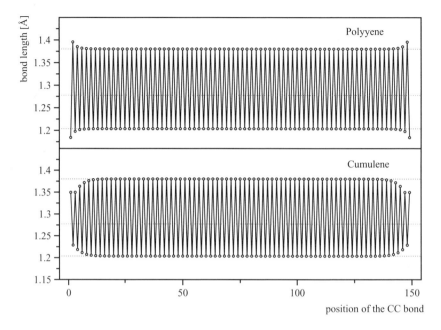

Fig. 8.   Bond length alternation pattern for the finite-length polyyne $C_{150}H_2$ (upper plot) and the finite-length cumulene $C_{150}H_4$ (lower plot). The bond lengths corresponding to the single and triple bonds in the infinite polyyne chain (thin black dotted line) and the bond length corresponding to the double bond in the infinite cumulene chain (thin red dotted line) are shown for comparison.

   Another striking feature apparent from the presented bond length distributions and the bond length alternation pattern is the similarity of the central regions of the finite cumulene and polyyne chains to the structure of the infinite polyyne chain. This regularity shows that the finite-border effect is visible only in the close vicinity of the terminal groups. This is quite a surprising observation. It is a common belief in organic chemistry that the geometric and electronic structure of a conjugated chain depends strongly on its length and the chemical character of the terminal groups. A slight alternation in the conjugation pattern, like in the case of the 11-*cis-trans* isomerisation of retinal — the molecule responsible for the vision process — can bring about enormous changes in the physical properties of the system. Similarly, one can expect that terminating a sequence of conjugated $\pi$ bonds by two different groups ($CH_2$ and CH, in our case) may introduce considerable changes to the geometric and electronic structure of the system. Our SCC-DFTB calculations show that this effect is pronounced only at the ends of finite chains, while the interior is almost completely insensitive to it.

## 4.2. *Electronic Structure of Finite Chains*

In the previous Section, we analyze the convergence of structural parameters for linear carbon chains of finite length. We have found that the equilibrium geometry of finite systems (i.e., $C_nH_4$ and $C_nH_2$) converges to the structure of infinite polyyne already for $n > 30$. It is tempting to investigate if a similar effect can be observed also for the electronic structure of finite-length chain. For this purpose, we construct the electronic densities of states (DOS) for $C_nH_4$ and $C_nH_2$ ($n = 2$–150) and compare them with the DOS of polyyne obtained from the SCC-DFTB solid-state calculations (see Fig. 4). The results are presented in Fig. 9(a) for polyyne chains and in Fig. 9(b) for cumulene chains. The solid-state DOS of polyyne is depicted in Fig. 9 using a thick solid line. The DOS plots for the finite systems are constructed as a superposition of Gaussians (half-width of $10^{-4}$ hartree) that correspond to the discrete molecular orbital energy levels computed using SCC-DFTB. The DOS plots for the solid-state calculations have been obtained by $k$-integration of the electronic energy bands shown in Fig. 3. It is clear that like in the case of structural parameters, the convergence of the shape of DOS with respect to $n$ is obtained very rapidly. In practice, the finite-length cumulene and polyyne chains containing more than 60 carbon atoms display a converged DOS almost identical to that obtained from solid-state SCC-DFTB calculations for polyyne. As expected, the convergence history for $C_nH_4$ and $C_nH_2$ is somewhat different, especially for shorter chains. Again, both sequences of the finite-length chain converge to the same limit, which shows that the $CH_2$ terminal groups have very little influence on the electronic properties of cumulene chains. A slight evidence of the terminal groups can be noticed from the two additional small bands appearing in Fig. 9(a) at approximately 0.4 and 0.8 hartree. We believe that for longer cumulene chains these two peaks diminish in intensity and in the limit of an infinite number of carbon atoms would eventually vanish.

## 4.3. *Vibrational Structure of Finite Chains*

We have shown that for finite-length cumulene and polyyne chains, the geometric and electronic properties converge to the analogous properties of an infinite polyyne chain. In this Section we show that similar effect is observed for the vibrational structure of polyyne and cumulene chains of

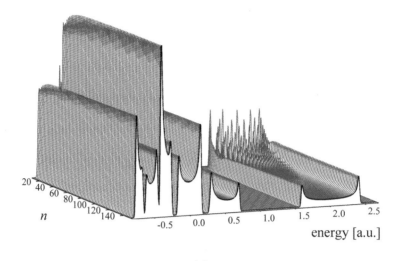

(a)

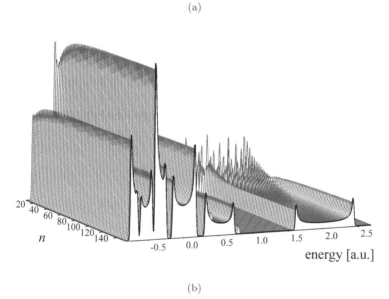

(b)

Fig. 9. Density of electronic states (DOS) for the finite-length polyyne chains $C_nH_2$ (a) and for the finite-length cumulene chains $C_nH_4$ (b) as a function of the number of carbon atoms $n$. Thick solid line gives the DOS obtained from solid-state SCC-DFTB calculations for polyyne.

finite length. The theoretical tool used to analyze the convergence of vibrational properties of finite-length chains is the vibrational density of states (VDOS). In standard solid-state calculations VDOS is defined as an integral of the phonon dispersion relation over the $k$-space. Here, we simulate the VDOS plots for finite-length molecules using the discrete vibrational levels of finite cumulene and polyyne chains computed with the SCC-DFTB method. Each VDOS is constructed as a superposition of Gaussian peaks with a half-width of 30 cm$^{-1}$; the positions of peaks are determined by the values of harmonic vibrational frequencies of the $C_nH_4$ and $C_nH_2$ molecules. The shape of the resultant VDOS plots is shown in Fig. 10 as a function of the number of carbon atoms in the chain ($n$). The converged VDOS plots are compared with the solid-state VDOS of polyyne obtained by integrating the phonon branches shown in Fig. 5.

The VDOSs corresponding to the longest analyzed finite-length cumulene and polyyne chains ($C_{150}H_4$ and $C_{150}H_2$, respectively) agree very well with the solid-state VDOS of polyyne. It shows that the vibrational structures of the finite and infinite chains are similar. In the VDOS plots for the finite-length chains a few additional bands can be observed that correspond to vibrations of the terminal groups. The additional bands in the VDOS of finite-length cumulene chains are located at 583, 992, and 1840 cm$^{-1}$. All of these modes correspond to the vibrations of the terminal $CH_2$ groups and can be characterized as the $CH_2$ wagging, rocking, and scissoring, respectively. The scissoring vibrational mode is partially coupled to the stretching of the double $C_2C_3$ bond, where $C_1$ denotes the carbon atom of the $CH_2$ group. The additional bands in the VDOS of finite-length polyyne chains are located at 424, 566, and 2031 cm$^{-1}$. Again, all these modes correspond to the vibrations of the terminal atoms and can be characterized as the $C-C\equiv C-H$ bending, the $C-H$ bending, and the stretching of the terminal $C\equiv C$ bonds, respectively. Here, we refer to the term "bending" as simultaneous out-of-phase perpendicular displacement of a group of atoms from the main axis of the molecule. We believe that for a long molecular chain the relative intensity of the additional bands diminishes and in the limit of an infinite number of carbon atoms these bands would disappear entirely. To verify this hypothesis, we present in Fig. 11 the VDOS of finite-length $C_nH_2$ polyynes computed up to $n = 300$. It is evident that for the longest chains the vibrational density of states in the region of the additional bands observed in Fig. 10(a) is already largely reduced, which confirms our analysis.

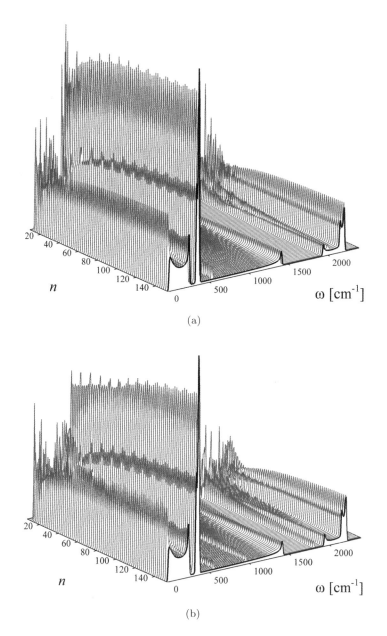

Fig. 10.    Density of vibrational states (VDOS) for the finite-length polyyne chains $C_nH_2$ (a) and for the finite-length cumulene chains $C_nH_4$ (b) as a function of the number of carbon atoms $n$. Thick solid line gives the VDOS obtained from solid-state SCC-DFTB calculations for polyyne.

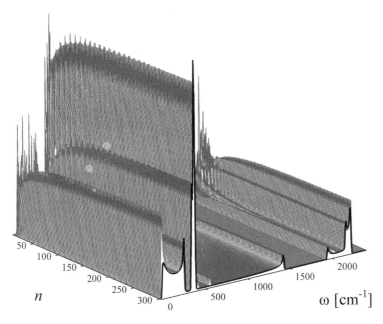

Fig. 11. Density of vibrational states (VDOS) for the finite-length polyyne chains $C_n H_2$ up to $n = 300$. Thick solid line gives the VDOS obtained from solid-state SCC-DFTB calculations for polyyne.

The presented results show that in analogy to the structural and electronic properties, the vibrational structures of the finite-length cumulene and polyyne chains also converge rapidly to the vibrational structure of the infinite polyyne chain. Already for chains containing 80 or more atoms, the vibrational density of states closely resembles the VDOS obtained from the solid-state calculations for polyyne. Again, both sequences of the finite-length chain converge to the same limit. However, there exist two issues showing a difference between the finite and infinite VDOSs. As we have discussed above, the first one concerns the presence of small additional bands in the VDOSs of the finite-length cumulene and polyyne chains. The additional bands are associated with various vibrations of the terminal CH and $CH_2$ groups. The second issue is more subtle and concerns the behavior of the vibrational densities of states around 0 $cm^{-1}$. The shape of VDOS in this region is determined by the frequencies of acoustic phonons, or, more precisely, by the $k$-dependence of the transversal acoustic phonon frequency in the vicinity of the $\Gamma$ point. As we have previously discussed (Sec. 1), the frequencies corresponding to the transversal acoustic phonon

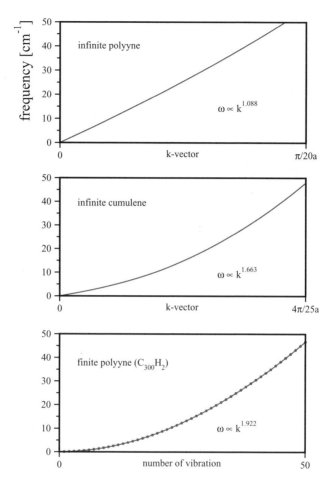

Fig. 12.   The behavior of the transversal acoustic (TA) phonon around the $\Gamma$ point for (a) an infinite polyyne chain, (b) an infinite cumulene chain, and (c) a finite-length polyyne chain represented by the $C_{300}H_2$ molecule. For the infinite chain, the dependence of the TA phonon's frequency on the wave vector is linear (polyyne) or superlinear (cumulene). For the finite-length polyyne, the dependence is nearly quadratic.

of polyyne should display nearly quadratic dependence on the wave vector for small values of $k$.[30]  A somewhat similar trend can be found in our solid-state SCC-DFTB calculations of the cumulene chain, where we get $\omega \propto k^{1.663}$ (see Fig. 12). However, our solid-state SCC-DFTB calculations for polyyne do not confirm this regularity giving the frequencies of low-energy transversal acoustic phonons proportional to $k^{1.088}$. A nearly quadratic dependence can be observed for the finite-length polyyne chains;

fitting the 50 lowest doubly degenerated frequencies of $C_{300}H_2$ as a function of the number of a harmonic $k$ yields the relation $\omega \propto k^{1.922}$ (also shown in Fig. 12). A very similar result was obtained earlier[34] on the base of HF, MP2, and DFT/B3LYP calculations. An analysis of these results showed that the quadratic dependence of $\omega$ on the wave vector can be explained on the grounds of the classical theory of a vibrating string.[34] In this context the nearly linear dependence of the phonon frequency on $k$ for the infinite polyyne chains is rather surprising. The difference may be attributed to the loose ends present in the finite-length polyynes or it may originate from the shortcomings of the SCC DFTB method. This issue is currently under investigation in our laboratory and we expect to give a conclusive explanation of this phenomenon in the future.

## 5. Vibrational Spectra

The analysis given in the last few sections shows that the structural, electronic, and vibrational properties of an infinite-length polyyne chain can be simulated using a series of finite-size $C_nH_2$ molecules. The presented results suggest that the convergence of various physical properties of finite-length carbon chains may be achieved for relatively small values of $n$. Stimulated by this finding, we apply in this section a similar strategy for modeling vibrational infrared (IR) and Raman spectra of linear carbon chains. It would certainly be beneficial to compare the converged IR and Raman spectra of very long, finite polyynes with analogous spectra computed for an infinite polyyne chain. Unfortunately, computing intensities in the IR and Raman spectra of periodic systems — associated here with differentiating the SCC-DFTB energy with respect to the external electric field — is not yet a well established procedure.[70,71] Therefore, no IR and Raman spectra of infinite chains are available within the SCC-DFTB framework yet. We present here the vibrational spectra computed for a series of finite-length polyynes, believing that the convergence of properties discussed in the previous sections justifies this approach. The positions of the bands in vibrational spectra depend predominantly on the equilibrium geometry and harmonic vibrational frequencies of a molecule, while the intensities are mainly associated with its electronic structure. We have shown that all three types of features computed for the finite-length polyyne chains converge rapidly to the corresponding properties of the infinite chain. In addition, the fast convergence of the presented vibrational spectra with $n$ (see Fig. 13) gives further justification of this approach.

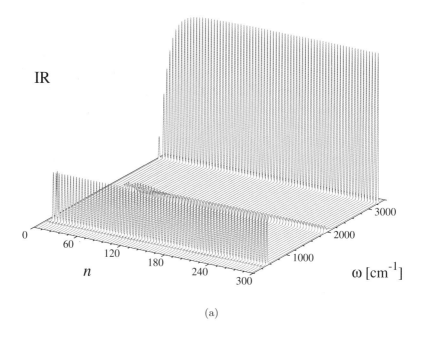

(a)

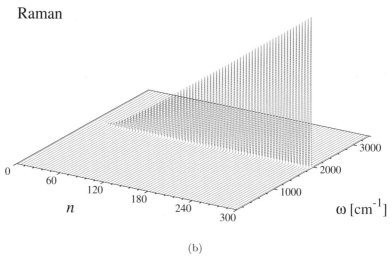

(b)

Fig. 13.   Vibrational IR (a) and Raman (b) spectra of the finite-length polyynes $C_nH_2$ as a function of the number of carbon atoms ($n$). Intensities are given in arbitrary units.

The series of IR spectra computed using SCC-DFTB for the finite-length $C_nH_2$ polyyne molecules is shown in Fig. 13(a). The analogous Raman spectra are given in Fig. 13(b). Both families of spectra have a simple structure. The convergence of the IR and Raman spectra can be observed for $n > 100$. The convergence has a different character for both vibrational spectra. For the IR ones, the intensity of the five distinct bands becomes constant, while for the Raman spectra, the intensity of the only observed band scales linearly for $n > 100$. Since the intensity in both plots corresponds to a single molecule, the physical consequences of this findings are quite different. Let us assume that the experimental samples, for which the IR or Raman spectrum is recorded, have a constant volume $V$. For polyynes, the volume of a single finite chain is proportional to the length of the system and can be written as $v \cdot n$. Assuming that the IR or Raman signal originating from each particle in the sample is identical, we see that the total IR or Raman intensity is proportional to the number of particles in the sample, which is inversely proportional to the number of carbon atoms in the chain. As we mentioned earlier, the IR signal from a single particle is constant, while the Raman signal scales as $n$. Therefore, we find that the total IR signal for a sample of $C_nH_2$ of volume $V$ scales as $1/n$ for large $n$. Similarly, the total Raman signal for a sample of $C_nH_2$ of volume $V$ scales as a constant with $n$. Extrapolating these results to $n = \infty$ (i.e., to an infinite polyyne chain), we see that an infinite polyyne would not be active in IR. Similar extrapolation shows that the Raman spectrum of an infinite polyyne would display a single narrow band (the so called $\alpha$ band) corresponding to the frequency of the longitudinal optical phonon at the $\Gamma$ point. Since in reality, experimental samples of polymers seldom contain perfect infinite chains, it is quite possible that the polymer polyyne would exhibit a similar IR spectrum to that shown in Fig. 13(a), except for the bands corresponding to vibrations involving hydrogen atom. A detailed discussion of features present in the computed IR and Raman spectra is given below.

The computed IR spectrum of the finite-length $C_{300}H_2$ polyyne molecule display five bands located at 424, 566, 1970, 2041, and 3254 cm$^{-1}$. The two most intensive bands correspond to the vibrations of the CH group: stretching at 3254 cm$^{-1}$ and bending at 566 cm$^{-1}$. Clearly these two bands would be missing in the IR spectrum of the polyyne chain not terminated with hydrogen atoms. Similar effect would be observed for the small satellite band located at 424 cm$^{-1}$, which originates from the terminal $C-C{\equiv}C-H$ bending. The band at 1970 cm$^{-1}$ is associated with a stretching vibration of the CC bond. It closely corresponds to the LO mode of solid-state polyyne at the Γ point (1951 cm$^{-1}$). This seems to be somewhat surprising, since the LO phonon has the *gerade* symmetry and should not be active in IR. The apparent paradox vanishes if one inspects the normal mode vector associated with this band. The whole long chain is divided into two domains with a border located in the center of the chain. Within each domain, the adjacent C≡C bonds vibrate in-phase, but the mutual vibration of both domains is out-of-phase yielding the effective *ungerade* symmetry. The small satellite band at 2041 cm$^{-1}$ has a similar origin with a more pronounced contribution from the terminal C≡C acetylenic units. Note that for the long polyyne chains not terminated by hydrogens, only the bands located around 2000 cm$^{-1}$ would be present in the IR spectrum, giving a similar overall shape to the Raman spectrum presented in Fig. 13(b).

The computed Raman spectrum of the finite-length $C_{300}H_2$ polyyne molecule has a very simple structure. It displays only a single band (so-called $\alpha$ band) located at 1951 cm$^{-1}$. A more detailed inspection of Fig. 13(b) shows a trace of an additional, very weak band located are 344 cm$^{-1}$. Both bands correspond to the optical phonon branches (LO and TO, respectively) of the infinite polyyne chain at the Γ point. The relative intensity of the $\alpha$ band grows linearly with the length of the chain. This finding agrees well with previous experimental observations for shorter chains.[23,24]

The quality of the simulated SCC-DFTB vibrational spectra of polyynes can be assessed by a comparison of the theoretical results to experiment. Unfortunately, no experimental Raman spectra are available for long polyyne chains. Raman spectra of relatively short, hydrogen-capped chains ($C_8H_2$–$C_{16}H_2$) were recorded in *n*-hexane.[23] Each spectrum consists of a main strong band and a few weak bands located in the region of 2000–2200 cm$^{-1}$. As the polyyne chain length increases, the strongest band shifts to lower frequencies. Our SCC-DFTB calculations are capable of reproducing the recorded Raman spectra with good accuracy. The number and positions of the bands agree well with the experiment. The largest observed deviation between the theoretical and experimental frequencies is only 37 cm$^{-1}$. Also the trend of frequency lowering for the most intensive Raman band with the growing size of the chain is correctly reproduced. The most serious discrepancy concerns the relative intensity of the satellite bands, which do not always match the experimental pattern. However, the intensities are not directly comparable as the experimental spectra were recorded in solution, while the theoretical simulations have been performed in the gas phase. Good performance of the SCC-DFTB method for the short polyyne chains lends credibility to the calculated vibrational spectra of long chains.

## 6. Conclusions

A detailed survey of available theoretical and experimental results on geometric, electronic, vibrational, and spectroscopic properties of linear carbon chains is presented. The experimental investigations mainly concern various properties (optical energy gap, vibrational frequencies, and spectroscopic properties in various environments) of finite chains containing up to a few dozens of carbon atoms. The theoretical efforts are focused on two main issues: 1) reconstructing and interpretation of the experimental results for short, finite-length chains and 2) constructing theoretical models capable of modeling physical properties of infinite chains. The presented survey is supplemented with extensive quantum calculations on the geometric, electronic, vibrational, and spectroscopic properties of finite and infinite chains. Two explicit models of the chains are considered: cumulenic with equidistant bond pattern and polyynic with alternated bond pattern. The models are studied using the SCC-DFTB method. The calculations for the infinite cumulene and polyyne chains are based on standard solid state physics techniques, while the calculations for the finite chains use molecular orbital (MO) based quantum chemical methods. The correspondence between both theoretical descriptions is established. The presented results show that the physical properties (equilibrium geometry, electronic and vibrational densities of states, and the IR and Raman spectra) of relatively short ($n > 100$) finite-length polyyne and cumulene chains ($C_nH_2$ and $C_nH_4$, respectively) are identical to those computed with the solid-state methodology. The instability of the cumulene chain is explicitly demonstrated at the molecular level, both in the solid-state and MO based calculations. The energy gap in the infinite polyyne chain is found to be 2.63 eV. The equilibrium bond length in cumulene is found to be 1.277 Å and those in polyyne, 1.204 and 1.380 Å. The band intensities in the IR spectra of long polyyne chains are found to decay as $1/n$ with the growing length of the chain, while the band intensities in the analogous Raman spectra are demonstrated to converge to a constant. Therefore, the IR spectrum of an infinite polyyne chain is expected to contain no intensive bands with residual IR intensity around 1970 cm$^{-1}$ corresponding to the signal from finite-length chains. The Raman spectrum of an infinite polyyne chain is expected to contain only a

single band ($\alpha$ band) around 1951 cm$^{-1}$ corresponding to the frequency of
the longitudinal optical phonon at the $\Gamma$ point.

## References

1. R. E. Peierls, *Quantum Theory of Solids.* (Clarendon, Oxford, 1955).
2. C. Glaser, Untersuchungen ueber einige derivate der zimmtsaeure, *Ann. Chem. Pharm.* **154**, 137, (1870).
3. M. B. Bell, P. A. Feldmann, S. Kwok, and H. E. Matthews, Detection of HC$_{11}$N in IRC + 10°216, *Nature.* **295**, 389, (1982).
4. H. W. Kroto, Buckminsterfullerene, the celestial sphere that fell to earth, *Angew. Chem.* **31**, 111, (1992).
5. M. T. W. Hearn, E. R. H. Jones, M. G. Pellatt, V. Thaller, and J. L. Turner, Natural acetylenes. Part XLII. Novel C$_7$, C$_8$, C$_9$, and C$_{10}$ polyacetylenes from fungal cultures, *J. Chem. Soc., Perkin Trans. 1.* p. 2785, (1973).
6. S. C. Shim and T. S. Lee, Photocycloaddition reaction of some conjugated hexatriynes with 2,3-dimethyl-2-butene, *J. Org. Chem.* **53**, 2410, (1988).
7. A. M. Sladkov and Y. P. Kudryavtsev, Polyynes, *Russian Chem. Rev.* **32**, 229, (1963).
8. E. R. H. Jones, H. H. Lee, and M. C. Whiting, Researches on acetylenic compounds. Part LXIV. The preparation of conjugated octa- and deca-acetylenic compounds, *J. Chem. Soc.* p. 3483, (1960).
9. T. R. Johnson and D. R. M. Walton, Silylation as a protective method in acetylene chemistry : Polyyne chain extensions using the reagents, Et$_3$Si(C≡C)$_m$ H ($m = 1, 2, 4$) in mixed oxidative couplings, *Tetrahedron.* **28**, 522, (1972).
10. R. J. Lagow, J. J. Kampa, H.-C. Wei, S. L. Battle, J. W. Genge, D. A. Laude, C. J. Harper, R. Bau, R. C. Stevens, J. F. Haw, and E. Munson, Synthesis of linear acetylenic carbon: The "sp" carbon allotrope, *Science.* **267**, 362, (1995).
11. T. Pino, H. Ding, F. Güthe, and J. P. Maier, Electronic spectra of the chains HC$_{2n}$H ($n = 8 - 13$) in the gas phase, *J. Chem. Phys.* **114**, 2208, (2000).
12. R. Dembinski, T. Bartik, B. Bartik, M. Jaeger, and J. A. Gladysz, Toward metal-capped one-dimensional carbon allotropes: Wirelike C$_6$–C$_{20}$ that span two redox-active ($\eta^5$-C$_5$Me$_5$)Re(NO)(PPh$_3$) endgroups, *J. Am. Chem. Soc.* **122**, 810, (2000).
13. M. Tsuji, S. Kuboyama, T. Matsuzaki, and T. Tsuji, Formation of hydrogen-capped polyynes by laser ablation of C$_{60}$ particles suspended in solution, *Carbon.* **41**, 2141, (2003).
14. S. Szafert and J. A. Gladysz, Carbon in one dimension: Structural analysis of the higher conjugated polyynes, *Chem. Rev.* **103**, 4175, (2003).

15. Q. Zheng and J. A. Gladysz, A synthetic breakthrough into an unanticipated stability regime: Readily isolable complexes in which $C_{16}$–$C_{28}$ polyynediyl chains span two platinum atoms, *J. Am. Chem. Soc.* **127**, 10508, (2005).

16. S. Eisler, A. Slepkov, E. Elliott, T. Luu, R. McDonald, F. A. Hegmann, and R. R. Tykwinski, Polyynes as a model for carbyne: Synthesis, physical properties, and nonlinear optical response, *J. Am. Chem. Soc.* **127**, 2666, (2005).

17. R. Matsutani, T. Kakimoto, K. Wada, T. Sanada, H. Tanaka, and K. Kojima, Preparation of long-chain polyynes $C_{18}H_2$ and $C_{20}H_2$ by laser ablation of pellets of graphite and perylene derivative in liquid phase, *Carbon.* **46**, 1091, (2008).

18. M. Grutter, M. Wyss, J. Fulara, and J. P. Maier, Electronic absorption spectra of the polyacetylene chains $HC_{2n}H$, $HC_{2n}H^-$, and $HC_{2n-1}N^-$ ($n = 6 - 12$) in neon matrices, *J. Phys. Chem. A.* **102**, 9785, (1998).

19. T. Wakabayashi, H. Nagayama, K. Daigoku, Y. Kiyooka, and K. Hashimoto, Laser induced emission spectra of polyyne molecules $C_{2n}H_2$ ($n = 5 - 8$), *Chem. Phys. Lett.* **446**, 65, (2007).

20. X. Zhao, Y. Ando, Y. Liu, M. Jinno, and T. Suzuki, Carbon nanowire made of a long linear carbon chain inserted inside a multiwalled carbon nanotube, *Phys. Rev. Lett.* **90**, (2003).

21. L. M. Malard, D. Nishide, L. G. Dias, R. B. Capaz, A. P. Gomes, A. Jorio, C. A. Achete, R. Saito, Y. Achiba, H. Shinohara, and M. A. Pimenta, Resonance raman study of polyynes encapsulated in single-wall carbon nanotubes, *Phys. Rev. B.* **76**, (2007).

22. A. Rusznyak, V. Zolyomi, J. Kürti, S. Yang, and M. Kertesz, Bond-length alternation and charge transfer in a linear carbon chain encapsulated within a single-walled carbon nanotube, *Phys. Rev. B.* **72**, (2005).

23. H. Tabata, M. Fujii, S. Hayashi, T. Doi, and T. Wakabayashi, Raman and surface-enhanced Raman scattering of a series of size-separated polyynes, *Carbon.* **44**, (2006).

24. T. Wakabayashi, H. Tabata, T. Doi, H. Nagayama, K. Okuda, R. Umeda, I. Hisaki, M. Sonoda, Y. Tobe, T. Minematsu, K. Hashimoto, and S. Hayashi, Resonance Raman spectra of polyyne molecules $C_{10}H_2$ and $C_{12}H_2$ in solution, *Chem. Phys. Lett.* **433**, (2007).

25. A. Lucotti, M. Tommasini, M. Del Zoppo, C. Castiglioni, G. Zerbi, F. Cataldo, C. S. Casari, A. Li Bassi, V. Russo, M. Bogana, and C. E. Bottani, Raman and SERS investigation of isolated sp carbon chains, *Chem. Phys. Lett.* **417**, (2006).

26. H. Tabata, M. Fujii, and S. Hayashi, Surface-enhanced Raman scattering from polyyne solutions, *Chem. Phys. Lett.* **420**, (2006).

27. M. Kertesz, J. Koller, and A. Azman, Ab initio Hartree-Fock crystal orbital studies. II. Energy bands of an infinite carbon chain, *J. Chem. Phys.* **68**,

(1978).

28. M. Kertesz, J. Koller, and A. Azman, Different orbitals for different spins for solids: Fully variational ab initio studies on hydrogen and carbon atomic chains, polyene, and poly(sulphur nitride), *Phys. Rev. B.* **19**, (1979).

29. A. Karpfen, Ab initio calculation on model chains, *Theoret. Chim. Acta.* **50**, (1978).

30. A. Karpfen, Ab initio studies on polymers. I. The linear infinite polyyne, *J. Phys. C.* **12**, (1979).

31. L. Brillouin, *Wave propagation in Periodic Structures.* (Dover, New York, 1953).

32. E. Cazzanelli, M. Castriota, L. Caputi, A. Cupolillo, C. Giallombardo, and L. Papagno, High-temperature evolution of linear carbon chains inside multiwalled nanotubes, *Phys. Rev. B.* **75**, (2007).

33. M. Tommasini, A. Milani, D. Fazzi, M. D. Zoppo, C. Castiglioni, and G. Zerbi, Modeling phonons of carbon nanowires, *Physica E.* **40**, (2008).

34. C. Seitz and A. L. L. East, Polyyne bending frequencies: why they vary with the square of the harmonic in the infinite limit, *Mol. Phys.* **101**, (2003).

35. M. Springborg, Self-consistent, first principles calculations of the electronic structure of a linear, infinite carbon chain, *J. Phys. C.* **19**, (1986).

36. M. Tommasini, D. Fazzi, A. Milani, M. D. Zoppo, C. Castiglioni, and G. Zerbi, Effective hamiltonian for $\pi$ electrons in linear carbon chains, *Chem. Phys. Lett.* **450**, (2007).

37. A. Abdurahman, A. Shukla, and M. Dolg, Ab initio many-body calculations on infinite carbon and boron-nitrogen chains, *Phys. Rev. B.* **65**, (2002).

38. E. J. Bylaska, R. Kawai, and J. H. Weare, From small to large behavior: The transition from the aromatic to the Peierls regime in carbon rings, *J. Chem. Phys.* **113**, (2000).

39. S. Yang and M. Kertesz, Bond length alternation and energy band gap of polyyne, *J. Phys. Chem. A.* **110**, (2006).

40. A. D. Becke, Density-functional exchange-energy approximation with correct asymptotic behavior, *J. Phys. Chem. A.* **38**, (1988).

41. C. Lee, W. Yang, and R. G. Parr, Development of the Colle-Salvetti correlation-energy formula into a functional of the electron density, *Phys. Rev. B.* **37**, (1988).

42. A. J. Cohen and N. C. Handy, Dynamic correlation, *Mol. Phys.* **99**, (2001).

43. A. D. Becke, Density-functional thermochemistry. III. The role of exact exchange, *J. Chem. Phys.* **98**, (1993).

44. J. P. Perdew, K. Burke, and M. Ernzerhof, Generalized gradient approximation made simple, *Phys. Rev. Lett.* **77**, (1996).

45. M. J. Frisch, G. W. Trucks, H. B. Schlegel, G. E. Scuseria, M. A. Robb, J. R. Cheeseman, J. A. Montgomery, Jr., T. Vreven, K. N. Kudin, J. C. Burant, J. M. Millam, S. S. Iyengar, J. Tomasi, V. Barone, B. Mennucci, M. Cossi,

G. Scalmani, N. Rega, G. A. Petersson, H. Nakatsuji, M. Hada, M. Ehara, K. Toyota, R. Fukuda, J. Hasegawa, M. Ishida, T. Nakajima, Y. Honda, O. Kitao, H. Nakai, M. Klene, X. Li, J. E. Knox, H. P. Hratchian, J. B. Cross, V. Bakken, C. Adamo, J. Jaramillo, R. Gomperts, R. E. Stratmann, O. Yazyev, A. J. Austin, R. Cammi, C. Pomelli, J. W. Ochterski, P. Y. Ayala, K. Morokuma, G. A. Voth, P. Salvador, J. J. Dannenberg, V. G. Zakrzewski, S. Dapprich, A. D. Daniels, M. C. Strain, O. Farkas, D. K. Malick, A. D. Rabuck, K. Raghavachari, J. B. Foresman, J. V. Ortiz, Q. Cui, A. G. Baboul, S. Clifford, J. Cioslowski, B. B. Stefanov, G. Liu, A. Liashenko, P. Piskorz, I. Komaromi, R. L. Martin, D. J. Fox, T. Keith, M. A. Al-Laham, C. Y. Peng, A. Nanayakkara, M. Challacombe, P. M. W. Gill, B. Johnson, W. Chen, M. W. Wong, C. Gonzalez, and J. A. Pople. Gaussian 03, Revision C.02. Gaussian, Inc., Wallingford, CT, 2004.

46. J. K. Kang and C. B. Musgrave, Prediction of transition state barriers and enthalpies of reaction by a new hybrid density-functional approximation, *J. Chem. Phys.* **115**, (2001).

47. S. Yang, M. Kertesz, V. Zolyomi, and J. Kürti, Application of a novel linear/exponential hybrid force field scaling scheme to the longitudinal Raman active mode of polyyne, *J. Phys. Chem. A.* **111**, (2007).

48. C. Castiglioni, M. Tommasini, and G. Zerbi, Raman spectroscopy of polyconjugated molecules and materials: confinement effect in one and two dimensions, *Philo. Trans. R. Soc. A.* **362**, (2004).

49. A. Milani, M. Tommasini, M. D. Zoppo, C. Castiglioni, and G. Zerbi, Carbon nanowires: Phonon and $\pi$-electron confinement, *Phys. Rev. B.* **74**, (2006).

50. A. Milani, D. Fazzi, M. Tommasini, M. D. Zoppo, C. Castiglioni, and G. Zerbi. Long range vibrational interactions in linear carbon chains. Research activities on high performance computing clusters at CILEA – 2006.

51. C. A. Coulson and H. C. Longuet-Higgins, The electronic structure of conjugated systems. I. General theory, *Proc. R. Soc. London Ser. A.* **191**, (1947).

52. C. A. Coulson and H. C. Longuet-Higgins, The electronic structure of conjugated systems. II. Unsaturated hydrocarbons and their hetero-derivatives, *Proc. R. Soc. London Ser. A.* **193**, (1948).

53. C. A. Coulson and H. C. Longuet-Higgins, The electronic structure of conjugated systems. III. Bond orders in unsaturated molecules; IV. Force constants and interaction constants in unsaturated hydrocarbons, *Proc. R. Soc. London Ser. A.* **193**, (1948).

54. W. Kohn, Image of the Fermi surface in the vibration spectrum of a metal, *Phys. Rev. Lett.* **2**, (1959).

55. M. Tommasini, D. Fazzi, A. Milani, M. D. Zoppo, C. Castiglioni, and G. Zerbi, Intramolecular vibrational force fields for linear carbon chains through an adaptative linear scaling scheme, *J. Phys. Chem. A.* **111**, (2007).

56. M. Elstner, D. Porezag, G. Jungnickel, J. Elsner, M. Haugk, T. Frauen-

heim, S. Suhai, and G. Seifert, Self-consistent-charge density-functional tight-binding method for simulations of complex material properties, *Phys. Rev. B.* **58**, 7260, (1998).

57. M. Elstner. *Weiterentwicklung quantenmechanischer Rechenverfahren für organische Moleküle und Polymere.* PhD thesis, Paderborm University, Germany, (1998).

58. T. Krüger, M. Elstner, P. Schiffels, and T. Frauenheim, Validation of the density-functional based tight-binding approximation method for the calculation of reaction energies and other data, *J. Chem. Phys.* **122**, 114110, (2005).

59. T. Frauenheim, G. Seifert, M. Elstner, Z. Hajnal, G. Jungnickel, D. Porezag, S. Suhai, and R. Scholz, A self-consistent charge density-functional based tight-binding method for predictive materials simulations in physics, chemistry, and biology, *Phys. Stat. Sol. B.* **217**, 41, (2000).

60. M. Elstner, Q.Cui, P. Munih, E. Kaxiras, T. Frauenheim, and M. Karplus, Modeling zinc in biomolecules with the self consistent charge-density functional tight binding SCC-DFTB method: Applications to structural and energetic analysis, *J. Comput. Chem.* **24**, 565, (2003).

61. H. A. Witek, S. Irle, G. Zheng, W. A. de Jong, and K. Morokuma, Modeling carbon nanostructures with the self-consistent charge density-functional based tight-binding method: Vibrational spectra and electronic structure of $C_{28}$, $C_{60}$, and $C_{70}$., *J. Chem. Phys.* **125**, 214706, (2006).

62. H. Witek, S. Irle, and K. Morokuma, Analytical second-order geometrical derivatives of energy for the self-consistent-charge density-functional tight-binding method, *J. Chem. Phys.* **121**, 5163, (2004).

63. H. Witek, K. Morokuma, and A. Stradomska, Modeling vibrational spectra using the self-consistent-charge density-functional tight-binding method. I. Raman spectra., *J. Chem. Phys.* **121**, 5171, (2004).

64. H. Witek, K. Morokuma, and A. Stradomska, Modeling vibrational spectra using the self-consistent-charge density-functional tight-binding method. II. Infrared spectra., *J. Theor. Comp. Chem.* **4**, 1, (2005).

65. B. Aradi, B. Hourahine, and T. Frauenheim, DFTB+, a sparse matrix-based implementation of the DFTB method, *J. Phys. Chem. A.* **111**, (2007).

66. H. Witek and K. Morokuma, Systematic study of vibrational frequencies calculated with the self-consistent-charge density-functional tight-binding method, *J. Comput. Chem.* **25**, 1858, (2004).

67. E. Małolepsza, H. Witek, and K. Morokuma, Accurate vibrational frequencies using the self-consistent-charge density-functional tight-binding method, *Chem. Phys. Lett.* **412**, 237, (2005).

68. K. Kuchitsu, Ed., *Structure of Free Polyatomic Molecules – Basic Data.* (Springer-Verlag, Berlin, 1998).

69. T. B. Lynge and T. G. Pedersen, Density-functional-based tight-binding approach to phonon spectra of conjugated polymers, *phys. stat. sol. b.* **241**, (2004).

70. Y. Dong and M. Springborg, Infinite polymers and electrostatic fields, *Synth. Met.* **135–136**, 349, (2003).

71. M. Springborg and B. Kirtman, Efficient vector potential method for calculating electronic and nuclear response of infinite periodic systems to finite electric fields, *J. Chem. Phys.* **126**, 104107, (2007).

## Chapter 13

# Dynamics Simulations of Fullerene and SWCNT Formation

Stephan Irle,[1,2,3] Guishan Zheng,[1] Zhi Wang,[1] and
Keiji Morokuma[1,2]

[1]*Cherry L. Emerson Center for Scientific Computation and Department
of Chemistry, Emory University, Atlanta, GA 30322, U.S.A.*
[2]*Fukui Institute for Fundamental Chemistry, Kyoto University,
Kyoto, 606-8103, Japan*
[3]*Institute for Advanced Research and Department of Chemistry,
Nagoya University, Nagoya, 464-8601, Japan*

Fullerene and carbon nanotube (CNT) formation are closely related processes, after all CNTs were first perceived as "buckytubes" meaning that one could think of them as elongated buckminsterfullerenes. Moreover, fullerenes are frequently accompanying the CNTs in the soot produced during carbon arc, laser evaporation, and HiPco processes, indicating a common origin. Recently, our theoretical simulations have indicated a possible solution to the mystery of fullerene formation, but the atomistic mechanisms governing metal catalyzed SWNT nucleation and growth processes remain almost completely unknown, making them all the more interesting for theoretical investigations. In this work we present experimental background on both fullerene and CNT syntheses, as well as results of our quantum chemical molecular dynamics (QM/MD) simulations for their formation mechanisms. Obviously, the presence of the metal catalyst prevents self-capping and allows a tube-growth behavior. We discuss in detail the relationship between the fullerene and Fe-catalyzed SWNT formation mechanisms from the point of view of bond partner statistics, ring creation statistics, and the role of charge transfer.

## 1. Introduction

Fullerene and carbon nanotube (CNT) formation are closely related processes, after all CNTs were first perceived as "buckytubes" [1] meaning that one could think of them as elongated buckminsterfullerenes. Moreover, fullerenes are frequently accompanying the CNTs in the soot produced during carbon arc, laser evaporation, and HiPco processes [2,3], indicating a common origin. Although the mystery of fullerene formation seems to be largely solved by a combination of experimental observations and quantum chemical molecular dynamics simulations [4-7], the mechanisms governing metal catalyzed SWNT nucleation and growth processes remain almost completely unknown, making them all the more interesting for theoretical investigations. Proper treatment of the metal catalyst is crucial both in theory and experiments, which is the essence of SWNT formation.

Experimentalists and theoreticians alike today agree that the basic structural parameters of CNTs such as number of layers, diameter, and chirality are most likely determined at the very early stages of CNT nucleation [8-12]. Transition metal catalysts are employed to promote growth of single-walled carbon nanotubes (SWNTs), double-walled carbon nanotubes (DWNTs), overall yield, and to control diameter distributions and growth rates [1,13-22]. It is therefore very important to gain atomic level understanding of the catalyst/carbon chemistry involved in the CNT nucleation and growth. In particular, gaining control over the tube structural parameters by means of critical nucleation parameters (temperature, carbon source, feeding rate, metal catalyst particle size, composition, and aggregation, catalyst interaction with substrate, etc.) is so to speak the Holy Grail of CNT and in particular SWNT synthetic research, as such knowledge would eliminate time consuming, expensive, and potentially tube damaging separation steps. Detailed atomic level knowledge about the CNT nucleation and growth mechanisms has therefore a very broad impact on CNT research in general, and is an imperative requirement for making carbon nanostructures integral part of future nanotechnology-based material science and engineering.

The plethora of sometimes confusing and even contradictory experimental SWNT yields, diameter distributions, purities, and tube lengths from various synthesis methods [18] that can even depend on different research groups trying to apply the same methods makes the development of a clear understanding of SWNT formation mechanisms extremely difficult. Experimentally, tube nucleation occurs too fast to be followed directly using microscopic techniques; the fastest HRTEM microscopes are operating at shutter speeds of approximately milliseconds [23,24], yet tube nucleation occurs many orders of magnitudes faster. For the lack of the atomistic details, hypothetical formation models have been proposed based mainly on experimental observations [21,25,26], and quantum chemical calculations of model compounds [10,14,27,28]. A few theoretical research groups have carried out atomic scale molecular dynamics (MD) simulations using either reactive empirical molecular force field approaches [29-38] or based on density functional theory (DFT) [39,40]. Obviously, before metal-catalyzed SWNT formation can be thoroughly understood by modeling, we have to properly understand the all-carbon systems without metal catalyst. Therefore, in this chapter, we are giving a review on both fullerene and SWNT formation studies in hot carbon vapor and try to shed light on the differences between them.

In Sec. 2, we will briefly summarize the present status on SWNT formation processes before we present results of MD simulations for fullerene formation and SWNT growth processes based on quantum chemical potential (QM/MD) in Sec. 3. Section 4 is a summary of essential ingredients for realistic atomistic simulations of high-temperature carbon systems and SWNT nucleation and growth processes in particular.

## 2. The Present Status of SWNT Formation Processes

### 2.1. *Experimental techniques for SWNT synthesis*

Smalley and Kroto *et al.* used the laser evaporation technique when they incidentally discovered buckminsterfullerene $C_{60}$ in 1985 [41], fifteen years after its existence had been proposed by Osawa based on "superaromaticity" arguments [42]. The laser evaporation technique can also be used to generate both SWNTs [14,43] and multi-walled carbon nanotubes (MWNTs) [44], as well as carbon arc discharge [1,13,16] and solar furnace methods [45]. Common to these methods, which can exist both in single pulsed and multi-pulsed versions, is the fact that a graphite target is used, which should contain a transition metal (TM) catalyst for the synthesis of SWNTs, added by sintering of graphite and metal powders. Multiple pulses are found to produce higher SWNT yields and greater purities in case of PLV [46], and to produce DWNTs in case of high-temperature arc discharge (HT-PAD) [47]. It is thought that the first pulse produces a lot of large graphene fragments, and that subsequent pulses further decompose these carbon clusters, but the processes found the direct ablation zone are without doubt very complex [48]. For CNT synthesis, subsequent heating after the initial pulse(s) is always required, and furnace temperatures vary typically between 1000 K and 1500 K. If on the other hand the carbon plumes are allowed to cool to room temperature immediately after a pulse, single-walled nanohorns (SWNHs) in case of laser evaporation [49] or amorphous carbon in case of HT-PAD are produced [47]. Ar and $N_2$ [50] are excellent carrier gases in both PLV and carbon arc syntheses, whose task is to evenly distribute kinetic energy and thus "thermostat" the carbon plumes. By-product of SWNTs in PLV synthesis are frequently fullerenes and giant fullerenes as well as graphite onions [2,3], but the amount of amorphous carbon produced in carbon arc is typically somewhat greater.

Apart from these "pure carbon" based CNT synthesis methods, catalytic CVD (CCVD) SWNT syntheses are nowadays widely used, in particular because they are more promising for industrial style continuous nanotube production. Most famous is the high-pressure CO conversion (HiPco) synthesis developed at Rice University, which relies on the Bouduard charcoal reaction: $2\ CO \rightarrow C + CO_2 + \Delta$. Frequently, $Fe(CO)_5$ and $Mo(CO)_6$ are used as gaseous source for both metal catalyst as well as carbon, but ferrocene, ethylene, acetylene, and other feedstock species have been used, in particular when the metal catalyst is deposited on a substrate like $Al_2O_3$ or $SiO_2$, in which cases vertically aligned forests of SWNTs and MWNTs are grown on the catalyst surface. Mixed in the carrier gas here is frequently $H_2$ to further reduce the metal catalyst. In CCVD synthesis, oven temperatures are typically lower than in the case of laser ablation or carbon arc syntheses, namely around 500 K to 900 K.

Following different approaches, Maruyama *et al.* [51] and Hata *et al.* [52] found that the presence of oxygen during the CCVD synthesis leads to increased yield and tube lengths, by the admixture of water vapor or the use of ethanol (alcohol CCVD, or short ACCVD) as carbon source, respectively. Also, sulfur has been successfully used to increase the yield of SWNTs [53] or prompt Y-junction formation [54]. It is believed that the role of the chalocogens is to burn away amorphous carbon [8,55], or to capture hydrogen radical atoms which would favor the formation of amorphous carbon by $sp^3$-carbon production [56]. As to the latter argument, it has recently been demonstrated that the presence of atomic hydrogen is in fact important for both nucleation and growth mechanisms [57].

A hybrid method combining both laser evaporation and CCVD syntheses is laser assisted CCVD (LCCVD), which is recently becoming a popular SWNT synthesis method [58].

## 2.2. *Hypothetical models of SWNT formation mechanisms*

The SWNT synthesis from graphite (pulsed laser vaporization (PLV) or electric arc discharge) or chemical vapor decomposition (CVD) of carbon-rich gases requires typically the presence of a TM catalyst such as Fe, Co, Ni, or Mo. The performance of these TM catalysts is different in different environments: Fe is typically a better performer for CVD synthesis but performs poorly during PLV or carbon arc synthesis, whereas Ni is a good catalyst for the latter methods and a bad performer for the former [59-61]. Moreover, adding small amounts of Y to Ni/Co

catalysts has been shown to increase yield and allow diameter control
in PLV and carbon arc SWNT synthesis [62,63], but pure Y catalyst
itself is a poor performer [64]. Hence, the yield, purity, and diameter
distributions vary to a large extent with different synthetic approaches,
even when the same chemical elements are used as catalyst [18]. The
exact mechanism of SWNT nucleation and growth on the atomic scale is
still completely unclear, although advances in time-resolved *in situ* high
resolution transmission electron microscopy (HRTEM) and its
application to SWNT growth have recently been reported in the case of
carbon fiber growth [24] and electron irradiation-induced growth reversal
[65]. In earlier times, a "scooter model" [14,27] was theoretically
investigated where single catalytic *TM atoms* are supposedly "scooting"
around open ends of growing SWNTs, preventing thereby tube closure.
Yudasaka *et al.* [19-21] proposed a "metal-particle" model, in which
*nanometer*-sized droplets of carbon-containing catalytic metal are first
formed and carbon atoms precipitate on the metal surface in the
subsequent cooling stage to form SWNTs. There are two possible
scenarios to this metal-particle model: the "root growth" [66-68] and the
"tip-growth" model [69-71], depending on whether the metal nanometer-
sized particle moves with the tip of the growing tube, or remains on the
substrate (see Fig. 1). Recent HRTEM images support actually both

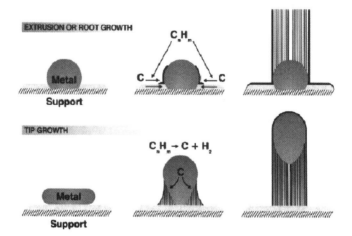

Fig. 1.  Base/root growth and tip growth models of SWNT formation (from Ref. 55).

versions of the metal-particle model [67,70]. Kataura *et al.* [26] proposed yet another hypothetical SWNT formation mechanism called "fullerene-cap model", where fullerene-like carbon clusters formed during laser-ablation attach to metal particles that serve as nuclei for SWNT growth. Such a mechanism would be very difficult to be distinguished from the metal particle (also called VLS) models. Remarkably, all three concepts share the notion that tube growth occurs at the metal/tube interface, and that the metal atoms serve to 1.) temporarily saturate the dangling bonds of the growing nanotube edges, and 2.) remove pentagon and other defects, before further growth of the honeycomb lattice in tube direction at the nanotube/TM interface can occur. This was already suspected in the ground-breaking "scooter" model papers by Tomanek *et al.* [14,27].

### 2.3. *Molecular force field modeling of TM-catalyzed SWNT formation*

In order to verify the hypothetical growth models, a few theoretical research groups have carried out atomic scale molecular dynamics (MD) simulations, using mainly Brenner's REBO type reactive empirical molecular force field approach [29-35,37]. Shibuta *et al.* have used periodic boundary systems containing carbon and Ni atoms [29] as well as different sized Ni clusters [30,31], while Ding *et al.* used Fe clusters of various sizes [33,37] and studied the role of temperature gradients [32,34,35]. An interesting study in the context of TM catalyst/carbon interaction is also the REBO simulation of molten FeC clusters [72,73]. While both groups have reported simulations of SWNT nucleation and growth processes occurring on the nanosecond time scale, the use of the REBO force field for sp- and $sp^2$-hybridized carbon in carbonaceous systems at high temperatures is questionable, in particular the neglect of aromaticity and $\pi$–conjugation effects. It was shown recently that electronic structure is a crucial ingredient for the simulation of high-temperature self-assembly processes of carbon nanostructures [6,74].

Recently Raty *et al.* carried out Car-Parinello MD (CPMD) simulations of carbon atoms on an $Fe_{55}$ cluster [39] (which is essentially *ab initio* DFT based molecular dynamics) and kept adding carbon atoms during the simulation at 1200 K. The time scale for SWNT nucleation in this case was on the order of 10 ps, which is markedly shorter than the aforementioned REBO based simulations by Ding *et al* or Shibuta *et al.* and consistent with out findings that carbon network self-healing processes are occurring faster when $\pi$–electron effects are included in the simulations. Unfortunately, no detailed analysis of the mechanism for SWNT nucleation was given, which might be difficult since only a single successful trajectory was reported. It is worth noting that, in order to avoid encapsulation of the metal particle by graphite, partial surface passivation had to be employed, achieved by adding hydrogen atoms to one side of the $Fe_{55}$ cluster and freezing the H-Fe bond distances.

Robertson *et al.* have recently presented a detailed density functional theory study about the structures and energetics of carbon nanotube caps on a regular hexagonal lattice representing a metal surface [10,11]. These simulations come close in spirit to the "fullerene cap model" of Kataura *et al.* [26], but no nucleation or growth was investigated in this otherwise interesting and important study.

So far, no detailed investigation and analysis of SWNT nucleation processes using QM/MD simulations have been reported, one reason for this being the enormous computer time associated with conventional CPMD simulations. An alternative approach is the density functional tight binding (DFTB) method, which combines the computational speed of tight binding approaches with the accuracy of DFT calculations. We have successfully employed DFTB in high-temperature QM/MD simulations of fullerene formation and SWNT growth processes and describe our results in the following Section.

## 3. QM/MD Simulations of Fullerene Formation and SWNT Growth Processes

After introducing the methodology (3.1), we present findings of DFTB based QM/MD simulations for high-temperature simulations of carbon (3.2), self-assembly of fullerenes (3.3), the importance of quantum potential vs. REBO (3.4), the importance of carbon flux for self-assembly processes (3.5), and Fe-catalyzed SWNT growth processes (3.6).

### 3.1. *Computational methodology for high-temperature QM/MD simulations*

The DFTB method is a popular method employed to compute on-the-fly potential energy surfaces (PESs) and energy gradients for direct trajectory calculations in the presented studies. All DFTB calculations were carried out with the program packages developed by Frauenheim, Seifert, and Elstner [75-77]. DFTB is an approximate density functional theory method based on the tight binding approach, and utilizes an optimized minimal LCAO Slater-type all-valence basis set in combination with a two-center approximation for Hamiltonian matrix elements. The authors have recently expanded the DFTB code by adding a routine to perform analytical geometrical second derivatives [78]. Within the DFTB formalism, there are three options available. One is the self-consistent charge (SCC) formalism, in which the atomic charge-atomic charge interaction is included in the energy and the atomic charges are determined self-consistently [77]. The second is the non-self-consistent charge (NCC) option, in which the atomic charge-atomic charge interaction is neglected [75]. The third is the most time consuming option, where both atomic charges as well as spin densities are self-consistently converged (SCC-sDFTB) [79]. The original NCC-DFTB method, approximately three to seven times faster than the SCC option which is itself much faster than the SCC-sDFTB method, is not a bad approximation for all-carbon systems like pure SWNTs and fullerenes like the author's benchmark comparison with B3LYP/6-31G(d) energetics showed [74,80,81], and they used NCC-DFTB in investigations on fullerene formation processes throughout. However, when transition metals are included, charge transfer plays an essential role in the chemistry of these systems, and the SCC-DFTB or SCC-sDFTB formalisms have to be applied.

Although typically NCC- and SCC-DFTB calculations are performed mainly for the closed shell state, when energy differences between occupied and vacant orbitals are smaller than $10^{-4}$ Hartree for a given geometry, these orbitals are considered to be degenerate, and an open shell occupancy is automatically adopted. While the total energy is not affected by the choice of occupancy of degenerate orbitals, the energy gradient depends on this choice. In the case of SCC-sDFTB, the different orbitals for different spin (DODS) approach is employed, and within this method, it is possible to compute energy splittings for different electronic spin states. In the derivation of the DFTB TM parameters for Fe-X, Ni-X, and Co-X (X = C, H, O, N) diatomic parameters, SCC-sDFTB was employed throughout (see below) [82].

The direct DFTB QM/MD simulations were performed by calculating analytical energy gradients on the fly and using them in a velocity Verlet integrator for time propagation. In case of hydrogen-free systems, we found that a time interval $\Delta t$ of 1.209 fs (= 50 atomic units) allows for energy conservation within 1 to 2 kcal/mol in Newtonian dynamics. The error introduced is negligible considering typical simulation temperatures of thousand Kelvin and higher. Temperature is kept constant throughout the system by scaling of atomic velocities an overall probability 20% scaling for the entire length of the simulations. Initial velocities are assigned randomly at the beginning of each simulation; however, when fragments are added during continued trajectories, only the velocities of these fragments are randomly chosen. We frequently monitor the accuracy of NCC- and SCC-DFTB trajectories by performing single point and geometry optimizations for snapshot geometries using the B3LYP hybrid or PBE density functional methods in combination with suitable polarized basis sets and effective core potentials (ECPs) as benchmark method. Figure 2 demonstrates such an approach for the monitoring of the DFTB PES during S3 giant fullerene growth [4], comparing single point DFTB energies with PBE/6-31G(d) results. We note that, while the total DFTB PES is shifted towards smaller relative energies, the landscape follows nicely the one predicted by the much more expensive *ab initio* PBE/3-21G method, and therefore we can expect that gradients and resulting trajectories are comparable.

With their collaborator Marcus Elstner in Germany the authors have developed M-X parameters for M = Fe, Co, Ni and X = C, H, O, N [82]. Concerning the accuracy of DFTB geometries and energetics involving TM atoms, a benchmark study was presented in reference [40] for Fe/C interactions. Figure 3 displays PBE/Lanl2DZ and SCC-sDFTB optimized geometries and energetics of the lowest electronic spin states of graphite

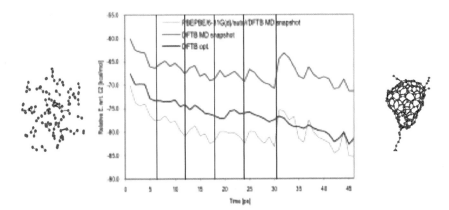

Fig. 2. DFTB PES (red) for direct MD simulation of S3 fullerene formation trajectory and corresponding PBE/6-31G(d) single point energies (blue), as well as energetics of DFTB optimized snapshot structures (black). Snapshots were taken at 1 ps intervals. Vertical red lines indicate time steps where 10 more $C_2$ molecules were added. After the last addition, temperature was raised from 2000 K to 3000 K.

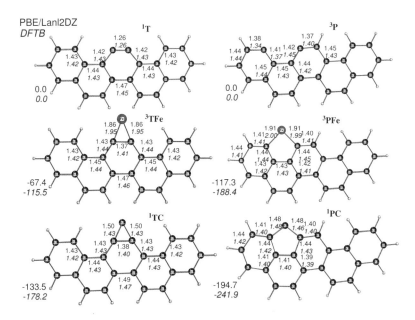

Fig. 3. PBE/Lanl2DZ (black and regular numbers) and DFTB (blue and italic numbers) bond lengths in [Å. and interaction energies in [kcal/mol. of graphite flakes T and P with respect to atomic electronic states $^3$P (carbon) and $^5$D (iron). From Ref. 38.

flake model systems that are similar in concept and spirit to the ones used by Tomanek et al. [27].

**T** indicates "triangle" and **P** "pentagon" final complexes of a $C_{22}H_{12}$ hydrocarbon fragment containing five hexagons and two unfilled valences ready to form 1,2- and 1,4-bonds with a single Fe or C atom to yield the corresponding complexes **TFe/TC** and **PFe/PC**, respectively. Spin states are indicated by superscripts preceding these labels, 1 for singlet, and 3 for triplet electronic states. The lowest energetic electronic states of each reactant species was used for the calculation of interaction energies; these are quintet for Fe and triplet for C atoms, and for the divalent graphite flakes singlet (**T**, a benzyne derivative) and triplet (**P**, a diradical species). After complexation with **T** and **P**, Fe complexes prefer triplet states with singlet being energetically higher in energy by about 15 kcal/mol, and carbene systems prefer singlet states; triplet states are higher by 34 kcal/mol and only 1 kcal/mol for **TC** and **PC**, respectively. As can be seen from the bond distances indicated in the Figure, DFTB C-C bonds deviate less than 0.05 Å from their respective DFT values even in the vicinity of the interaction region, and frequently are even much closer, indicating excellent performance of DFTB for C-C interactions in agreement with previous previous benchmarks [74,80,81,83]. DFTB Fe-C bond distances and the ${}^{1}$**TC** C-C bond of the carbenoid system show greater deviations from their respective DFT values, but are nevertheless in a better than 0.1 Å agreement. A similar general agreement between B3LYP and DFTB Fe-C bond lengths was observed before [82], which is a reasonable DFTB performance given the fact that even calculations using different density functionals and basis sets in standard DFT calculations can deviate by as much as 0.1 Å in transition metal chemistry.

DFTB interaction energies involving transition metal atoms are typically in less favorable agreement with standard DFT calculations [82], but very often the DFT relative energetic of related chemical systems in their electronic ground states is reasonably reproduced by DFTB. The interaction energies of Fe and C systems are very different, with Fe binding much weaker in TFe and PFe (−67 and −117 kcal/mol for DFT, −116 and −188 kcal/mol for DFTB) to the graphite flake than carbon in TC and PC (−134 and −195 kcal/mol for DFT, −178 and

−242 kcal/mol for DFTB). Both levels of theory predict the P complexes to be energetically much more favorable than T complexes by a constant amount of about 60 to 70 kcal/mol, which can be ascribed to the ring strain inherent to the T compounds. Although DFTB shows a systematic tendency to overbinding by about 60 kcal/mol compared to PBE/Lanl2DZ binding energies, the relative PBE energetic differences between P/T and Fe/C combinations are well reproduced at the DFTB level of theory at a small fraction of the computational cost. The DFTB overbinding tendency has been discussed and explained in more detail elsewhere [82], and the fact that the deviations of DFTB energetics from elaborate and computationally expensive DFT are of systematic nature in this example is indicative of the reliability of future QM/MD simulations of SWNT nucleation and growth processes using this economical quantum chemical method.

### 3.2. *High-temperature QM/MD simulations of carbon: Polyyne chains are everywhere*

The authors have performed quantum chemical molecular dynamics (QM/MD) high-temperature simulations on all-carbon self-capping processes of open-ended single-walled carbon nanotubes [74,83] as well as on fullerene formation from ensembles of randomly oriented $C_2$ molecules with various densities [4-7,80]. In the investigation of high-temperature dynamics of open-ended carbon nanotubes, the formation of long-lived "wobbling $C_2$-units" at the open ends was a predominant feature in these simulations, in noticeable agreement with the observations of Car *et al.* in their computationally much more expensive CPMD studies on similar systems [28]. In elaborate studies on open-ended carbon nanotubes of different lengths, diameter, and chirality, they also found the formation of "wobbling $C_4$ units", in particular for chiral and armchair nanotubes. These studies confirm that linear, sp-hybridized carbon polyyne chains are favorable species under high temperature conditions, and indicate that at initial stages of fullerene formation, the party line mechanism is most likely to dominate [84,85]. It is reported that Prof. Smalley mentioned during a conference that high-temperature carbon nanotube defects would resemble the threads coming out of a worn old sweater. If it is true, the results of DFTB-based QM/MD simulations and his anticipation of such structural defects are in perfect agreement. Recently, experimental evidence for the existence of polyyne chains inside MWNTs [86] and at high-temperature coalesced DWNTs [87] was observed in Raman spectra, but the nature of these high-frequency Raman peaks is still debated. Preliminary QM/MD studies of Fe-catalyzed SWNT have demonstrated that the formation of polyyne chains in carbon vapor is suppressed in the presence of Fe atoms attached to growing SWNT openings, favoring long-lived Fe-$C_2$-Fe units instead, which cannot form pentagons by themselves (see Fig. 4) [40].

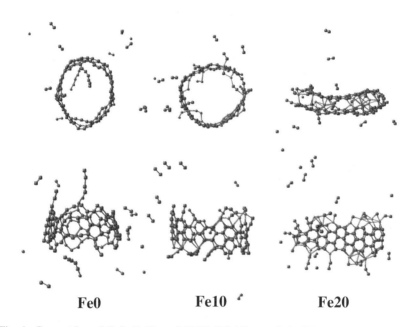

**Fe0**               **Fe10**               **Fe20**

Fig. 4. Geometries of Fe0, Fe10, and Fe20 (10,10) armchair SWNT model structures with 20 $C_2$ molecules after 9.6 ps simulation at 2000 K. View along tube axis (top) and viw perpendicular to tube axis (bottom). From Ref. 38.

### 3.3. The self-assembly mechanism of fullerenes in QM/MD simulations

The concept of "self-assembly" in fullerene formation emerged somewhat surprisingly but naturally as a consequence of such non-equilibrium dynamics [4-7,80], and the authors found that fullerenes might in fact represent "frozen" dissipative dynamic carbon structures commonly found in non-linear dynamic systems [88], trapped by rapid cooling thanks to their kinetic stability. Figure 5 displays schematically the self-assembly mechanism of giant fullerenes from ensembles of randomly oriented $C_2$ molecules in QM/MD simulations. The authors actually found that giant fullerene growth occurs in three stages under non-equilibrium dynamic conditions, utilizing the aforementioned polyyne chains heavily: *nucleation* of polycyclic structures from entangled carbon chains (irreversible carbon $sp^2$ hybdrization from sp-hybridized polyynes, formed by the party line mechanism), *growth* by ring condensation of carbon chains attached to the hexagon and pentagon containing nucleus (similar to Rubin's proposed ring condensation mechanism [89]), and at final stages *cage closure* similar to the

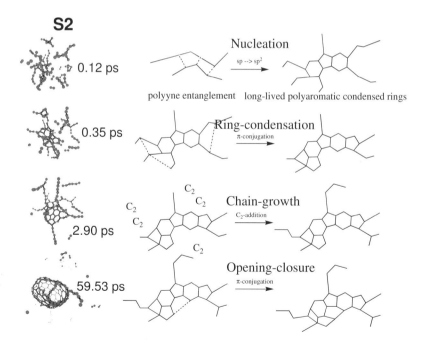

Fig. 5. Trajectory snapshots from **S2** giant fullerene formation trajectory from Ref. 4 and schematic explanations of events during the simulation.

mechanisms observed for the transformation of open-ended carbon nanotubes to fullerenes, where polyyne chains reach over the opening and "zip" them closed.

During the nucleation stage, long sp-hybridized carbon polyyne chains entangle and form initial clusters of condensed "small" rings (which we call nucleus), which can attain lifetimes long enough at around 2000 K temperature to form bonds with neighboring carbon atoms of attached chains, thus creating larger condensed ring systems by the ring-collapse mechanism [89]. It appears that due to its exothermicity, this sp to $sp^2$ re-hybridization step is irreversible, that means the only way structural conformations can be achieved from this point forward are ring size transformation similar to the 6/6 to 5/7 Stone-Wales mechanism. In particular we found that the possibility for ring destruction is greatly reduced once two to three rings are forming a condensed ring system. In the entanglement process, four-membered rings are frequently formed, which quickly isomerize to pentagons and hexagons. At temperatures of 2000 K and higher, the energetic difference between pentagons and hexagons does not matter very much, and we found that both 5- and 6-membered rings are readily created with a ratio of about 1:1, respectively. Over time, this ratio approaches approximately 1:2 at 2000 K, favoring hexagons as a consequence of continued annealing [6]. While the growing condensed ring system gains energy non-linearly by growing $\pi$–delocalization, the embedded pentagons force the growing slab to adapt a uniform curvature. At the same time, attached polyyne chains are growing by catching additional $C_n$ molecules from the carbon mixture of the environment, and retain thereby their flexibility to bend so much that ring condensation between slab border atoms and the nearest chain neighbor atoms can continue to proceed. The result is typically a basket-shaped carbon cluster with several long polyyne chains attached to its opening, and these chains are able to reach over to the other ends, forcing the system to eventually close to a fullerene cage. The energy profile associated with this mechanism is constantly downhill, because catching high-energetic small carbon fragment molecules and forming bonds to attach them at the attached carbon chains releases constantly energy, and the curvature of giant fullerenes is less steep than that of $C_{60}$ or $C_{70}$. Starting at higher carbon densities reduced the size of resulting fullerenes down to $C_{72}$, but somewhat surprisingly, neither $C_{60}$ nor $C_{70}$, the most important and abundant representative of fullerenes, were found in these simulations [4-7,80]. Continued heating of giant fullerene cages however reveals a slow but continuous loss of $C_2$ fragments on the order of several ps per carbon atom as a result of kink and octagon formation, where kink strain in these "wobbly" giant fullerenes is reduced by expelling carbon fragments [5,7]. $C_{60}$ as a kinetically very stable, spherical molecule represents presumably the end point of a continuously heated giant fullerene, and the author's "Shrinking Hot Giant" road of fullerene formation explains naturally a majority of reported experimental findings of $C_{60}$ synthesis [5-7], including the characteristic mass spectrum of soot product carbon fragments, containing large numbers of higher and giant fullerenes [90].

### 3.4. *Importance of quantum chemical potential in carbon nanochemistry*

Computationally very cheap reactive empirical bond order (REBO) semiclassical molecular mechanics potentials [91,92] have already been applied in molecular dynamics studies of fullerene formation mechanisms [93-96] and SWNT nucleation and growth [29-35,37]. Their finding was that fullerene formation and SWNT nucleation occurs on the order of nanoseconds, without observing the presence of polyyne chains, which are omnipresent in QM/MD simulations as major player in high temperature carbon chemistry, as described above.

The authors have shown in their previous quantum chemical molecular dynamics (QM/MD) studies on the self-capping mechanism of open-ended SWNTs [74,83] and on the growth mechanism of fullerenes [4,5,80] that the inclusion of full quantum chemical potential is crucial to describe essential features of carbon nanochemistry such as $\pi$–delocalization and aromatic stabilization. Figure 6 demonstrates that the use of REBO as potential in MD simulations can lead to qualitatively wrong intermediate structures that contain many sp$^3$ defects, and the fact that purely local potentials do not consider the effect of delocalized $\pi$–electronic states leads to a poor description of graphite's amazing self-

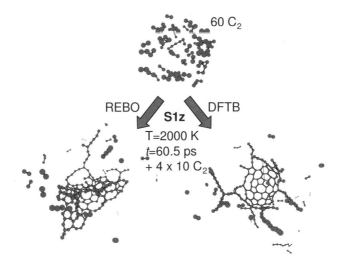

Fig. 6. S1z trajectory from Ref. 6 of the same random ensemble of 60 $C_2$ molecules in a 30 Å$^3$ periodic boundary box after heating to 2000 K and 4 times addition of 10 $C_2$ units in 12.1 ps intervals. Total simulation time at this stage is 60.5 ps.

healing capabilities, and consequently to significantly overestimated formation time scales [74].

This discrepancy is apparent when one compares the time scale of Maruyama's fullerene formation REBO MD simulations, which is on the order of nanoseconds [96] with QM/MD fullerene formation simulation, where giant fullerenes are grown within 40 to 100 ps, i.e. about 100-1000 times faster. A similar difference between REBO and QM/MD simulations can be seen when one compares the time scales for SWNT formation simulations of Ding or Shibuta with the only available density functional theory based Car-Parinello MD simulation study of this kind by Raty *et al.* [39]. It appears that also in case of SWNT nucleation, in the quantum world, events are occurring within tens of picoseconds, as opposed to nanoseconds in the classical REBO world.

### 3.5. Importance of carbon flux for self-assembly processes

Under experimental conditions of fullerene and SWNT formation, a constant flow of carbon material is certainly present during the microsecond long processes, forcing initially formed carbon clusters constantly out of thermodynamic equilibrium. In particular, as temperatures are decreasing in the case of laser- or arc-generated carbon plumes, condensation processes are setting in, leading to increasing densities in localized spatial areas over time. We chose to model such an open environment [4] by adding more $C_2$ units to a trajectory once an equilibrium state is reached, where no apparent structural reorganizations take place. Strictly speaking, we are leaving the domain of a well-defined potential energy surface (PES) in this way, but on the other hand a particle-based relative energy can still be defined. Moreover, we are keeping all velocities and only use random velocities for newly added $C_2$ fragments. We found that the gradual addition of $C_2$ is essential for the formation of giant fullerenes, and that if one starts simulations with the final number of $C_2$ molecules of our successful trajectories from the beginning without adding more $C_2$ molecules subsequently, only very few linear chains grow very long, as opposed to many shorter chains created by the subsequent addition of $C_2$ molecules. The ultra-long carbon chains created by the "all-at-once" approach are likely to lead eventually to large graphene sheets with lesser curvature due to the lack of initial entanglement than when starting from more but shorter polyyne chains.

The author's idea of adding carbon material during MD simulations has been applied in the REBO-based SWNT nucleation simulations of Ding *et al.* [32-35,37] as well as in the CPMD study by Raty *et al.* [39] While Ding chose to add carbon atoms in the center of $Fe_n$ clusters (which is a highly unphysical choice), Raty selected the outside of the $Fe_n$ surface as place to randomly add carbon atoms during the simulations.

### 3.6. *Fe/C interactions during SWNT growth in carbon vapor*

We have performed isokinetic QM/MD simulations using the DFTB quantum chemical potential on SWNT growth using 20 $C_2$ feedstock molecules randomly distributed around an open-ended (10,10) armchair tube model consisting of 100 carbon atoms [40]. The simulations were performed under periodic boundary conditions for three series that differ in the initial orientation of the $C_2$ ensembles, at temperatures of 1500 K, 2000 K, 2500 K, and 3000 K (see also Fig. 4). The goal was to investigate a chemically well-defined system where DFTB geometries and energetics can reasonably be verified by suitable *ab initio* PBE/Lanl2DZ model calculations, and to use DFTB-based QM/MD simulations to study the delicate balance between Fe/C interactions in dependence on Fe concentrations and temperature. To this end, we added 10 (Fe10 systems) and 20 (Fe20 systems) Fe atoms in 1,4-positions on the rims of the open-ended armchair tube, as shown in Fig. 4. In a total of 36 DFTB QM/MD trajectories on the open-ended SWNT tube models we noticed the following trends, summarizing the results presented in Tables 1 and 2:

Table 1. Stochiometry of SWNT model after 9.6 ps simulations for 3 trajectories **TX1**, **TX2**, and **TX3** at temperatures 1500 K, 2000 K, 2500 K, and 3000 K. Initial values are Fe0C100, Fe10C100, and Fe20C100 for the Fe0, Fe10, and Fe20 systems, respectively. From Ref. 38.

| | SWNT model stochiometry | | | Average $Cn$ |
|---|---|---|---|---|
| $T = 1500$ | **TX1** | **TX2** | **TX3** | |
| Fe0 | C110 | C124 | C127 | 120.3 |
| Fe10 | Fe10C120 | Fe10C118 | Fe10C120 | 119.3 |
| Fe20 | Fe20C124[a] | Fe20C124[a] | Fe20C124[a] | 124.0 |
| | | | | |
| $T = 2000$ | **TX1** | **TX2** | **TX3** | |
| Fe0 | C118 | C121 | C116 | 118.3 |
| Fe10 | Fe10C118 | Fe10C119 | Fe10C111 | 116.0 |
| Fe20 | Fe20C117[a] | Fe20C114[a] | Fe20C125[a] | 118.7 |
| | | | | |
| $T = 2500$ | **TX1** | **TX2** | **TX3** | |
| Fe0 | C125 | C126 | C121 | 124.0 |
| Fe10 | Fe10C111[b] | destroyed[c] | Fe10C119[b] | |
| Fe20 | destroyed[c] | Fe20C107[a] | Fe20C102[a] | |
| | | | | |
| $T = 3000$ | **TX1** | **TX2** | **TX3** | |
| Fe0 | C122 | C130 | C127 | 127.3 |
| Fe10 | Fe10C112[b] | destroyed[c] | destroyed[c] | |
| Fe20 | destroyed[c] | destroyed[c] | destroyed[c] | |

[a] Category II: tube collapsed to barbell shape.
[b] Category III: ring completely opened but largest cluster consists of 100 or more carbon atoms.
[c] Category IV: largest cluster consists of less than 100 carbon atoms.

Table 2. Average number of Fe-C bonds per Fe atom and Fe-Fe bonds per Fe atom and total number of Fe valences after 9.6 ps simulations for 3 trajectories **TX1**, **TX2**, and **TX3** at temperatures 1500 K, 2000 K, 2500 K, and 3000 K. Initial values are 2.00 for the number of Fe-C bonds per Fe atom and 0.00 for the number of Fe-Fe bonds per Fe atom and 2.00 for the total number of Fe valences. From Ref. 38.

| Average # of Fe-C bonds/Fe | | | Avg. | Average # of Fe-Fe bonds/Fe | | | | Avg. | Total |
|---|---|---|---|---|---|---|---|---|---|
| $T = 1500$ | **TX1** | **TX2** | **TX3** | $T = 1500$ | **TX1** | **TX2** | **TX3** | | |
| Fe10 | 3.60 | 3.60 | 3.70 | 3.63 | Fe10 | 0.00 | 0.00 | 0.00 | 0.00 | 3.63 |
| Fe20 | 3.35 | 3.60 | 3.45 | 3.47 | Fe20 | 1.70 | 1.60 | 1.50 | 1.60 | 5.07 |
| | | | | | | | | | | |
| $T = 2000$ | **TX1** | **TX2** | **TX3** | $T = 2000$ | **TX1** | **TX2** | **TX3** | | |
| Fe10 | 3.80 | 4.10 | 3.50 | 3.80 | Fe10 | 0.00 | 0.00 | 0.00 | 0.00 | 3.80 |
| Fe20 | 3.35 | 3.40 | 3.55 | 3.43 | Fe20 | 2.40 | 2.70 | 1.50 | 2.20 | 5.63 |
| | | | | | | | | | | |
| $T = 2500$ | **TX1** | **TX2** | **TX3** | $T = 2500$ | **TX1** | **TX2** | **TX3** | | |
| Fe10 | 3.70 | 3.70 | 4.00 | 3.80 | Fe10 | 0.00 | 0.40 | 0.40 | 0.27 | 4.07 |
| Fe20 | 3.20 | 3.75 | 4.10 | 3.68 | Fe20 | 1.80 | 1.80 | 0.60 | 1.40 | 5.08 |
| | | | | | | | | | | |
| $T = 3000$ | **TX1** | **TX2** | **TX3** | $T = 3000$ | **TX1** | **TX2** | **TX3** | | |
| Fe10 | 4.20 | 3.90 | 4.00 | 4.03 | Fe10 | 0.00 | 0.40 | 0.40 | 0.27 | 4.30 |
| Fe20 | 3.55 | 3.75 | 3.75 | 3.68 | Fe20 | 0.60 | 0.90 | 1.30 | 0.93 | 4.62 |

1. Fe atoms bind strongly to the rims of open-ended tubes, possibly much stronger than Ni, but less strongly than atomic carbon. Fe atoms do not frequently leave their position on the opening rim, except in order to form Fe-Fe bonds or larger Fe aggregates,

2. Pure carbon tubes withstand much higher temperatures without breaking apart, and continue to grow into the precursors of giant fullerenes in the absence of the metal catalyst, featuring the familiar polyyne ring condensation growth that incorporates pentagons as well as hexagons in the growing carbon cluster,

3. Fe-containing SWNT models are less stable with respect to thermal destruction, and we attribute this finding to the partial occupation of $\pi^*$-orbitals of the tube sidewalls, based on the results of Mulliken population analysis of the DFTB wavefunctions underlying the author's QM/MD simulations (see Fig. 7),

4. High concentration of Fe atoms on both sides of the opening (Fe20 systems) leads to a structural collapse of the tube from round to barbell shape, caused by predominant intermetallic bonds between Fe

atoms on opposite sides of the same openings; the barbell-shaped structures are slightly less stable than pure carbon clusters with respect to increased temperatures,

5. Only partial saturation of dangling bonds by Fe atoms on the open rims cannot enforce the barbell shape, and lead to structures that are even more frail than the barbell-shaped Fe20 systems,

6. Fe tries to maximize the number of chemical binding partners by interacting with $C_2$ feedstock molecules, leading to the formation of Fe-$C_2$-Fe bridge structures with lifetimes of several picoseconds which cannot form pentagon defects as opposed to longer polyyne chains,

7. Fe prefers to form intermetallic Fe-Fe bonds, indicating a strong tendency to form metal aggregates instead of Fe-C bonds, corroborating the metal-particle model of SWNT growth.

These studies confirm the original suspicion by Tomanek *et al.* [14,27] that the presence of metal atoms "softens" carbon clusters in the interaction regions due to charge transfer into $\pi^*$ molecular orbitals, making them more susceptible for annealing to perfect hexagon lattices, and that transition metal catalyzed SWNT growth occurs preferably at lower temperatures than fullerene formation. The authors have also shown in QM/MD simulations that Fe is not a very good catalyst for $C_2$ addition due its very strong Fe/C interactions.

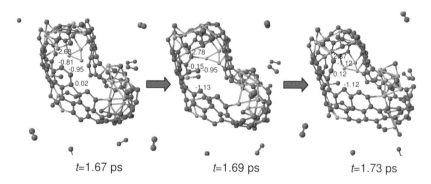

Fig. 7. Sequence of snapshot structures for trajectory (Fe20, 2500 K, TX1) at 1.67 ps, 1.69 ps, and 1.73 ps. The images trace the formation of a wobbling $C_2$ unit with immediate re-attachment to a 4-coordinated Fe atom. Mulliken atomic charges for the involved atomic centers are given in [e⁻]. From Ref. 38.

## 4. Summary

In this review we have presented reliable QM/MD high-temperature simulations of carbon, self-capping of open-ended SWNTs, and self-assembly of giant fullerenes. Using these all-carbon systems, we have also documented the importance of quantum vs. semiempirical REBO potential underlying such MD simulations, and pointed out the importance of carbon flux for carbon nanostructure self-assembly processes. In the case of Fe-catalyzed SWNT growth processes we have shown that electron transfer from Fe to the open ends of growing SWNT rims can make the carbon skeleton softer and more susceptible towards structural reorganizations, i.e. annealing to perfect hexagon structures.

In detail, we found that REBO based MD simulations are describing events too slow by a factor of approximately 1000 compared with QM/MD simulations, both in case of fullerene as well as SWNT formation processes. We attribute this lack of events during REBO simulations mainly to the fact that it does not include the delocalized nature of the quantum world, since it is a strictly local potential where bonds across ring openings for instance do not know of each other, which is not correct since the associated π–bonding orbitals are delocalized over the entire region. Similarly, π–stabilization increases non-additively with growing $sp^2$-carbon slabs, leading to synergistic enhancement of the cluster strengths. Moreover, in SWNT nucleation studies where metal particles are present, charge transfer weakens the growing carbon $sp^2$ skeleton, and makes it more flexible to reorganize into a regular hexagon network, thus pushing the curved SWNT cap faster away from the particle as described by Raty *et al.* [39].

Accordingly, essential ingredients for realistic atomistic simulations of high-temperature carbon systems and SWNT nucleation and growth processes in particular are the use of an accurate quantum chemical method that can describe carbon-metal interactions and charge transfer properly, as well as metal-metal and carbon-carbon interactions, where for the latter, proper description of sp to $sp^2$ hybridization processes and associated π–conjugation stabilization is required. Whether one uses DFTB or CPMD is perhaps only a question of computer time, but given the complexity of the problem with many environmental variables, a faster computational method is clearly preferable to probe each possible avenue pointed out by experiments.

## Acknowledgements

S. Irle is acknowledging a short term visiting fellowship from JSPS to Nagoya University. This work was partially supported by a grant from the Mitsubishi Chemical Corporation, and computer resources were provided in part by the Air Force Office of Scientific Research DURIP grant (FA9550-04-1-0321) as well as by the Cherry L. Emerson Center of Emory University. We also thank the Pacific Northwest Laboratory's EMSL for valuable computer time.

## References

1. S. Iijima, *Nature* **363**, 603 (1993).
2. S. Ramesh, B. Brinson, M. P. Johnson, Z. Gu, R. K. Saini, P. Willis, T. Marriott, W. E. Billups, J. L. Margrave, R. H. Hauge, R. E. Smalley, *J. Phys. Chem. B.* **107**, 1360 (2003).
3. A. K. Sadana, F. Liang, B. Brinson, S. Arepalli, S. Farhat, R. H. Hauge, R. E. Smalley, W. E. Billups, *J. Phys. Chem. B.* **109**, 4416 (2005).
4. S. Irle, G. Zheng, M. Elstner, K. Morokuma, *Nano Lett.* **3**, 1657 (2003).
5. G. Zheng, S. Irle, K. Morokuma, *J. Chem. Phys.* **122**, 014708 (2005).
6. S. Irle, G. Zheng, Z. Wang, K. Morokuma, *J. Phys. Chem. B* **110**, 14531 (2006).
7. G. Zheng, Z. Wang, S. Irle, K. Morokuma, *J. Nanosci. Nanotechnol.* **7**, 1662 (2007).
8. C. D. Scott, S. Arepalli, 2nd NASA/Rice University Workshop on SWNT Nucleation and Growth Mechanisms, 2nd NASA/Rice University Workshop on SWNT Nucleation and Growth Mechanisms, Boerne, TX, 2005.
9. Y. Miyauchi, S. Chiashi, Y. Murakami, Y. Hayashida, S. Maruyama, *Chem. Phys. Lett.* **387**, 198 (2004).
10. S. Reich, L. Li, J. Robertson, *Phys. Rev. B* **72**, 165423 (2005).
11. L. Li, S. Reich, J. Robertson, *J. Nanosci. Nanotechnol.* **6**, 1290 (2006).
12. S. Irle, Z. Wang, G. Zheng, K. Morokuma, M. Kusunoki, *J. Chem. Phys.* **125**, 044702 (2006).
13. D. S. Bethune, C. H. Kiang, M. S. DeVries, G. Gorman, R. Savoy, R. Beyers, *Nature* **363**, 605 (1993).
14. A. Thess, R. Lee, P. Nikolaev, H. Dai, P. Petit, J. Robert, C. Xu, Y. H. Lee, S. G. Kim, A. G. Rinzler, D. T. Colbert, G. E. Scuseria, D. Tomanek, J. E. Fischer, R. E. Smalley, *Science* **273**, 483 (1996).

15. H. Dai, A. G. Rinzler, P. Nikolaev, A. Thess, D. T. Colbert, R.E. Smalley, *Chem. Phys. Lett.* **260**, 471 (1996).

16. C. Journet, W. K. Maser, P. Bernier, A. Loiseau, M. Lamy de la Chapelle, S. Lefrant, P. Deniard, R. Lee, J. E. Fischer, *Nature* **388**, 756 (1997).

17. J. A. Hafner, M. J. Bronikowski, B. R. Azamian, P. Nikolaev, A. G. Rinzler, D. T. Colbert, K. A. Smith, R. E. Smalley, *Chem. Phys. Lett.* **296**, 195 (1998).

18. C. Journet, P. Bernier, Appl. Phys. A. 67, 1 (1998).

19. M. Yudasaka, T. Ichihashi, S. Iijima, *J. Phys. Chem. B* **102**, 10201 (1998).

20. M. Yudasaka, T. Komatsu, T. Ichihashi, Y. Achiba, S. Iijima, *J. Phys. Chem. B* **102**, 4892 (1998).

21. M. Yudasaka, R. Yamada, N. Sensui, T. Wilkins, T. Ichihashi, S. Iijima, *J. Phys. Chem. B* **103**, 6224 (1999).

22. T. Hiraoka, T. Kawakubo, J. Kimura, R. Taniguchi, A. Okamoto, T. Okazaki, T. Sugai, Y. Ozeki, M. Yoshikawa, H. Shinohara, *Chem. Phys. Lett.* **382**, 679 (2003).

23. K. Suenaga, R. Taniguchi, T. Shimada, T. Okazaki, H. Shinohara, S. Iijima, *Nano Lett.* **3**, 1395 (2003).

24. S. Helveg, C. Lopez-Cartes, J. Sehested, P. L. Hansen, B. S. Clausen, J. R. Rostrup-Nielsen, F. Abild-Pedersen, J. K. Nørskov, *Nature* **427**, 426 (2004).

25. Y. Saito, T. Yoshikawa, M. Inagaki, M. Tomita, T. Hayashi, *Chem. Phys. Lett.* **204**, 277 (1993).

26. H. Kataura, Y. Kumuzawa, Y. Maniwa, Y. Ohtsuka, R. Sen, S. Suzuki, Y. Achiba. *Carbon* **38**, 1691 (2000).

27. Y. H. Lee, S. G. Kim, D. Tomanek, *Phys. Rev. Lett.* **78**, 2393 (1997).

28. J.-C. Charlier, A. De Vita, X. Blase, R. Car, *Science* **275**, 646 (1997).

29. Y. Shibuta, S. Maruyama, *Physica B* **323**, 187 (2002) .

30. Y. Shibuta, S. Maruyama, *Chem. Phys. Lett.* **382**, 381 (2003).

31. Y. Shibuta, S. Maruyama, *Thermal Science & Engineering* **12**, 79 (2004).

32. F. Ding, K. Bolton, *J. Vac. Sci. Technol. A* **22**, 1471 (2004).

33. F. Ding, A. Rosen, K. Bolton, *J. Chem. Phys.* **121**, 2775 (2004).

34. F. Ding, A. Rosen, K. Bolton, *Chem. Phys. Lett.* **393**, 309 (2004).

35. F. Ding, K. Bolton, A. Rosen, *J. Phys. Chem. B* **108**, 17369 (2004).

36. K. D. Nielson, A. C. T. van Duin, J. Oxgaard, W. Q. Deng, W. A. Goddard III, *J. Phys. Chem. A* **109**, 493 (2005).

37. F. Ding, A. Rosen, K. Bolton, *Carbon* **43**, 2215 (2005).

38. F. Ding, A. Rosen, C. J. Campbell, L. K. L. Falk, K. Bolton, *J. Phys. Chem. B* **110**, 7666 (2006).

39. J.-Y. Raty, F. Gygi, G. Galli, *Phys. Rev. Lett.* **95**, 096103 (2005).

40. G. Zheng, S. Irle, K. Morokuma, *J. Nanosci. Nanotechnol.* **6**, 1259 (2006).

41. H. W. Kroto, J. R. Heath, S. C. O'Brien, R. F. Curl, R. E. Smalley, *Nature* **318**, 162 (1985).

42. E. Osawa, *Kagaku* **25**, 854 (1970).

43. T. Guo, P. Nikolaev, A. Thess, D. T. Colbert, R. E. Smalley, *Chem. Phys. Lett.* **243**, 49 (1995).

44. T. Guo, P. Nikolaev, A. G. Rinzler, D. Tomanek, D. T. Colbert, R.E. Smalley, *J. Phys. Chem.* **99**, 10694 (1995).

45. T. Guillard, S. Cetout, L. Alvarez, J. L. Sauvajol, E. Anglaret, P. Bernier, G. Flamant, D. Laplaze. *Eur. Phys. J.* **5** (1999).

46. A. G. Rinzler, J. Liu, H. Dai, P. Nikolaev, C. B. Huffman, F. J. Rodriguez-Marcias, P. J. Boul, A. H. Lu, D. Heymann, D. T. Colbert, R. S. Lee, J. E. Fischer, A. M. Rao, P. C. Eklund, R. E. Smalley, *Appl. Phys. A* **67**, 29 (1998).

47. T. Sugai, T. Okazaki, H. Yoshida, H. Shinohara, *New J. Phys.* **43**, 431 (2004).

48. C. D. Scott, S. Arepalli, P. Nikolaev, R. E. Smalley, *Appl. Phys. A* **72**, 573 (2001).

49. S. Iijima, M. Yudasaka, R. Yamada, S. Bandow, K. Suenaga, F. Kokai, K. Takahashi, *Chem. Phys. Lett.* **309**, 165 (1999).

50. D. Nishide, H. Kataura, S. Suzuki, K. Tsukagoshi, Y. Aoyagi, Y. Achiba, *Chem. Phys. Lett.* **372**, 45 (2003).

51. S. Maruyama, R. Kojima, Y. Miyauchi, S. Chiashi, M. Kohno, *Chem. Phys. Lett.* **360**, 229 (2002).

52. K. Hata, D. N. Futaba, K. Mizuno, T. Namai, M. Yumura, S. Iijima, *Science* **306**, 1362 (2004).

53. W. C. Ren, F. Li, H. M. Cheng, Shell control and growth mechanism of carbon nanotubes by sulfur-enhanced floating catalyst method, NASA/Rice University 2nd Workshop on Nucleation and Growth of Single-Wall Carbon Nanotubes, Boerne, TX, 2005.

54. C. Valles, M. Perez-Mendoza, P. Castell, M. T. Martinez, W. K. Maser, A. M. Benito, *Nanotechnology* **17**, 4292 (2006).

55. H. Ago, N. Uehara, N. Yoshihara, M. Tsuji, M. Yumura, N. Tomonaga, T. Setoguchi, *Carbon* **44**, 2912 (2006).

56. G. Zhang, D. Mann, L. Zhang, A. Javey, Y. Li, E. Yenilmez, Q. Wang, J. P. McVittie, Y. Nishi, J. Gibbons, H. Dai, *Proc. Natl. Acad. Sci.* **102**, 16141 (2005).

57. Y.-Q. Xu, E. Flor, H. Schmidt, R. E. Smalley, R. H. Hauge, *Appl. Phys. Lett.* **89**, 123116 (2006).

58. R. Alexandrescu, A. Crunteanu, R.-E. Morjan, F. Morjan, F. Rohmund, L. K. L. Falk, G. Ledoux, F. Huisken, *Infrared Phys. Technol.* **44**, 43 (2003).

59. H. Dai, Nanotube Growth and Characterization, Springer Verlag GmbH, New York (2001).

60. J.-C. Charlier, S. Iijima, Growth Mechanisms of Carbon Nanotubes, Springer Verlag GmbH, New York (2001).

61. M. Daenen, R. D. de Fouw, B. Hamers, P. G. A. Janssen, K. Schouteden, M.A.J. Veld, Wondrous Word of Carbon Nanotubes. Eindhoven University of Technology, Eindhoven (2003).

62. M. Yudasaka, N. Sensui, M. Takizawa, S. Bandow, T. Ichihashi, S. Iijima, *Chem. Phys. Lett.* **312**, 155 (1999).

63. M. Takizawa, S. Bandow, M. Yudasaka, Y. Ando, H. Shimoyama, S. Iijima, *Chem. Phys. Lett.* **326**, 351 (2000).

64. T. Saito, K. Kawabata, M. Okuda, *J. Phys. Chem.* **99**, 16076 (1995).

65. V. Stolojan, Y. Tison, G. Y. Chen, R. Silva, *Nano Lett.* **6**, 1837 (2006).

66. A. Maiti, C. J. Brabec, J. Bernholc, *Phys. Rev. B* **55**, R6097 (1997).

67. J. Gavillet, A. Loiseau, C. Journet, F. Willaime, F. Ducastelle, J.-C. Charlier, *Phys. Rev. Lett.* **87**, 275504 (2001).

68. C. Lu, J. Liu, *J. Phys. Chem. B* **110**, 20254 (2006).

69. O. A. Louchev, T. Laude, Y. Sato, H. Kanda, *J. Chem. Phys.* **118**, 7622 (2003).

70. S. Huang, M. Woodson, R. Smalley, J. Liu, *Nano Lett.* **4**, 1025 (2004).

71. L. Zheng, M. O'Connel, S. Doorn, X. Liao, Y. Zhao, E. Akhadov, M. Hoffbauer, B. Roop, Q. Jia, R. Dye, D. Peterson, S. Huang, J. Liu, Y. Zhu, *Nature Materials* **3**, 673 (2004).

72. F. Ding, A. Rosen, K. Bolton, *Phys. Rev. B* **70**, 075415 (2004).

73. F. Ding, A. Rosen, S. Curtarolo, K. Bolton, *Appl. Phys. Lett.* **88**, 133110 (2006).

74. G. Zheng, S. Irle, M. Elstner, K. Morokuma, *J. Phys. Chem. A* **108**, 3182 (2004).

75. D. Porezag, T. Frauenheim, T. Köhler, G. Seifert, R. Kaschner, *Phys. Rev. B* **51**, 12947 (1995).

76. G. Seifert, D. Porezag, T. Frauenheim, *Int. J. Quantum Chem.* **58**, 185 (1996).

77. M. Elstner, D. Porezag, G. Jungnickel, J. Elsner, M. Haugk, T. Frauenheim, S. Suhai, G. Seifert, *Phys. Rev. B* **58**, 7260 (1998).

78. H. Witek, S. Irle, K. Morokuma, *J. Chem. Phys.* **121**, 5163 (2004).

79. C. Kohler, G. Seifert, U. Gerstmann, M. Elstner, H. Overhof, T. Frauenheim, *Phys. Chem. Chem. Phys.* **3**, 5109 (2001).

80. S. Irle, G. Zheng, M. Elstner, K. Morokuma, High Temperature Quantum Chemical Molecular Dynamics Simulations of Carbon Nanostructure Self-Assembly Processes in C.E. Dykstra (Ed.), Theory and Applications of Computational Chemistry: The First 40 Years, Elsevier (2005).

81. G. Zheng, S. Irle, K. Morokuma, *Chem. Phys. Lett.* **412**, 210 (2005).

82. G. Zheng, H. A. Witek, P. Bobadova-Parvanova, S. Irle, D. G. Musaev, R. Prabhakar, K. Morokuma, M. Lundberg, M. Elstner, C. Kohler, T. Frauenheim, *J. Chem. Theor. Comput.* **3**, 1349 (2007).

83. S. Irle, G. Zheng, M. Elstner, K. Morokuma, *Nano Lett.* **3**, 465 (2003).

84. H. W. Kroto, J. R. Heath, S. C. O'Brien, R. F. Curl, R. E. Smalley, *Astrophys. Journal* **314**, 352 (1987).

85. H. W. Kroto, *Carbon* **30**, 1139 (1992).

86. X. Zhao, Y. Ando, Y. Liu, M. Jinno, T. Suzuki, *Phys. Rev. Lett.* **90**, 187401 (2003).

87. C. Fantini, E. Cruz, A. Jorio, M. Terrones, H. Terrones, G. Van Lier, J.-C. Charlier, M. S. Dresselhaus, R. Saito, Y. A. Kim, T. Hayashi, H. Muramatsu, M. Endo, M. A. Pimenta, *Phys. Rev. B* **73**, 193408 (2006).

88. I. Prigogine, I. Stengers, Order out of Chaos: Man's new dialogue with nature, Bantam Books, Toronto (1984).

89. Y. Rubin, F. Diederich, in F. Vögtle, J.F. Stoddart, S. Masakatsu (Eds.), Stimulating Concepts in Chemistry. Wiley-VCH, Weinheim (2000).

90. H. Shinohara, H. Sato, Y. Saito, M. Takayama, A. Izuoka, T. Sugawara, *J. Phys. Chem.* **95**, 8449 (1991).

91. D. W. Brenner, *Phys. Rev. B* **42**, 9458 (1990).

92. D. W. Brenner, *Phys. Rev. B* **46**, 1948 (1992).

93. J. R. Chelikowsky, *Phys. Rev. Lett.* **67**, 2970 (1991).

94. J. R. Chelikowsky, *Phys. Rev. B* **45**, 12062 (1992).

95. S. Maruyama, Y. Yamaguchi, *Chem. Phys. Lett.* **286**, 343 (1998).

96. Y. Yamaguchi, S. Maruyama, *Chem. Phys. Lett.* **286**, 336 (1998).

Chapter 14

# Mechanisms of Carbon Gasification Reactions Using Electronic Structure Methods

Juan F. Espinal,[1] Thanh N. Truong[2] and Fanor Mondragón[1]

[1]*Institute of Chemistry, University of Antioquia, Medellín, Colombia*
[2]*Department of Chemistry, University of Utah, Salt Lake City, U.S.A.*

A comprehensive review is given on theoretical studies of carbon gasification reactions. Such studies rely on the exploration of potential energy surfaces, vibrational spectra, transition states, and minimum energy reaction pathways. We focus on *ab initio* and density functional theory electronic structure calculations that have been used to understand key problems in the carbon gasification system, otherwise impossible to address using available experimental techniques.

## 1. Introduction

Fossil fuels have been the most important source of primary energy in the world. This tendency will probably continue without significant changes, at least during the next few decades. The use of this energy source has associated a highly negative impact on the environment due to the generation of different pollutant gases, liquids, and solids. Climatic changes induced by these contaminants have led to the search of alternative energy sources as well as to the optimization of existent processes oriented to the reduction of the environmental impact caused by their utilization. However, this task has been very difficult to implement partly due to the lack of a detailed knowledge at the molecular level of the available processes.

Among the fossil fuels utilization processes, coal combustion is the one that produces the highest degree of environmental pollution. An alternative process is coal gasification which reduces pollutant emissions and allows more efficient use of coal than combustion. Efficiency of conventional coal combustion plants is typically between 30-35% efficient (fuel-to-electricity). With improved technologies, this type of plants can achieve efficiencies of around 45%. However, coal gasification offers the prospects of boosting efficiencies up to 55% in the short-term and potentially to nearly 65% with technological advancements (Buskies 1996). Higher efficiencies translate into better economy and environmental impacts due to inherent reductions in greenhouse gases. In addition coal gasification is a process that allows the so called pre-combustion $CO_2$ capture. All of these characteristics make gasification to be considered as one of the cleanest technologies of the near future for energy production. A coal gasification process basically consists of two stages. The first one is the pyrolysis consisting of many reactions that take place very fast and during which moisture and volatile matter are eliminated leaving behind a low reactivity carbon rich solid product known as char. The char is an array of carbon atoms organized in clusters of graphenes of different sizes, mainly 3-7 six-member carbon rings randomly joint together by single C–C bonding. The second stage is the char gasification with a gasifying agent which can be oxygen, carbon dioxide, steam or hydrogen. These solid-gas reactions are known to be chemically controlled and are the rate-limiting step of the process. The main overall gasification reactions are presented in Table 1.

Table 1. Main reactions that take place during char gasification.

| Reaction | Enthalpy (kJ/mol) |
|---|---|
| $C_f + \frac{1}{2} O_2 \rightarrow CO$ | −123.1 |
| $C_f + CO_2 \rightarrow 2\,CO$ | 159.7 |
| $C_f + H_2O \rightarrow CO + H_2$ | 118.9 |
| $C_f + 2\,H_2 \rightarrow CH_4$ | −87.4 |

Therefore, detailed knowledge of the thermodynamic and kinetic parameters of the gasification reactions as well as of those reactions that can inhibit carbon conversion are important in order to improve efficiencies and decrease environmental impact. Furthermore, this knowledge is important to solve problems in other areas, such as coal storage, nanotubes purification, and combustion of other carbonaceous-based materials (i.e. obtained from biomass and/or waste). Reactions of carbon gasification by oxygen, carbon dioxide, steam, hydrogen and NOx have been studied for several decades using experimental techniques, such as Temperature Programmed Desorption (TPD) (Hermann and Hüttinger 1986; Marchon, Tysoe et al. 1988; Hall and Calo 1989; Zhuang, Kyotani et al. 1994), Thermo-Gravimetric Analysis (TGA) (Chen and Yang 1993), Transient Kinetics (TK) (Radovic, Jiang et al. 1991; Zhuang, Kyotani et al. 1994; Zhuang, Kyotani et al. 1995), X-ray Photoelectron Spectroscopy (XPS) (Marchon, Carrazza et al. 1988; García, Espinal et al. 2004), Scanning Electron Microscopy (SEM) (Tomita, Higashiyama et al. 1981), Scanning Tunneling Microscopy (STM) (Huang and Yang 1999), Transmission electron microscopy (TEM) (Yang and Duan 1985; Chen and Yang 1993), and by isotopic tracing (Vastola, Hart et al. 1964), among others. However, current molecular-level knowledge of these processes is limited due to the complexity of carbonaceous structures making it impossible to obtain experimentally necessary information for molecular-level detailed simulations of the gasification process.

Electronic structure methods are a powerful tool that helps us to improve our understanding of the carbon gasification processes and their use has been increasing for this kind of systems in the last few years. The reliability of the data obtained from molecular modeling has increased with the continuing development of more accurate quantum chemistry methods and more powerful computers. Numerous quantum chemistry software are available for calculating potential energy surfaces, vibrational spectra, transition states, reaction mechanisms, etc.

Even though numerous publications can be found for molecular modeling using semi-empirical methods, such as Hückel Molecular Orbital Theory (HMO) (Stein and Brown 1987), Extended Hückel Theory (EHT) (Bennett, McCarroll et al. 1971), Pariser-Parr-Pople (PPP)

(Radovic, Karra et al. 1998), Complete Neglect of Differential Overlap (CNDO) (Bennett, McCarroll et al. 1971), Intermediate Neglect of Differential Overlap (INDO) (Chen, Yang et al. 1993), Neglect of Diatomic Differential Overlap (NDDO) (Dybala-Defratyka, Paneth et al. 2004), Modified Intermediate Neglect of Differential Overlap (MINDO) (Casanas, Illas et al. 1983), Modified Neglect of Diatomic Overlap (MNDO) (Barone, Lelj et al. 1987), Austin Model 1 (AM1) (Shimizu and Tachikawa 2002), and Parametric Method 3 (PM3) (Aksenenko and Tarasevich 1996). This chapter will mainly focus on applications of more accurate methods, such as *ab initio* Molecular Orbital and/or Density Functional Theory (DFT), which are commonly used in more recent publications.

## 2. Molecular Systems and Model Chemistries

The first step in a computational work to study the basic reactions that take place in gasification, combustion, or liquefaction processes is to define a physical model that can be used to study chemical properties of the full system. The selection of this physical model is arbitrary and different researchers have chosen different physical models (Chen and Yang 1998; Kyotani and Tomita 1999; Montoya, Truong et al. 2000; Radovic and Bockrath 2005). In the case of coal gasification, since the controlling step is the gasification of the char, the preferred ones are aromatic units comprising from 4-9 aromatic rings with carbon active sites in the zigzag and armchair configurations.

Once the physical model has been established, it is necessary to define the theoretical model, also known as model chemistry. Model chemistries consist of a combination of theoretical procedures (theory level) and the mathematical representation of the orbitals within a molecule, known as basis set. Details about the selection of the model chemistries can be found in many Computational Chemistry textbooks (Foresman and Frisch 1996; Jensen 1999; Young 2001). Selection of the model chemistry often depends on the available computing resources. To reduce the computational demand, it is possible to take advantage of a multilevel approach that uses different levels of theory for different parts of the same physical model. An example of a multi-layer approach is to treat the reactive site with the highest level of theory and the rest of the structure with a lower level of theory (Humbel, Sieber et al. 1996; Svensson, Humbel et al. 1996). However, in the case of aromatic structures, the partition of the molecular structure can induce computational errors due to the disruption of aromaticity (Montoya, Mondragón et al. 2002). Overall, the selections of the physical models and model chemistries are considered together to form the cost-effective strategy for carrying out the computational study.

It is known that semi-empirical molecular orbital theories are useful for providing insight into the carbon gasification mechanisms. However, it is necessary to use more accurate methods (such as *ab initio* and DFT) in order to get more reliable energetic and thermodynamic data. An *ab initio* molecular orbital study on graphenes as models of graphite surface was first performed by Chen and Yang (Chen and Yang 1998). Six graphene models with sizes up to seven rings were used (see Fig. 1). The authors assumed the active sites to be the unsaturated carbon atoms while other edge carbon atoms are capped with hydrogen atoms. Three levels of *ab initio* calculations (HF/STO-3G, HF/3-21G(d), and B3LYP/ 6-31G(d)) were used for all graphite models. It was found that the

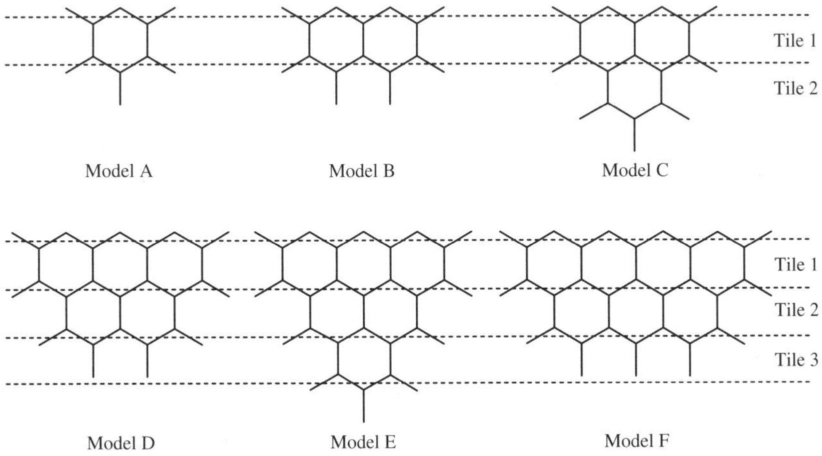

Fig. 1. Graphite models. Note all edge atoms are saturated with hydrogen atoms except the active surface (on top tile). (Reprinted from (Chen and Yang 1998), Copyright 1998, with permission from Elsevier).

B3LYP/6-31G(d)//HF/3-21G(d) model chemistry, which means single point energy at the B3LYP/6-31G(d) using geometry optimized at the HF/3-21G(d) level (Foresman and Frisch 1996), can reproduce results that compare well with experimental data but at a reasonable computational cost. Among the models shown in Fig. 1, the authors reported model F ($C_{25}H_9$) as the most appropriate model to represent a graphite structure, it yielded parameters in good agreement with the experimental data. In addition, at the selected model chemistry, model F can represent the unbalanced graphite edge sites, which are the active sites for carbon gasification reactions.

As mentioned above, one way to represent an active site in the models is to remove a hydrogen atom from an aromatic carbon atom. This leads to the existence of an unpaired electron thus either unrestricted or restricted open-shell wavefunction must be used. Montoya et al. (Montoya, Truong et al. 2000) examined the effect of spin contamination in the Unrestricted Hartree-Fock (UHF) and DFT levels of theory in modeling NO adsorption on graphite using the model shown in Fig. 2. It was found that there is a significant large spin contamination in the UHF wave function due to the nature of the active sites. This leads to an error of 46 kcal/mol in the NO binding energy, equivalent to 51% of the restricted open-shell HF binding energy. Such a spin contamination would be more severe in the transition state region. Thus, care should be taken in using UHF-based methods in modeling reaction mechanisms on carbonaceous surfaces. Montoya's results also showed that the B3LYP method has a much smaller spin contamination. Consequently, there are smaller differences in both adsorption geometry and binding energy between the unrestricted and restricted open-shell B3LYP wave function. Due to the computational efficiency of the unrestricted open-shell

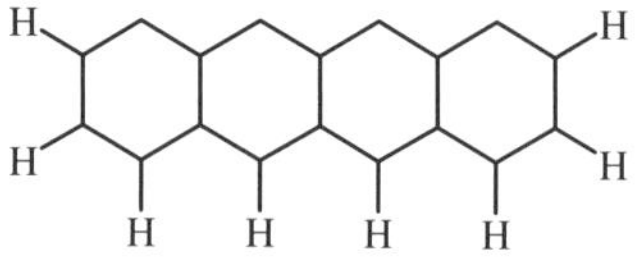

Fig. 2. Single graphene layer used as a model of char to investigate the spin contamination. (Reprinted with permission from (Montoya, Truong et al. 2000), Copyright 2000 American Chemical Society).

methods compared to the restricted open-shell ones, it was recommended that nonlocal DFT methods should be used in modeling adsorption processes or reactions on carbonaceous surfaces.

Recently, Radovic and Bockrath (Radovic and Bockrath 2005) questioned the above models used to represent active sites in char and suggested a different model. They proposed that in the zigzag models the active sites are carbene-like structures with the triplet ground state being the most common. In the armchair models the active sites are of the carbyne type with singlet ground state as the most abundant. The ground state of the carbyne structure is seen to be more stable than the ground state of the carbene structures. It is interesting to point out that all the model structures used to investigate the carbene structures have only one active site. No analysis was done for the case where more than one active site is present in a model such as the ones used by Chen and Yang (Chen and Yang 1998), Montoya et al. (Montoya, Truong et al. 2000), and Kyotani and Tomita (Kyotani and Tomita 1999).

In the multi-layer approach, the active site is treated with the highest possible level of theory and the rest of the structure with a lower theory level. A systematic analysis of the accuracy and efficiency of several of such multi-layer model chemistries for studying reactions involving aromatic systems was presented by Montoya et al. (Montoya, Mondragón et al. 2002). Different multi-layer ONIOM (Own N-layered Integrated molecular Orbital and molecular Mechanics) models in which the whole system is divided into subsystems that can be treated at different levels of theory as explained above, see Fig. 3.

Average calculation time per optimization cycle (hour/cycle) for the carbonaceous models and their corresponding surface carbon-oxygen complexes are listed in Table 2. The nomenclature used for the level of theory is: OL2-BH(L): O for ONIOM model; L2, two-layer ONIOM method; B, B3LYP; H, Hartree-Fock; (L), linear chain in the active region. It corresponds to the B3LYP/6-31G(d):HF/3-21G level of theory. OL3-B(L): it corresponds to the B3LYP/6-31G(d):B3LYP/3-21G:B3LYP/STO-3G level of theory. OL2-BA(L): it corresponds to the B3LYP/6-31G(d):AM1 level of theory and the model system is shown in Fig. 3a. OL2-BA(R): it corresponds to the B3LYP/6-31G(d):AM1 level of theory where the model system is the central six-member ring as shown in Fig. 3b. The results are relative to the least expensive method, i.e., the OL2-BA(R) method. It was found that the computational cost increases in the order OL2-BA(R) ≅ OL2-BA(L) < OL2-BH(L) < OL3-B(L) < B3LYP/6-31G(d).

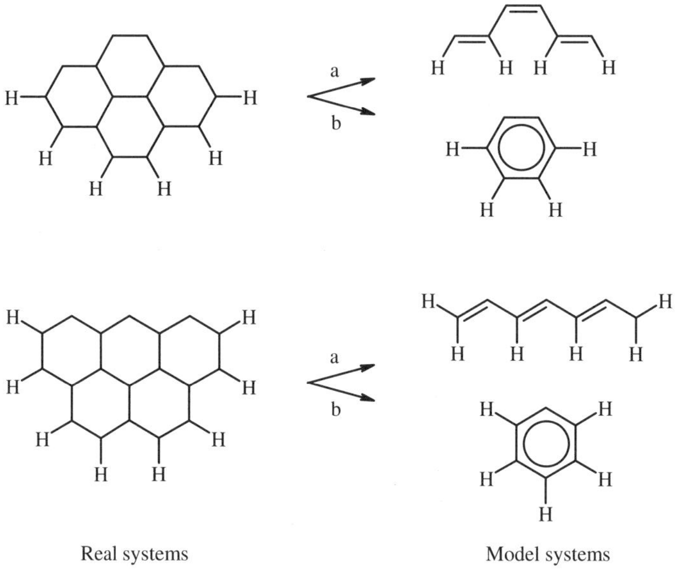

Real systems                                    Model systems

Fig. 3. Schematic representation of the different systems in the multiple-layer ONIOM model for the armchair and zigzag carbonaceous models. (Reprinted from (Montoya, Mondragón et al. 2002), Copyright 2002, with permission from Elsevier).

Table 2. Relative time per cycle during the optimization of the armchair and zigzag models at different levels of theory.[a] (Reprinted from (Montoya, Mondragón et al. 2002), Copyright 2002, with permission from Elsevier.)

| Level of theory | Armchair | | | Zigzag | | |
|---|---|---|---|---|---|---|
| | Carbon model | Semiquinone | Carbonyl | Carbon model | Semiquinone | Carbonyl |
| OL2-BA(R) | 1.0 | 1.0 | 1.0 | 1.0 | 1.0 | 1.0 |
| OL2-BA(L) | 1.2 | 1.0 | 0.9 | 1.0 | 1.0 | 1.0 |
| OL2-BH(L) | 1.7 | 6.4 | 1.6 | 2.0 | 1.7 | 2.5 |
| OL3-B(L) | 3.4 | 6.0 | 4.4 | 5.8 | 2.5 | 3.6 |
| B3LYP/6-31G(d) | 9.1 | 10.2 | 7.8 | 12.6 | 3.9 | 10.6 |

[a] All values are relative to the OL2-BA(R) level.

It was found that any attempt to partition the system into subsystems and to treat them at different levels of theory would lead to large errors. This is because the carbonaceous surfaces are highly aromatic. Such a partition would disrupt the delocalization of the π-bond network. However, it was found that it is possible to treat different regions with different basis sets and still maintain a reasonable level of accuracy. This, in fact, gives a noticeable decrease in computational cost and thus the approach is recommended for future studies of adsorption on larger carbonaceous surfaces.

## 3. Oxidation Reactions

Combustion and gasification of carbonaceous materials are processes that involve oxidation reactions. In the case of combustion the oxidation takes place by molecular oxygen or by atomic oxygen, while for gasification the oxidation is carried out mainly by molecular oxygen, water, and/or carbon dioxide. Many aspects of the oxidation reactions still remain uncertain. Therefore, molecular level details on the mechanism of the gasification and combustion reactions are still not complete. As will be discussed below, most of the available information is on the thermodynamics of the reactions, very little is known on the dynamics of these reactions.

Before 1990 there were just very few publications on molecular modeling of reactions between carbon and oxidant agents. As these studies used semi-empirical methods and only the reaction on the basal plane was discussed (Bennett, McCarroll et al. 1971; Hayns 1975), they will not be considered in this review.

Strongly bonded oxygen in graphite was studied by Pan and Yang (Pan and Yang 1992) using detection by high-temperature TPD and semiempirical molecular orbital MINDO method. The molecular orbital calculations considered oxygen chemisorptions on four possible graphite surface sites (see Fig. 4). The results are shown in Table 3. Structures A and B in Fig. 4 represent the most stable chemisorbed oxygen structures on the zigzag face, structure C represents the most stable chemisorbed oxygen on the armchair face, structure D represents that in between two basal planes. Structure D shows only one basal plane because the oxygen atom was closer to that plane, while its bonding with the plane on the opposite side was negligible. For structures A, B, and C, the C–O bonds were stronger than the neighboring C–C bonds. This result indicates that it was more favorable to break the C–C bonds rather than the C–O bonds; hence CO should be desorbed rather than $O_2$. The opposite was

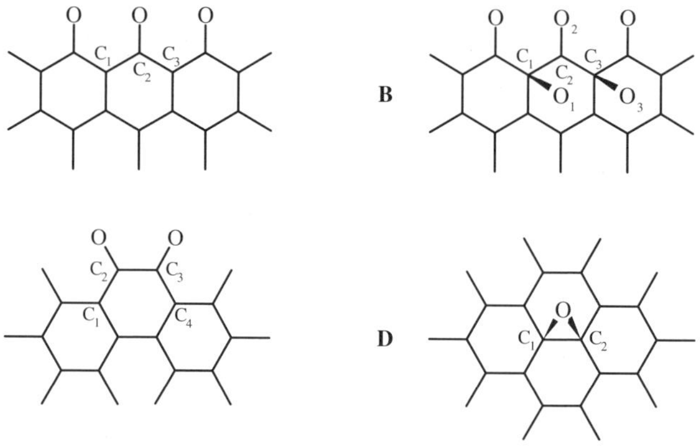

Fig. 4. Structures of chemisorbed oxygen on different faces of graphite. (Reprinted with permission from (Pan and Yang 1992), Copyright 1992 American Chemical Society).

Table 3. Summary of molecular orbital calculations by MINDO method (structures A-D shown in Fig. 4). (Reprinted with permission from (Pan and Yang 1992), Copyright 1992 American Chemical Society.)

| Structure | Bond | Bond length, Å | Wiberg indices* | Net charge | |
|---|---|---|---|---|---|
| A | $C_1$–$C_2$ | 1.46 | 0.904 | $C_1$, $C_3$ | −0.208 |
| | $C_2$–O | 1.18 | 1.812 | $C_2$ | 0.616 |
| | - | - | - | O | −0.211 |
| B | $C_1$–$C_2$ | 1.63 | 0.584 | $C_1$ | 0.621 |
| | $C_1$–$O_1$ | 1.30 | 1.242 | $C_2$ | 0.634 |
| | $C_2$–$O_2$ | 1.19 | 1.791 | $O_1$ | −0.186 |
| | - | - | - | $O_2$ | −0.201 |
| C | $C_1$–$C_2$ | 1.53 | 0.672 | $C_1$ | −0.189 |
| | $C_2$–O | 1.18 | 1.841 | $C_2$ | 0.412 |
| | - | - | - | O | −0.205 |
| D | $C_1$–$C_2$ | 1.42 | 0.814 | $C_1$ | 0.491 |
| | $C_2$–O | 1.41 | 0.752 | O | −0.203 |

* Wiberg indices are an estimation of the bond strength between two atoms.

found for structure D, where the C–O bond was weaker than the neighboring C–C bonds. Consequently, the C–O bonds were more favorable to breakage, which was likely followed by diffusion of oxygen atoms between the graphite layers to the edge carbon atoms where a semiquinone group was formed and desorbed as CO. In structure A, the net charge densities of carbon atoms $C_1$ and $C_3$ were highly negative making oxygen chemisorption on these sites a favorable process that produces structure B. From Table 3, the weakest bonds were the C–C bonds neighboring the C–O bond in structures B and C. Therefore, these C–C bonds were broken first at low temperatures (below 1000 °C) in TPD to yield CO. Bonds in structure D were relatively stronger and hence could be broken only at higher temperatures (above 1000 °C).

Semi empirical molecular orbital theory calculations using the INDO method were done by Chen et al. (Chen, Yang et al. 1993). Table 4 lists C–C and C–O bond strengths that were calculated for the structures shown in Fig. 5. Results given in Table 4 show that the C–C bond energies (for CO desorption from graphite) are lowered approximately 30% by bonding of oxygen on the saturated carbon atoms, i.e., by transforming from structure B to D or from structure C to E in Fig. 5. Structures D and E are formed due to the high net electron charges of the saturated carbon sites. Structures D and E are abundant in the $C-O_2$ reaction but not in the $C-CO_2$ and $C-H_2O$ reactions. This more active form of oxygen complex is a main contributor to the $C-O_2$ reaction. From the above discussion the following mechanism for carbon gasification by oxygen-containing gases is proposed. Taking $CO_2$ as example, the mechanism is expressed as follows:

$$CO_2 + C_f \leftrightarrow C_f(O) + CO \qquad K_1$$

$$CO_2 + C_f(O) \leftrightarrow C(O)C_f(O) + CO \qquad K_2$$

$$C(O)C_f(O) \rightarrow CO + C_f(O) \qquad k_3$$

$$C_f(O) \rightarrow CO + C_f \qquad k_4$$

where K stands for equilibrium constant and k for rate constant. The symbols $C_f(O)$ and $C(O)C_f(O)$ represent the complexes in substrates B, C and D, E, respectively (or to be more exact, they represent the two main groups of oxygen containing structures, i.e., in-plane group and off-plane group). $C_f$ is the edge carbon site with a free sp$^2$ electron and $C$ is the saturated carbon atom. It is concluded that this unified mechanism is consistent with all TPD, TK, and kinetic results for gas-carbon reactions reported in the literature.

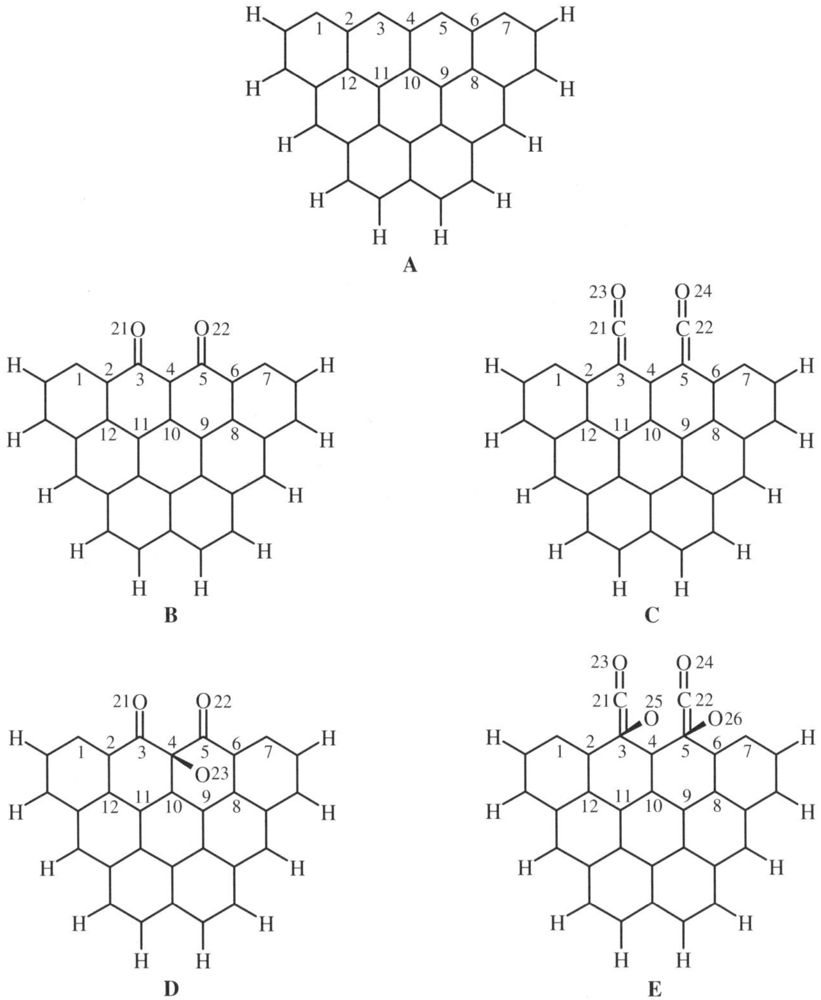

Fig. 5. Structures of chemisorbed oxygen on graphite. (Reprinted with permission from (Chen, Yang et al. 1993), Copyright 1993 American Chemical Society).

Table 4. Carbon-carbon bond strength in different substrates (structures A-E shown in Fig. 5) in terms of diatomic energy (kcal/mol).[a] (Reprinted with permission from (Chen, Yang et al. 1993), Copyright 1993 American Chemical Society.)

| No. | Structure A | Structure B | Structure C | Structure D | Structure E |
|-----|-------------|-------------|-------------|-------------|-------------|
| $C_{2-3}$ | 569.20 | 507.31 | 493.42 | 416.92 | 402.33 |
| $C_{3-4}$ | 578.42 | 515.89 | 505.89 | 360.80 | 386.98 |
| $C_{4-5}$ | 578.42 | 515.89 | 505.89 | 360.80 | 386.98 |
| $C_{5-6}$ | 569.20 | 507.31 | 493.42 | 416.92 | 402.33 |
| $C_{3-21}$ | - | 660.49 (C–O) | 572.56 | 861.03 (C–O) | 434.88 |
| $C_{5-22}$ | - | 660.49 (C–O) | 572.56 | 861.03 (C–O) | 434.88 |
| $C_{21-23}$ | - | - | 864.81 (C–O) | - | 789.82 (C–O) |
| $C_{22-24}$ | - | - | 864.81 (C–O) | - | 789.82 (C–O) |
| $C_{3-25}$ | - | - | - | - | 480.06 (C–O) |
| $C_{5-26}$ | - | - | - | - | 480.06 (C–O) |
| $C_{4-23}$ | - | - | - | 456.61 (C–O) | - |

[a] Empirically, the bond energies are of the order of 1/5 of the diatomic energies.

In a following study, Chen and Yang (Chen and Yang 1998) performed *ab initio* molecular orbital calculations at the B3LYP/6-31G(d)//HF/3-21G(d) model chemistry on graphite and oxygen intermediates on and near the edges of graphite. Stable wave functions were achieved for all graphite models shown in Fig. 6, indicating that there is a certain minimum on the potential surface that corresponds to the expected structure for each of the models. It was found that the results are in agreement with experimental data for structural geometry parameters, vibrational frequencies, and, more importantly, bond energies. Two reaction pathways were deduced from the calculation results, one for the $C + O_2$ reaction and one for the $C+ CO_2$ and $C + H_2O$ reactions. These two pathways are shown in Fig. 7. The zigzag edge is used as an example, while the same pathways also apply to the armchair edge. The rate-limiting step for all gasification reactions by oxygen-containing gases is the breakage of C–C bonds to free CO from the semiquinone intermediate. The energy for this C–C bond is about 80 kcal/mol, which is close to the experimental activation energy for the reactions with $CO_2$ and $H_2O$. The C–C bond is weakened by 33% by the effect of an adjacent epoxy oxygen, to a C–C bond with a bond energy of nearly 53 kcal/mol. This extent of weakening in the C–C bond energy

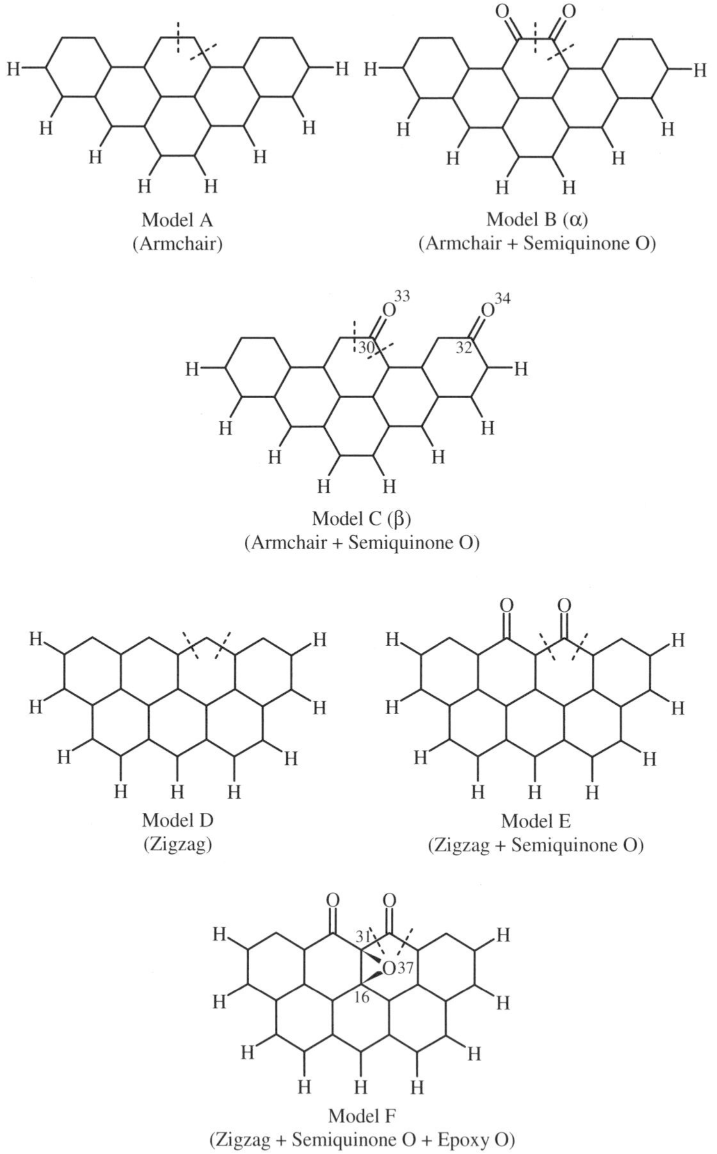

*J. F. Espinal, T. N. Truong and F. Mondragón*

Fig. 6. Selected graphite models showing oxygen intermediates. The dotted line indicates where the C–C bond breakage takes place to free CO. Epoxy oxygen, $O_{37}$, is perpendicular to the basal plane. Semiquinone and carbonyl oxygen are bonded to carbon with a double bond. (Reprinted with permission from (Chen and Yang 1998), Copyright 1998 American Chemical Society).

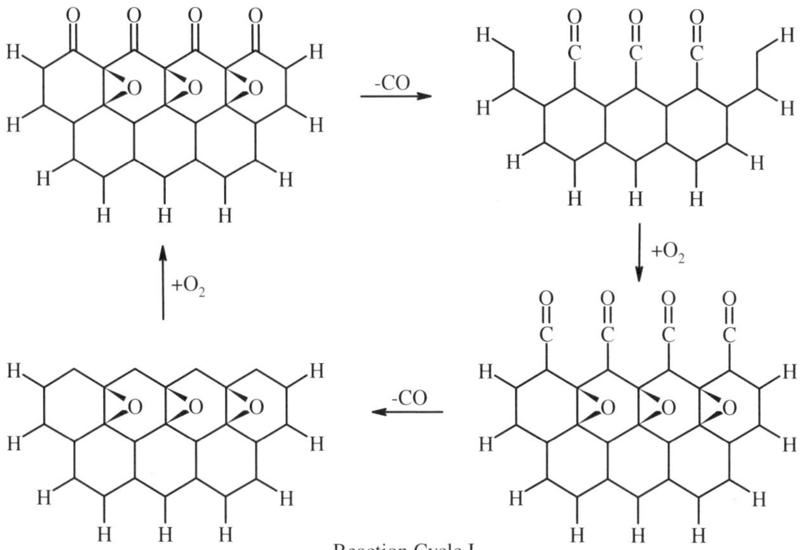

Reaction Cycle I
For reaction: $C + O_2$ (Activation energy: 58 kcal/mol)

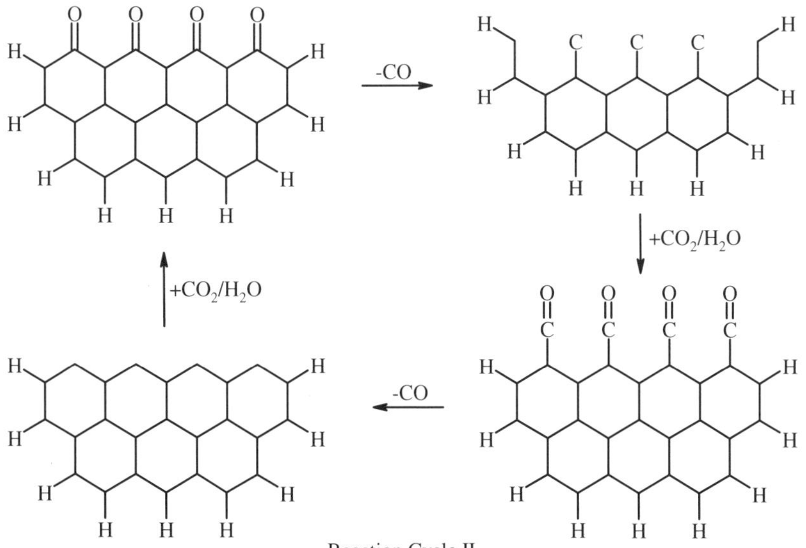

Reaction Cycle II
For reactions: $C + CO_2$ & $C + H_2O$ (Activation energy: 85 kcal/mol)

Fig. 7. Reaction pathways for graphite gasification, indicating three oxygen intermediates. (Reprinted with permission from (Chen and Yang 1998), Copyright 1998 American Chemical Society).

coincides with the decrease in the experimental activation energy from 85 kcal/mol for the reactions with $CO_2$ and $H_2O$ to 58 kcal/mol for the reaction with $O_2$.

The reaction of oxygen with graphite was studied by Backreedy et al. (Backreedy, Jones et al. 2001) by means of *ab initio* modeling. The molecular system used is the graphene model $C_{25}H_9$ based on the study of Chen and Yang (Chen and Yang 1998), see Model F in Fig. 1. The unrestricted Hartree-Fock (UHF) method was used with the 3-21G(d) basis set for geometry optimization followed by a single point energy calculation at the B3LYP DFT method with the 6-31G(d) basis set. A reaction scheme that replicates the general features that describe the reactivity of graphite was proposed for high temperatures where carbon monoxide is the dominant product. Active sites, formation of carbon monoxide, and formation of new active sites as the reaction proceeds were modeled. Dangling carbon atoms were designated as super-reactive sites. This work indicates that single activation energy assigned to the oxidation of coal char and graphite is insufficient and should have a distribution of activation energies.

Xu and Li (Xu and Li 2004) also studied the interaction of molecular oxygen with graphite by means of DFT coupled with cluster models. In particular, the interaction with the basal surface and the active sites of graphite was investigated. All the calculations were carried out at the B3LYP/6-31G(d)//3-21G(d) level. For geometry optimizations the 6-31G(d) basis set was used for oxygen atoms and the 3-21G(d) for the other atoms. It was concluded that the calculated adsorption energy for $O_2$ physisorbed on the clean basal surface (0.26 kcal/mol) is in good agreement with the experimental value. The edge sites on the zigzag and armchair surfaces exhibit high reactivity toward the adsorption and dissociation of $O_2$. Not only high adsorption energies have been obtained, but the obvious O–O bond strength weakening was found with respect to the adsorption of $O_2$ at the edge active sites. It was noted that the local detailed arrangement of edge carbon atoms can play an important effect on $O_2$ adsorption behavior.

An experimental and theoretical study about the role of substitutional boron in carbon oxidation was carried out by Radovic et al. (Radovic, Karra et al. 1998). However, simple Hückel molecular orbital (SHMO) theory was used. In a follow-up study, the effect of substitutional boron on the electronic structure and reactivity of carbon model structures has been investigated by Wu and Radovic (Wu and Radovic 2004) using *ab initio* molecular orbital calculations and frontier orbital theory. The model chemistry selected for the calculations was HF/6-31G(d)//B3LYP/6-31G(d). Boron substitution was found to decrease the global model cluster stability and to affect the local reactivity of its edge sites. For a zigzag model cluster, the reactivity of carbon active sites may be increased or decreased by boron substitution and the exact effect appears to be dependent on substituents position: in general, the reactivity of unsaturated edge sites decreases, but substitution at certain basal-plane sites may increase the reactivity of some active sites which in turn suggests a catalytic effect. For an armchair model cluster, boron substitution increases the reactivity of one or more armchair edge sites. Single atom substitution in the zigzag cluster may result in thermodynamically favorable or unfavorable $O_2$ chemisorption, the exact effect was found to be site-dependent. It also increases the energy barrier for CO desorption. Such an intriguing dual effect provides an explanation for the experimentally observed conflicting effects of boron doping in carbon oxidation.

A theoretical study using DFT to provide molecular-level understanding on the desorption of carbon monoxide from surface oxygen complexes, such as carbonyl, that are formed in the gasification and combustion of coal was performed by Montoya et al. (Montoya, Truong et al. 2001). Carbonyl models of different sizes in the zigzag, armchair, and tip shapes of the active sites were selected, as can be seen in Fig. 8. B3LYP/6-31G(d) level of theory was used in all calculations. It was found that the shape of the local active site has a strong effect on the CO desorption energy and that this behavior can contribute to the broaden feature of the CO molecule desorption in TPD experiments of oxidized carbonaceous materials. The desorption activation energy does not depend strongly on the size of the models. The calculated desorption activation energy ranging from 31 to 49 kcal/mol is in good agreement with experimental data. Molecular size convergence analyses on the carbonyl models suggest that the smallest graphene molecular system for accurate desorption structure on char is a three-ring molecule. The activation energy and normal-mode analyses for selected carbonyl complexes suggest that carbonyl surface oxygen complexes are stable structures and that they can be considered as labile surface oxygen complexes. The CO molecule desorption energy is affected by the influence of different neighboring surface oxygen groups on the carbon surface as well as the aromatic character of the molecular models.

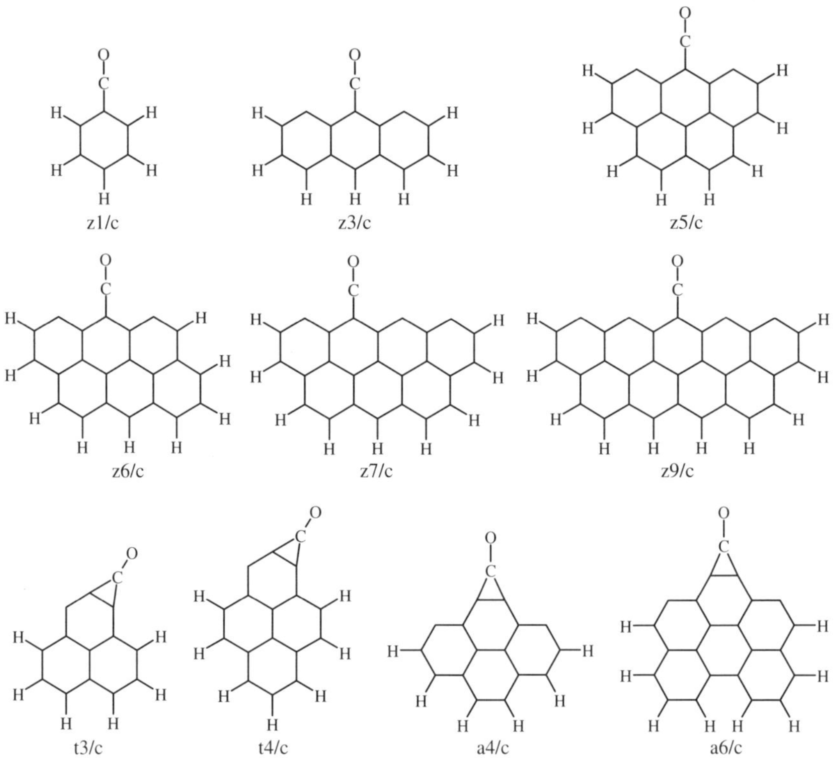

Fig. 8. Geometry representation of selected carbonyl models on a single graphene layer. (Reprinted with permission from (Montoya, Truong et al. 2001), Copyright 2001 American Chemical Society).

Further work by Montoya et al. (Montoya, Mondragón et al. 2002) presented an *ab initio* study on the kinetics of CO desorption from semiquinone carbon-oxygen species in carbonaceous surfaces. DFT, in particular B3LYP/6-31G(d) level, was used to calculate the potential energy surface information. A five six-member ring in zigzag shape was used, as shown in Fig. 9. It was found that because of the unpaired electrons located at unsaturated carbon atoms representing the active sites of char, the molecular system has several low-lying electronic states. The authors suggested that as CO desorbs, these electronic states can cross. Careful examinations of such crossing allow determining the potential energy profile for CO desorption from the semiquinone surface oxygen species. It was obtained for the first time the temperature-dependent rate constant for this process using a model surface oxygen complex. The high CO desorption activation energy suggests that semiquinone complex has an important effect on the rate-limiting step on the char gasification mechanism. The fitted Arrhenius expression for the calculated rate constants is $k(T) = 1.81 \times 10^{17} \exp[-47682/T(K)](s^{-1})$, which is within the experimental uncertainty for char gasification, indicating that the approach used in this study can be used to provide necessary thermodynamic and kinetic properties of important reactions in coal gasification processes. In the above investigation the CO desorption is assumed to take place by the simultaneous cleaving of the two C–C bonds that holds the semiquinone group.

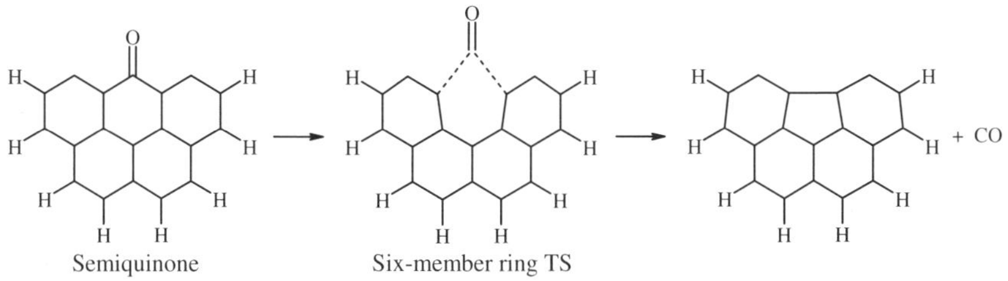

Fig. 9. CO desorption from semiquinone surface complex. (Reprinted with permission from (Montoya, Mondragón et al. 2002), Copyright 2002 American Chemical Society).

The study about kinetics of CO desorption from semiquinone groups reported by Montoya et al. (Montoya, Mondragón et al. 2002) was further undertaken by Frankcombe and Smith (Frankcombe and Smith 2004) by considering alternative paths to direct gasification of the CO fragment that were not taken into account by Montoya et al. when studying the microscopic mechanism of carbon gasification. Frankcombe and Smith analyzed reaction paths which involved the initial breaking of a single C–C bond of the carbon atom of the desorbed CO fragment in addition to the path that involves simultaneous breaking of two C–C bonds (direct gasification). These alternative mechanisms provide step-wise paths to the gasification product with significantly lower energy barriers. It was also shown that the methodology used by Montoya et al. (Montoya, Mondragón et al. 2002) had several shortcomings, the inappropriate use of unrestricted molecular orbital theory and imposing high symmetry. All molecular orbital calculations were performed at the B3LYP level of theory using the 6-31G(d,p) basis set. Relaxed scan calculations indicate that picking a simple reaction coordinate that does not adequately reflect the true minimum energy path, or imposing too high symmetry on the calculation can lead to quantitatively incorrect conclusions. Even though the calculated reaction rates are reasonably consistent with experimental data, experimentally derived rates are effectively averages over many microscopic competing paths rather than the lone oxygen, pure zigzag site desorption modeled here. Slower reactions from more stable oxygen-containing structures would reduce the observed rate.

The effect of free active sites next to the leaving semiquinone group, as those shown in Fig. 9, was recently investigated by Sendt and Haynes (Sendt and Haynes 2005) who studied the reactions of a ketone surface oxide group on the zigzag and armchair edges of a model char using density functional theory at the B3LYP/6-31G(d) level of theory in order to provide a kinetic understanding of the behavior of the ketone group and also resolving the confusion surrounding the issue of gasification from semiquinone groups on zigzag edges that exists among the studies of Montoya (Montoya, Mondragón et al. 2002) and Frankcombe (Frankcombe and Smith 2004). Both zigzag and armchair edges were characterized, with the armchair edge being the most stable due to its benzyne-like structure, followed by the Kekulé zigzag edge, with the non-Kekulé zigzag edge highest in energy. In addition, it was found that the difference between the two zigzag edges (Kekulé and non-Kekulé) is significant for chemisorption reactions, as well as oxides bound with only single bonds. A number of desorption, rearrangement, and surface migration reactions that are energetically accessible under normal reaction conditions were characterized. The rearrangement and surface migration reactions are proposed to take place on much shorter time scales than the desorption reactions, and consideration should be given to this when developing mechanisms for carbon gasification. The experimentally observed broad activation energy profile for desorption can be partially explained by the range of reactions presented in this study.

Sendt and Haynes also studied the chemisorption of molecular oxygen on the armchair (Sendt and Haynes 2005) and the zigzag (Sendt and Haynes 2005) surfaces of graphite using density functional calculations at the B3LYP/6-31G(d) level of theory. Equilibrium and transition state geometries were optimized for the chemisorption, desorption, rearrangement and migration reactions in order to provide a mechanistic understanding of the processes occurring during carbon gasification. From the computational results, two types of chemisorption behaviors were observed: Type A, predominating at low temperatures and leading to the reaction of both oxygen atoms in the adsorbing $O_2$ molecule; and Type B, where one oxygen atom is retained and one is desorbed as CO. It was concluded that the energetics and even mechanistic details of a single ketone group were found to be perturbed by both an adjacent ketone group and an adjacent 5-member ring, indicating that the chemical environment of a surface oxide has significant effect on its reaction behavior.

Zhu et al. (Zhu, Finnerty et al. 2002) conducted a comparative study of carbon gasification with $O_2$ and $CO_2$ by DFT calculations. The proposed mechanisms for carbon gasification with $O_2$ and $CO_2$ are shown in Fig. 10 using a randomly cut piece of monolayer graphite (model A in Fig. 10). It was found that $O_2$ has a strong adsorption capacity and that the dissociative chemisorption of $O_2$ is thermodynamically favorable on either bare carbon surface (consecutive edge sites) or even isolated edge sites. As a result, a large number of semiquinone and o-quinone oxygen can be formed indicating a significant increase in the number of active sites. Moreover, the weaker o-quinone C–C bonds can also drive the reaction forward at (ca. 30%) lower activation energy. Epoxy oxygen forms under relatively high $O_2$ pressure, and it increases the number of active sites. $CO_2$ has a lower adsorption capacity. Dissociative chemisorption of $CO_2$ can only occur on two consecutive edge sites and o-quinone oxygen formed from $CO_2$ chemisorption is negligible. Therefore, $CO_2$–carbon reaction needs

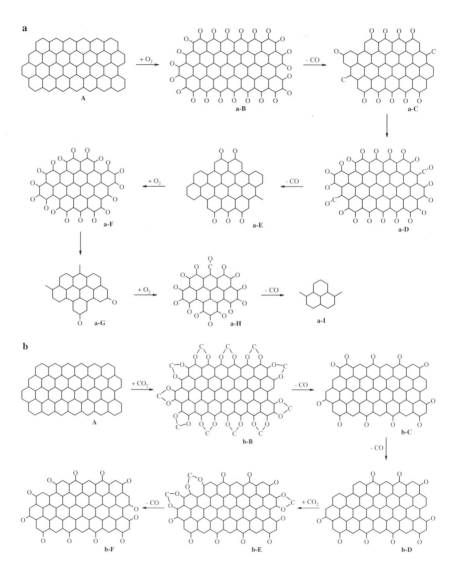

Fig. 10. Models for different mechanisms of $O_2$ and $CO_2$ reactions. (Reprinted with permission from (Zhu, Finnerty et al. 2002), Copyright 2002 American Chemical Society).

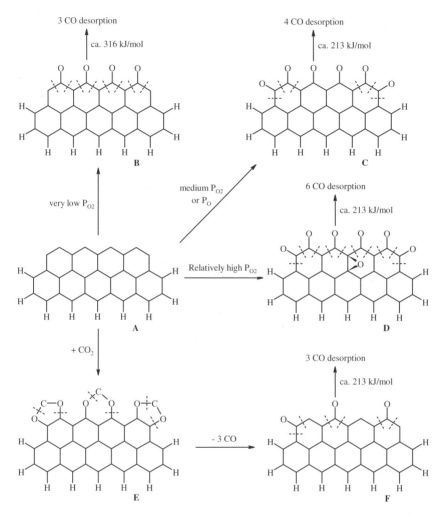

Fig. 11. Models showing the new unified mechanism. (Reprinted with permission from (Zhu, Finnerty et al. 2002), Copyright 2002 American Chemical Society).

(ca. 30%) higher activation energy. A combination of the higher activation energy and the fewer active sites leads to the much lower reaction rate of $CO_2$–carbon reaction. From results thus obtained and the previous unified mechanism proposed by Chen and Yang (Chen and Yang 1998), a new unified mechanism is proposed (see Fig. 11). On the basis of these results, the real reason for the lower activation energy of the $O_2$–carbon reaction is the formation of o-quinone rather than epoxy oxygen (as in the old unified mechanism), the formation of the latter only increases the number of active sites, but not further decreases the activation energy. However, in the proposed mechanism there is no explanation on how the epoxy group can be formed from molecular oxygen.

In a following study, Zhu et al. (Zhu, Finnerty et al. 2002) carried out molecular orbital theory calculations of the $H_2O$-carbon reactions. Although the experimental gasification research is vast, the mechanism for $H_2$ formation during carbon gasification with steam is still unclear. In this work, DFT method (B3LYP/6-31G(d,p)//B3LYP/3-21G(d,p)) was used to demonstrate a favorable energetic pathway where $H_2O$ is first physically adsorbed on the active sites of the graphite surface with negligible change in molecular structure (see Fig. 12). Chemisorption occurs via the O atom approaching the carbon edge site with one H atom stretching away from the O atom in the transition state. This is followed by a local minimum state, in which the stretching H atom is further disconnected from the O atom and the remaining OH group is still on the

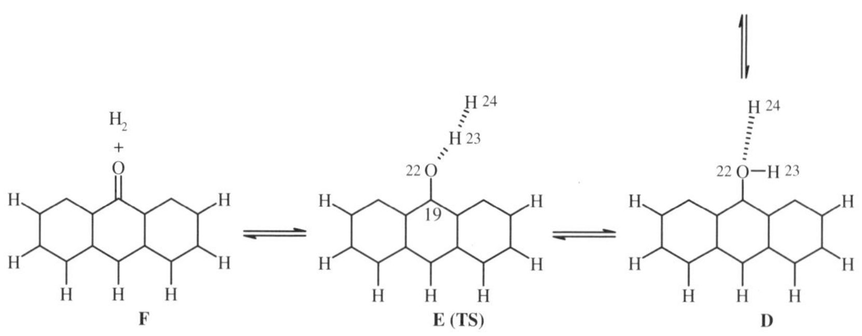

Fig. 12. Pathway from $H_2O$ adsorption to $H_2$ formation. (Reprinted with permission from (Zhu, Finnerty et al. 2002), Copyright 2002 American Chemical Society).

carbon edge site. The disconnected H atom then pivots around the OH group to bond with the H atom of the OH group and forms $H_2$. The O atom remaining on the carbon edge site is subsequently desorbed as CO. The reverse pathway occurs when $H_2$ reacts with the surface oxygen to produce $H_2O$. It is concluded that the proposed mechanism is in good agreement with all the experimental observations. First, it explains that $H_2O$ is first physically adsorbed, which is a slightly exothermic process. Second, it shows good agreement with the experimentally obtained $\Delta H$ for the total reaction $H_2O + C \rightarrow H_2 + CO$. Third, the mechanism is consistent with the reverse reaction, i.e., $H_2 + C(O) \rightarrow H_2O + C$, being possible. Finally, this study can do further justice to results, such as inhibition and anisotropy in the $H_2O$ carbon reaction.

Formation of CO precursors during char gasification with $O_2$, $CO_2$, and $H_2O$ was analyzed by Montoya et al. (Montoya, Mondragón et al. 2002) using DFT at the B3LYP/6-31G(d) level of theory. Four different oxygen groups were obtained by the chemisorption process. $CO_2$ can form lactones and heterocyclic complexes while $H_2O$ and $O_2$ can form phenols and peroxide surface complexes, respectively (see Fig. 13). Formation of all four carbon-oxygen complexes is exothermic. The transformations of the carbon-oxygen complexes described above into stable semiquinone groups were systematically examined and are presented in Fig. 13 along with their corresponding heats of reaction. It is seen that $CO_2$ complexes (lactone and heterocyclic compound) can transform into semiquinone groups by an endothermic process. However, the energy required for such transformation is much smaller than the energy released during the complex formation, thus such transformations are thermodynamically possible. Transformations of $H_2O$ and $O_2$ complexes into semiquinone groups are thermodynamically possible via exothermic processes. Based on the above results it can be said the semiquinone groups can be formed from different surface complexes, since this group is highly stable, its decomposition becomes a common rate-limiting step during gasification. These results suggest that it is possible to develop an unified mechanism for carbon gasification involving all of the oxidizing agents.

Recently, Sanchez and Mondragon (Sánchez and Mondragón 2007) carried out a DFT study to evaluate $CO_2$ desorption in combustion/

gasification reactions. All calculations were performed at the B3LYP level of theory using the 6-31G(d) basis set. Two possible routes for heterogeneous $CO_2$ desorption during oxidation of carbon materials with molecular oxygen were proposed. The first one takes place by CO

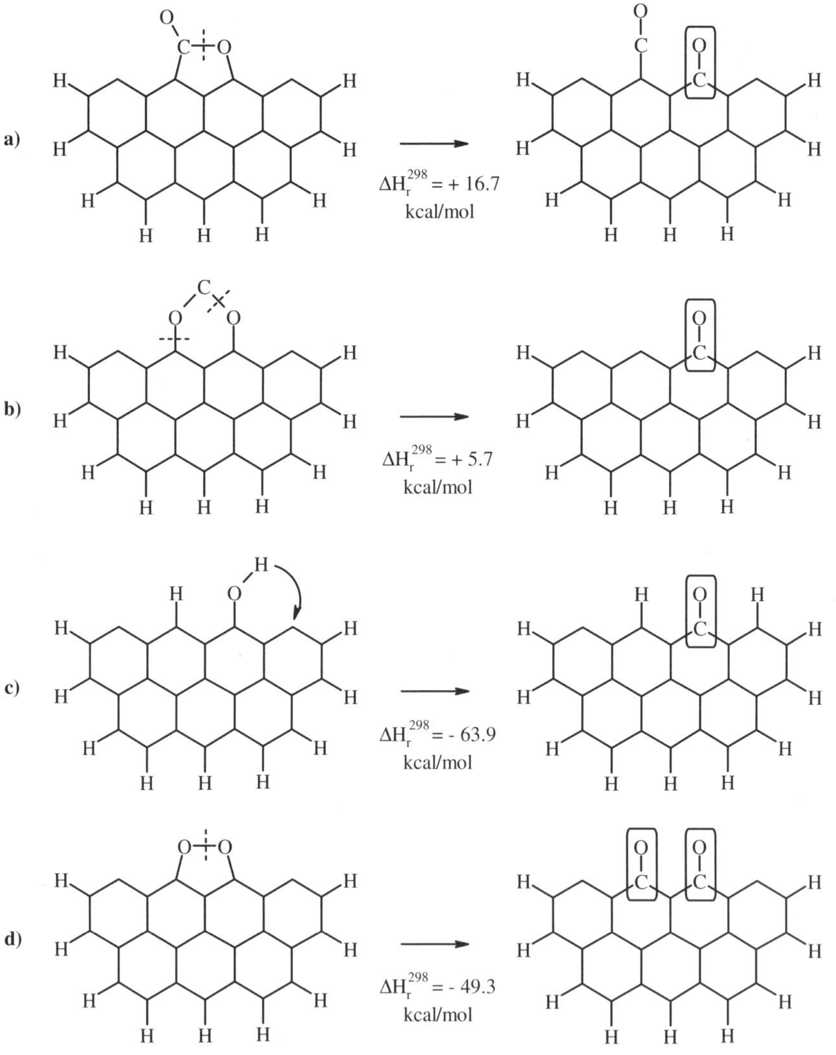

Fig. 13. Surface transformation reactions to semiquinone groups. (Reprinted from (Montoya, Mondragón et al. 2002), Copyright 2002, with permission from Elsevier).

re-adsorption on an oxidized surface that produces a carbonate group which decomposes as $CO_2$ with an energy barrier of 60 kcal/mol. The second route involves $O_2$ molecular chemisorption on an oxidized surface that leads to the formation of a dioxiranyl complex which can evolve as $CO_2$ with an energy barrier of 48 kcal/mol. Additionally, it was found that the presence of an epoxy complex near the edge of the graphene layer facilitates the $CO_2$ desorption leaving a cyclic ether complex on the edge after the dioxiranyl decomposition. It was concluded that the suggested mechanisms are useful for explaining several experimental observations concerning CO and $CO_2$ desorption at high temperatures as well as the role of epoxy groups in combustion and gasificaition reactions. Detailed analysis of bond order changes facilitates the understanding of formation, transformation, and cleavage of bonds involved in a given reaction.

## 4. $CO_2$ Interaction with Carbon Surfaces

Carbon gasification with $CO_2$ is a well documented reaction in the scientific literature (Sakawa, Sakurai et al. 1982; Kapteijn and Moulijn 1983; DeGroot and Shafizadeh 1984; Kapteijn, Abbel et al. 1984; Freund 1985; Kelemen and Freund 1986; Koenig, Squires et al. 1986; Kwon, Kim et al. 1988; Meijer, Weeda et al. 1991; Suzuki, Ohme et al. 1992; Suzuki, Ohme et al. 1994; Molina, Montoya et al. 1999) to produce gaseous fuels or to tailor the porosity properties of the char (Rodriguez-Mirasol, Cordero et al. 1993; Linares-Solano, Salinas-Martinez de Lecea et al. 2000). Some of these studies have proposed mechanisms for the reactions (Suzuki, Ohme et al. 1992; Suzuki, Ohme et al. 1994) but the system is very complex and more exhaustive investigations need to be done. The detailed knowledge of the mechanism of this reaction is important in the design of gasification and combustion reactors. An approach to this mechanism can be achieved by means of molecular modeling. The interaction of $CO_2$ with carbon models has been studied in some of the investigations already discussed in the previous section (Chen, Yang et al. 1993; Chen and Yang 1998; Montoya, Mondragón et al. 2002; Zhu, Finnerty et al. 2002).

Chen and Yang (Chen and Yang 1993) performed molecular orbital theory calculations using the INDO method in order to compare the relative adsorption strengths of $CO_2$ and/or $SO_2$ on the edge plane and the basal plane of graphite. It was found that, for $CO_2$, the total energy change on the edge plane is six times that on the basal plane. For $SO_2$, the change is only slightly stronger on the edge plane. This result is in agreement with the experimental fact that $CO_2$ has a strong preference for the edge plane whereas $SO_2$ adsorbs on both planes (edge and basal) with almost equal strengths. It should be noted that the absolute values of the energy change from these calculations are overestimated and the equilibrium positions are underestimated. However, the relative values are generally correct and are valuable for comparison purposes.

The adsorption of $CO_2$ on carbonaceous surfaces was evaluated by Montoya et al. (Montoya, Mondragón et al. 2003) using DFT at the B3LYP/6-31G(d) level of theory to study the $CO_2$ interaction on the edge of the carbonaceous models. The interaction on the basal plane was studied at the MP2/6-31G(d)//B3LYP/6-31G(d) level of theory since dispersion interactions are not accounted for in DFT. At low surface coverage it was found that $CO_2$ forms stable surface oxygen complexes with the active sites of the char. Three stable carbon-oxygen complexes are formed due to the interaction of $CO_2$ with a clean carbon model, i.e. lactone, heterocyclic, and furan-type complexes. As the surface coverage increases, the $CO_2$ adsorption energy decreases due to the presence of surface oxygen atoms and by the formation of new surface oxygen complexes, such as carbonates. In the high-coverage region, the theoretical result shows low adsorption energy, suggesting that $CO_2$ molecules chemisorb on the surface oxygen complexes and on the graphene planes. The calculated results for heats of adsorption are consistent with experimental values.

The mechanism of $CO_2$ chemisorption on zigzag carbon active sites was also studied by Radovic (Radovic 2005) using a graphene layer model (see Fig. 14) in order to illustrate the sensitivity of the results obtained to the assumptions made regarding the electronic configuration of the graphene layer. DFT (B3LYP) was used with the 6-31G(d) basis set. Various electronic environments of the four ring model, with and

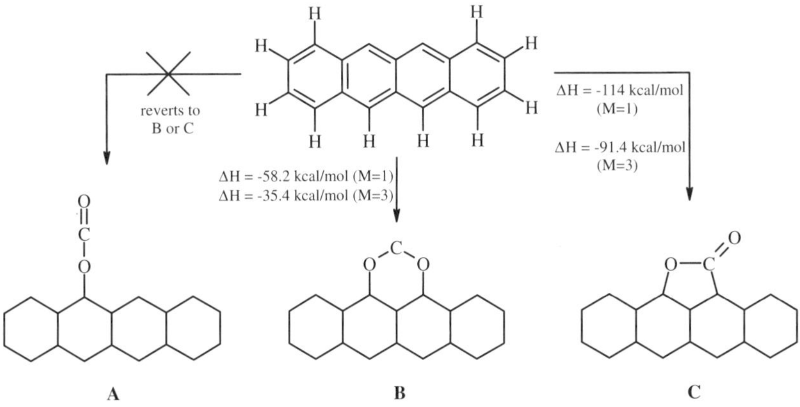

Fig. 14. Three hypothetical pathways for chemisorption of $CO_2$ on the graphene layer having two free and adjacent zigzag sites. (Reprinted from (Radovic 2005), Copyright 2005, with permission from Elsevier).

without surface oxygen and with special emphasis on carbene-like structures were analyzed. It was found that the resulting optimized geometries were quite sensitive to such electronic configurations. In agreement with experimental evidence, dissociative $CO_2$ adsorption was found to be particularly favorable. Furthermore, dissociation was found to be favored on isolated carbene-like zigzag sites, as can be seen in Fig. 15. It is recommended that such a pathway, rather than the Montoya et al. (Montoya, Mondragón et al. 2003) dual site adsorption and C–$CO_2$ complex formation, deserves the dominant attention in further theoretical studies of adsorption, reaction, and desorption processes during $CO_2$ gasification of carbons.

Xu et al. (Xu, Irle et al. 2006) studied dissociative adsorption reactions of $CO_x$ and $NO_x$ ($x = 1$, 2) molecules on the basal graphite surface based on potential energy surfaces obtained using ONIOM. The B3LYP/6-31+G(d) method was selected as high-level theory for a coronene model system and the DFTB-B method was employed at the low level lo include the horizontal bulk effect using a larger dicircumcoronene real system. This ONIOM approach and model systems have been successfully tested in a previous study of the interaction of water clusters on the basal plane of graphite (Xu, Irle et al.

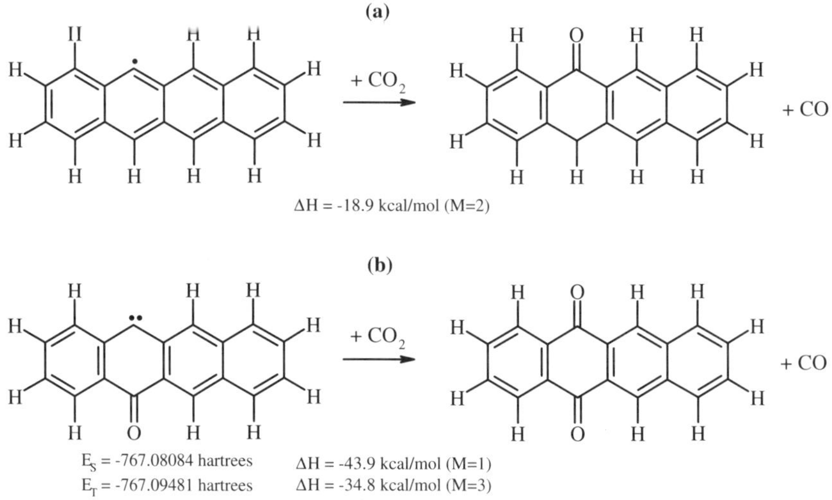

Fig. 15. Proposed principal pathway for $CO_2$ chemisorption on zigzag sites in a graphene layer: (a) advanced stages of gasification (burnoff level > 0%); (b) initial stage of gasification (burnoff level = 0%). (Reprinted from (Radovic 2005), Copyright 2005, with permission from Elsevier).

2005). For the adsorption of $CO_2$ on graphite, the CO + O dissociation was found to be the primary irreversible dissociation path. For the adsorption of NO on graphite, it was found that it is possible to create an N defect on graphite with overall exothermicity involving elimination of CO species. Rate constants were predicted for the dissociative adsorption reactions of $CO_x$ and $NO_x$ on graphite using RRKM. Overall, it was concluded that it is very difficult to attack the (0001) graphite surface even by radical species, but pathways for erosion feasible at high temperatures were found.

## 5. Hydrogenation Reactions

Carbon hydrogenation reactions are important in several processes such as gasification and liquefaction. Hydrogasification of carbonaceous materials leads to hydrogenated structures that could be the precursors of hydrocarbons evolution, mainly methane (also called: synthetic natural gas). Therefore, this process has been studied as a source of substitute natural gas. Even though carbon-hydrogen reactions are relatively simple, their mechanisms are still not completely clear.

Most molecular modeling studies of the carbon-hydrogen reactions were focused on the reaction on the basal plane of graphite (Bennett, McCarroll et al. 1971; Bennett, McCarroll et al. 1971; Dovesi, Pisani et al. 1976; Dovesi, Pisani et al. 1981; Casanas, Illas et al. 1983; Caballol, Igual et al. 1985; Barone, Lelj et al. 1987; Jeloaica and Sidis 1999), in spite of the fact that graphite edges have much higher reactivity. Moreover, nearly all of these studies were carried out using semi-empirical methods. Therefore, those studies will not be included in this review and we will center our attention in reactions taking place on the edges of graphene layers.

There are not many publications about the reaction of hydrogen on the edge sites of graphite. To the best of our knowledge, Pan and Yang (Pan and Yang 1990) were the first to study the mechanism of methane formation from the reaction between graphite and hydrogen using EHT (Extended Hückel Theory). The proposed mechanism is shown in Fig. 16. By evaluating the total overlap population (TOP) around edge carbon atoms it was found that on a clean surface the TOP on the zigzag edge atom is lower than that of the armchair atom, hence the zigzag face is more reactive. After one hydrogen atom is added, the same comparison holds but with a smaller difference. After two hydrogen additions, the situation is reversed and the armchair edge atom becomes more reactive. For the two edge planes, surface C–C bond cleavage is

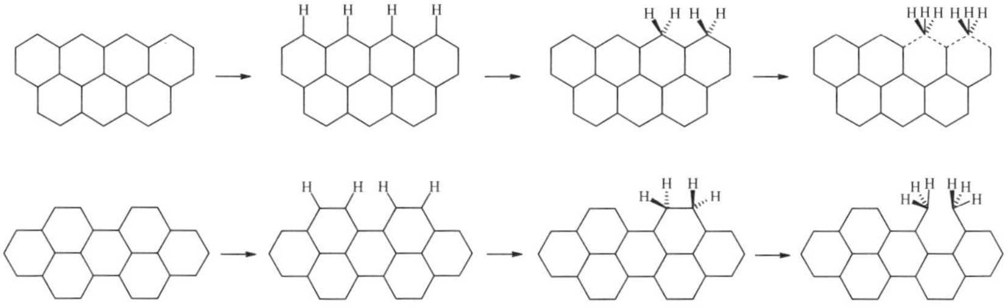

Fig. 16. Mechanism for methane formation on zigzag (upper) and armchair (lower) faces by successive H addition. (Reprinted from (Pan and Yang 1990), Copyright 1990, with permission from Elsevier).

required for the third hydrogen addition, after which dangling bonds remain and methane formation is energetically very favorable. The edge surface carbon atoms become saturated and inactive for further chemisorption after two hydrogen addition. Consecutively, the third hydrogen addition which requires C–C bond breakage is the rate limiting step in methane formation. Unfortunately, no thermodynamic or kinetic information was presented for this study.

The mechanisms for methane and ethane formation in the reaction of hydrogen with carbonaceous materials were recently studied by Espinal et al. (Espinal, Mondragón et al. 2005) using DFT. Figures 17 and 18 show the proposed mechanisms for the initial stages of methane and ethane formation on zigzag and armchair edges, respectively. These mechanisms are similar to the ones proposed by Pan and Yang (Pan and Yang 1990). However, hydrogen migration involving C–C bond cleavage is assumed to take place after the second hydrogen molecule addition instead of a third hydrogen molecule addition. Since the single C–C bonds are the weakest C–C bonds in the system and the reaction is taking place at rather high temperatures, a reasonable route for formation

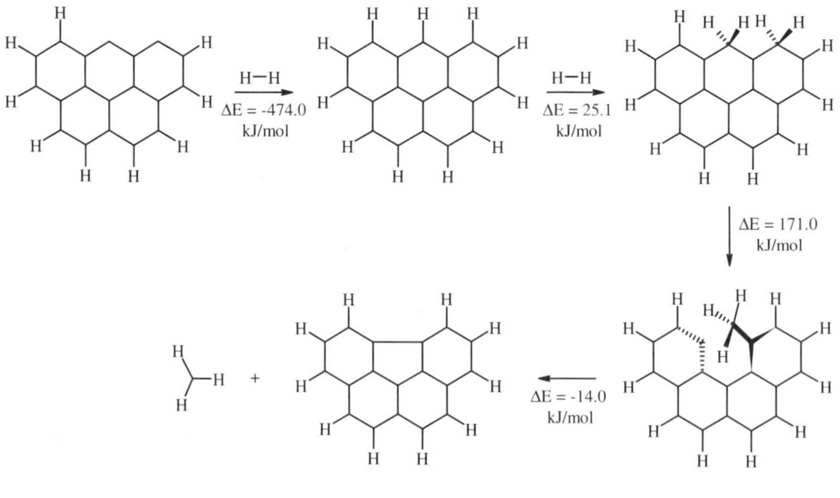

Fig. 17. Proposed mechanism for the initial stages of methane formation from the zigzag edges of a hypothetical graphene layer. (Reprinted from (Espinal, Mondragón et al. 2005), Copyright 2005, with permission from Elsevier).

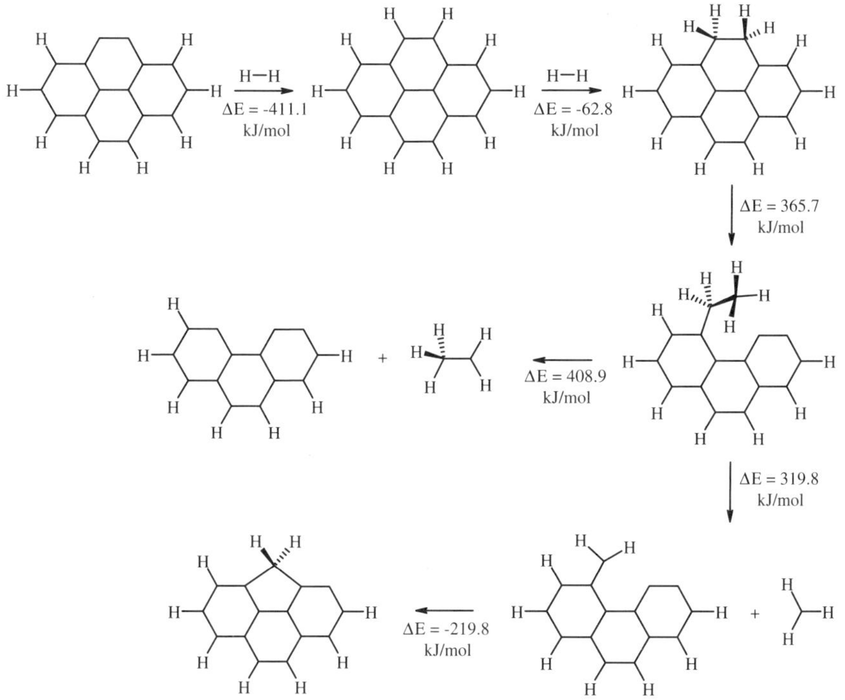

Fig. 18. Proposed mechanism for the initial stages of methane and ethane formation from the armchair edges of a hypothetical graphene layer. (Reprinted from (Espinal, Mondragón et al. 2005), Copyright 2005, with permission from Elsevier).

of methane and ethane is from dissociation of these single C–C bonds to form methyl and ethyl radicals. Once methyl and ethyl radicals are in the gas phase, they can form methane and ethane by hydrogen abstraction from $H_2$ or radical recombination with available hydrogen atoms or other methyl radicals. It was determined that the reaction rate limiting step is the hydrogen migration. Transition states and rate constants were calculated for the rate limiting steps. It was found that from a thermodynamic and a kinetic point of view it is more favorable to produce methane from zigzag edges.

A DFT study of adsorption of molecular hydrogen on graphene layers was carried out by Arellano et al. (Arellano, Molina et al. 2000). In addition, adsorption of atomic hydrogen on graphite was studied by

Yang and Yang (Yang and Yang 2002) performing *ab initio* molecular orbital calculations. Nevertheless, the main purpose of both studies was to investigate hydrogen storage on carbon nanotubes instead of the hydrogasification reaction and further details of those studies will not be explained in this review.

On the other hand, there is a lack of published information on the hydrogenation of heteroatoms on carbon materials such as nitrogen and sulfur functionalities. There is a recent publication in this area carried out by Espinal, et al. (Espinal, Truong et al. 2007) who studied mechanisms of $NH_3$ formation during the interaction of $H_2$ with nitrogen containing carbonaceous materials using DFT. The proposed mechanisms (see Figs. 19 and 20) involve consecutive hydrogenation steps and rearrangements that produce $C-NH_2$ groups, which can be released to the gas phase by dissociation of the $C-N$ bond producing $NH_2$ radicals. As a last step, $NH_3$ can be produced by hydrogen abstraction from the hydrogenated carbonaceous structures (heterogeneous reaction) or by hydrogen abstraction from $H_2$ molecules or by radical recombination with available hydrogen atoms (homogeneous reactions), as it is shown in Fig. 21.

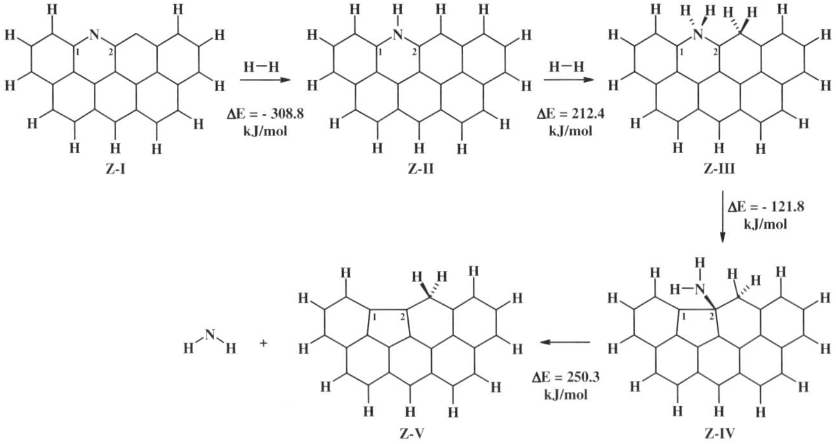

Fig. 19. Reaction of the $H_2$ molecule with a nitrogen containing carbonaceous model on zigzag configuration and thermodynamic data. (Reprinted from (Espinal, Truong et al. 2007), Copyright 2007, with permission from Elsevier).

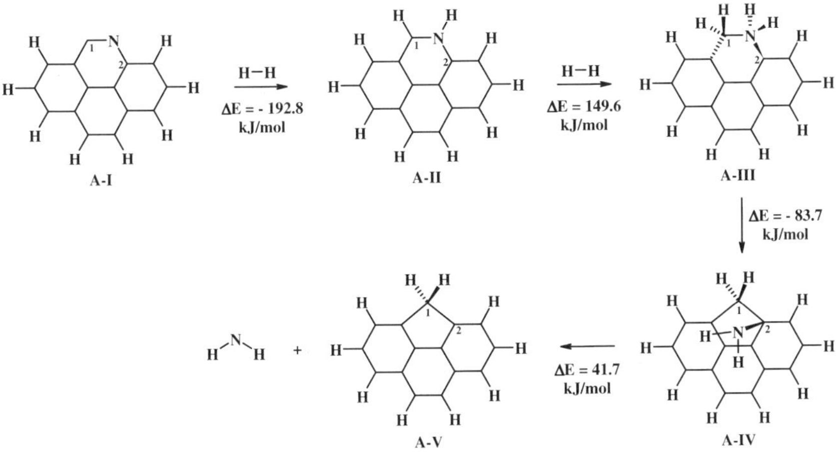

Fig. 20. Reaction of the $H_2$ molecule with a nitrogen containing carbonaceous model on armchair configuration and thermodynamic data. (Reprinted from (Espinal, Truong et al. 2007), Copyright 2007, with permission from Elsevier).

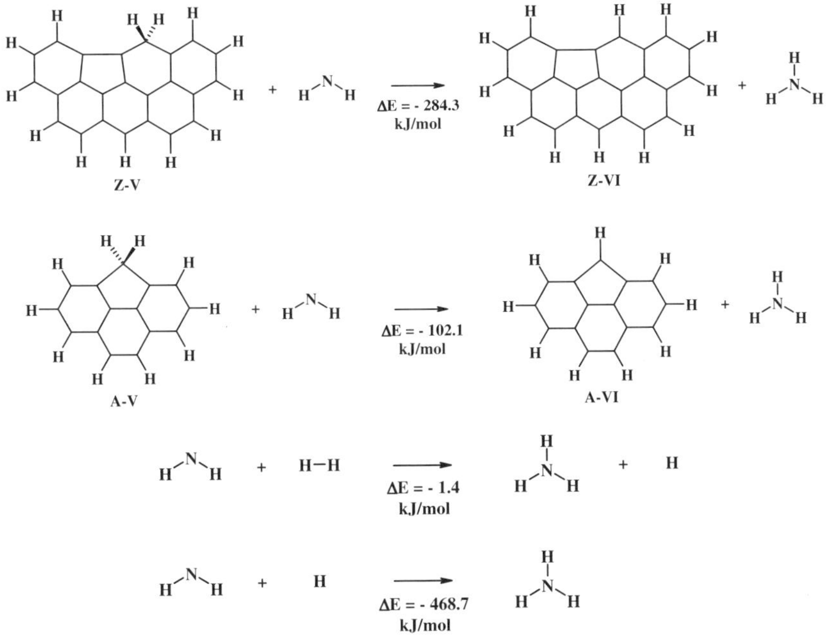

Fig. 21. Heterogeneous and homogeneous reactions that produce $NH_3$. (Reprinted from (Espinal, Truong et al. 2007), Copyright 2007, with permission from Elsevier).

## 6. CO Interaction with Carbon Surfaces

During gasification and combustion of carbon materials there is a high concentration of CO in the environment surrounding the carbon particles. Therefore, re-adsorption reactions can take place. Marchon et al. (Marchon, Tysoe et al. 1988) based on experimental results proposed that CO can adsorb on a five-member ring to produce a semiquinone group. Hall and Calo (Hall and Calo 1989) carried out TPD experiments to investigate secondary reactions of CO on the carbon surface. They found that some of the $CO_2$ produced during the reaction can be ascribed to the attack of the CO molecule on oxygenated surface complexes. However, no mechanistic explanation was given for the results.

A more detailed analysis of the possible reactions that can take place with a carbon surface was investigated by Espinal et al. (Espinal, Montoya et al. 2004) using DFT (B3LYP) level of theory with the 6-31G(d) basis set. It was found that CO can be adsorbed exothermically on the active sites of zigzag, armchair, and tip carbonaceous models to yield stable intermediates, such as cyclic ether, carbonyl, ketone, and semiquinone (on clean surfaces, see Fig. 22); lactone and carbonate (on oxidized surfaces). Adsorption of CO during the gasification process blocks the active sites of the carbonaceous material and thus can reduce the efficiency of the process. Furthermore, it was found that when CO is adsorbed in a carbonyl type structure (C=C=O) there is a reversible interconversion process by ring closure with a neighbor active site to produce a cyclic ether (furan type, see Fig. 23). Consequently, the available number of active sites for gasification reaction is decreased and therefore the gasification reaction is inhibited. The presence of oxygenated groups, such as semiquinone, on a carbonaceous surface before CO adsorption makes the adsorption energy less exothermic, except for lactone formation. Several pathways for $CO_2$ desorption after consecutive CO adsorption on clean and oxidized surfaces were proposed. This reaction can either reduce the surface by removing a surface oxygen atom or leaving a carbon atom on the surface. Both effects lead to inhibition or retardation of the gasification reaction. The $CO_2$ desorption through CO disproportionation reaction also provides useful insight into the mechanism for single-walled nanotubes (SWNT) growth; see proposed reactions in Fig. 24. The formation of five member rings induces the formation of curved surfaces.

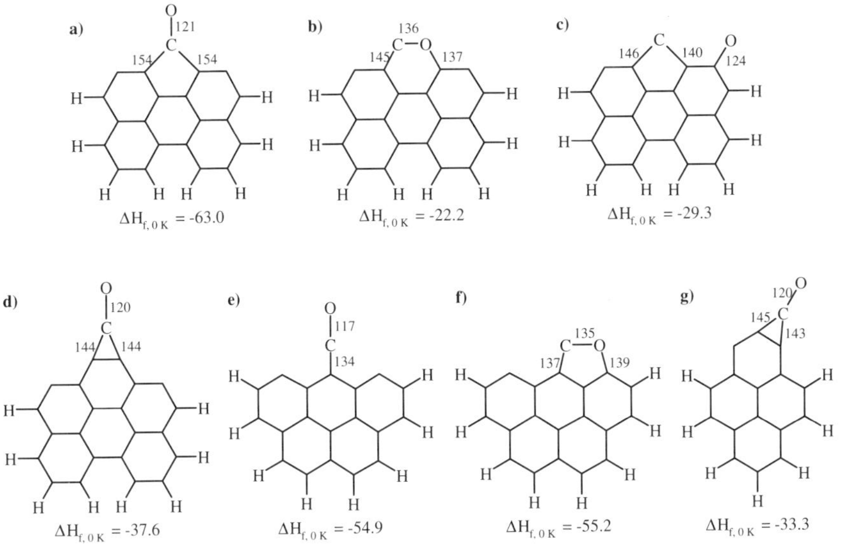

Fig. 22. Structures of surface oxygen complexes formed after interaction of the CO molecule with carbonaceous models. Selected bond lengths are in pm. Formation enthalpies are also shown for each complex in kcal/mol. (Reprinted with permission from (Espinal, Montoya et al. 2004), Copyright 2004 American Chemical Society).

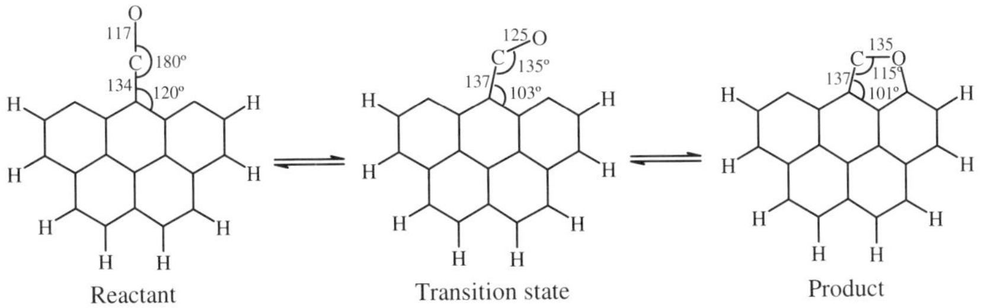

Fig. 23. Reaction for transformation of carbonyl into ether model. Selected bond lengths are in pm and angles are in degrees. (Reprinted with permission from (Espinal, Montoya et al. 2004), Copyright 2004 American Chemical Society).

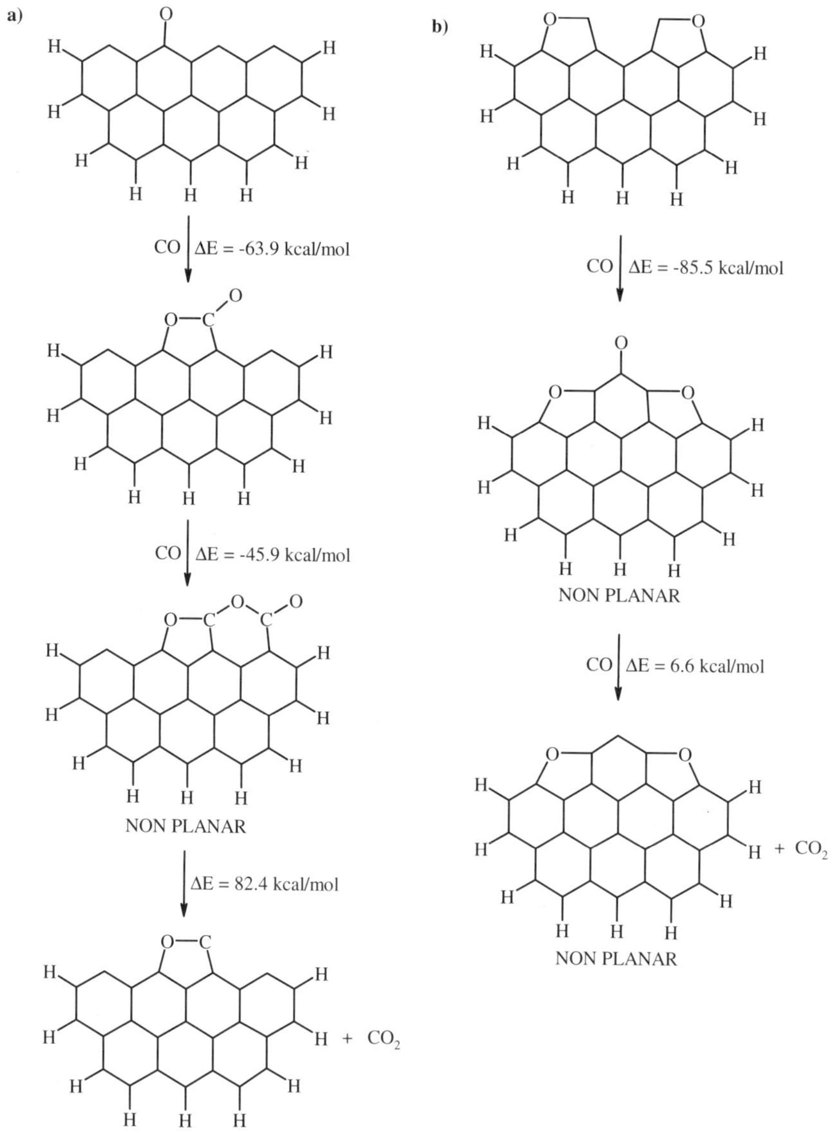

Fig. 24. CO consecutive adsorption that favors: a) $CO_2$ desorption and b) six-member rings formation. (Reprinted with permission from (Espinal, Montoya et al. 2004), Copyright 2004 American Chemical Society).

## 7. Catalyzed Reactions

It is well known that carbon gasification reactions are catalyzed by, i.e., alkali and alkaline earth oxides and salts as well as transition-metal oxides added to the carbon substrate. Voluminous literature has been devoted to this area. Many types of intermediates and mechanisms have been proposed for the catalyzed gasification reactions. However, those mechanisms only account for some of the experimental observations. Therefore, the mechanisms of this reaction are still unclear. Molecular modeling has proved to be a useful tool for answering some of the questions that experimentalists have not been able to address.

There are not many publications on molecular modeling for catalysis of the carbon gasification reactions. Janiak et al. (Janiak, Hoffmann et al. 1993) studied the potassium promoter function in the oxidation of graphite using the tight-binding extended Hückel method on an infinite, two dimensional surface. In addition to the reaction on the basal plane of graphite with and without catalyst, the reaction of $O_2$ on surface defects, such as vacancies, was also studied. It was found that in the presence of potassium on the surface the $O_2$ dissociation barrier is significantly reduced, when compared to the dissociation in the absence of the catalyst. The $O_2$ molecule strongly adsorbs on top of a potassium atom, with no interaction on the graphite layer, and can dissociate into oxygen atoms. The promoter effect of potassium surpasses that of surface defects and distortions. The potassium catalyst's role can be described as enabling the sticking of $O_2$ and subsequent dissociation on graphite at a surface-$O_2$ distance where normally only physisorption occurs.

The interaction of potassium and oxygen with the basal plane of graphite, individually and as coadsorbates was performed by Lamoen and Persson (Lamoen and Persson 1998) by first-principles methods within the density functional formalism. It was shown that although the main physics is correctly described by a single graphite layer, non-negligible corrections arise from the second graphite layer. When potassium is adsorbed on the graphite surface, $O_2$ chemisorbs at the potassium which is consistent with the large sticking coefficient observed for $O_2$ on a potassium covered surface, whereas the $O_2$-graphite interaction is found to be repulsive on a clean surface. The $O_2$ molecule binds not on-top of the K atom, as has been suggested earlier (Janiak, Hoffmann et al. 1993), but rather side on. The energy barrier towards dissociation of $O_2$ on the clean graphite surface is estimated to be similar to that of gas phase $O_2$. For $O_2$ on K/graphite the barrier for dissociation is much smaller than for $O_2$ on the clean graphite surface.

The studies of Janiak (Janiak, Hoffmann et al. 1993) and Lamoen (Lamoen and Persson 1998) concentrated on the basal plane of graphite, they are not practical since graphite edges are much more reactive. The active surface species in alkali-catalyzed carbon gasification was studied by Chen and Yang (Chen and Yang 1993) performing CNDO semi-empirical molecular orbital calculations on model graphite substrates with –O and –O–K groups bonded to the zigzag and armchair faces, as shown in Figs. 25 and 26. On the zigzag face, the carbon atom bridging two C–O–K groups (C28) gains a large negative charge (–0.486) and

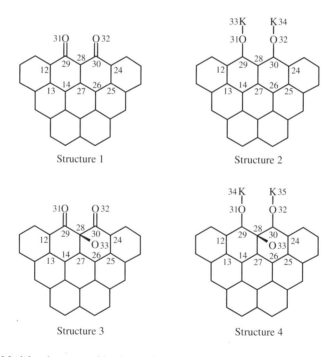

Fig. 25. Model substrates with zigzag face on graphite for CNDO molecular orbital calculation. (Reprinted from (Chen and Yang 1993), Copyright 1993, with permission from Elsevier).

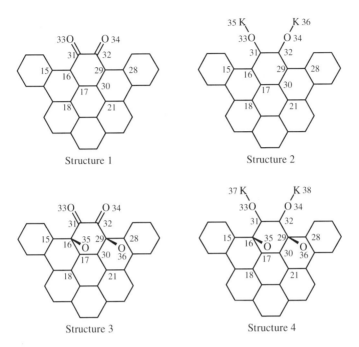

Fig. 26. Model substrates with armchair face on graphite for CNDO molecular orbital calculation. (Reprinted from (Chen and Yang 1993), Copyright 1993, with permission from Elsevier).

hence is a favorable site for binding an O atom. Consequently, more oxygen chemisorption on the bridge carbon atom would occur with the presence of potassium. The surface C–C bonds in this structure are substantially weakened by adding O atoms on the bridging C atoms, leading to CO release. The O atoms can be supplied by the dissociation of $H_2O$ and $CO_2$. The rate of carbon gasification should be directly proportional to the concentration of the CO species on the bridge carbon. Therefore, the presence of potassium increases the reaction rate. This catalytic mechanism can also explain the observation that the activation energies were not changed in the alkali and alkaline earth catalyzed carbon gasification reactions by $H_2O$ and $CO_2$. The CNDO results showed that with or without potassium, the C–C bonds on the surface (Structures 3 and 4 in Fig. 25) were equally weakened. Therefore, one would not expect a decrease in the activation energy by potassium. The

CNDO results also predict that the C–O–K groups have an inhibiting effect on the armchair face since the C–C bonds were actually strengthened.

In a follow-up study, Chen and Yang (Chen and Yang 1997) carried out molecular orbital calculations to propose a unified mechanism of alkali and alkaline earth catalyzed gasification reactions of carbon by $CO_2$ and $H_2O$ using MNDO type of approximation methods since CNDO is not accurate enough for calculating bond energies. The substrates used in this study are the same as the ones used previously (Chen and Yang 1993). By studying net charges, bond energies, and bond orders, a more comprehensive mechanism for alkali and alkaline earth catalyzed gasification reactions of carbon by $H_2O$ and $CO_2$ was proposed on the basis of the off-plane oxygen species. The following mechanism was proposed using $CO_2$ as the reactant:

$$CO_2 + * \leftrightarrow CO + O* \qquad K'$$

Diffusion of O*

$$O* + C_f \leftrightarrow C_f(O*) \qquad K_1''$$

$$O* + C_f(O*) \leftrightarrow C(O*)C_f(O*) \qquad K_2''$$

$$C(O*)C_f(O*) \rightarrow CO + CC_f(O*) + * \qquad i_3$$

$$C_f(O*) \rightarrow CO + C_f* \qquad i_4$$

where $K$ stands for equilibrium constant and $i$ for rate constant. * represents the catalyst cluster which is a nonstoichiometric compound and could be described as $M_xO_y$ (where M is the alkali or alkaline earth metal atom and $x$ and $y$ are changeable during the gasification reaction). The symbols $C_f(O*)$ and $C(O*)C_f(O*)$ represent two main groups of oxygen-containing structures, i.e., the in-plane group (semiquinone) and off-plane group (bonded to two saturated carbon atoms that are adjacent to the semiquinone species). The two oxygen intermediates are shown in Fig. 27. In this mechanism the first step takes place at the catalyst surface. The second step, diffusion, may be accomplished by electron transfer between oxygen atoms in the cluster. So the migration of O* is very fast and the diffusion step may be omitted. Like the uncatalyzed reactions, the C–C bond breakage is the rate-limiting step. The proposed

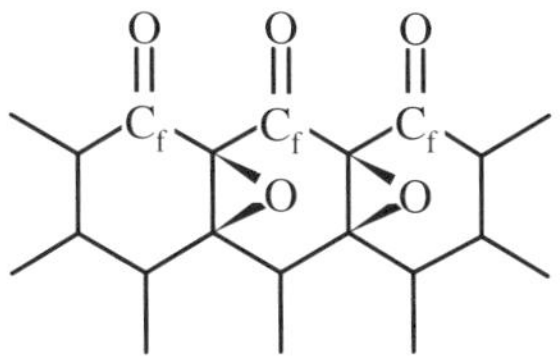

Fig. 27. Schematic showing two surface oxygen intermediates. $C_f(O)$ is the semiquinone type species. $C(O)C_f(O)$ is the species with an adjacent off-plane oxygen which weakens the $C_f$–$C$ bonds. (Reprinted with permission from (Chen and Yang 1997), Copyright 1997 American Chemical Society).

mechanism not only provides an explanation of why catalysts can change gasification rates without changing the activation energy but also gives an explanation of how the catalysts change the concentration of active complexes. However, only zigzag structures were studied.

Backreedy et al. (Backreedy, Jones et al. 2002) studied the reaction of oxygen with carbon catalyzed by metals inherent within the char matrix. Empirical and *ab initio* modeling were used to calculate geometries for the structures shown in Fig. 28. Geometry optimizations were carried out at the UHF/3-21G(d) followed by B3LYP/6-31G(d) single point energy calculations. The catalysis of metallic atoms in a graphene matrix was simulated with a model structure where an iron atom is bonded to the graphene molecule as shown in Fig. 28. The results show that the structure in Fig. 28b is sterically more probable. It was found that iron weakens the $C_5$–$C_6$ and $C_6$–$C_7$ bonds providing an easy route for CO desorption which can further react to form $CO_2$. However, this study was based on geometries and no thermodynamic information is provided.

Recently, Domazetis et al. (Domazetis, Raoarun et al. 2007; Domazetis, Raoarun et al. 2008) carried out molecular modeling studies on steam gasification of coal catalyzed by iron species performing semiempirical and DFT calculations. The formation of $H_2$ was evaluated via H abstraction from O–H and C–H groups by iron clusters. It was found that H abstraction and $H_2$ formation were energetically favored. Two routes were evaluated for CO formation: via the adsorption of FeO and $Fe_2O$ on graphite followed by decomposition of the formed C–O–Fe group that produces CO (Domazetis, Raoarun et al. 2007). Additionally,

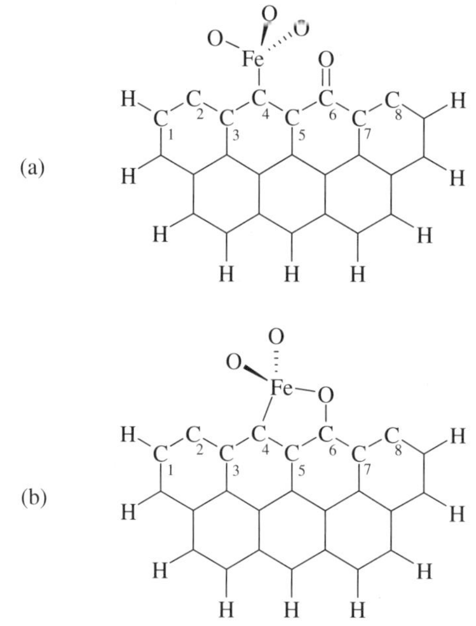

(a)

(b)

Fig. 28. Molecular models of carbon structure with Fe metal additive to simulate the effect of catalyses on combustion. (Reprinted with permission from (Backreedy, Jones et al. 2002), Copyright 2002).

the active site for gasification was suggested to be Fe–C system. The mechanism of catalytic steam gasification involves the chemisorption of $H_2O$ on Fe–C, formation of the Fe←$OH_2$ coordination bond, with production of $H_2$ via iron hydride complexes. Formation of CO takes place via oxygen insertion into Fe–C to form Fe–O–C that decomposes producing CO and another Fe–C site. It was also found that large sized iron species were not catalytically active but caused formation of large pores in char (Domazetis, Raoarun et al. 2008).

## 8. Carbon-NOx reactions

Nitrogen oxides (NOx) are formed in several industrial processes such as power generation units, waste incinerators, as well as in mobile sources. Improvement of these industrial processes and engine designs are necessary to mitigate emissions of nitrogen oxides. Catalyzed and uncatalyzed processes have been studied to reduce nitrogen oxides to $N_2$. However, detailed mechanisms (at molecular level) cannot be determined with available experimental techniques. Therefore, electronic structure methods are very helpful for the understanding of the reaction mechanisms. This reaction was already addressed by Xu et al. (Xu, Irle et al. 2006) and discussed in the $CO_2$ interaction section.

To our knowledge, Kyotani and Tomita (Kyotani and Tomita 1999) were the first to analyze the reaction of zigzag and armchair edge sites of carbon with NO and $N_2O$ by using an *ab initio* molecular orbital theory. The B3LYP/6-31G(d)//HF/3-21G(d) model chemistry was employed. From the thermal stability of the formed nitrogen-containing complexes, it was concluded that for NO adsorption the N-down approach toward the edge site is thermally more favorable than the O-down one, while the O-down mode is more favorable than the N-down one for $N_2O$ adsorption because the former process releases a stable $N_2$ molecule to form a surface oxygen complex. The chemisorption of the NO or $N_2O$ molecule with its bond axis parallel to the edge plane (side-on mode) gave the most stable chemisorbed species. The presence of surface oxygen complexes (quinone-type carbonyl group) on the edge sites reduced the heat of NO chemisorption to some extent. The calculation of the system including both NO molecule and a nitrogen chemisorbed species led to the formation of a six-membered ring complex including NNO bonding. The bond population analysis predicted that $N_2$ desorption from such a complex is very probable. A similar analysis was done about $N_2$ formation either from the reaction of pyridinic or pyridone type nitrogen with NO or from two neighboring C(N) species on a zigzag site; in the case of the C–$N_2O$ reaction, the calculation predicted the release of $N_2$ and the formation of C(O) when $N_2O$ reacted on the edge site in both the O-down and side-on modes. Although the effect of surface oxygen complexes on the C–NO reaction was investigated, the situation is still far from the real one in combustion because it involves the presence of low levels of NO/$N_2O$ and a high concentration of $O_2$.

The reaction of NO with char-bound nitrogen during combustion was studied by Montoya et al. (Montoya, Truong et al. 2000) using DFT at the B3LYP/6-31G(d) level for all calculations. Different zigzag models of nitrogen-containing char were studied in order to establish the feasibility of nitrous oxide evolution from a heterogeneous process. It was found that the reaction of a NO molecule with char yields predominantly CO and $N_2$ to the gas phase. However, reaction of a NO molecule with char-nitrogen in the presence of chemisorbed oxygen can also release $N_2O$ but as a minor product. It also was mentioned that additional calculations need to be carried out on different model compounds and for different pathways to $N_2$ and $N_2O$, particularly those involving sequential heterogeneous and homogeneous reactions. Unfortunately, no information is presented for the armchair configuration.

Zhu et al. (Zhu, Finnerty et al. 2001) studied the opposite roles of $O_2$ in NO– and $N_2O$–carbon reactions by using *ab initio* calculations. Only zigzag models are considered since the mechanism obtained for the zigzag models is not significantly different from that obtained for armchair models as it was suggested by Yang (Chen, Yang et al. 1993; Chen and Yang 1998) and Tomita (Kyotani and Tomita 1999). It was found that the opposite roles of $O_2$ are caused by the different adsorption paths of $N_2O$ and NO on the carbon surface. In the presence of excess $O_2$, most of the active sites are occupied by oxygen groups. In the competition for the remaining active sites, NO is more likely to chemisorb in the form of $NO_2$ while NO chemisorption is thermodynamically more favorable than $O_2$ chemisorption. By contrast, the presence of excess $O_2$ makes $N_2O$ chemisorption much less thermally stable either on the consecutive edge sites or edge sites isolated by semiquinone oxygen. Therefore, the $N_2O$ carbon reaction rate is depressed by the presence of excess $O_2$. For the NO-carbon reaction, $N_2$ can be formed by reaction of the –C–ONO–C– group with another NO molecule. For the $N_2O$ carbon reaction with the presence of excess oxygen, the transfer of the oxygen atom of $N_2O$ to the carbon sites is followed by the formation of $N_2$. It is not necessary to weaken or to break the adjacent C–C bonds during the formation of $N_2$. After the release of $N_2$, oxygen still remains on the carbon edge sites.

The kinetics of nitric oxide desorption from carbonaceous surfaces was studied by Montoya et al. (Montoya, Mondragón et al. 2002) carrying out molecular modeling. DFT at the B3LYP/6-31G(d) model chemistry was used to provide potential energy surface information.

Transition state theory was used to provide temperature dependant rate constants. Major discrepancies between theoretical and experimental activation energies for the NO desorption were found in this study. However, desorption rate constants of NO and CO are not considerably different as it has been assumed in single particle models. The NO desorption activation energy was found to be 10 kcal/mol lower than that of the CO desorption. The hypothesis that NO can be released by a different mechanism from that of CO is not supported from this study. Then, it is necessary to consider other types of NO precursors to have a complete insight into the overall NO desorption process.

Surface nitrogen complexes formation upon reaction of coal char with NO in the presence or absence of $O_2$ and/or $H_2O$ was studied by García et al. (García, Espinal et al. 2004) using XPS. Complexes such as pyrrolic-N type, pyridine-N-oxide, pyridinic-N, and –$NO_2$ were observed. From the experimental results several mechanisms are proposed to explain the formation of pyridinic and –$NO_2$ complexes at low temperatures. In addition, a complementary molecular modeling study was carried out at the B3LYP/6-31G(d) model chemistry in order to obtain thermodynamic data for the proposed mechanisms. From the theoretical calculations it was concluded that five member ring structures (formed by CO desorption from semiquinone complexes) are susceptible to chemisorb NO giving pyridine-N-oxide at temperatures where gasification is not relevant. –$NO_2$ complexes can be formed by NO chemisorption on adjacent sites to oxygen complexes or by chemisorption of $NO_2$ present in the gas mixtures that contain oxygen. These complexes are responsible for the NO reversible desorption.

Huang and Yang (Huang and Yang 1999) studied the catalyzed carbon-NO reaction by STM and *ab initio* molecular orbital calculations on model graphite substrates with –O and –O–M groups (where M=metal) bonded to the zigzag face, as can be seen in Fig. 29. The B3LYP/3-21G//HF/STO-3G model chemistry was used for all calculations. From the C–C bond energies of the different studied models (see Table 5) it is clear that the C–C bonds on the zigzag face were substantially weakened when an O atom was added to the active carbon site, $C_4$, and was further weakened by adding M to the O atom. It can be seen in Table 5 that the bond energies are very close, indicating that different elements do not have a large influence on the total bond energies of the $C_4$–$C_3$ and the $C_4$–$C_5$ bonds. In addition, it was found that the relative order of the extent of the C–C bond weakening by different –O–M groups is in general agreement with the order of catalytic activity results. Cu catalyst is an exception.

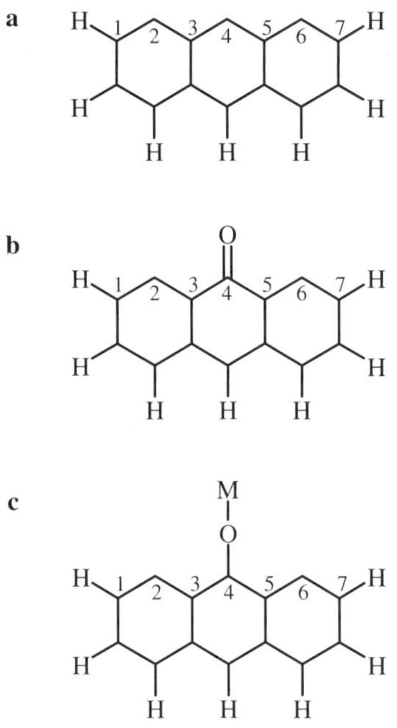

Fig. 29. Graphite models for molecular orbital calculations. (a) model with hydrogen saturating the edge carbon sites, (b) model of chemisorbed oxygen on active site, (c) model with –O–M group attached on the active site (M = metal). (Reprinted from (Huang and Yang 1999), Copyright 1999, with permission from Elsevier).

Table 5. Calculated C–C bond ($C_3$–$C_4$ and $C_4$–$C_5$, see Fig. 29) energies for different models. The bond energies (BE) are in kcal/mol. The experimental C–C bond energy is about 170 kcal/mol. (Reprinted from (Huang and Yang 1999), Copyright 1999, with permission from Elsevier.)

| Model | $C_{14}H_7$ | $C_{14}H_7O$ | $C_{14}H_7OLi$ | $C_{14}H_7ONa$ | $C_{14}H_7OK$ | $C_{14}H_7OCa$ | $C_{14}H_7OCu$ |
|---|---|---|---|---|---|---|---|
| C–C BE | 216.5 | 203.9 | 200.8 | 197.7 | 193.9 | 196.1 | 197.6 |

## 9. Concluding Remarks

Molecular modeling of the gasification reactions allows to estimate changes in enthalpy ($\Delta H$), entropy ($\Delta S$), and free energy ($\Delta G$), as well as transition states (energy barriers), rate constants (k), and infrared spectra, of the elemental reactions between carbon materials and different oxidizing agents. However, chemical intuition and knowledge of the studied system is required in order to analyze the results and realize whether they are reasonable or not. In addition, it is worth to mention that finding transition states is a rather difficult task, and because of that reason, most of the existing information is on the thermodynamics of the elemental gas-carbon reactions.

Even though useful information has been obtained that helps us in to rationalize the experimental data, there are still many questions to be answered. For example, molecular modeling studies of the carbon gasification reactions have been focused mainly at the first stages of the reaction (low oxygen coverage), studies of the reactions that take place at high oxygen coverage are yet to be done. It has been proposed from experimental observations that the activity of metal clusters are higher than that of the C–O–M groups. However, to the best of our knowledge, only a few investigations have been carried out to explore the chemistry of metal clusters catalysis in carbon gasification reactions. On the other hand, molecular dynamics studies of the carbon gasification reactions need to be carried out, incorporating various reaction conditions, such as temperature, pressure, and gasifying agents.

The main conclusion of this review is that it has been proved that electronic structure calculations are a powerful tool which helps us to understand key problems in the carbon gasification system that would not be possible to address using available experimental techniques.

## Acknowledgments

The authors wish to thank the University of Antioquia for the financial support of the "SOSTENIBILIDAD" program. JFE acknowledges "COLCIENCIAS" and the University of Antioquia for the PhD scholarship.

# References

Aksenenko, E. V. and Tarasevich, Y. I. (1996). Quantum chemical study of the interaction of water molecules with a partially oxidized graphite surface, *Adsorption Science & Technology*, 14, pp. 383-391.

Arellano, J. S., Molina, L. M., Rubio, A. and Alonso, J. A. (2000). Density functional study of adsorption of molecular hydrogen on graphene layers, *Journal of Chemical Physics*, 112, pp. 8114-8119.

Backreedy, R., Jones, J. M., Pourkashanian, M. and Williams, A. (2001). A study of the reaction of oxygen with graphite: Model chemistry, *Faraday Discussions*, 119, pp. 385-394.

Backreedy, R. I., Jones, J. M., Pourkashanian, M. and Williams, A. (2002). Modeling the reaction of oxygen with coal and biomass chars, *Proceedings of the Combustion Institute*, 29, pp. 415-421.

Barone, V., Lelj, F., Minichino, C., Russo, N. and Toscano, M. (1987). Cluster model study of the chemisorption of atomic hydrogen on the basal plane of graphite, *Surface Science*, 189, pp. 185-189.

Bennett, A. J., McCarroll, B. and Messmer, R. P. (1971). Molecular orbital approach to chemisorption. I. Atomic hydrogen on graphite, *Surface Science*, 24, pp. 191-208.

Bennett, A. J., McCarroll, B. and Messmer, R. P. (1971). Molecular orbital approach to chemisorption. II. Atomic H, C, N, O, and F on graphite, *Physical Review B: Solid State*, 3, pp. 1397-1406.

Buskies, U. (1996). The efficiency of coal-fired combined-cycle power plants, *Applied Thermal Engineering*, 16, pp. 959-974.

Caballol, R., Igual, J., Illas, F. and Rubio, J. (1985). MINDO/3 potential energy surface for hydrogen-graphite system: active sites and migration, *Surface Science*, 149, pp. 621-629.

Casanas, J., Illas, F., Sanz, F. and Virgili, J. (1983). Dissociative chemisorption of molecular hydrogen on graphite: a MINDO/3 study, *Surface Science*, 133, pp. 29-37.

Chen, N. and Yang, R. T. (1998). Ab-initio molecular orbital calculation on graphite: selection of molecular system and model chemistry, *Carbon*, 36, pp. 1061-1070.

Chen, N. and Yang, R. T. (1998). Ab-initio molecular orbital study of the unified mechanism and pathways for gas-carbon reactions, *Journal of Physical Chemistry A*, 102, pp. 6348-6356.

Chen, S. G. and Yang, R. T. (1993). The active surface species in alkali-catalyzed carbon gasification: phenolate (C-O-M) groups vs clusters (particles), *Journal of Catalysis*, 141, pp. 102-113.

Chen, S. G. and Yang, R. T. (1993). Titration for basal plane versus edge plane surface on graphitic carbons by adsorption, *Langmuir*, 9, pp. 3259-63.

Chen, S. G. and Yang, R. T. (1997). Unified mechanism of alkali and alkaline earth catalyzed gasification reactions of carbon by $CO_2$ and $H_2O$, *Energy & Fuels*, 11, pp. 421-427.

Chen, S. G., Yang, R. T., Kapteijn, F. and Moulijn, J. A. (1993). A new surface oxygen complex on carbon: toward a unified Mechanism for carbon gasification reactions, *Industrial and Engineering Chemistry Research*, 32, pp. 2835-2840.

DeGroot, W. F. and Shafizadeh, F. (1984). Kinetics of gasification of Douglas fir and cottonwood chars by carbon dioxide, *Fuel*, 63, pp. 210-216.

Domazetis, G., Raoarun, M. and James, B. D. (2007). Semiempirical and density functional theory molecular modeling of brown coal chars with iron species and $H_2$, CO formation, *Energy & Fuels*, 21, pp. 2531-2542.

Domazetis, G., Raoarun, M., James, B. D. and Liesegang, J. (2008). Molecular modelling and experimental studies on steam gasification of low-rank coals catalysed by iron species, *Applied Catalysis A: General*, 340, pp. 105-118.

Dovesi, R., Pisani, C., Ricca, F. and Roetti, C. (1976). Regular chemisorption of hydrogen on graphite in the crystalline orbital NDO approximation, *Journal of Chemical Physics*, 65, pp. 3075-3084.

Dovesi, R., Pisani, C. and Roetti, C. (1981). Ab initio HF versus semi-empirical results of chemisorption calculations of hydrogen on graphite, *Chemical Physics Letters*, 81, pp. 498-502.

Dybala-Defratyka, A., Paneth, P., Pu, J. and Truhlar, D. G. (2004). Benchmark Results for Hydrogen Atom Transfer between Carbon Centers and Validation of Electronic Structure Methods for Bond Energies and Barrier Heights, *Journal of Physical Chemistry A*, 108, pp. 2475-2486.

Espinal, J. F., Mondragón, F. and Truong, T. N. (2005). Mechanisms for methane and ethane formation in the reaction of hydrogen with carbonaceous materials, *Carbon*, 43, pp. 1820-1827.

Espinal, J. F., Montoya, A., Mondragón, F. and Truong, T. N. (2004). A DFT study of interaction of carbon monoxide with carbonaceous materials, *Journal of Physical Chemistry B*, 108, pp. 1003-1008.

Espinal, J. F., Truong, T. N. and Mondragon, F. (2007). Mechanisms of $NH_3$ formation during the reaction of $H_2$ with nitrogen containing carbonaceous materials, *Carbon*, 45, pp. 2273-2279.

Foresman, J. B. and Frisch, A. (1996). Exploring Chemistry with Electronic Structure Methods. Pittsburg, PA, Gaussian, Inc.

Frankcombe, T. J. and Smith, S. C. (2004). On the microscopic mechanism of carbon gasification: A theoretical study, *Carbon*, 42, pp. 2921-2928.

Freund, H. (1985). Kinetics of carbon gasification by carbon dioxide, *Fuel*, 64, pp. 657-660.

García, P., Espinal, J. F., Salinas Martinez de Lecea, C. and Mondragón, F. (2004). Experimental characterization and molecular simulation of nitrogen complexes formed upon NO-char reaction at 270 °C in the presence of $H_2O$ and $O_2$, *Carbon*, 42, pp. 1507-1515.

Hall, P. J. and Calo, J. M. (1989). Secondary interactions upon thermal desorption of surface oxides from coal chars, *Energy & Fuels*, 3, pp. 370-376.

Hayns, M. R. (1975). Molecular orbital calculations of the chemisorption and diffusion of oxygen and water on a graphite substrate, *Theoretica Chimica Acta*, 39, pp. 61-74.

Hermann, G. and Hüttinger, K. J. (1986). Mechanism of water vapour gasification of carbon - a new model, *Carbon*, 24, pp. 705-713.

Huang, H. Y. and Yang, R. T. (1999). Catalyzed carbon-NO reaction studied by scanning tunneling microscopy and ab Initio molecular orbital calculations, *Journal of Catalysis*, 185, pp. 286-296.

Humbel, S., Sieber, S. and Morokuma, K. (1996). The IMOMO (integrated MO MO) method: integration of different levels of molecular orbital approximations for geometry optimization of large systems: test for n-butane conformation and SN2 reaction: RCl + Cl, *Journal of Chemical Physics*, 105, pp. 1959-1967.

Janiak, C., Hoffmann, R., Sjovall, P. and Kasemo, B. (1993). The potassium promoter function in the oxidation of graphite: an experimental and theoretical study, *Langmuir*, 9, pp. 3427-3440.

Jeloaica, L. and Sidis, V. (1999). DFT investigation of the adsorption of atomic hydrogen on a cluster-model graphite surface, *Chemical Physics Letters*, 300, pp. 157-162.

Jensen, F. (1999). Introduction to Computational Chemistry. New York, John Wiley & Sons.

Kapteijn, F., Abbel, G. and Moulijn, J. A. (1984). Carbon dioxide gasification of carbon catalyzed by alkali metals. Reactivity and mechanism, *Fuel*, 63, pp. 1036-1042.

Kapteijn, F. and Moulijn, J. A. (1983). Kinetics of the potassium carbonate-catalyzed carbon dioxide gasification of activated carbon, *Fuel*, 62, pp. 221-225.

Kelemen, S. R. and Freund, H. (1986). Model carbon dioxide gasification reactions on uncatalyzed and potassium catalyzed glassy carbon surfaces, *Journal of Catalysis*, 102, pp. 80-91.

Koenig, P. C., Squires, R. G. and Laurendeau, N. M. (1986). Char gasification by carbon dioxide. Further evidence for a two-site model, *Fuel*, 65, pp. 412-416.

Kwon, T. W., Kim, S. D. and Fung, D. P. C. (1988). Reaction kinetics of char-carbon dioxide gasification, *Fuel*, 67, pp. 530-535.

Kyotani, T. and Tomita, A. (1999). Analysis of the reaction of carbon with $NO/N_2O$ using ab initio molecular orbital theory, *Journal of Physical Chemistry B*, 103, pp. 3434-3441.

Lamoen, D. and Persson, B. N. J. (1998). Adsorption of potassium and oxygen on graphite: A theoretical study, *Journal of Chemical Physics*, 108, pp. 3332-3341.

Linares-Solano, A., Salinas-Martinez de Lecea, C., Cazorla-Amoros, D. and Martin-Gullon, I. (2000). Porosity development during $CO_2$ and steam activation in a fluidized bed reactor, *Energy & Fuels*, 14, pp. 142-149.

Marchon, B., Carrazza, J., Heinemann, H. and Somorjai, G. A. (1988). TPD and XPS studies of $O_2$, $CO_2$, and $H_2O$ adsorption on clean polycrystalline graphite, *Carbon*, 26, pp. 507-514.

Marchon, B., Tysoe, W. T., Carrazza, J., Heinemann, H. and Somorjai, G. A. (1988). Reactive and kinetic properties of carbon monoxide and carbon dioxide on a graphite surface, *Journal of Physical Chemistry*, 92, pp. 5744-5749.

Meijer, R., Weeda, M., Kapteijn, F. and Moulijn, J. A. (1991). Catalyst loss and retention during alkali-catalyzed carbon gasification in carbon dioxide, *Carbon*, 29, pp. 929-941.

Molina, A., Montoya, A. and Mondragon, F. (1999). $CO_2$ strong chemisorption as an estimate of coal char gasification reactivity, *Fuel*, 78, pp. 971-977.

Montoya, A., Mondragón, F. and Truong, T. N. (2002). Adsorption on carbonaceous surfaces: cost-effective computational strategies for quantum chemistry studies of aromatic systems, *Carbon*, 40, pp. 1863-1872.

Montoya, A., Mondragón, F. and Truong, T. N. (2002). First-principles kinetics of CO desorption from oxygen species on carbonaceous surface, *Journal of Physical Chemistry A*, 106, pp. 4236-4239.

Montoya, A., Mondragón, F. and Truong, T. N. (2002). Formation of CO precursors during char gasification with $O_2$, $CO_2$ and $H_2O$, *Fuel Processing Technology*, 77-78, pp. 125-130.

Montoya, A., Mondragón, F. and Truong, T. N. (2002). Kinetics of nitric oxide desorption from carbonaceous surfaces, *Fuel Processing Technology*, 77-78, pp. 453-458.

Montoya, A., Mondragón, F. and Truong, T. N. (2003). $CO_2$ adsorption on carbonaceous surfaces: a combined experimental and theoretical study, *Carbon*, 41, pp. 29-39.

Montoya, A., Truong, T.-T. T., Mondragón, F. and Truong, T. N. (2001). CO desorption from oxygen species on carbonaceous surface: 1. Effects of the local structure of the active site and the surface coverage, *Journal of Physical Chemistry A*, 105, pp. 6757-6764.

Montoya, A., Truong, T. N. and Sarofim, A. F. (2000). Application of density functional theory to the study of the reaction of NO with char-bound nitrogen during combustion, *Journal of Physical Chemistry A*, 104, pp. 8409-8417.

Montoya, A., Truong, T. N. and Sarofim, A. F. (2000). Spin contamination in Hartree-Fock and density functional theory wavefunctions in modeling of adsorption on graphite, *Journal of Physical Chemistry A*, 104, pp. 6108-6110.

Pan, Z. and Yang, R. T. (1992). Strongly bonded oxygen in graphite: detection by high-temperature TPD and characterization, *Industrial and Engineering Chemistry Research*, 31, pp. 2675-2680.

Pan, Z. J. and Yang, R. T. (1990). The mechanism of methane formation from the reaction between graphite and hydrogen, *Journal of Catalysis*, 123, pp. 206-214.

Radovic, L. R. (2005). The mechanism of $CO_2$ chemisorption on zigzag carbon active sites: A computational chemistry study, *Carbon*, 43, pp. 907-915.

Radovic, L. R. and Bockrath, B. (2005). On the chemical nature of graphene edges: origin of stability and potential for magnetism in carbon materials, *Journal of the American Chemical Society*, 127, pp. 5917-5927.

Radovic, L. R., Jiang, H. and Lizzio, A. A. (1991). A transient kinetics study of char gasification in carbon dioxide and oxygen, *Energy & Fuels*, 5, pp. 68-74.

Radovic, L. R., Karra, M., Skokova, K. and Thrower, P. A. (1998). The role of substitutional boron in carbon oxidation, *Carbon*, 36, pp. 1841-1854.

Rodriguez-Mirasol, J., Cordero, T. and Rodriguez, J. J. (1993). Activated carbons from carbon dioxide partial gasification of eucalyptus kraft lignin, *Energy & Fuels*, 7, pp. 133-138.

Sakawa, M., Sakurai, Y. and Hara, Y. (1982). Influence of coal characteristics on carbon dioxide gasification, *Fuel*, 61, pp. 717-720.

Sánchez, A. and Mondragón, F. (2007). Role of the epoxy group in the heterogeneous $CO_2$ evolution in carbon oxidation reactions, *Journal of Physical Chemistry C*, 111, pp. 612-617.

Sendt, K. and Haynes, B. S. (2005). Density functional study of the chemisorption of $O_2$ on the armchair surface of graphite, *Proceedings of the Combustion Institute*, 30, pp. 2141-2149.

Sendt, K. and Haynes, B. S. (2005). Density functional study of the chemisorption of $O_2$ on the zig-zag surface of graphite, *Combustion and Flame*, 143, pp. 629-643.

Sendt, K. and Haynes, B. S. (2005). Density functional study of the reaction of carbon surface oxides: The behavior of ketones, *Journal of Physical Chemistry A*, 109, pp. 3438-3447.

Shimizu, A. and Tachikawa, H. (2002). The direct molecular orbital dynamics study on the hydrogen species adsorbed on the surface of planar graphite cluster model, *Journal of Physics and Chemistry of Solids*, 63, pp. 759-763.

Stein, S. E. and Brown, R. L. (1987). π-Electron properties of large condensed polyaromatic hydrocarbons, *Journal of the American Chemical Society*, 109, pp. 3721-3729.

Suzuki, T., Ohme, H. and Watanabe, Y. (1992). A mechanism of sodium-catalyzed carbon dioxide gasification of carbon investigation by pulse and TPD techniques, *Energy & Fuels*, 6, pp. 336-343.

Suzuki, T., Ohme, H. and Watanabe, Y. (1994). Mechanisms of alkaline-earth metals catalyzed $CO_2$ gasification of carbon, *Energy & Fuels*, 8, pp. 649-658.

Svensson, M., Humbel, S., et al. (1996). ONIOM: A multi-layered integrated MO + MM method for geometry optimizations and single point energy predictions. A test for Diels-Alder reactions and $Pt(P(t-Bu)_3)_2 + H_2$ oxidative addition, *Journal of Physical Chemistry*, 100, pp. 19357-19363.

Tomita, A., Higashiyama, K. and Tamai, Y. (1981). Scanning electron microscopic study on the catalytic gasification of coal, *Fuel*, 60, pp. 103-114.

Vastola, F. J., Hart, P. J. and Walker, P. L., Jr. (1964). A study of carbon-oxygen surface complexes using $^{18}O$ as a tracer, *Carbon*, 2, pp. 65-71.

Wu, X. and Radovic, L. R. (2004). Ab initio molecular orbital study on the electronic structures and reactivity of boron-substituted carbon, *Journal of Physical Chemistry A*, 108, pp. 9180-9187.

Xu, S., Irle, S., Musaev, D. G. and Lin, M. C. (2005). Water clusters on graphite: Methodology for quantum chemical a priori prediction of reaction rate constants, *Juornal of Physical Chemistry A*, 109, pp. 9563-9572.

Xu, S. C., Irle, S., Musaev, D. G. and Lin, M. C. (2006). Quantum chemical prediction of reaction pathways and rate constants for dissociative adsorption of COx and NOx on the graphite (0001) surface, *Journal of Physical Chemistry B*, 110, pp. 21135-21144.

Xu, Y.-J. and Li, J.-Q. (2004). The interaction of molecular oxygen with active sites of graphite: a theoretical study, *Chemical Physics Letters*, 400, pp. 406-412.

Yang, F. H. and Yang, R. T. (2002). Ab initio molecular orbital study of adsorption of atomic hydrogen on graphite: Insight into hydrogen storage in carbon nanotubes, *Carbon*, 40, pp. 437-444.

Yang, R. T. and Duan, R. Z. (1985). Kinetics and mechanism of gas-carbon reactions: conformation of etch pits, hydrogen inhibition and anisotropy in reactivity, *Carbon*, 23, pp. 325-331.

Young, D. C. (2001). Computational Chemistry: A Practical Guide for Applying Techniques to Real-World Problems. New York, John Wiley & Sons, Inc.

Zhu, Z. H., Finnerty, J., Lu, G. Q., Wilson, M. A. and Yang, R. T. (2002). Molecular orbital theory calculations of the $H_2O$-carbon reaction, *Energy & Fuels*, 16, pp. 847-854.

Zhu, Z. H., Finnerty, J., Lu, G. Q. and Yang, R. T. (2001). Opposite roles of $O_2$ in NO- and $N_2O$-carbon reactions; an ab initio study, *Journal of Physical Chemistry B*, 105, pp. 821-830.

Zhu, Z. H., Finnerty, J., Lu, G. Q. and Yang, R. T. (2002). A comparative study of carbon gasification with $O_2$ and $CO_2$ by density functional theory calculations, *Energy & Fuels*, 16, pp. 1359-1368.

Zhuang, Q.-L., Kyotani, T. and Tomita, A. (1994). DRIFT and TK/TPD analyses of surface oxygen complexes formed during carbon gasification, *Energy & Fuels*, 8, pp. 714-718.

Zhuang, Q., Kyotani, T. and Tomita, A. (1995). Dynamics of Surface Oxygen Complexes during Carbon Gasification with Oxygen, *Energy & Fuels*, 9, pp. 630-634.

## Appendix (Calculations output)

In this section we will illustrate some of the information that can be obtained from molecular orbital calculations, such as optimizations (see Table 6) and harmonic frequencies (see Table 7). As an example, we list information for an armchair carbonaceous model (first structure in the mechanism shown in Fig. 6) calculated using Gaussian 03.

Table 6. Optimized Cartesian coordinates, Mulliken spin densities and Mulliken[*] atomic charges. Temperature = 0 K.

| Atom | Cartesian coordinates (angstroms) | | | Spin densities | Atomic charges |
|---|---|---|---|---|---|
| | X | Y | Z | | |
| C | 0.030495 | 0.000000 | −0.006954 | 0.000000 | −0.002488 |
| C | 0.030467 | 0.000000 | 1.433410 | 0.000000 | −0.002483 |
| C | 1.258448 | 0.000000 | 2.211676 | 0.000000 | 0.101864 |
| C | −1.210065 | 0.000000 | −0.715478 | 0.000000 | 0.148522 |
| C | −1.210121 | 0.000000 | 2.141886 | 0.000000 | 0.148522 |
| C | −2.437947 | 0.000000 | 1.393324 | 0.000000 | −0.190725 |
| C | −2.437919 | 0.000000 | 0.033035 | 0.000000 | −0.190725 |
| H | −3.375234 | 0.000000 | 1.944354 | 0.000000 | 0.133114 |
| C | 1.218747 | 0.000000 | 3.606408 | 0.000000 | −0.114856 |
| C | −1.203767 | 0.000000 | 3.549429 | 0.000000 | −0.210864 |
| H | −2.154380 | 0.000000 | 4.076963 | 0.000000 | 0.131311 |
| C | −0.012204 | 0.000000 | 4.267462 | 0.000000 | −0.146897 |
| H | −0.038580 | 0.000000 | 5.353554 | 0.000000 | 0.134839 |
| C | 1.258506 | 0.000000 | −0.785171 | 0.000000 | 0.101868 |
| C | 1.218859 | 0.000000 | −2.179906 | 0.000000 | −0.114856 |
| C | −0.012065 | 0.000000 | −2.841008 | 0.000000 | −0.146897 |
| C | −1.203656 | 0.000000 | −2.123021 | 0.000000 | −0.210864 |
| H | −2.154249 | 0.000000 | −2.650591 | 0.000000 | 0.131311 |
| H | −3.375184 | 0.000000 | −0.518033 | 0.000000 | 0.133114 |
| H | −0.038398 | 0.000000 | −3.927101 | 0.000000 | 0.134839 |
| C | 2.372461 | 0.000000 | 1.330017 | 0.000000 | −0.081907 |
| C | 2.372474 | 0.000000 | 0.096537 | 0.000000 | −0.081901 |
| H | 2.149088 | 0.000000 | −2.738069 | 0.000000 | 0.098079 |
| H | 2.148951 | 0.000000 | 4.164613 | 0.000000 | 0.098079 |

[*] Note that spin densities are zero because the optimization was performed for singlet state multiplicity. Any other value for multiplicity would produce spin densities different than zero.

In addition to the information already listed in Table 6, optimization calculations also give:

Rotational constants (GHz), 1.0921, 0.5458, 0.3639

Full point group, C1

Total molecule energy (Hartrees), −614.4210

Table 7. Zero-point vibrational energy (ZPE), moments of inertia (I), heat capacity at constant volume (Cv), entropy (S), enthalpy (H), and harmonic vibrational frequencies (ν) obtained at a temperature = 298.15 K and pressure = 1 atmosphere.

| ZPE (kcal/mol) | I (atomic units) | Cv (cal/molK) | S (cal/molK) | H (kcal/mol) |
|---|---|---|---|---|
| 114.58 | 1652.53<br>3306.66<br>4959.18 | 44.16 | 97.57 | 385439.4 |
| Harmonic frequencies, ν (cm⁻¹) | | | | |
| 99, 157, 229, 259, 270, 362, 381, 382, 408, 452, 497, 503, 522, 543, 546, 578, 587, 604, 632, 717, 727, 781, 785, 798, 817, 822, 849, 902, 914, 968, 977, 980, 984, 1004, 1081, 1094, 1126, 1179, 1201, 1209, 1211, 1248, 1258, 1307, 1308, 1351, 1381, 1419, 1445, 1460, 1470, 1544, 4552, 1609, 1631, 1652, 1672, 2111, 3181, 3186, 3187, 3200, 3202, 3203, 3217, 3218 | | | | |

Besides to the information already listed in Table 7, frequency calculations also give electronic (0.00), translational (0.89), rotational (0.89), and vibrational (119.16) energies in kcal/mol.

# Index

1D systems, 376
2D distribution, 69, 72

$\alpha$ band, 380, 384, 407, 408, 410
$\Gamma$ point, 380, 381, 383, 393, 394, 403, 404, 407, 408, 410
$\pi$ system
    conjugation, 384, 423, 439
    delocalization, 352, 355, 361, 363, 364, 366
    electrons, 376, 382
    stabilization, 439
$\Omega$ quantum number, 117

*ab initio*, 380, 382, 387
    calculations, 96, 261
    methods, 445, 448, 449, 460, 461, 478, 487, 489-491
ablated material, 201, 213
ablation, 267-269, 272, 273
    graphite, 167, 169, 172
    plasma, 210
    plume, 167, 168
    threshold, 172
absolute concentrations, 77, 78, 105
    densities of radicals, 87
    number densities, 248
    OH concentrations, 87
absorption, 18, 32, 45, 48
    coefficient, 79, 93, 94, 208
    length, 205, 207, 210
    profiles, 205
    spectroscopy, 77, 80, 82, 90, 98-100, 168

FTIR, 3, 32, 44, 45
computation of synthetic spectrum, 159, 160
    algorithm, 160
    example, 161
UV, 16, 32, 45
abundance spectra, 344, 345, 347, 348, 350, 364
acetylene, 258, 259, 264, 272, 275, 421
acetylenic units, 377-380, 382, 385, 408
actinometer, 37-40
activation energy, 457, 460-463, 465, 467, 485, 487, 491
active region, 228
active site, 451, 461, 480, 488, 492
adiabatic expansion, 72
afterglow, 210, 213, 217, 265
    region, 210
aggregation, 200
AIREBO, 319
allyl ($C_3H_5$), 267
    bromide, 267
    iodide, 267
AM1, 326, 328
ambient atmosphere, 173, 228, 239, 240, 243
    conditions, 239
    nitrogen pressures, 250
ambipolar characteristics, 299, 300
amorphous carbon, 285, 288, 289, 295, 296, 304, 318, 335, 420, 421
amplification, 207
angle strain, 201

angular momentum
    nuclear orbital **R**, 118, 129
    total electronic orbital **L**, 118, 129
    total electronic spin **S**, 118
    total nuclear spin **T**, 118, 122
    total orbital **N**, 118
    total with spin **F**, 118
    total without nuclear spin **J**, 118, 124
annulenes, 346
arc discharge, 284, 288, 293, 377
ArF laser, 172
argon, 203, 211, 213, 263, 264, 268
armchair edge, 429, 430, 435, 436
aromaticity, 423
    anti-aromatic, 346, 347, 356, 361, 364, 371
    aromatic, 346, 347, 352, 358, 361, 362, 364, 371
    aromatic stabilization, 433
    aromatic units, 448
astigmatic mirror cells, 264
astronomical applications, 255
astronomical sources, 93
astrophysical investigations, 200
    observations, 218
astrophysics, 97, 204, 225, 257
asymmetric vibration, 272
atmospheres of stars, 97
atomic carbon recombination, 177, 234, 235
atomic force microscopy (AFM), 286, 289, 298, 299, 306

B3LYP, 326, 328, 349, 351, 353-359, 361-368, 370, 381, 383, 385, 405
background gas, 213
    gas pressures, 239
backward propagation, 70

Balmer series, 33, 40
band gap energy, 376, 378, 379, 381-383, 387, 390, 391, 409
    metallic, 376, 390
    semiconducting, 377
band heads, 228, 229, 231, 232, 234
    intensity, 237
    sequence, 232
band origin, 205, 206, 214
bath gas, 88
Beer-Lambert law, 79, 208
benzene, 258, 259, 389
bi-dimensional imaging, 212, 213
bimodal distribution, 344
binding energy, 326, 328, 331, 335, 336
biomass, 447
blackbody emission, 178, 179
    radiation, 60, 62
blue satellite, 217
Boltzmann, 10, 26, 40-43
    constant, 237, 240
Boltzmann-Plot
    example, 157, 158
    modified, 153
bombardment, 333-337
bond
    angle, 318
    covalent bond, 318-320, 327, 329, 332, 335, 338
    dangling, 201, 304, 423, 438, 460, 476
    order, 317, 318, 337, 338
    oriented, 305
bond length difference (BLD), 379, 382-385, 389
bond strength, 454, 457, 460
bonding patterns, 283
Born-Oppenheimer approximation, 124
Bouduard reaction, 421
boundary conditions, 243

box-car, 269, 271, 273
    integrator, 57, 59
Bremsstrahlung radiation, 174
Brillouin zone, 380, 382, 383, 385,
    390, 393, 394
broadband absorption, 207
    intra-cavity spectra, 218
broadened profiles, 209
broadening mechanisms, 209
broadening of spectral lines, 262
    absorption lines, 101
buckminsterfullerene, 199, 417, 418,
    420
buckytubes, 417, 418
buffer gas, 201, 211, 284
bundles, 289
butadiyne, 377
butatriene, 376
$B_v$, 130

$C_2$, 9-12, 14-16, 18-22, 24, 26-29, 33,
    35, 36, 38, 41, 48, 49, 56, 58-60,
    62-69, 72-74, 345, 361, 365, 367,
    427, 429-431, 433-436, 438
        Swan Bands, 9, 10, 12-16, 18, 19,
        21, 22, 24, 27, 33, 41, 48
$C_2$ addition, 261
$C_2$ and $C_3$, 167-169, 171, 172, 174,
    179, 180, 183, 185, 187, 190, 191
        densities, 55
$C_2H$, 256, 258, 267, 271, 272, 275, 277
    radical, 258
$C_2H_2$, 82-84, 86, 90-92, 96
$C_2H_3$, 266, 271, 272
$C_2H_4$, 83, 84, 86, 103
$C_2H_5$, 271, 276
$C_2H_6$, 83, 84, 86
$C_2H_x$, 333
$C_{2n}$, 345-347, 361, 362, 364, 368, 371
$C_2N_2$, 84
$C_2$ Swan band, 233, 250

$C_3$, 11, 12, 18, 21, 22, 27-29, 43, 49,
    345, 376, 389, 401
$C_3H_3$, 256, 259, 260, 267, 269, 271
    radical, 259
$C_3H_3^+$, 256, 259, 260
$C_3H_5$, 267, 269, 271, 277
$C_3$ radicals, 58, 65, 66, 73
$C_4$, 273, 274, 345, 347-352, 371, 376,
    377, 429
$C_5$, 350, 365, 366, 371, 376
$C_5$, $C_6$, $C_8$ and $C_{10}$, 213, 217
$C_6$, 273, 345, 347, 352-356, 358, 360,
    361, 364, 371
$C_7$, 274
$C_8$, 346, 352, 356-358, 361, 371, 380,
    389, 408
$C_9$, 274
$C_{10}H_2$, 379
$C_{11}$, 274
$C_{12}$, 361-363, 371
$C_{12}H_2$, 379
$C_{13}$, 274
$C_{14}$, 345, 361-364, 371
$C_{16}$, 361-363, 371, 380, 408
$C_{18}$, 345, 361-364, 371
$C_{28}$, 326-328
$C_{60}$, 199, 420, 432, 433
calibration factor, 249
capsules, 284
carbanions, 260
carbene, 451, 473
carbide, 286, 302-304
carbon arc, 417, 418, 420-422
    technique, 224
carbon nitride (CN), 228
carbon atoms, 283, 287, 304
carbon
        conversion, 447
        dioxide, 275, 446
        gasification reaction, 445
        materials, 478, 480, 493

monoxide, 83, 87, 88, 90, 91,
  258, 263, 275, 457, 460
particles, 199, 218
planar, 283
plumes, 56
species, 202, 203, 217
tetrahedral, 283
vapor, 55, 57, 59, 61, 63, 65, 67,
  69, 71, 73, 199, 200, 203
carbon chain ion, 260
carbon chain radicals, 255
carbon chains, 239, 375, 377, 379,
  381, 383, 385, 387-389, 391, 393,
  395, 397, 399, 401, 403, 405,
  407, 409
carbon clusters, 55, 56, 60, 72-74,
  199-201, 204-206, 217, 218, 256,
  257, 259-261, 267-269, 272-274,
  277, 283, 287, 304, 305
  cyclic form, 350-352, 355-358,
    365, 368-371
  even-numbered, 343-345, 347,
    364, 371
  linear, 283
  linear form, 343-352, 354-359,
    361, 364-371
  monocyclic, 343-348, 350, 354,
    358, 361, 364, 371
  odd-numbered, 343, 345, 346,
    364, 368, 371
carbon dust, 170
  nitride $(CN_x)$, 167
carbon nanofiber (CNF), 32, 33, 35
carbon nanostructures, 419, 423
carbon nanotube (CNT), 3, 4, 7-14, 16,
  17, 20, 22-24, 27, 29, 32-36, 44-46,
  48, 49, 417, 418, 429, 432
  chirality, 285, 429
  diameter, 283, 284, 292, 294,
    295, 297-301, 304, 418, 419,
    422, 429

double-walled (DWCNTs),
  285-290, 292-301, 306, 418,
  420, 430
durability, 285, 286, 297
hollow space, 286
length, 284, 299
metallic, 286, 299, 300
multi-walled carbon nanotubes,
  420, 421, 430
nucleation and growth, 417-419,
  421-423, 429, 433, 439
purity, 422
root growth mechanism, 422
semiconducting, 299, 300
separation, 418
single-walled carbon nanotubes
  (SWCNTs), 284, 285, 288,
  289, 292-301, 304, 417-424,
  429, 430, 433-439
solubility, 285
tip-growth mechanism, 422
tips, 286, 289, 298, 299, 306
carbon skeleton, 439
carbon source, 418, 421
carbon stars, 257, 344, 377
carbonaceous
  materials, 453, 461, 474, 476, 478
  structures, 447, 478
carrier gas, 420
CARS, 89, 276
cascade laser absorption spectroscopy
  (QCLAS), 99
CASSCF, 348, 352, 353
catalyst, 284, 288, 292, 294, 295,
  417-423, 437, 438
cavity enhanced absorption
  spectroscopy (CEAS), 100
cavity length, 208, 210
  losses, 207
  ring-down spectroscopy, 199,
    204-206, 214, 271

cavity mode, 102

cavity-enhanced methods, 271

CCD array, 212
    photocamera, 213

CCD camera, 57, 58, 174

CCSD and CCSD(T), 349-351, 353, 354, 356-362, 364-370, 382, 389

centrifugal corrections, 134

CH, 259, 269, 271, 275, 276

$CH_2$, 256, 266, 271, 275, 276

$CH_2O$, 83, 84, 86

$CH_3$, 82, 83, 86, 88, 91-95, 99, 267, 271, 277

$CH_3OH$, 83, 84, 86

$CH_4$, 82-84, 86-88, 90-92, 101, 102

$CH_4$-$H_2$, 82

chain lengthening, 259, 261
    species, 203

char, 446, 448, 450, 451, 460, 461, 463, 464, 469, 471, 472, 487-491

characteristic lifetime, 266

chemical modelling, 85, 105
    processes, 213
    reactions, 328, 333
    sputtering, 333

chemical vapor deposition (CVD), 256, 284, 420, 421
    alcohol CCVD (ACCVD), 421
    catalytic (CCVD), 420, 421

chemiionization reaction, 259, 260

chemiions, 259

chemiluminescent emission lines, 275

chemisorption, 317, 453, 455, 456, 461, 464-466, 469, 471-474, 476, 485, 488-492

chirp, 102

chromatography, 285

$CH_x$, 333

classical MD, 317, 330, 338

classical string theory, 381

Clebsch-Gordan coefficient
    building $JM$ states from $MN_M$ and $SM_S$ states, 124
    Clebsch-Gordan series inverse, 125, 132
    in case (b) basis, 126
    swap $j_1m_1$ and $j_2m_2$, 122
    total nuclear spin, 122

climatic changes, 445

cluster absorption, 211
    bands, 217
    dynamics, 206
    ejection, 202
    formation, 56, 73, 74, 167-170, 173, 199, 201, 204, 213, 218
    source, 201

clustering reactions, 60

clusters, 284, 286, 291, 292, 301, 303, 446, 473, 487, 493

$C_n$, 60, 68, 73, 432, 436

CN, 97, 98

CN bands, 228, 229

CNT, 3, 4, 7-14, 16, 17, 20, 22-24, 27, 29, 32-36, 44-46, 48, 49

C-O bonds, 454, 455

CO, 83, 87, 88, 90, 91, 263, 275

$CO_2$, 275

$CO_2$ capture, 446

coagulation, 257, 259

coal combustion, 446
    gasification, 446, 448, 463
    storage, 447

coherent radiation, 206

collinear beam configuration, 187

collision, 66, 68, 72, 74

collisional and collision-less, 239
    cross-sections, 248
    process, 235

collisional interactions, 169
    processes, 180, 186, 189

color center lasers, 271

Coulomb explosion (CE), 347, 348

combustion, 344, 446-448, 453, 461,
    469, 471, 480, 488, 489

combustion chemistry, 257, 261
    intermediates, 256
    research, 255, 257

comet tails, 344

comets, 200, 266

commutator
    $J_z$ and $J_{z'}$, 131

competitive chemical reactions, 258

Complete Neglect of Differential
    Overlap (CNDO), 448, 484-486

complex carbon stuctures, 199

compressed front, 240

computational cost, 429, 448, 450

concentration modulation, 265, 266,
    272

concentration monitoring, 104

condensation, 201, 213, 217

cone cutoff method, 319, 329, 330,
    334, 337

confinement effect, 73
    of the plume, 73, 247

conservation of mass, 242

continual (d.c.) discharges, 265

continuous background, 178
    light sources, 79

continuum, 55, 56, 59-62, 73, 74
    light source, 207
    optical emission, 55

continuum emission, 169, 174, 178

controllable delay, 211

conversion path, 88

cooling, 420
    rapid, 431

cooling mechanism, 72
    of molecules, 263

crater, 213, 214

CRDS, 89, 100-102

critical pressure, 247

cross-sections, 286, 298

cryogenic cooling, 80
    matrices, 274

cumulative double bonds, 376

cumulene, 343, 345-364, 371,
    376-378, 382, 386, 388-404, 409

cumulenic, 380, 384-386, 396, 409

current, 287, 288, 290, 299, 300

curvature, 432, 435

cutoff functions, 318, 319, 321, 325,
    329

CVD diamond coating, 89
    growth, 82

CVD reactor, 89

cyclic carbon isomers, 218
    cluster cation, 202, 218

cyclic isomers, 274

cyclization reactions, 259

cyclopentadiene, 259

cylindrical symmetry, 66


data acquisition, 87, 101, 103
    processing, 87, 89

DCCD in He, 206, 217

decay curve, 71, 72

degenerate four-wave mixing
    (DFWM), 276

delay, 227, 229-232, 238-241, 248-250
    generator, 227

density functional theory (DFT), 348,
    366, 381-387, 389, 405, 419, 423,
    424, 428, 429, 445, 448-451,
    460-462, 464, 465, 468, 469, 472,
    476-478, 480, 487, 489, 490

density of the particles, 242

density-functional tight-binding
    (DFTB), 326, 328, 375, 377,
    386-393, 395, 396, 398-405,
    407-409, 424-430, 434-437, 439

deposition of diamond, 84

Deslandres-D'Azambuja bands, 177
desorption, 286, 301-304
    activation energy, 461
detection limits, 102
deuterium, 333
devices, 284, 298-301
    Field Effect Transistor, 299
diacetylene, 258-260
diagnostic techniques, 56
diamond coating, 90
    growth rate, 91, 92
    growth species, 90
    hot-filament CVD, 81
diamond infiltration (CVI), 92
    quality, 91, 92
diamond-like carbon (DLC), 167
diatomic eigenfunction exact, 117
diazomethane ($CH_2N_2$), 266
difference frequency generation (DFG),
    271
diffraction device, 207
    losses, 207
    spectrograph, 212
diffuse, 200, 204, 218
    interstellar bands, 200, 256
direct absorption, 204, 271
dispersion curves, 380
dissipative structures, 431
dissociation, 201, 213, 218, 260, 264,
    266, 267
dissociation of higher clusters, 234, 235
    of larger clusters, 185
dissociation processes, 80
DLC films, 168
Doppler broadening, 97, 262
    line width, 95
"Doppler-free" technique, 276
double peak structure, 65
double pulse laser ablation, 185
drag force, 242, 243
    models, 240

drain bias voltages, 290
drift tube, 290, 291
dual-laser plasma, 186
    configuration, 187-191
dye laser, 207, 209-212, 227
dynamic correlation, 348, 349, 352, 356
    fast photography, 225
    phenomena, 56
    range, 207, 208
dynamical measurements, 210

echelle grating, 207
economy, 446
effective line width, 102
EHT, 447, 475
Einstein transition probabililty in terms
    of line strength, 145
electric discharge, 96, 263, 265, 267,
    268
electrodes, 284, 287, 288
electron collision, 85
        cyclotron resonance plasmas, 82
        induced plasma reactions, 81
        irradiation, 422
        kinetic equation, 85
electron spin resonance (ESR)
    spectroscopy, 348, 356
electron temperature, 240
electronic band structure, 375
electronic excitations, 382
electronic gas-phase spectra, 218
    spectra, 199, 203
    transitions, 202, 204, 213
electronic structure, 343, 345, 346
    energies, 381
    excitations, 379
    states, 463
    structure calculations, 445, 493
    structure methods, 445, 447
    transition moment, 143
electronically excited molecules, 243

electrostatic probes, 225
electro-thermal atomizer, 212
elemental gas-carbon reactions, 493
emission intensities, 59
emission intensity enhancement, 186, 189
emission spectrum
example algorithm, 146
example synthetic, 148-150
free spontaneous, 145
inferring temperature from, 149
example, 154
energy levels, 58, 263
profile, 432
source, 445
entanglement, 431, 432, 435
enthalpy, 446
environment, 445, 465, 480
environment-dependent interaction potential (EDIP), 318
environmental impact, 445, 447
environmental monitoring, 100
pollution, 446
epoxy oxygen, 457
equation of motion, 242
equilibrium
constant, 455
geometry, 200, 399, 409
internuclear distance, 98
erosion, 333, 336, 337
etalon fringes, 273
etching plasmas, 80
ethanol, 421
ethylene, 421
ethyne, 377
evaporating, 344
even-numbered clusters, 201
exchange symmetry
nuclear
alternation of intensities, 150

eigenvalues, 122
eigenvalues for exchange of identical spins, 123
homonuclear alternation of intensities, 123
homonuclear ratio of statistical weights, 123
operator, 122
operator for identical nuclei, 122
operator to exchange identical spins, 123
excimer, 266-268
excited electronic states, 204
excited-states lifetimes, 183
exothermic addition, 260
chemical reactions, 185
expanding plasma, 223, 227, 235, 240, 246, 248, 250
plume, 202, 212
expansion front, 240
expansion of the plume, 62-64, 68
explosion front, 241
explosive detection, 100
exponential functions, 318
Extended Hückel Theory, 447
external magnetic field, 246
pressure force, 243, 251

Far Infra-Red (FIR), 79
fast photography, 223, 225, 239
feeding rate, 418
feedstock, 421, 435, 438
ferrocene, 421
field emission devices, 224
fine structure, 130
finite systems, 375
flame, 46, 50
spectroscopy, 225
flash-lamp pumped, 210
fluorescence emissions, 58
fluorocarbons, 77, 79

force constants, 380, 384, 385

forced gas flow, 91

formaldehyde (H₂CO), 88

formation, 417-419, 421-425, 427,
    429-431, 433-435, 438, 439
    mechanisms, 55, 417
    models, 419
        of C₂ and CN, 223, 250
        of CN, 228, 234, 235, 250
        of soot, 257
        processes, 199, 200

fossil fuels, 445
    utilization, 446

Fourier Transform Infrared (FTIR)
    spectroscopy, 79

four-wave mixing, 276

fragmentation, 170, 172, 173, 199

Franck-Condon factor, 143, 237

free electrons, 260
    radicals, 77, 79, 80

frequencies, 378-380, 383, 385-387,
    392, 401, 403-405, 408, 409

frequency shift, 209

full width at half maximum, 245

fullerene, 55, 74, 167, 283-288, 295,
    301, 302, 304, 344, 417, 418,
    424
    discovery of fullerenes, 269
    endohedrally encapsulated metal
        nitride, 286
    formation, 417- 419, 429, 433, 434
    fullerene-cap model, 423
    giant fullerene, 420, 431-435,
        437, 439
    metallofullerenes, 290, 301, 303

fundamental, 210, 218
    wavelength, 210

fundamental and second harmonic
    lines, 180

fusion physics, 84
    reactors, 317

gas ion chromatography, 274

gas phase, 378, 381, 409
    temperatures, 80

gasification, 445-450, 453, 455, 457,
    459, 461, 463-465, 468-471, 473,
    474, 480, 483-488, 491, 493
    reactions, 445-447, 450, 457,
        470, 483, 486, 493

gasifying agent, 446

gas-matrix shifts, 203

gas-phase experiments, 203
    investigations, 218

gate bias, 290

gate width, 61

gated ICCD, 227, 228

gated image intensifier, 57, 58

Gaussian limited mode structure, 226

geometry optimization, 380, 383

germanium, 318

glow discharge, 30
    atmospheric, 32
    DC, 9, 30-33
    microwave, 31
    RF, 30, 31, 33, 40

graphene, 224, 446, 449, 450, 460-462,
    471-477, 487
    armchair edge, 333, 334, 336,
        337, 448, 451-453, 457, 460,
        461, 464, 465, 475-477, 479,
        480, 484-486, 489, 490
    boron substitution, 461
    chemical reactions, 446-448, 451,
        453, 455-457, 459, 460,
        463-466, 468, 470, 471,
        473-475, 478-481, 483, 485,
        486, 488, 490, 493
    edge carbon atoms, 449, 455,
        460, 475
    zigzag edge, 333, 334, 336, 337,
        448, 451-453, 457, 460, 461,
        463-465, 472-478, 480, 484,
        487, 489-491

graphene flake, 428
   sheet, 295
graphite, 317-319, 327-329, 331-337,
   344, 420, 421, 424, 426-428, 433
      hydrogen interaction with, 317,
         318, 322, 323, 326, 333-337
      interlayer interaction, 317, 319,
         327, 331, 332, 334-337
      melting point, 184
      onions, 420
      peeling, 317, 334-336
      surface, 449, 453, 468, 474, 483,
         484
            basal surface, 460
            NO adsorption, 450
      target, 55, 56, 69, 72, 74
growth mechanisms, 204
   clusters, 73
   processes, 55, 168
         rates, 418
         time of instability, 247

$H_2$, 421
$H_2O$, 83, 87, 88, 101, 275
Hamiltonian, 425
      $\chi$-dependent term, 129
      diagonal matrix elements, 133
      example matrix, 140, 141
      fine structure, 130
      offdiagonal elements, 134
      rotational, 128, 129
      spin-orbit term, 130
      spin-rotation fine structure term,
         130
      spin-spin term, 130
      total diatomic, 116, 127
Hankel functions, 381
Hartree-Fock (HF), 381-383, 405
HCN, 84, 86-88, 275
He, 55-57, 59-61, 63-74, 263, 264, 268
heat-induced dissociation, 267
heating, 420, 433, 434

heavier clusters, 73
hemispherical expansion, 243, 251
      model, 250
hermetic chamber, 211
Herriot cells, 264
heterogeneous target, 211
heterostructures, 223
hexagon, 431, 438, 439
hexatriyne, 377
HF, 275
HF-CVD, 89-91
      diamond coating, 90
HgCdTe detector, 273
high finesse, 100, 101, 207
      irradiance, 218
      -resolution spectrograph, 210
high pressure (HP) bands, 171
highly sensitive techniques, 278
high-resolution laser spectroscopy, 270
      spectroscopy, 263, 277
high-resolution transmission electron
      microscopy (HRTEM), 419, 422
high-temperature, 258
      pulsed-arc discharge (HTPAD),
         284, 285, 287, 292, 306
HiPco synthesis, 417, 418, 420
hollow cathode, 265, 272
honeycomb lattice, 423
hopping integrals, 384
hot band, 94, 96, 97
      -filament reactors, 89, 90
hot carbon particles, 178
Hückel, 381, 384
      theory, 345, 348, 447, 460, 475,
         483
Hund's basis function, 124
      (a)↔(b) transformations, 127
      case (a), 125
      case (a) diagonal matrix
         elements, 133
      case (b), 125
      example, 138

parity
case (a), 126
case (b), 126
hydrocarbons, 77, 79-81, 83, 84, 344, 376-378, 386-388
cations, 256
combustion, 269
flames, 257, 261
plasma, 256, 261
radicals, 256, 258-260, 266, 267, 269-271, 275
molecules, 333
plasmas, 84
precursors, 81
hydrodynamical effects, 173
hydrogen, 421, 424, 426, 446, 447, 449, 450, 475-478, 492
abstraction, 259
hydrogenated carbon clusters, 255
hydroxyl radical, 87

imaging spectroscopy, 55-58, 72
immiscible fluids, 246
in signal-noise ratio (SNR), 265
*in situ* diagnostics, 256
monitoring, 277
incandescence, 174
induced surface reactions, 103
inert environment, 203
infinite limit, 375
influence of rare gases, 83
infrared absorption spectra, 202
diode lasers, 79
spectrum, 271
spectroscopy (IR), 347, 348, 352, 356, 358, 375, 386-388, 405-409, 493
TDLAS, 80, 82
Infrared Laser Absorption Spectroscopy (IRLAS), 77
initial conditions, 245
explosion mass, 241

initiation reaction, 258
in-situ, 199, 204
measurements, 100
integrated absorption coefficients, 97
cavity output spectroscopy (ICOS), 100
integrated intensity, 208
intensified charge-coupled device (ICCD), 227
interaction energies, 427, 428
interband cascade lasers (ICL), 100
interdependent gas-phase reactions, 89
interference filters, 57, 58
interferometer, 209
intergalactic dust, 344
inter-layer, 217
bond breakage, 170
dissociation, 170
intermediate molecules, 88
species, 286
intermolecular collisions, 267
interstellar clouds, 200, 204, 257
exploration, 204
medium, 93, 97
intra-cavity, 199, 201, 203, 205, 207, 209, 211, 213, 215, 217, 219
intra-layer, 217
ion chromatography, 202
ion detection, 286
ion drift velocity, 286
ion mobility, 286, 290, 291, 301-303, 306
combined with mass spectroscopy (IMS/MS), 286, 287, 290, 301, 303-306
observation of reactions, 286
spectrometry (IMS), 286
ionization, 286, 288, 301-304
ionized species, 175
ion-molecular reactions, 265
IR and VIS spectroscopy, 261
IR beam, 103

IRMA, 80, 89, 90, 93
irradiated target, 202
irradiation conditions, 172, 180-182, 202
ir-vibrational frequencies, 203
isoenergetic, 350, 356, 361, 371
isotopic tracing, 447
isotopomers, 97, 98
ITER, 333

$J$ quantum number, 117
Jahn-Teller effect, 346

Kekulé structure, 464
ketene ($CH_2O_2$), 266
ketone group, 464, 465
key intermediate, 258
kinetic energy, 420
    factors, 447, 456, 463, 464, 476, 477
    stability, 433
    studies, 268
kinetics of plasma processes, 78
Knudsen layer, 237
Kohn anomaly, 375, 394
KrF excimer laser, 172

laminar flow, 92
lanthanides, 295
large-amplitude bending modes, 261
laser
    fluence, 286, 303
    furnace, 284
    irradiation, 286
    pumped, 210
    solid interaction, 201
    spectroscopy, 199, 206
    vaporization, 284, 287, 291, 304
laser ablation, 55, 56, 60, 61, 70, 72, 74, 167, 169, 172, 174, 180, 185, 186, 199, 201, 206, 212-218, 377, 417, 418, 420, 421

fluence, 56, 60, 169, 173, 175, 179, 180, 182, 184
gain medium, 207
gasification, 268
    -induced fluorescence, 275
    -initiated plasma, 269
    -magnetic resonance, 274
    spectroscopy, 255, 262, 270, 274, 275
induced breakdown spectroscopy (LIBS), 186, 211
induced fluorescence, 168, 207
irradiance, 182, 185
-laser delay time, 187-191
photoionization, 73
plume, 17
power density, 170
pulse duration, 172, 173
Laser Induced Fluorescence (LIF), 18, 20, 26, 27, 29, 55, 225, 248
layer interaction, 285, 289, 297, 299, 301
lead salt diode lasers, 77, 80
    laser systems, 98
leading edge, 64, 70, 240
level populations, 183
LIF, 89, 275
    images, 248, 249
    measurement, 58
    spectra
        computation of synthetic, 160
        example, 163
        simple algorithm, 162
lifetime, 235, 262
light scattering, 207
lightning discharge, 224
limiting density, 242
line emissions, 60, 61
    profiles, 78, 97

line strength, 93-96
  definition, 116
  diatomic, 143
    as a product, 144
    electronic-vibrational, 144
    Hönl-London factor, 144
    rotational factor, 144
  example table, 152
    in case (a), 142
    in emission computation, 146
linear chains, 204, 274, 375, 377, 379, 381, 383, 385, 387, 389, 391, 393, 395, 397, 399, 401, 403, 405, 407, 409
linear combination of atomic orbitals (LCAO), 425
liquefaction, 448, 474
liquid particles, 213
LMR, 275
local ion sound speed, 240
local reactivity, 461
lock-in amplifiers, 271
long carbon chain, 264
Lorentzian line shapes, 209
low pressure flames, 258
low temperature plasmas, 80, 81
low-frequency (a.c.) discharges, 265
low-irradiance, 218
low-temperature plasma, 265

*M* quantum number, 117
magic numbers, 345
magnetic lines of force, 246
  moment, 275
magnetically enhanced reactive ion etcher (MERIE), 103
mass production, 285
mass spectrometry (MS), 82, 286
  transport, 92
mass spectroscopy (M), 168, 261, 276, 344

-selected clusters, 261
-selected species, 269
materials, 283-287, 295, 304-306, 318
matrix, 378, 391, 392
  spectra, 203, 214, 217, 218
Maxwell-Boltzmann distribution, 242
McKenna burner, 269
mean free path, 235
medical diagnostics, 100
metal, 284, 286-288, 295, 301, 303
  powders, 420
metastable, 58
  materials, 168
methane plasmas, 79, 85
methyl radical (CH3), 82
micro-instabilities, 239
microsensor, 224
microwave, 81-87, 93, 97-99
microwave (MW) discharges, 265
microwave-plasma, 224
mid infrared fibers, 104
  spectrometer, 99
  spectroscopy, 105
minimum energy reaction pathways, 445, 464
MIR absorption spectroscopy, 100
model chemistries, 448
Modified Neglect of Diatomic Overlap (MNDO), 448, 486
moisture, 446
molecular band spectra, 60, 62, 73
molecular dynamics (MD), 317, 418, 419, 423, 424, 427, 429, 430, 433-435, 439
  Car-Parinello (CPMD), 423, 424, 429, 434, 435, 439
  non-equilibrium, 431
  quantum chemical molecular dynamics (QM/MD), 417, 419, 424-426, 429-431, 433-439

molecular dynamics simulations, 202
molecular fingerprint region, 101
    gas concentrations, 90
    ions, 77, 80
    plasmas, 78, 80
molecular modeling, 447, 453, 471,
    475, 483, 487, 490, 491, 493
molecular nanowires, 377
molecular orbital (MO), 447, 448
molecular structures, 343, 344, 348,
    358, 368, 371
molecular symmetry, 346
momentum-transfer, 173
monochromator, 227
Morse potential, 318
MP2, 349, 350, 353-359, 361-366,
    368-370, 381-383, 389, 405
MP4, 349, 350, 353-355, 359, 361
MRCI, 349-355, 357
Mulliken band ($D^1\sum_u - X_1\sum_g^+$), 178
multi-configuration coupled pair
    functional (MCPF), 349, 350,
    353-357, 359-361
multi-pass absorption cells, 80
multi-ring clusters, 200
multimode, 209, 210
multipass, 263, 271, 273
    optical set-up, 273
multiphoton, 276
multiple pass optics, 84
multiply-bonded chain, 201
multireflection cells, 263
multiterm method, 85

$N_2O$, 275
nanobuds, 224
nanocages, 224
nanocarbon, 284-287, 304-306
nanocarbon materials, 283
nanochemistry, 433
nano-crystaline diamonds (NCD), 256

nanomaterials, 200, 224, 225
    -particle, 200
nanoparticle formation, 236
nanoparticles, 19, 21-24, 26, 28, 29, 48,
    49
nanorods, 223
nanosecond ablation, 173
    laser ablation, 169
nanostructures, 167, 168, 187
nanotubes, 55, 167, 168, 187, 379, 382,
    447, 478, 481
nanowires, 223, 224
Nd, 210, 211
Ne, 264
near-degeneracy, 352, 355, 357, 358
Near infrared (NIR), 100
negative ion, 343, 345, 347, 350, 358,
    361, 364, 368, 371
Nelder-Mead algorithm, 149
neon matrix, 202, 205, 218
Neptune, 93
neutral atoms and ions, 235
    carbon chains, 203, 218
        clusters, 202
    compounds, 343
        radicals, 266
$NF_3$, 103
$NH_3$, 84, 86, 87
NMR, 286
noble gas matrices, 261
noble gases, 264
non-emissive species, 225
non-equilibrium, 78, 80, 81
    distributions, 183
non-invasive detection, 255
    studies, 270
non-linear competition, 209
    optical devices, 224
        processes, 276
    spectroscopy, 275, 276
non-overlapping spheres, 381

non-reactive collisions, 169
non-selective cavity, 209
nucleation, 168, 200, 201
nucleogenesis, 260
number density, 82, 225, 242, 248-251
NVE, 334

$O_2$, 345
observation area, 63, 66
OH, 87, 88
oligomers, 375, 379, 383
ONIOM calculations, 451, 452, 473
open-shell, 256
    wavefunction, 450
OPO laser, 57, 68, 69
optical absorption spectroscopy, 225
    emission spectroscopy (OES), 18,
        22, 31-33, 41, 168, 225
    fiber cable, 226
optical fiber, 57
    parametric oscillator (OPO), 58
optical multichannel analyzer (OMA),
    212
optical multi-pass cells, 101
optical parametric oscillators (OPO),
    271
optical path extension, 263
    length, 208
optical phonon, 380, 385, 392, 394,
    407, 408, 410
organic electronics, 377
organo-silicon compounds, 77, 79
out-of-phase motion, 379
oxidation, 285, 288, 289, 296, 297
    processes, 258
oxygen, 421, 446, 447, 451, 453-466,
    468-473, 480, 481, 483-493

PAH formation, 257, 260
parameters
    determining from recorded
        spectrum, 135

example determination, 138
example table, 139
parametrization, 381
parity
    algorithm, 121
    case (a), 126
    case (b), 126
    definition, 118
    e/f designation, 120
    eigenvalues, 119
    example matrix, 141
    operator
        electronic, 122
        nuclear, 122
particle density, 189
    formation, 201
    size, 418
particulates, 60, 62, 73
partition functions, 97
passive cavity, 205
Peierls distortion, 375, 376, 385, 386,
    391, 393, 394, 396
Penning ionization, 264
pentagon, 423, 428, 431, 438
Perdew-Berke-Ernzerhof (PBE), 385,
    426, 427, 429, 436
periodic lattice deformations, 376
perturbation theory, 354, 361
perturbations, 246
phase-sensitive (lock-in) amplifier,
    265
phenyl radical, 259
phenylacetylene, 258, 259
phonon, 376, 381, 384, 385, 403, 404
    branches, 380, 391, 392, 394,
        401, 408
    dispersion curves, 375
photochemistry, 172
photo-detachment spectroscopy, 204
photodissociation, 170
photoelectron spectroscopy (XPS), 274,
    447

photofragmentation, 170
photoionization diagnostics, 73
photolysis, 263, 264, 266
physisorption, 460
picosecond and femtosecond ablation,
    173
pixel-wise LIF intensity, 249
planar, 205, 212
    laser beam, 58, 69
Planck's constant, 237
    radiation distribution, 60
plasma, 317, 333, 337
    confinement, 179, 333
    parameters, 242
        -gas boundary, 246
        pressure, 242
    plasma surface interaction (PSI),
        337
    plasma-wall interaction (PWI),
        333
    processing, 255
    processing reactors, 218
        -chemical reactors, 204
        expansion, 213
        initiation, 213
        processing, 200
        species, 85
plasma absorption cell, 82
    chemical conversion, 84
    chemistry, 81, 82, 102, 103
    diagnostics, 80, 99, 100
plasma enhanced CVD (PECVD), 6,
    30, 46, 49, 81
plume, 420, 435
    confinement, 230
    cooling, 73, 74
    cooling temperature, 55, 56,
        68-74
    front position, 227, 239
    length, 227, 228, 243
    linear momentum, 243, 251

shape, 243
volume, 72
PLV synthesis, 420
PM3, 326, 328
point explosion, 240
polarizability, 384
Polarization Spectroscopy (PS), 276
pollutant
    emissions, 446
    gases, 445
polyacetylene, 343, 345, 346, 348,
    351-356, 359-364
polyaromatic hydrocarbons (PAH),
    255, 256
polymer, 328
polymerization, 257
poly-peptides, 286
polyyne, 377-386, 388-392, 394-405,
    407-409, 429-433, 435, 437,
    438
polyynic chain, 379, 381-385
population densities, 77, 78
    modulation, 265
positive ion, 343, 351, 361, 364, 366,
    368, 370, 371
potential
    attractive, 318, 320, 327
    barrier, 326
    Brenner, 317-319, 324-326, 330,
        331, 334, 337
    long range interactions, 327
    model, 317, 319, 321, 323, 325,
        327, 329, 331, 333, 335, 337
    repulsive, 318, 320, 327
    Tersoff, 318, 319, 337
    three-body, 321
    two-body, 317, 321, 327, 331
potential curve crossing, 171
potential energy surface (PES), 425,
    445, 447, 473
pre-breakdown, 213

pre-combustion, 446

precursor clusters, 199

    molecules, 81, 84

pre-mixed flames, 269

pressure (collision) broadening, 262

    regime, 247, 250

probe dye-laser, 210

profile narrowing, 191

propagation, 239, 241, 243

propane, 258

propargyl, 260

pulse decay-time, 205

    repetition rate, 211

pulsed arc discharge (PAD), 284, 306

pump laser fluence, 248

purification, 283, 285, 288, 296, 297, 447

    burning-assisted, 285

pyrolysis, 257, 258, 263, 264, 267, 446

    -generated radicals, 258

    of hydrocarbons, 257


*Q*-branch, 82, 94, 96

QCL, 77, 80, 98, 100-103

QCLAS, 78, 99, 103

Q-MACS, 99, 103, 104

quadrupole mass spectrometer, 291

quantum cascade (QC), 271

    lasers, 77, 80, 100, 101, 105

    yields, 275

quantum chemistry, 200

    chemical method, 317, 338

    dots, 223

    effects, 284

quartic distortion constants, 97

"quasi-continuum", 178

quasimomentum, 376, 391

quenching, 56, 66, 68


R branch, 68, 69

*R* mode, 383

radial breathing mode (RBM), 292, 295, 297

radial plasma expansion, 237

radiating plume, 214

radiation initiated chemical reactions, 263

radiative recombination, 174

radical atoms, 421

    densities, 56, 62-64, 74

radicals and ions, 255, 256, 265, 267, 268, 271, 278

radio frequency (RF), 81

    discharges, 265

raising and lowering operators, 131

    molecule-fixed, 131

    on electronic spin states, 132

    on rotation matrix element, 132

Raman spectroscopy, 3, 18, 25, 35, 45-48, 289, 292-296, 375, 376, 378-380, 382, 383, 386-388, 405-409, 430

    Coherent Anti-Stokes (CARS), 18, 25, 27

    surface-enhanced (SERS), 378, 380

rare gas atmosphere, 56

    matrices, 348

rate coefficients, 85

    constant, 94, 96

rate-limiting step, 446, 457, 463, 469, 486

Rayleigh-Taylor (RT) instability, 246

*r*-centroids, 143

reaction kinetics, 77, 81

    pathways, 85

reactive empirical bond order (REBO), 423, 424, 433-435, 439

reactive force field, 317

    free radicals, 256

REBO, 317-319, 324-328, 330, 331, 334, 337

recombination interactions, 239
  mechanism, 217
  of carbon atoms, 177, 184
  reactions, 258
    with electrons, 260
recombinational (nucleation)
  mechanism, 171
recombinative processes, 169
red satellites, 215
red-shifted, 205
relative intensities, 78, 82
  population, 237
  stabilities, 351, 355, 359, 362,
    366, 368, 370, 371
REMPED, 277
REMPI, 89, 276, 277
Renner-Teller effect, 178
resonance energy, 379
resonant multi-photon methods, 277
re-splitting, 237-239, 245, 247
restricted open-shell wavefunction,
  450, 451
ring condensation mechanism, 431
ring-collapse mechanism, 432
RM2PD, 277
rotation matrix element, 117
  eigenfunction of $J_z$ and $J_z$, 131
  raising and lowering operators
    on, 132
  temperature, 267
rotation-vibration spectroscopy, 271,
  276
rotational quantum numbers, 69
  temperature, 69
rotational temperatures, 82
  transitions, 93
rotationally resolved spectra, 271
rovibration spectra, 262, 271
rovibronic lines, 93
ruby, 210

saddle point, 354, 361, 362, 364

satellite bands, 376, 380, 383, 385, 408
saturated absorption, 209
  carbon atoms, 455, 486
  excitation, 69
saturation parameter, 209
Saturn, 93
$Sc_2 C_2 C_{84}$, 286
scanning electron microscopy (SEM),
  289, 289, 447
scanning tunneling microscopy (STM),
  447, 491
Schrödinger equation, 124, 128, 134,
  144, 147, 318
scooter model, 422
segregation effects, 202
selective excitation, 248
self recombination, 94
self-absorption, 9-11, 13, 14, 42, 48
self-assembly, 285, 423, 424, 430, 431,
  434, 439
self-emissions, 58
semiconductor, 77, 78, 100, 103, 104
  diode lasers (TDLs), 271
semi-empirical method, 348, 447, 453
  Austin Model 1 (AM1), 448, 451
  Intermediate Neglect of
    Differential Overlap (INDO),
    448, 455, 472
  Modified Intermediate Neglect of
    Diatomic Overlap (MINDO),
    448, 453, 454
  Neglect of Diatomic Differential
    Overlap (NDDO), 448
  Parametric Method 3 (PM3), 448
  Pariser-Parr-Pople (PPP), 447
  simple Hückel MO (SHMO), 460
semiquinone, 455, 457, 462-465, 469,
  470, 480, 486, 487, 490, 491
shock front, 240, 242
  /blast wave, 239, 240
  wave, 64, 70, 239
  wave model, 243

short-lived, 256, 266, 278
$SiF_4$, 103, 104
signal to noise, 102
    ratio, 63
silane plasmas, 80
silicon, 318, 319
    wafer, 103
simulations, 202
single and double pulse, 167, 169
single-reference methods, 350
singles and doubles CI (SDCI), 349,
    350, 352-357, 359, 360
singlet, 346, 356, 367, 428
sintering, 420
$SiO_2$, 103
size-extensivity, 354, 355, 360
snowplow model, 247
solid samples, 209
    soot particles, 270
    target, 213
solid-gas reactions, 446
solid-state theory, 375, 377, 386-388,
    390-395, 399-404, 408, 409
solvent, 285, 301, 302
soot, 255-261, 269, 270, 277, 417, 418,
    433
    formation chemistry, 255, 257
    formation rates, 259
    precursor, 256
spatial and temporal distribution, 228
spatial distributions, 55, 61
    resolution, 275
spatiotemporal variation, 68
species densities, 211
spectral band-profiles, 207
    distribution, 59
    interference, 217
    line positions, 78
    multiplexing, 207
spectrochemical analysis, 208, 210, 211
spectroscopic diagnostic, 55

spherical blast waves, 240
    molecule, 433
sp-hybridized carbon, 429
spike emission, 181
spiky emission, 59, 60
spin contamination, 450
spin rotation interaction, 98
spin states, 426
spin-orbit fine structure term, 130
    diagonal matrix elements, 133
spin-rotation fine structure term, 130
    parameter $\gamma_v$, 133
spin-spin fine structure term, 130
    parameter $\lambda_v$, 133
splitting of the plume, 237, 238
spontaneous emission, 207
    coefficients, 183
stability of wave functions, 457
stable isomers, 201
    species, 213
stationary and moving components, 250
steam, 446, 447, 468, 487, 488
stellar atmospheres, 84
stereo-chemical structure, 204
stray light, 58
strip-off (dissociative) mechanism, 169
structural identification, 286
    isomers, 200
styrene, 259
sub-Doppler, 271
substrate, 418, 421, 422
sulfur, 292, 421, 478
superaromaticity, 420
supercapacitors, 224
superlattices, 223
supersonic cluster beam, 224
    expansion, 263, 268
        jet, 262, 267, 268, 271, 273
    gas pulse, 206
        planar plasma, 205, 206
        slit/nozzle, 215
        spectra, 217

surface activation, 78
Swan band emissions, 225
    system, 232, 236
Swan system ($d\,{}^3\Pi_g$ - $a^3\Pi_u$), 177
sweater, 430
(SWNTs), 187
synchronized lasers, 210
synergy effect, 187

target surface, 56, 57, 64
Taylor-Sedov (T-S) theory, 240
TDL, 77, 84-86
TDLAS, 80, 82-84, 86, 87, 89-92, 97, 271
temperature, 3, 4, 6, 7, 9-13, 15, 16, 19, 20, 23-26, 29, 30, 33, 37-46, 48, 284, 287, 288, 290, 292-295, 299, 305, 306, 418-420, 423-425, 427, 429, 430, 432, 433, 436, 439
    dependence of SWNT vs DWNT growth, 292
    electronic, 20, 26, 27, 40, 42, 44, 48
    rotational, 10, 11, 20, 26, 41-45, 48
    vibrational, 13, 19, 25, 27, 41-43
temperature programmed desorption (TPD), 447, 453, 455, 456, 461, 480
temporal evolution, 55, 56
    profiles of emission, 180, 191
    pulse shape, 208
    variation, 59, 69
temporally resolved spectrum, 236
*tert*-butyl nitrite, 267
theoretical modeling, 89
    predictions, 199
thermal ablation mechanisms, 217
    conduction, 72
        kinetics, 72
    equilibrium, 183, 237
        velocity, 173
    target ablation, 217

thermodynamics, 447, 449, 453, 463, 476-479, 487, 491, 493
thermo-gravimetric Analysis (TGA), 447
thin films, 167, 168, 187, 191
    deposition, 77, 78
three body reaction, 94
    collisions, 170
    recombination, 234
three-dimensional structures, 201
threshold laser irradiance, 182
tight-binding, 375, 386, 424, 425
time dependent measurement, 93
    Boltzmann equation, 85
time integrated plasma emission spectra, 174
    resolved OES, 168
time resolved spectroscopy, 225
    ICCD, 243
    LIF signal, 227
time scale, 423, 434
    step, 324, 325, 334
time-domain measurement, 57
time-of-flight, 269, 276
time-of-flight (TOF) measurements, 291, 292
TOBI, 80
TOF, 277
trace constituents, 99
    gas detection, 80, 100
    level, 209
trajectory calculations, 425
transient carbon species, 255-257, 262-265, 267-272, 275-277
    kinetics (TK), 447, 456
    molecular species, 77, 80
    plasma, 199, 206, 207, 217
    properties, 56
transition dipole moment, 83, 93-97
    moment case (a), 143
    operator, 116

electric dipole in case (a), 142

electric dipole moment, 143

electronic transition moment, 143

Franck-Condon factor, 143

*r*-centroids, 143

spherical tensor form, 142

transition state, 445, 447, 450, 465, 468, 493

translating target, 268

translation degrees of freedom, 170

motion of molecules, 263

temperature, 69

transmission electron microscopy (TEM), 286, 289, 292-298, 306, 447

profile, 208

transport mechanism, 228

triplet, 347, 348, 367, 368, 428, 451

tritium, 333

tropical weeds, 377

tubular furnace, 211

tunable diode laser, 255, 271, 273

laser pulses, 58

laser sources, 271

semiconductor lasers, 77

turbulence, 247

turbulent interactions, 240

twin peak distribution, 182

two-dimensional images, 58, 228

two-photon ionization, 277

two-ring clusters, 200

ultra-high vacuum scanning tunnel microscopy (UHV-STM), 286

ultra-sensitive spectroscopy, 199

ultraviolet (UV) laser, 172

un-damped cavity, 207

unpaired electron, 256, 275, 450

unrestricted Hartree-Fock (UHF), 450, 460, 487

unstable neutral species, 256

UV radiation, 240

UV-irradiation, 203

-visible region, 200

UV-Vis, 378

vacant orbitals, 425

Vacuum Ultra Violet (VUV), 79

vapor diagnostics, 207

vaporization, 213

vapor-liquid-solid (VLS) mechanism, 423

velocity distribution, 167, 168

modulation, 266

vibration frequencies, 274

vibration rotation spectrum, 97

vibrational and rotational relaxation, 171

densities of states (VDOS), 409

distribution of $C_2$, 185

levels, 237

temperature, 236, 237

spectra, 445, 447

spectroscopy, 375

temperatures, 82, 95, 96, 183, 184

temperatures of $C_2$, 184

vibrations, 376, 378-381, 391, 392, 401, 403, 407

vinylacetylene, 258

violet $B^2 \Sigma^+ - X^2 \Sigma^+$ band, 235

violet $B^2 \Sigma^+ - X^2 \Sigma^+$ band system, 228, 250

volatile matter, 446

waste treatment, 78

water vapor, 83

wave vector, 380, 405
White cell, 84, 85, 87, 263

XeCl$_2$ excimer laser, 172
X-ray diffraction, 286
        photoelectron spectroscopy
            (XPS), 447, 491

YAG laser, 56, 172, 180, 224, 226, 227
yield, 418, 421, 422, 428
yttrium, 295

zoom lens, 227